VOLUME SIX HUNDRED AND FORTY TWO

METHODS IN ENZYMOLOGY

Odorant Binding and Chemosensory Proteins

METHODS IN ENZYMOLOGY

Editors-in-Chief

ANNA MARIE PYLE
Departments of Molecular, Cellular and Developmental Biology and Department of Chemistry
Investigator, Howard Hughes Medical Institute
Yale University

DAVID W. CHRISTIANSON
Roy and Diana Vagelos Laboratories
Department of Chemistry
University of Pennsylvania
Philadelphia, PA

Founding Editors

SIDNEY P. COLOWICK and NATHAN O. KAPLAN

VOLUME SIX HUNDRED AND FORTY TWO

METHODS IN ENZYMOLOGY

Odorant Binding and Chemosensory Proteins

Edited by

PAOLO PELOSI

AIT Austrian Institute of Technology GmbH,
Biosensor Technologies,
Tulln, Austria

WOLFGANG KNOLL

AIT Austrian Institute of Technology GmbH,
Biosensor Technologies; CEST Competence Center
for Electrochemical Surface Technology,
Tulln, Austria

ELSEVIER

ACADEMIC PRESS
An imprint of Elsevier

Academic Press is an imprint of Elsevier
50 Hampshire Street, 5th Floor, Cambridge, MA 02139, United States
525 B Street, Suite 1650, San Diego, CA 92101, United States
The Boulevard, Langford Lane, Kidlington, Oxford OX5 1GB, United Kingdom
125 London Wall, London, EC2Y 5AS, United Kingdom

First edition 2020

Notices
Knowledge and best practice in this field are constantly changing. As new research and experience broaden our understanding, changes in research methods, professional practices, or medical treatment may become necessary.

Practitioners and researchers must always rely on their own experience and knowledge in evaluating and using any information, methods, compounds, or experiments described herein. In using such information or methods they should be mindful of their own safety and the safety of others, including parties for whom they have a professional responsibility.

To the fullest extent of the law, neither the Publisher nor the authors, contributors, or editors, assume any liability for any injury and/or damage to persons or property as a matter of products liability, negligence or otherwise, or from any use or operation of any methods, products, instructions, or ideas contained in the material herein.

ISBN: 978-0-12-821157-1
ISSN: 0076-6879

Publisher: Zoe Kruze
Editorial Project Manager: Shellie Bryant
Production Project Manager: James Selvam
Cover Designer: Greg Harris

Typeset by SPi Global, India

Contents

Contributors

Béatrice Amigues
Architecture et Fonction des Macromolécules Biologiques, Centre National de la Recherche Scientifique (CNRS) & Aix-Marseille Université, Marseille, France

Patrik Aspermair
CEST Competence Center for Electrochemical Surface Technology; AIT Austrian Institute of Technology GmbH, Biosensor Technologies, Tulln, Austria; University of Lille, CNRS, Centrale, Lille, Université Polytechnique Hauts-de-France, UMR 8520—IEMN, Lille, France

Christine Belloir
Centre des Sciences du Goût et de l'Alimentation, AgroSup Dijon, CNRS, INRAE, Université Bourgogne Franche-Comté, Dijon, France

Rabah Boukherroub
University of Lille, CNRS, Centrale Lille, Université Polytechnique Hauts-de-France, UMR 8520—IEMN, Lille, France

Anil Bozdogan
AIT-Austrian Institute of Technology GmbH, Tulln; CEST Kompetenzzentrum für elektrochemische Oberflächentechnologie GmbH TFZ, Wiener Neustadt, Austria

Loïc Briand
Centre des Sciences du Goût et de l'Alimentation, AgroSup Dijon, CNRS, INRAE, Université Bourgogne Franche-Comté, Dijon, France

Marine Brulé
Centre des Sciences du Goût et de l'Alimentation, AgroSup Dijon, CNRS, INRAE, Université Bourgogne Franche-Comté, Dijon, France

Khasim Cali
Department of Instrumentation and Analytical Science, The University of Manchester, Manchester, United Kingdom

Christian Cambillau
Architecture et Fonction des Macromolécules Biologiques, Centre National de la Recherche Scientifique (CNRS) & Aix-Marseille Université, Marseille, France

Song Cao
State Key Laboratory for Biology of Plant Diseases and Insect Pests, Institute of Plant Protection, Chinese Academy of Agricultural Sciences, Beijing, China

Francesca R. Dani
Department of Biology, University of Firenze, Firenze, Italy

Chiara D'Onofrio
AIT Austrian Institute of Technology GmbH, Biosensor Technologies, Tulln; Department of Nanobiotechnology, University of Natural Resources and Life Sciences, Vienna, Austria

Jakub Dostalek
AIT-Austrian Institute of Technology GmbH, Tulln, Austria

Alessandro Emendato
Department of Chemical Sciences, University of Naples "Federico II", Complesso Universitario Monte S. Angelo, Napoli, Italy

Paula Escuer
Departament de Genètica, Microbiologia i Estadística and Institut de Recerca de la Biodiversitat (IRBio), Universitat de Barcelona, Barcelona, Spain

Anaïs Gaubert
Architecture et Fonction des Macromolécules Biologiques, Centre National de la Recherche Scientifique (CNRS) & Aix-Marseille Université, Marseille, France

Margot Glaz
Centre des Sciences du Goût et de l'Alimentation, AgroSup Dijon, CNRS, INRAE, Université Bourgogne Franche-Comté, Dijon, France

Simone Hageneder
AIT-Austrian Institute of Technology GmbH, Tulln, Austria

Stefanie Hartmann
Evolutionary Adaptive Genomics, Institute for Biochemistry and Biology, University of Potsdam, Potsdam, Germany

Nadja Hellmann
Department of Chemistry/Biochemistry, Johannes Gutenberg-University Mainz, Mainz, Germany

Michael Hofreiter
Evolutionary Adaptive Genomics, Institute for Biochemistry and Biology, University of Potsdam, Potsdam, Germany

Wolfgang Knoll
AIT Austrian Institute of Technology GmbH, Biosensor Technologies; CEST Competence Center for Electrochemical Surface Technology, Tulln, Austria

Serena Leone
Department of Chemical Sciences, University of Naples "Federico II", Complesso Universitario Monte S. Angelo, Napoli, Italy

Wei Liu
State Key Laboratory for Biology of Plant Diseases and Insect Pests, Institute of Plant Protection, Chinese Academy of Agricultural Sciences, Beijing, China

Kyriaki Manoli
Dipartimento di Chimica - Università degli Studi di Bari "A. Moro"; CSGI (Centre for Colloid and Surface Science), Unit of Bari, Bari, Italy

Vladyslav Mishyn
University of Lille, CNRS, Centrale Lille, Université Polytechnique Hauts-de-France, UMR 8520—IEMN, Lille, France

Lucie Moitrier
Centre des Sciences du Goût et de l'Alimentation, AgroSup Dijon, CNRS, INRAE, Université Bourgogne Franche-Comté, Dijon, France

Mohammad Y. Mulla
Research Institutes of Sweden (RISE), Norrköping, Sweden

Fabrice Neiers
Centre des Sciences du Goût et de l'Alimentation, AgroSup Dijon, CNRS, INRAE, Université Bourgogne Franche-Comté, Dijon, France

Paolo Pelosi
AIT Austrian Institute of Technology GmbH, Biosensor Technologies, Tulln, Austria

Krishna C. Persaud
Department of Instrumentation and Analytical Science, The University of Manchester, Manchester, United Kingdom

Jean-François Picimbon
School of Bioengineering, QILU University of Technology, Jinan, Shandong, China

Delia Picone
Department of Chemical Sciences, University of Naples "Federico II", Complesso Universitario Monte S. Angelo, Napoli, Italy

Giuseppe Pieraccini
CISM, Mass Spectrometry Centre, Department of Health Sciences, University of Firenze, Firenze, Italy

Nicolas Poirier
Centre des Sciences du Goût et de l'Alimentation, AgroSup Dijon, CNRS, INRAE, Université Bourgogne Franche-Comté, Dijon, France

Julio Rozas
Departament de Genètica, Microbiologia i Estadística and Institut de Recerca de la Biodiversitat (IRBio), Universitat de Barcelona, Barcelona, Spain

Takeshi Sakurai
Faculty of Agriculture, Tokyo University of Agriculture, Kanagawa, Japan

Alejandro Sánchez-Gracia
Departament de Genètica, Microbiologia i Estadística and Institut de Recerca de la Biodiversitat (IRBio), Universitat de Barcelona, Barcelona, Spain

Andrea Scaloni
Proteomics and Mass Spectrometry Laboratory, ISPAAM, National Research Council, Naples, Italy

Yusuke Shiota
Department of Evolutionary Neuroethology, Max Planck Institute for Chemical Ecology, Jena, Germany; Research Center for Advanced Science and Technology, The University of Tokyo, Tokyo, Japan

Roberta Spadaccini
Department of Science and Technology, Università degli Studi del Sannio, Benevento, Italy

Silvia Spinelli
Architecture et Fonction des Macromolécules Biologiques, Centre National de la Recherche Scientifique (CNRS) & Aix-Marseille Université, Marseille, France

Rudolf Alexander Steinbrecht
Max-Planck-Institute for Ornithology, Seewiesen, Germany

Sabine Szunerits
University of Lille, CNRS, Centrale Lille, Université Polytechnique Hauts-de-France, UMR 8520—IEMN, Lille, France

Luisa Torsi
Dipartimento di Chimica - Università degli Studi di Bari "A. Moro"; CSGI (Centre for Colloid and Surface Science), Unit of Bari, Bari, Italy; Center for Functional materials, The Faculty of Science and Engineering, Åbo Akademi University, Turku, Finland

Elena Tuccori
Pall Corporation, Portsmouth, United Kingdom

Alina Vasilescu
International Centre of Biodynamics, Bucharest, Romania

Joel Vizueta
Departament de Genètica, Microbiologia i Estadística and Institut de Recerca de la Biodiversitat (IRBio), Universitat de Barcelona, Barcelona, Spain

Bing Wang
State Key Laboratory for Biology of Plant Diseases and Insect Pests, Institute of Plant Protection, Chinese Academy of Agricultural Sciences, Beijing, China

Guirong Wang
State Key Laboratory for Biology of Plant Diseases and Insect Pests, Institute of Plant Protection, Chinese Academy of Agricultural Sciences, Beijing; Guangdong Laboratory of Lingnan Modern Agriculture, Shenzhen; Genome Analysis Laboratory of the Ministry of Agriculture, Agricultural Genomics Institute at Shenzhen, Chinese Academy of Agricultural Sciences, Shenzhen, China

Valeriia Zaremska
AIT Austrian Institute of Technology GmbH, Biosensor Technologies, Tulln; Department of Nanobiotechnology, University of Natural Resources and Life Sciences, Vienna, Austria

Jiao Zhu
AIT Austrian Institute of Technology GmbH, Biosensor Technologies, Tulln, Austria; Faculty of Biology, Institute of Molecular Physiology, Johannes Gutenberg-Universität, Mainz, Germany

Preface

The discovery of odorant-binding proteins (OBPs) nearly 4 decades ago marked a turning point in the study of olfaction. The coincidental isolation of proteins able to bind odorants and pheromones from the nasal mucosa of mammals and from the antennae of insects showed for the first time that biochemistry could be applied to investigate the sense of smell. This prompted scientists to start a search for olfactory receptors, successfully accomplished 10 years later by uncovering the specific biochemical sensors for detecting and discriminating all the different types of smells.

OBPs, at first, were regarded only as accessory polypeptides performing a role of carriers for hydrophobic volatiles, and the previous interest for these proteins rapidly decreased. Recently, however, we have witnessed a renewed attention in the study of OBPs (and other family of odorant carriers discovered in the meantime), because of their structural characteristics, that have determined a strong evolutionary success for these proteins. Such novel interest is developing along two lines: (a) OBPs are active not only in olfaction, but in several different physiological processes, such as development, inflammatory processes, transport of nutrients and resistance of insects to pesticides; (b) OBPs are increasingly being regarded as ideal biosensors for odors to be used in artificial (biomimetic) olfaction.

These new aspects of OBPs are based on unique structural characteristics, that made such proteins highly successful across evolution (and therefore adopted for different tasks), as well as extremely stable to challenging environmental conditions. The OBPs of mammals and those of insects represent two distinct classes of proteins, markedly different in amino acid sequence and three-dimensional structure. However, they share (also with other soluble binding proteins) a compact structure and a well-defined hydrophobic binding site. These characteristics make OBPs ideal carriers for different types of ligands, responsible for different functions in various parts of the body. Due to their compact structure, OBPs are also exceptionally resistant to high temperatures (can even be boiled for several minutes), to solvents and, to a certain extent, to the action of proteases. These unique qualities have prompted their use in biosensors, where the binding to volatile compounds is monitored by variations in electric parameters, optical properties or mass.

All these studies, in particular technological applications, require large amounts of OBPs (and other ligand-binding proteins) that need to be prepared using simple, inexpensive and reliable protocols. In this volume we have collected well established and tested methods for the study of OBPs and CSPs, from genome mining and sequence analysis to heterologous expression, structural characterization and ligand-binding assays. We have also added applications of OBPs and CSPs in sensing devices for odor monitoring, using different approaches.

The first three chapters deal with sequence identification and analysis. The availability of an increasing number of genomes represents a rich database to search for proteins that might be useful for technological applications. At present, using bioinformatic tools and based on the wide information available, reliable predictions can be formulated just from the amino acid sequence of a protein. Moreover, thanks to automated and inexpensive production, synthetic genes encoding small proteins, such as OBPs and CSPs can be easily obtained and used for heterologous expression of the proteins. This enables us to synthesize and study OBPs of endangered or even extinct species, and to formulate hypotheses on the long-gone pheromones that once occupied their binding pockets.

The following two chapters illustrate proteomic approaches and the application of mass spectrometry techniques to the detection of OBPs and CSPs in biological samples and the identification of post-translational modifications.

Chapters 6–10 describe protocols for the expression and characterization of soluble olfactory proteins. Bacterial expression is currently the method of choice, but also yeast has been successfully used to prepare OBPs for structural and functional studies. Determination of the three-dimensional structure is a key step both for functional investigations and for technological applications. Whether we want to formulate hypotheses on ligand pheromones of OBPs from extinct species, or we wish to design mutated proteins to be used as sensors for odors, the starting point is always the three-dimensional structure. Chapters 7 and 8 describe how to solve the structure of OBPs and CSPs using X-ray crystallography or NMR, respectively. Thermal stability is another interesting characteristic of OBPs, particularly when these proteins are incorporated in sensing elements for environmental monitoring. How to measure such feature is illustrated in Chapter 9. Finally, ligand-binding assays (Chapter 10) are essential in the characterization of OBPs and other olfactory proteins, both when studying these proteins within a biological/physiological context to understand their role in chemical ecology, and when selecting or modifying OBPs to be used in biosensors for odors.

Chapter 11 describes the techniques used to detect the presence of OBPs at the sub-cellular level, using immunocytochemistry and electron microscopy. Given the wide variety of functions recently discovered for OBPs, identifying their sites of action represents a great help in understanding their specific roles.

In Chapter 12 we look at the effect of OBPs on the response of olfactory receptors. Although it is generally accepted that odor detection and discrimination occur mainly at the level of olfactory receptors, OBPs can modulate the signals by making the system more sensitive and specific.

Chapters 13 and 14 describe protocols for introducing mutations in the genes encoding OBPs and for inserting mutants into an insect or deleting genes, with the purpose of studying the effect of mutations or of specific genes on the performance of the protein or on the behavior of an insect. Introducing mutations into specific position of an OBP binding site and studying their effect on the ligand characteristics of the protein is essential when designing OBPs for specific applications as biosensing elements.

The last five chapters illustrate the use of OBPs as odor detecting elements in artificial devices. Ligand-binding events can produce variations in electrochemical (Chapter 15) or electronic devices, such as field-effect transistors (Chapters 16 and 17), changes in the mass-load and hence the resonant frequencies of quartz crystal microbalances (Chapter 18) or generate optical signals that can be detected in by surface plasmon resonance instruments or other devices (Chapter 19). The transducer concepts described in these last five chapters offer novel and label-free approaches for the quantitative evaluation of the molecular interactions between the typically small odorant molecules and OBPs, either in wild-type form or with various mutations, thus complementing the classical protocols (cf. Chapter 10).

On the other hand, these transduction platforms offer novel, biomimetic solution for the development of artificial noses. The instruments currently used are still bulky, complex, expensive, and can only be used in a well-equipped lab by experienced personnel. But the new concepts described in Chapters 15–19 are the principles from which new chemical sensors for practical uses will be built. Research on artificial olfaction (electronic noses) currently is very active and rapidly expanding.

Paolo Pelosi
AIT Austrian Institute of Technology GmbH,
Biosensor Technologies, Tulln, Austria

Wolfgang Knoll
AIT Austrian Institute of Technology GmbH,
Biosensor Technologies; CEST Competence Center
for Electrochemical Surface Technology, Tulln, Austria

CHAPTER ONE

Genome mining and sequence analysis of chemosensory soluble proteins in arthropods

Joel Vizueta, Paula Escuer, Alejandro Sánchez-Gracia, Julio Rozas*
Departament de Genètica, Microbiologia i Estadística and Institut de Recerca de la Biodiversitat (IRBio), Universitat de Barcelona, Barcelona, Spain
*Corresponding author: e-mail address: jrozas@ub.edu

Contents

Abstract

Identifying protein-coding genes from genome and transcriptome data is the first and one of the most important steps towards their comprehensive study. This chapter introduces both general procedures for sequence mining, and specific approaches for recognizing characteristic motives and chemical properties in soluble proteins potentially involved in arthropod chemical communication. We describe (i) the workflow to identify members of the OBP (Odorant-Binding Proteins) and CSP (Chemosensory Proteins) families in genomic and transcriptomic sequences using our recently developed bioinformatic solution, BITACORA, and (ii) the main further steps to visualize and to accurately annotate these genes in the Apollo genome browser. The success of further biochemical, functional and evolutionary analyses largely depends on the quality of these initial steps.

Methods in Enzymology, Volume 642
ISSN 0076-6879
https://doi.org/10.1016/bs.mie.2020.05.015

Abbreviations

BED browser extensible data
CSP chemosensory protein
GFF general feature format
HMM hidden Markov model
HTS high-throughput sequencing.
OBP odorant-binding protein

1. Introduction

Chemoreception in animals is a physiological process capable of detecting and discriminating a wide range of chemical signals from the external environment, playing a fundamental role in organisms' survival, reproduction and social communication (Asahina, Pavlenkovich, & Vosshall, 2008; Meyerhof & Korsching, 2009; Smadja & Butlin, 2009). Although the chemosensory system has been traditionally divided into two main modalities, the smell and taste, this distinction is not always clear, especially in the case of some invertebrates. The classical view (Mori, Takahashi, Igarashi, & Yamaguchi, 2006; Touhara & Vosshall, 2009) emphasize in the range of the detection, separating long (i.e., smell) from short-range (or direct contact) chemoreception (i.e., taste) distances. Other authors (Mollo et al., 2014) however, define smell as the ability to detect largely hydrophobic volatile compounds, while taste would mediate the detection of hydrophilic highly soluble molecules.

The chemoreception process is mediated by proteins encoded by medium to large-sized gene families (Vieira & Rozas, 2011; Vizueta, Rozas, & Sánchez-Gracia, 2018) belonging to two main functional categories, chemoreceptors, located in the dendritic membranes of chemosensory neurons, and soluble proteins, small globular proteins secreted to the aqueous space surrounding these dendrites. In arthropods, the latter category includes a number of unrelated families encoding abundantly secreted soluble proteins, such as the Odorant-Binding Protein (OBP), Chemosensory Protein (CSP), Chemosensory Protein A and B (CheA and CheB) and Niemann-Pick protein type C2 (NPC2) families (Angeli et al., 1999; Ishida et al., 2014; Nomura, Kawasaki, Kubo, & Natori, 1992; Pelosi, Iovinella, Felicioli, & Dani, 2014; Sánchez-Gracia, Vieira, & Rozas, 2009; Vieira & Rozas, 2011; Vogt & Riddiford, 1981). The genomes of vertebrates also encode a set of proteins called Odorant Binding Proteins, but they are structurally different and non-homologous to insect OBPs. Despite their different evolutionary origin, vertebrate and insect OBPs would have an analogous functional role (Pelosi, 1990; Pelosi & Maida, 1995), although far from being understood, especially in insects. The classic idea is that in insects these molecules have a relevant function in olfaction by solubilizing and transporting hydrophobic molecules to the proximity of transmembrane receptors. Despite some recent results (Larter, Sun, & Carlson, 2016; Xiao, Sun, & Carlson, 2019) would suggest that many of the OBPs expressed in the Drosophila antenna would not be essential for triggering olfactory signal transduction (but see also Xu, Atkinson, Jones, & Smith, 2005), they would be important in modulating olfactory physiology and behavioral responses. In flies, furthermore, many of such proteins are expressed in tissues not related with the chemosensory function, likely playing other physiological roles as carriers. Similarly, the putative chemosensory and non-chemosensory roles of the CSP have been only studied in insects (Pelosi, Iovinella, Zhu, Wang, & Dani, 2018; Robinson, 2011).

It is worth noting that most functional knowledge of the role of these soluble proteins in invertebrates is primarily based on studies in insects. Current high-throughput sequencing (HTS) technologies are undoubtedly providing novel genomics data from other arthropods, highly valuable to get insights into the origin, evolution and function of soluble proteins in this highly diverse animal group. Indeed, initiatives such as the i5k project (Robinson, 2011) have greatly boosted the sequencing and functional annotation of complete genomes from many non-model arthropod species. Remarkably, these studies have permitted the identification in chelicerates and myriapods of candidate homologs to insect OBPs, named as OBP-like, placing the origin of this gene family much older than previously thought (Renthal et al., 2017; Vizueta et al., 2017, 2018).

The accurate identification and annotation of gene family members in genome assemblies of non-model organisms is a major challenge in evolutionary and functional genomics studies, particularly in the absence of a properly annotated reference genome from closely related species. This drawback is especially relevant in older gene families where divergent and newly originated (very similar) copies coexist in the same genomes, as in the OBPs. In these cases, the oldest divergent copies are very difficult to detect using sequence similarity-based searches, whereas newly arisen duplicates (originated by unequal crossing-over) usually appears physically close in the genome in long tandem gene arrays (Clifton et al., 2020; Vieira, Sánchez-Gracia, & Rozas, 2007), precluding the prediction of correct gene models (e.g., producing incomplete or chimeric genes).

In this chapter, we summarize the main steps and bioinformatics methods applied to the structural and functional annotation of gene family members in the genome drafts of non-model organisms, using the OBP family as an example. To illustrate the process, we will use BITACORA (Vizueta, Sánchez-Gracia, & Rozas, 2020) a bioinformatics pipeline that implements many of the algorithms and methods to assist the comprehensive annotation of gene families in genome assemblies. This software performs a semi-automatic mining of genome sequences that permits the characterization of gene family copies, including both recent and ancient members. In this chapter, we also describe some commonly used methods to verify that the identified genes contain the protein domains and the structural features characteristic of the OBP and CSP proteins, such as the signal peptide, the specific α-helical domains that assemble into a stable and compact structure with a hydrophobic binding pocket, or the conserved cysteine pattern forming the characteristic disulfide bridges.

2. Identification and visualization of OBP genes in genome assemblies

The correct identification and gene model annotation of all the members of a gene family present in a genome is critical for the success of further analyses, both for conducting evolutionary genomic analyses (estimating gene birth-and-death rates, detecting the footprint of positive selection), or to perform further functional studies. In this section, we provide a brief guide of the appropriate workflow and bioinformatics tools to identify and annotate the members of a gene family of interest. Specifically, we center on how to identify members of the OBP and CSP gene families from genomic data. We will focus only on the technical aspects; the main concepts and theoretical framework associated with gene identification and annotation can be consulted in the suggested literature.

In this review, we use the genome sequence information of the tobacco hornworm moth *Manduca sexta* (BioProject number PRJNA81037; Kanost et al., 2016) as an example of a genome draft from a non-model organism. This species has a typical insect genome size (419 Mb), an intermediate assembly quality (N50 = 40.4 kb; 20,871 scaffolds), and encodes for a medium size OBP gene repertoire (49 proteins) (Kanost et al., 2016; Vogt, Grosse-Wilde, & Zhou, 2015), which include members of the Classic (24 proteins), Minus-C (18 proteins) and Plus-C (7 proteins) subfamilies (Hekmat-Scafe, Scafe, McKinney, & Tanouye, 2002; Vieira et al., 2007; Zhou, Huang, Zhang, Pickett, & Field, 2004), as well as 22 CSPs.

2.1 Identifying structural annotations

In this section we have mainly used the pipeline BITACORA to conduct an exhaustive and assisted identification and annotation of chemosensory gene family members. Although we describe the details for the OBP, the protocol could be straightforward applied for the CSP or any other gene family. BITACORA can be downloaded from https://github.com/molevol-ub/bitacora. See also the publication and documentation of this software (http://www.ub.edu/softevol/bitacora/) for specific methodological details and installation requirements.

2.1.1 Procedure

1. Prepare the genome data. Download the genome sequences and structural gene annotations of the surveyed species. The last release of the *Manduca sexta* genome can be found in https://www.ncbi.nlm.nih.gov/genome/?term=manduca (under BioProject PRJNA81037; Assembly ID GCA_000262585.1). BITACORA needs a FASTA file with the DNA sequences (referred to as genome in NCBI), the GFF (general feature format) file with annotation features, and the file with predicted protein sequences (also in FASTA format). The GFF file contains the genome coordinates (including the chromosome or scaffold) of all gene features (e.g., genes, exons, introns or repetitive regions) in the genome assembly (see Section 2.1.1.1 note 4 below for specific details about the NCBI GFF format).
2. Prepare the Focal Protein Data Base (FPDB). Build the database with the OBP sequences to be used as queries to mine the genome of *M. sexta.* The quality of this dataset is critical for the successful identification and annotation of the members of the focal gene family. The FPDB (see the BITACORA documentation) is composed of two types of data, the protein sequence database (in FASTA format), and the HMM (Hidden Markov Model) profile of the focal gene family. The FPDB data used in our example was entirely obtained from Pfam database (El-Gebali et al., 2019) and includes the PBP_GOBP (PF01395) profile and all OBP sequences included to build this profile (http://pfam.xfam.org/family/PF01395#tabview=tab0). Protein sequences can be downloaded from "Alignments" subsection (tab 3 in Pfam webpage), in "Format an alignment" box, and selecting the preferred dataset (in this case the full set) in FASTA format with no gaps (unaligned). The file with the HMM profile can be obtained from the subsection "Curation & model" in Pfam (tab 6) in "HMM information." In the present example we have renamed the FPDB files as "OBP_db.fasta" and "OBP_db.hmm."

3. Execute the BITACORA pipeline. Specify the path of the files with genome sequences and the FPDB datasets in the script "runBITACORA.sh" and execute it in the command prompt or terminal ($ > bash runBITACORA.sh). BITACORA will run automatically the next two steps.
4. Identify and curate proteins with structural gene annotation. This corresponds with Step 1 in the workflow of the BITACORA manual. In this step the software launches BLASTP and HMMER programs (Altschul, Gish, Miller, Myers, & Lipman, 1990; Eddy, 2011) to identify all putative OBP members among the already predicted peptides in the *M. sexta* genome, using as input for similarity-based searches the FPDB prepared for the OBP gene family. The positive hits from these searches are then filtered for quality (removing very small hits and identifying chimeric models and other incorrect annotations) and combined into a single integrated result. BITACORA reports the gene coordinates (in a new GFF file), and the curated protein sequences (in a FASTA file) of all identified OBP, which are incorporated into the updated FPDB (uFPDB) to be used in the following step.
5. Identify structurally unannotated OBP genes. This corresponds to Step 2 in the workflow of the BITACORA manual. In this step, the software uses TBLASTN to search for OBP genes in the genome (DNA sequence), using as query the OBP information included in the uFPDB. Then, the GeMoMa algorithm (Keilwagen et al., 2016; Keilwagen, Hartung, Paulini, Twardziok, & Grau, 2018) is used to reconstruct gene models based on the TBLASTN hits (see Section 2.1.1.1 step 7 below). The results from this step are also curated for erroneous gene models originated by the incorrect merging of exons from different genes generating gene fusions. We use the specific OBP protein domain to delimit individual genes (see Section 2.1.1.1 note 8 below).
6. Generation of the output files. BITACORA generates three output files ("*.txt") and one folder. The files include information about the number of OBP genes identified and annotated, whereas the output folder (e.g., the OBP folder in our case) includes files with the relevant data: (1) information of all OBP members identified in the genome (a GFF and FASTA file with both curated and novel identified gene models), (2) all non-redundant proteins encoded by the models in the GFF (a FASTA file), and (3) two BED files with the genomic coordinates of all OBPs in *M. sexta*. These BED files are especially useful to visualize (and if needed manually editing) annotations in genome editors such as Apollo (See Fig. 1). Moreover, BITACORA can also generate a second

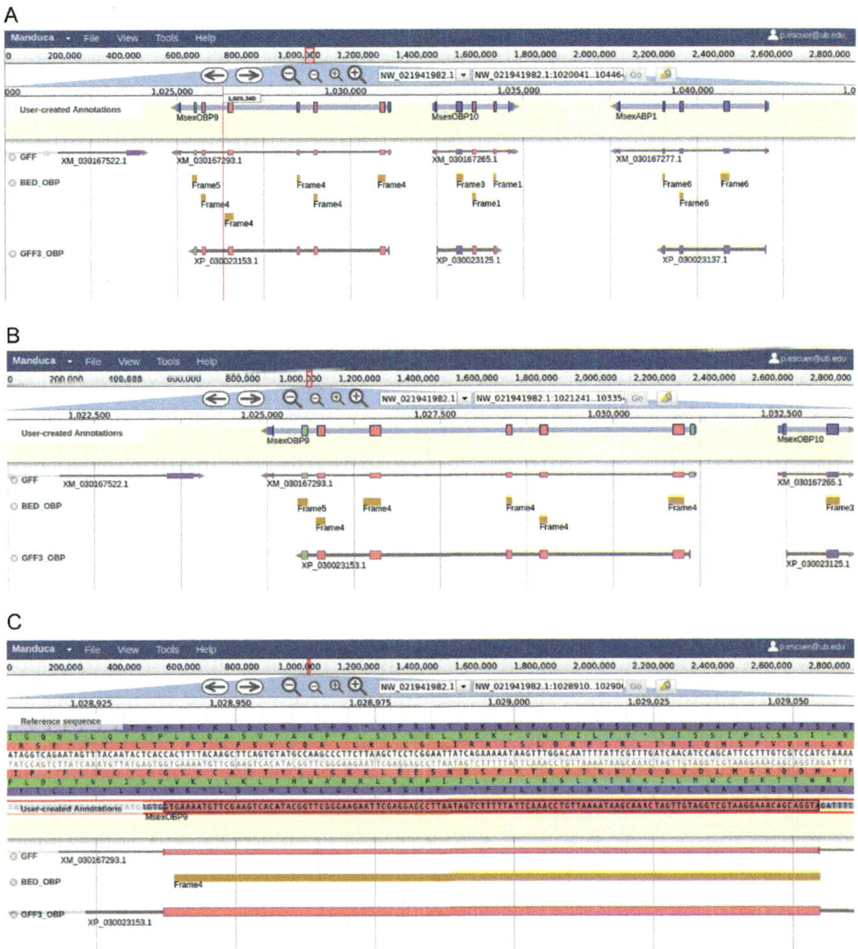

Fig. 1 Apollo genome browser visualization of the annotations in *Manduca sexta* genome. The screenshot shows some of the OBP annotated on the scaffold 0005 in Vogt, Grosse-Wilde, and Zhou (2015), which correspond to the scaffold NW_021941982.1 in the BioProject number PRJNA81037 (Kanost et al., 2016). The GFF file for the genome annotation track was obtained from the *Manduca* genome project (point 1 in Section 2.2.1). The BED_OBP and GFF3_OBP tracks were generated using the GFF obtained with BITACORA for the OBP gene family. (A) Image showing a cluster of 3 OBP genes in this genomic region. XM_030167293.1, XM_030167265.1 and XM_030167277.1 genes (in the GFF track) corresponds to MsexObp9, MsexObp10 and MsexAbp1 in Vogt, Grosse-Wilde, and Zhou (2015), respectively. The gene XM_030167522.1 does not encode an OBP protein. (B) Zoom of panel A showing the structure of the MsexObp9 gene (which is encoded in the reverse strand). (C) Zoom of panel B showing the DNA sequence of the third exon of this gene (i.e., the second frame in the BED track of this gene model). At this maximum zoom level, the Apollo browser shows the six possible open reading frames (in colors) across the two strands.

folder (selecting the "cleaning" option) with all intermediate files and information generated during the run, including unedited gene models for comparisons.

7. Perform additional iterative searches. The user can also perform an additional iterative search round (Step 3 in the workflow of the BITACORA manual). The input for this step is a new OBP database that includes all OBPs identified in points 4 and 5 above (referred as sFPDB database). This step could be especially useful for searching remote homologs or in case of having used a FPDB file with OBP from evolutionary distant species. In the present example all *M. sexta* OBP sequences were properly identified in the first round (points 4 and 5 above). Nevertheless, we recommend running one iterative step to ensure a comprehensive identification of all OBP members in the genomes of non-model organisms (see Note 6 below).

2.1.1.1 Notes

(1) Although the current protocol focuses on the OBP gene family, it could be extended to any gene family.

(2) It is very important that the FPDB must include highly curated gene family members, including available data from closely related species. Otherwise, the iterative searches might identify as a member of the focal family other evolutionary unrelated sequences, especially in very old and/or fast-evolving gene families.

(3) In the absence of publicly available HMM profiles, the user can generate their own profiles using the proteins already available in databases for a particular family with the program *hmmbuild* (Finn, Clements, & Eddy, 2011). This software generates HMM profiles from multiple protein sequence alignments. Also, the InterPro tool (http://www.ebi.ac.uk/interpro/) can be used to identify the characteristic protein domains of a gene family (to be used in the point 5 above)

(4) The NCBI GFF files must be converted to a format compatible with BITACORA (to correctly assign gene features IDs with the protein sequences); this task can be done with the script "reformat_ncbi_gff.pl" (included in the Tools folder of the BITACORA package). Occasionally, some proteins (included in the FASTA file downloaded from the NCBI web) are not encoded in the GFF; in this case it is convenient to exclude them from the annotation step; this task is performed by the script "get_proteins_notfound_ingff.pl" (also included in the Tools folder of BITACORA).

(5) BITACORA pipeline can also be used only with protein evidence, denoted as the "protein" mode in BITACORA. For instance, there are several RNA-seq data experiments (mainly in non-model organisms) without reference genome information. In these cases, using the "protein" mode of BITACORA, the pipeline can characterize the focal gene families from the conceptually translated proteins from the transcriptome.

(6) The absence of structural gene annotation in a genome project is not common. However, BITACORA can also be used to perform a de novo annotation ("genome" mode of BITACORA). In this case, it is highly recommended to run the optional iterative search round (point 7 of this section).

(7) BITACORA can construct novel gene models using two different algorithms (point 5). By default, BITACORA applies the GeMoMa algorithm (Keilwagen et al., 2016, 2018) to build accurate gene models. This method, nevertheless, could not be sensitive enough to identify members under putative sequencing or assembly errors. Alternatively, using the "close proximity method" allows identifying all putative exon sequences (obtained through TBLASTN), which are merged into genes based on their physical distance in the scaffold. Although this method consistently recovers most gene family members, it requires the further validation by the user, especially for clustered copies, and the verification of exon-intron delimitation; BITACORA facilitates this first task by labeling each predicted protein with the ID of the identified protein domain (e.g., OBP domains).

(8) BITACORA offers two options for the validation and curation of novel gene family members obtained in point 5 (BITACORA "genomicblastp" option). For the one hand (default option), BITACORA will assess for the presence in the annotated proteins of the FPDB specific domain (using HMMER), and only those gene models having the domain will be retained and curated. Alternatively, BITACORA will validate the quality of the new identified proteins both using positive HMMER (as in the abovementioned option) or also BLASTP (against the proteins in the FPDB) hits. In this case, although the sensitivity of the annotations increases considerably, the specificity could decrease, especially if the FPDB contains poor quality annotations (or repetitive regions) which might generate wrong gene models (with non-homologous fragments).

(9) Specific details and more advanced specifications of the protocol can be found in (Vizueta et al., 2020), and references therein.

2.2 Visualization of OBP genes using the Apollo genome editor

Proteins identified using the steps specified in the previous section can be directly used for prospective evolutionary and functional analyses, however, they might need further manual curation if the objective is a comprehensive study. In this section, we describe how to carry out that with the Apollo genome browser, a user-friendly genome visualization and edition tool (Fig. 1). This genome browser is especially prone for researchers working simultaneously in the annotation of the same genome project (Lee et al., 2013). With Apollo, the user can visualize OBP members annotated using the protocol explained in the Section 2.1 (BED and GFF files generated by BITACORA) along with other external evidence, for example transcriptome data. Such additional information is incorporated as different genome annotation tracks, which permits a visual inspection of the distribution of different genomic features across the scaffold (e.g., gene lengths, gene clustering), but also assists the manual curation and editing of gene models. The Apollo software can be downloaded from: https://genomearchitect.readthedocs.io/en/latest/index.html.

2.2.1 Procedure

1. Local installation. Install Apollo on your personal computer; the specific details can be obtained from: https://genomearchitect.readthedocs.io/en/latest/Setup.html. A correct installation may require the modification of some of the setup configuration files (see the "Apollo configuration" section). Some other parameters, however, can be configured directly in the browser interface.
2. Prepare the track with the reference genome sequence data (the genome of *M. sexta* in our example). Genome data (FASTA file used in the point 1 of the Section 2.1.1) should be formatted before being used in the Apollo browser. The "data generation pipeline" section of Apollo includes a script, "prepare-refseqs.pl," specially set for this task. The output of this script can be found in the user defined directory.
3. Prepare the track with the GFF data of *M. sexta*. Again, the GFF file (the file used in the point 1 of the Section 2.1.1) should be correctly formatted to be used in Apollo. In this case, the script "flatfile-to-json.pl," available in the "data generation pipeline" section of Apollo, can be used. The output of this script is sent to the user defined data directory.
4. Generate a searchable index of names (optional). Although this option is not mandatory, it is very useful to facilitate the search for terms in the

browser navigation. The script "generate-names.pl" available in the "data generation pipeline" section of Apollo can be used for this task.

5. Add a BAM file track (optional). Apollo can also read alignment BAM files (http://samtools.github.io/hts-specs). These files can incorporate some complementary evidence, such as transcriptomic data, which helps annotating and curating gene models. Use the "add-bam-track.pl" script available in the "data generation pipeline" section of Apollo to prepare the BAM file track.
6. Add a BigWig track with raw read data (optional). BigWig is a file format to display continuous data in a genome browser as a graph. To visualize more efficiently these data, it is recommended to convert the BAM file into BigWig format. Use the "add-bw-track.pl" script available in the "data generation pipeline" section of Apollo to perform this task.
7. Add the tracks with annotations of the OBP gene family. The BED/GFF files generated by BITACORA (files including the coordinates and gene structure of *M. sexta* OBPs) can be directly uploaded (without the need to be formatted) as a temporary file in the Apollo browser, permitting a faster visualization. Use the "File > Open track file" option to upload or drag these files.
8. Edit structural annotations of genes. To edit incorrectly annotated genes, select the pre-annotated gene from the GFF track (it can be any track with annotations or evidence of our gene of interest), and drag it to the "User Annotation" area. The user can modify gene structure using the right-click menu, and selecting the desired options, such as adding or removing exons, UTRs, etc., and also having the option of changing the reading frame (click at the exon and extend the region).
9. Edit gene information. To add or modify the "name," "description," "GO annotation," or some of the other comments or attributes of a gene, the user should use the "information editor." To edit, select the desired element, use the right-click option, and under "Edit information" introduce the new information.
10. Extracting the gene features of genome sequence. The user can get this information using the right-click menu at the focal gene and selecting "Get GFF3" or "Get Sequence" options. For the later, the user can choose protein, cDNA, CDS or genomic sequences for download.

2.2.1.1 Notes

(1) If tracks are correctly formatted and uploaded, the user should see the "Navigation Panel," with specific options to browse the genome, and a yellow track (on the top of the working area) to incorporate specific annotations ("User-created Annotations" area).

(2) Although it is not mandatory, using transcriptome data is extremely useful for correctly defining exon-intron boundaries. Visualizing this information from a BAM file is less friendly than from BigWig file format. Therefore, it is highly recommended to convert BAM files to BigWig using scripts such as those included in the python suite deepTools ("BamCoverage.py" script).

(3) Most eukaryotic splice sites have the canonical GT/AG sequences. Nevertheless, Apollo also shows the "non-canonical" splice sites. It is recommended to check these features to assess for putative errors.

(4) Since all editing steps performed in Apollo are recorded, it is possible to undo and redo all actions. Right-click to the annotated gene > Show history to see this information.

3. Predicting the characteristic biochemical properties of soluble proteins

3.1 Characteristic motives of insect soluble proteins

In this section, we introduce a number of bioinformatics tools used to recognize the characteristic motives (motives for structural features) encoded by the OBPs, such as (1) the presence of a signal peptide, i.e., a highly hydrophobic motif that tag these proteins for secretion. (2) A characteristic secondary structure conformed by four to eight α-helices assembled in a compact globular and stable structure. (3) A distinctive pattern of conserved cysteines (mostly six, although it can vary from 4 to 12) connecting α-helices with disulfide bridges. (4) A characteristic 3D structure a hydrophobic ligand-binding pocket.

3.1.1 Procedure

1. Software used in this section:

SignalP (Almagro Armenteros et al., 2019) http://www.cbs.dtu.dk/services/SignalP-5.0/

MAFFT (Katoh & Standley, 2013) https://mafft.cbrc.jp/alignment/server/

WebLogo 3 (Crooks, Hon, Chandonia, & Brenner, 2004) http://weblogo.threeplusone.com/create.cgi

PSIPRED (McGuffin, Bryson, & Jones, 2000) http://bioinf.cs.ucl.ac.uk/psipred/

DiANNA (Ferre & Clote, 2005) http://clavius.bc.edu/~clotelab/DiANNA/
Protein-Sol (Hebditch, Carballo-Amador, Charonis, Curtis, & Warwicker, 2017) https://protein-sol.manchester.ac.uk/
PHYRE2 (Kelley, Mezulis, Yates, Wass, & Sternberg, 2015) http://www.sbg.bio.ic.ac.uk/~phyre2/html/page.cgi?id=index

2. Identify the signal peptide (N-terminal part of the protein) in the putative OBP precursor sequences using SignalP (Fig. 2). Upload or paste the query sequences in the SignalP webpage and submit the work. Alternatively, the local version of the software can also be downloaded and ran on a desktop PC. This software reports the probability that the protein contains a signal peptide, the predicted amino acid positions, and the putative specific cleavage site.
3. Identify conserved cysteine patterns. Align all predicted OBP proteins using for example MAFFT web application (with default options; see note 3) or downloading and running this program on your desktop PC. The output (a multiple sequence alignment) can be visualized selecting the "view" option at the top of the MAFFT page and saved in a simple format alignment file by selecting "Fasta format." Next, use this FASTA file to create a logo of the amino acid pattern in the alignment in WebLogo 3 webpage, from which we should clearly see the conserved cysteine pattern characteristic of soluble proteins (Fig. 2).
4. Infer the secondary structure of the putative OBP using software like PSIPRED. Paste the protein sequence in the PSIPRED web page and submit the work. The web application outputs the predicted secondary structure for the query protein, which in the case of a real OBP member includes the characteristic α-helices (Fig. 3).
5. Predict disulfide bonds using DiANNA. Enter a predicted OBP sequence in the DiANNA web server and select the option "Disulfide bond connectivity prediction". The program shows all predicted disulfide bonds among cysteines in our query protein along with their putative prediction score.
6. Infer protein solubility using Protein-Sol. Submit the predicted OBP protein sequence in the Protein-Sol web server. The software predicts protein solubility, fold propensity and net segment charge per amino acid.
7. Predict the 3D structure of candidate OBP proteins by homology modeling using PHYRE2. Submit the protein sequence into the PHYRE2 web application. The program models the 3D structure of

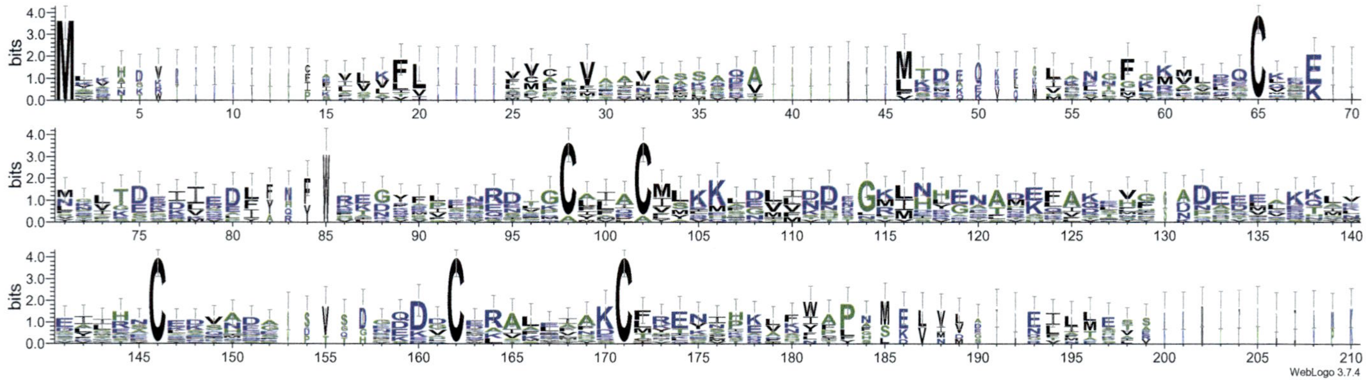

Fig. 2 OBP sequence logo. Degree of amino acid sequence conservation across the 24 sequences of classic OBP from *M. sexta*, showing the six characteristic cysteines conserved in this subfamily. The first amino acids (∼20) corresponds to the signal peptide as identified by SignalP.

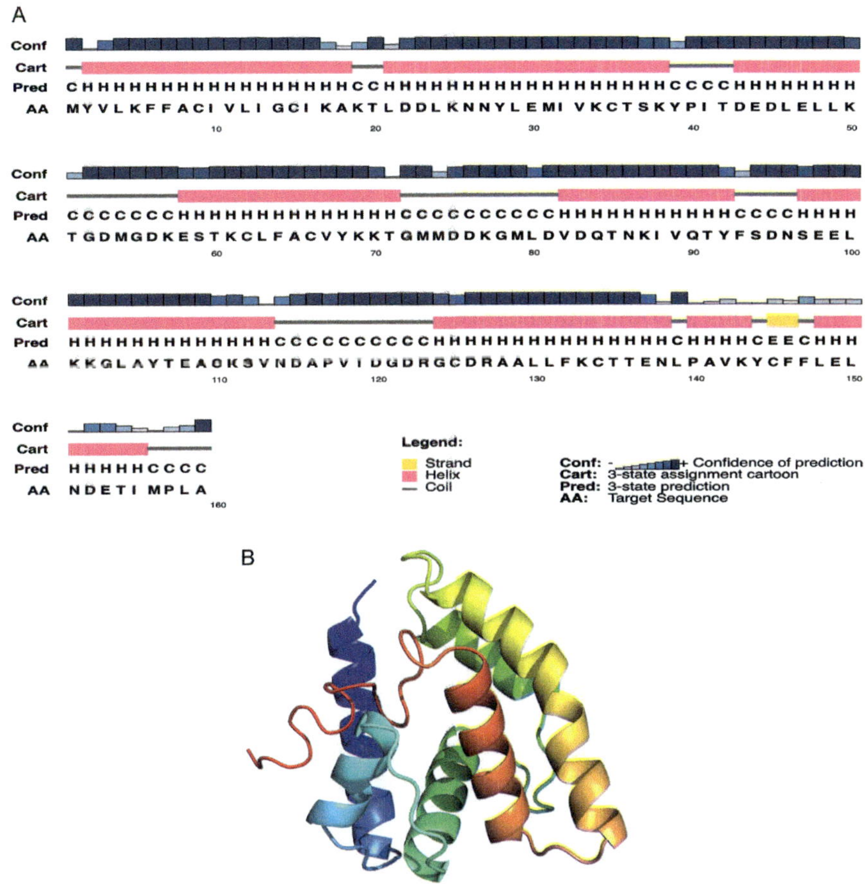

Fig. 3 (A) Secondary structure of MsexOBP9 precursor predicted with PSIPRED (159 amino acids, where the first 19 amino acids correspond to the signal peptide). The mature protein has six cysteines, which form three disulfide bridges (C1–C6; C2–C4; C5–C7) as determined by DiANNA. (B) Predicted three-dimensional structure of MsexOBP9 based on homology modeling in PHYRE2. The OBP4 of the mosquito *Anopheles gambiae* OBP4 was used as a template (PDB ID 3Q8I). The image of predicted structure is generated in the JSmol Viewer included in PHYRE2 server. Color ramp ranges from blue (amino-terminal) to red (carboxyl-terminal).

the query sequence (the candidate OBP) in base of the sequence similarity and predicted secondary structure against some experimentally determined structures (the template) available in the Protein DataBase (PDB; Berman et al., 2000). For each model, the server reports the alignment coverage between the query and the template, a score of the global quality of the model, and information about the structure of the template (Fig. 3). Since the 3D structure of several OBP proteins (in several species) have been determined experimentally (and are available in PDB), they can be used as possible templates. The results include a prediction of the putative ligand-binding sites using 3DLigandSite (Wass, Kelley, & Sternberg, 2010).

3.1.1.1 Notes

(1) The absence of a predicted signal peptide in a gene identified as an OBP or CSP member (or other soluble protein families) could indicate erroneous genome annotations.
(2) Since the signal peptide of soluble proteins evolves rapidly, it is recommended to exclude this part of the protein in downstream evolutionary analyses.
(3) Given that different OBP subfamilies can have different lengths, it is important to construct multiple sequence alignments based solely on clear homologous fragments (i.e., separate the double OBPs in two sequences, ensure the cysteines of the minus-C and plus-C subfamilies are aligned with the corresponding homologous positions in the other families, etc.). This checking can also help to identify artifactual proteins (chimeric gene models) that should be properly curated or discarded from the analyses.
(4) It is worth noting that all these computational predictions should be experimentally validated using proper biochemical and biophysical analyses.

4. Concluding remarks

Current development and falling costs of HTS technologies are promoting the genome sequencing of many non-model organisms and, in many cases, with low or medium quality data. However, the structural and functional annotation of these genomes is a big challenge for evolutionary and functional analysis, especially for gene families. Remote members of gene families are difficult to identify in genome sequences, while very recent members are often organized in clusters, making them difficult to annotate without errors and often failing to identify some of the copies. In this chapter we have described the commonly used bioinformatics protocols to identify all gene family members encoded in a genome, using the OBP family of *M. sexta* as an example and the software BITACORA as a guide. We have also summarized the main workflow steps to recognize and interpret the characteristic biochemical and structural properties of insect soluble protein families. All these tasks are the first and the necessary steps for further functional studies in chemosensory soluble protein gene families such as OBPs or CSPs.

Acknowledgments

This work was supported by the Ministerio de Economía y Competitividad of Spain (CGL2013-45211, CGL2016-75255) and the Comissió Interdepartamental de Recerca I Innovació Tecnològica of Catalonia, Spain (2017SGR1287). A.S.-G. is a Serra Húnter Fellow. J.V. and P.E. were supported by FPI grants from the Ministerio de Economía y Competitividad of Spain (BES-2014-068437 and BES-2017-081740, respectively).

References

Almagro Armenteros, J. J., Tsirigos, K. D., Sonderby, C. K., Petersen, T. N., Winther, O., Brunak, S., et al. (2019). SignalP 5.0 improves signal peptide predictions using deep neural networks. *Nature Biotechnology*, *37*(4), 420–423. https://doi.org/10.1038/s41587-019-0036-z.

Altschul, S. F., Gish, W., Miller, W., Myers, E. W., & Lipman, D. J. (1990). Basic local alignment search tool. *Journal of Molecular Biology*, *215*(3), 403–410. https://doi.org/10.1016/S0022-2836(05)80360-2.

Angeli, S., Ceron, F., Scaloni, A., Monti, M., Monteforti, G., Minnocci, A., et al. (1999). Purification, structural characterization, cloning and immunocytochemical localization of chemoreception proteins from Schistocerca gregaria. *European Journal of Biochemistry*, *262*(3), 745–754. https://doi.org/10.1046/j.1432-1327.1999.00438.x.

Asahina, K., Pavlenkovich, V., & Vosshall, L. B. (2008). The survival advantage of olfaction in a competitive environment. *Current Biology*, *18*(15), 1153–1155. https://doi.org/10.1016/j.cub.2008.06.075.

Berman, H. M., Westbrook, J., Feng, Z., Gilliland, G., Bhat, T. N., Weissig, H., et al. (2000). The protein data bank. *Nucleic Acids Research*, *28*(1), 235–242. https://doi.org/10.1093/nar/28.1.235.

Clifton, B., Jimenez, J., Kimura, A., Chahine, Z., Librado, P., Sánchez-Gracia, A., et al. (2020). Understanding the early evolutionary stages of a tandem D. melanogaster-specific gene family: A structural and functional population study. *Molecular Biology and Evolution*, in press. https://doi.org/10.1093/molbev/msaa109.

Crooks, G. E., Hon, G., Chandonia, J. M., & Brenner, S. E. (2004). WebLogo: A sequence logo generator. *Genome Research*, *14*(6), 1188–1190. https://doi.org/10.1101/gr.849004.

Eddy, S. R. (2011). Accelerated profile HMM searches. *PLoS Computational Biology*, 7(10), e1002195. https://doi.org/10.1371/journal.pcbi.1002195.

El-Gebali, S., Mistry, J., Bateman, A., Eddy, S. R., Luciani, A., Potter, S. C., et al. (2019). The Pfam protein families database in 2019. *Nucleic Acids Research*, *47*(D1), D427–D432. https://doi.org/10.1093/nar/gky995.

Ferre, F., & Clote, P. (2005). DiANNA: A web server for disulfide connectivity prediction. *Nucleic Acids Research*, *33*(Web Server issue), W230–W232. https://doi.org/10.1093/nar/gki412.

Finn, R. D., Clements, J., & Eddy, S. R. (2011). HMMER web server: Interactive sequence similarity searching. *Nucleic Acids Research*, *39*(Web Server issue), W29–W37. https://doi.org/10.1093/nar/gkr367.

Hebditch, M., Carballo-Amador, M. A., Charonis, S., Curtis, R., & Warwicker, J. (2017). Protein-sol: A web tool for predicting protein solubility from sequence. *Bioinformatics*, *33*(19), 3098–3100. https://doi.org/10.1093/bioinformatics/btx345.

Hekmat-Scafe, D. S., Scafe, C. R., McKinney, A. J., & Tanouye, M. A. (2002). Genome-wide analysis of the odorant-binding protein gene family in *Drosophila melanogaster*. *Genome Research*, *12*(9), 1357–1369. https://doi.org/10.1101/gr.239402.

Ishida, Y., Tsuchiya, W., Fujii, T., Fujimoto, Z., Miyazawa, M., Ishibashi, J., et al. (2014). Niemann-pick type C2 protein mediating chemical communication in the worker ant. *Proceedings of the National Academy of Sciences of the United States of America*, *111*(10), 3847–3852. https://doi.org/10.1073/pnas.1323928111.

Kanost, M. R., Arrese, E. L., Cao, X., Chen, Y. R., Chellapilla, S., Goldsmith, M. R., et al. (2016). Multifaceted biological insights from a draft genome sequence of the tobacco hornworm moth, Manduca sexta. *Insect Biochemistry and Molecular Biology*, *76*, 118–147. https://doi.org/10.1016/j.ibmb.2016.07.005.

Katoh, K., & Standley, D. M. (2013). MAFFT multiple sequence alignment software version 7: Improvements in performance and usability. *Molecular Biology and Evolution*, *30*(4), 772–780. https://doi.org/10.1093/molbev/mst010.

Keilwagen, J., Hartung, F., Paulini, M., Twardziok, S. O., & Grau, J. (2018). Combining RNA-seq data and homology-based gene prediction for plants, animals and fungi. *BMC Bioinformatics*, *19*(1), 189. https://doi.org/10.1186/s12859-018-2203-5.

Keilwagen, J., Wenk, M., Erickson, J. L., Schattat, M. H., Grau, J., & Hartung, F. (2016). Using intron position conservation for homology-based gene prediction. *Nucleic Acids Research*, *44*(9), e89. https://doi.org/10.1093/nar/gkw092.

Kelley, L. A., Mezulis, S., Yates, C. M., Wass, M. N., & Sternberg, M. J. (2015). The Phyre2 web portal for protein modeling, prediction and analysis. *Nature Protocols*, *10*(6), 845–858. https://doi.org/10.1038/nprot.2015.053.

Larter, N. K., Sun, J. S., & Carlson, J. R. (2016). Organization and function of Drosophila odorant binding proteins. *eLife*, *5*, e20242. https://doi.org/10.7554/eLife.20242.

Lee, E., Helt, G. A., Reese, J. T., Munoz-Torres, M. C., Childers, C. P., Buels, R. M., et al. (2013). Web Apollo: A web-based genomic annotation editing platform. *Genome Biology*, *14*(8), R93. https://doi.org/10.1186/gb-2013-14-8-r93.

McGuffin, L. J., Bryson, K., & Jones, D. T. (2000). The PSIPRED protein structure prediction server. *Bioinformatics*, *16*(4), 404–405. https://doi.org/10.1093/bioinformatics/16.4.404.

Meyerhof, W., & Korsching, S. (2009). Chemosensory systems in mammals, fishes, and insects. *Results and Problems in Cell Differentiation*, *47*, v–xi. Retrieved from https://www.ncbi.nlm.nih.gov/pubmed/19655435.

Mollo, E., Fontana, A., Roussis, V., Polese, G., Amodeo, P., & Ghiselin, M. T. (2014). Sensing marine biomolecules: Smell, taste, and the evolutionary transition from aquatic to terrestrial life. *Frontiers in Chemistry*, *2*, 92. https://doi.org/10.3389/fchem.2014.00092.

Mori, K., Takahashi, Y. K., Igarashi, K. M., & Yamaguchi, M. (2006). Maps of odorant molecular features in the mammalian olfactory bulb. *Physiological Reviews*, *86*(2), 409–433. https://doi.org/10.1152/physrev.00021.2005.

Nomura, A., Kawasaki, K., Kubo, T., & Natori, S. (1992). Purification and localization of p10, a novel protein that increases in nymphal regenerating legs of Periplaneta americana (American cockroach). *The International Journal of Developmental Biology*, *36*(3), 391–398. Retrieved from https://www.ncbi.nlm.nih.gov/pubmed/1445782.

Pelosi, P. (1990). Odorant-binding proteins in vertebrates and insects: Similarities and possible common function. *Chemical Senses*, *15*(2), 205–215. https://doi.org/10.1093/chemse/15.2.205.

Pelosi, P., Iovinella, I., Felicioli, A., & Dani, F. R. (2014). Soluble proteins of chemical communication: An overview across arthropods. *Frontiers in Physiology*, *5*, 320. https://doi.org/10.3389/fphys.2014.00320.

Pelosi, P., Iovinella, I., Zhu, J., Wang, G., & Dani, F. R. (2018). Beyond chemoreception: Diverse tasks of soluble olfactory proteins in insects. *Biological Reviews of the Cambridge Philosophical Society*, *93*(1), 184–200. https://doi.org/10.1111/brv.12339.

Pelosi, P., & Maida, R. (1995). Odorant-binding proteins in insects. *Comparative Biochemistry and Physiology Part B: Biochemistry and Molecular Biology*, *111*(3), 503–514. https://doi.org/10.1016/0305-0491(95)00019-5.

Renthal, R., Manghnani, L., Bernal, S., Qu, Y., Griffith, W. P., Lohmeyer, K., et al. (2017). The chemosensory appendage proteome of Amblyomma Americanum (Acari: Ixodidae) reveals putative odorant-binding and other chemoreception-related proteins. *Insect Sci*, *24*(5), 730–742. https://doi.org/10.1111/1744-7917.12368.

Robinson, R. (2011). For genomes, repetition breeds diversity. *PLoS Biology*, *9*(6), e1001079. https://doi.org/10.1371/journal.pbio.1001079.

Sánchez-Gracia, A., Vieira, F. G., & Rozas, J. (2009). Molecular evolution of the major chemosensory gene families in insects. *Heredity*, *103*(3), 208–216. https://doi.org/10.1038/hdy.2009.55.

Smadja, C., & Butlin, R. K. (2009). On the scent of speciation: The chemosensory system and its role in premating isolation. *Heredity*, *102*(1), 77–97. https://doi.org/10.1038/hdy.2008.55.

Touhara, K., & Vosshall, L. B. (2009). Sensing odorants and pheromones with chemosensory receptors. *Annual Review of Physiology*, *71*, 307–332. https://doi.org/10.1146/annurev.physiol.010908.163209.

Vieira, F. G., & Rozas, J. (2011). Comparative genomics of the odorant-binding and chemosensory protein gene families across the Arthropoda: Origin and evolutionary history of the chemosensory system. *Genome Biology and Evolution*, *3*, 476–490. https://doi.org/10.1093/gbe/evr033.

Vieira, F. G., Sánchez-Gracia, A., & Rozas, J. (2007). Comparative genomic analysis of the odorant-binding protein family in 12 Drosophila genomes: Purifying selection and birth-and-death evolution. *Genome Biology*, *8*(11), R235. https://doi.org/10.1186/gb-2007-8-11-r235.

Vizueta, J., Frias-Lopez, C., Macias-Hernandez, N., Arnedo, M. A., Sánchez-Gracia, A., & Rozas, J. (2017). Evolution of chemosensory gene families in arthropods: Insight from the first inclusive comparative transcriptome analysis across spider appendages. *Genome Biology and Evolution*, *9*(1), 178–196. https://doi.org/10.1093/gbe/evw296.

Vizueta, J., Rozas, J., & Sánchez-Gracia, A. (2018). Comparative genomics reveals thousands of novel chemosensory genes and massive changes in chemoreceptor repertories across chelicerates. *Genome Biology and Evolution*, *10*(5), 1221–1236. https://doi.org/10.1093/gbe/evy081.

Vizueta, J., Sánchez-Gracia, A., & Rozas, J. (2020). BITACORA: A comprehensive tool for the identification and annotation of gene families in genome assemblies. *Molecular Ecology Resources*, in press. https://doi.org/10.1111/1755-0998.13202.

Vogt, R. G., Grosse-Wilde, E., & Zhou, J. J. (2015). The Lepidoptera odorant binding protein gene family: Gene gain and loss within the GOBP/PBP complex of moths and butterflies. *Insect Biochemistry and Molecular Biology*, *62*, 142–153. https://doi.org/10.1016/j.ibmb.2015.03.003.

Vogt, R. G., & Riddiford, L. M. (1981). Pheromone binding and inactivation by moth antennae. *Nature*, *293*(5828), 161–163. https://doi.org/10.1038/293161a0.

Wass, M. N., Kelley, L. A., & Sternberg, M. J. (2010). 3DLigandSite: Predicting ligand-binding sites using similar structures. *Nucleic Acids Research*, *38*(Web Server issue), W469–W473. https://doi.org/10.1093/nar/gkq406.

Xiao, S., Sun, J. S., & Carlson, J. R. (2019). Robust olfactory responses in the absence of odorant binding proteins. *eLife*, *8*, e51040. https://doi.org/10.7554/eLife.51040.

Xu, P., Atkinson, R., Jones, D. N., & Smith, D. P. (2005). Drosophila OBP LUSH is required for activity of pheromone-sensitive neurons. *Neuron*, *45*(2), 193–200. https://doi.org/10.1016/j.neuron.2004.12.031.

Zhou, J. J., Huang, W., Zhang, G. A., Pickett, J. A., & Field, L. M. (2004). "Plus-C" odorant-binding protein genes in two Drosophila species and the malaria mosquito *Anopheles gambiae*. *Gene*, *327*(1), 117–129. https://doi.org/10.1016/j.gene.2003.11.007.

CHAPTER TWO

Reconstructing protein-coding sequences from ancient DNA

Michael Hofreiter*, Stefanie Hartmann
Evolutionary Adaptive Genomics, Institute for Biochemistry and Biology, University of Potsdam, Potsdam, Germany
*Corresponding author: e-mail address: mhofreit@uni-potsdam.de

Contents

Abstract

Obtaining information about functional details of proteins of extinct species is of critical importance for a better understanding of the real-life appearance, behavior and ecology of these lost entries in the book of life. In this chapter, we discuss the possibilities to retrieve the necessary DNA sequence information from paleogenomic data obtained from fossil specimens, which can then be used to express and subsequently analyze the protein of interest. We discuss the problems specific to ancient DNA, including mis-coding lesions, short read length and incomplete paleogenome assemblies. Finally, we discuss an alternative, but currently rarely used approach, direct PCR amplification, which is especially useful for comparatively short proteins.

Methods in Enzymology, Volume 642
ISSN 0076-6879
https://doi.org/10.1016/bs.mie.2020.05.008

1. Introduction

The majority of species that have ever lived on earth is now extinct. In fact, many scientists argue that extinct species account for more than 99% of the species that have existed on earth so far (Jablonski, 2004). Yet, we know only a tiny fraction of these, often from only partial specimens. For even fewer, enough fossils have been recovered to make some informed guesses about their ecology. And only a tiny fraction died out recently enough that their genome can be reconstructed. However, even though it can be applied to only a small number of species, the field of paleogenomics has made enormous progress in recent years (Shapiro & Hofreiter, 2014). Paleogenomes of varying quality—mostly partial, but also occasionally of high coverage—are being reconstructed from an increasing number of species. Although at the moment, most paleogenomes available have been reconstructed from fossils of modern humans or domestic animals (McHugo, Dover, & MacHugh, 2019; Skoglund & Mathieson, 2018), the number of paleogenomes obtained from extinct taxa is also on the rise. If we ignore taxonomic issues and just focus on distinct evolutionary lineages, genomes have been reconstructed for our closest relatives, the Denisovans and Neanderthals (Green et al., 2010; Reich et al., 2010), as well as for the European aurochs (Park et al., 2015), cave bears (Barlow et al., 2018), mammoths, straight tusked elephant, American mastodon (Palkopoulou et al., 2018) and several other taxa.

Such paleogenomic data can be used for many types of analyses, including the inference of phylogenies, gene flow or population dynamics (Shapiro & Hofreiter, 2014). However, in this chapter we focus on the possibility of learning more about the functioning of specific proteins of extinct species. There are in principle two major approaches to identify the sequence of a specific protein of an extinct evolutionary unit. First, and this is also chronologically the first approach that was taken, one can use the polymerase chain reaction (PCR) to amplify the DNA sequence of interest from a DNA extract obtained from a fossil. And second, one can sequence a complete paleogenome to sufficient coverage and then use computational approaches to assemble the coding sequence of the gene of interest. In the methods section, we will describe this second approach, but for completeness, we will treat the first one here.

PCR was the first method that allowed obtaining reliable ancient DNA sequences. For a long time, researchers focused mainly on the mitochondrial genome (Paijmans, Gilbert, & Hofreiter, 2013), due to its higher copy number in a cell compared to the nuclear genome. However, nuclear sequences were also regularly analyzed, and the first nuclear gene segments from Pleistocene specimens were reported as early as 1999 (Greenwood, Capelli, Possnert, & Pääbo, 1999). It took another 7 years, however, before the first functional analysis of an extinct gene, the MC1R gene of the woolly mammoth, was published (Römpler, Rohland, et al., 2006). Although a few more genes were analyzed this way (e.g., Campbell et al., 2010; Lalueza-Fox et al., 2007), the fragmented nature of ancient DNA makes this approach difficult for longer genes, and it never became a strongly pursued field of research.

Interest in the functional effect of genetic variants from extinct evolutionary lineages got renewed with the availability of paleogenomes of our closest relatives, Neanderthals and Denisovans, especially since ancient admixture between these lineages and our own one resulted in a substantial number of their alleles having survived in the modern gene pool (Green et al., 2010; Reich et al., 2010). One early and spectacular result of a closer look at these introgressed alleles was the realization that one key allele for high-altitude adaptation in Tibetans was in fact derived from introgression from Denisovans (Huerta-Sánchez, Jin, Asan, et al., 2014). Interestingly, despite a growing number of publications investigating the functional consequences of genetic variation introgressed from these archaic human lineages into the modern human gene pool (e.g., Dannemann & Kelso, 2017; Dannemann, Prüfer, & Kelso, 2017), all these studies work with association analyses. This approach is possible since many archaic alleles are preserved in living modern humans, and a lot is known about the functional implications of different alleles of many genes in our own species.

However, for species whose genetic variants went physically extinct together with the species that carried them, a different approach is necessary. In this chapter, we discuss how to obtain complete gene sequences from paleogenomes of extinct species. These sequences can subsequently be used to express the extinct gene variants in order to understand their function in as much detail as in-vitro approaches allow. Although rarely used, they can even be analyzed in-vivo, by creating transgenic animals carrying the extinct variant (Huynen et al., 2014).

2. Assembling protein-coding sequences from paleogenomic data

We will not go into the details of assembling genomes de-novo, as this approach is normally not performed on paleogenomic data. Although genome assemblies were initially attempted from shotgun genome sequencing data alone (Li et al., 2010), and this approach has been used to assemble the genome of the recently extinct marsupial wolf (Feigin et al., 2018), this is not commonly done. The main reason is that the resulting genomes are of low contiguity, and the structure of many proteins in such assemblies is incomplete or incorrect, even for modern samples. Currently used approaches for de-novo genome assemblies from modern DNA use a combination of techniques, of which several rely on very long DNA sequencing reads, something that cannot be obtained from ancient DNA due to its highly fragmented nature. We refer readers interested in this process to relevant reviews on this topic (e.g., Ghurye & Pop, 2019; Simpson & Pop, 2015). Independent of the way the assembly is performed, a de-novo genome assembly then needs to be annotated, i.e., the sections of the genome sequence that code for genes have to be identified. For protein-coding genes, this includes the identification of the locations and precise boundaries of exons and introns. Moreover, the function of these genes has to be predicted. Both structural and functional genome annotations are not trivial processes, and close inspection of published genomes often reveals errors in assembly as well as annotation.

Using modern samples, it is also possible to apply RNAseq, i.e., sequencing of cDNA that is obtained via reverse transcription from cellular RNA and assembly of genes using these transcriptomic data. The gene sets obtained this way are usually quite fragmented, and reconstructed genes are often chimeric. Moreover, although RNA data can be obtained from a minority of ancient samples (Smith & Gilbert, 2018), reconstructing a full transcriptome from these data is not possible.

2.1 Obtaining paleogenomic data

Compared to modern DNA, ancient DNA presents researchers with a number of additional challenges. Samples from which ancient DNA is obtained have been exposed to the environment for thousands to tens, in rare cases even hundreds, of thousands of years. This results in ancient DNA being highly fragmented (average read length often below 50 bp), being present in relatively small amounts (at least compared to modern DNA), and being contaminated with DNA (often the majority of the DNA extracted) from microbes that have lived on, in and from the fossil investigated. Finally, ancient DNA extracts often contain additional substances such as humid acids that inhibit enzymatic steps that are required for processing the DNA in order to obtain sequence data.

We cannot detail all these experimental aspects, but we will give a short overview with references to well established methods for the main steps. A key step in obtaining paleogenomic data is DNA extraction. Numerous protocols have been described for maximizing DNA amounts while simultaneously minimizing co-extraction of inhibitors. However, in recent years, two related protocols using binding of DNA to silica columns have been established as the most successful approaches (Dabney et al., 2013; Glocke & Meyer, 2017). Interestingly, recent research also showed that selection of the right skeletal element (almost all ancient DNA work on animals is performed using bone samples) prior to DNA extraction plays an important role for the success of a project. Here, the petrous bone from the inner ear was found to perform exceedingly well compared to all other skeletal elements (Pinhasi et al., 2015). If no petrous bone is available, the cementum of teeth (Damgaard et al., 2015) or the hard, outer surface of other bones, which can be determined using CT-scanning (Alberti et al., 2018) provide promising, albeit less reliable sources for ancient DNA.

The next key step is the construction of a massively parallel sequencing (MPS) library from the ancient DNA extract. The sequencing platform currently used almost exclusively in paleogenomic research is Illumina sequencing. Therefore, we only refer to this MPS technology. There are two principles for constructing Illumina sequencing libraries, based on double-stranded and single-stranded DNA, respectively. There is a variety of different double-stranded library construction approaches, including commercial kits. However, for obtaining optimal results, it is best to use a variant that is optimized for use of ancient DNA such as described in (Henneberger, Barlow, & Paijmans, 2019). Even better results are obtained with approaches based on single stranded DNA (Meyer et al., 2012), at least when these are implemented competently. Finally, before the library can be sequenced, it needs to be quantified and amplified adequately (Paijmans et al., 2017). Depending on genome size, the required genome coverage, endogenous content and average read length, several hundred million to several billion reads are then obtained using one of the Illumina platforms. For example, for obtaining a 30 × genome for a mammalian species with 2.5 Gb genome size, 50 bp average read length and 50%

endogenous content (which is higher than in most samples), 3 billion reads are required. Higher coverage, larger genome size, shorter average read length and/or lower endogenous content accordingly result in even higher read numbers required.

2.2 Assembling meaningful sequence information from ancient DNA short read data

As noted above, most modern genome sequencing approaches today use a combination of long sequencing reads and shotgun sequencing in order to obtain genomes with high contiguity, ideally chromosome-level resolution (Jaworski, Allan, & Matzkin, 2020; Michael & VanBuren, 2020). These approaches not only require DNA of high quality, i.e., fragment lengths >50,000 base pairs, but also large amounts of this DNA in high purity. Ancient DNA does not fulfill any of these requirements. First, as also noted above, it is fragmented to short lengths, with average fragment lengths often being below 50 bp. Moreover, reads below 30 bp cannot be mapped confidently, resulting in a relatively narrow length distribution of usable reads. Second, ancient DNA extracts invariably contain DNA of other sources, mainly of the microbes that live and lived in and of the tissue since the death of the target organism. Often this microbial DNA represents the majority of DNA in an ancient DNA extract (Noonan et al., 2005). And third, even though our knowledge which parts of an animal preserve DNA best as well as how to extract ancient DNA has improved considerably (see above), the amounts of ancient DNA we can obtain from a fossil are still—and will probably always remain—relatively limited.

Due to these factors, standard de-novo genome assembly approaches do not work with paleogenomic data. Instead, mapping approaches are generally used in paleogenomics, i.e., the sequencing reads obtained are mapped against a reference genome of the same or an as closely related species as possible. In a next and separate step, either the positions that are variable between reference and sample are saved for further analysis, or the entire sequence of the sample, based on the mapping results, is generated. If—and this is a big if—sufficient endogenous sequencing reads, i.e., reads that originate from the species the fossil belongs to rather than from microorganisms that have lived on the fossil, can be obtained, the quality of these results depends mainly on two factors: (1) the completeness of the reference used, and (2) the sequence similarity between the DNA sequence of the reference and the sampled species. Of course, there are additional factors that play a smaller, albeit not necessarily unimportant role. Thus, samples with longer average and/or maximum read lengths yield better results than those with very short read lengths, even for ancient DNA. Also, samples with high endogenous content more readily allow obtaining sufficient sequence data, and at the same time, the problem of incorrectly mapping reads from microbial contaminants with sequence similarity simply by chance is reduced. Finally, the program used for mapping and the values chosen for different parameters play a role. For example, using a longer minimum read length for mapping improves mapping quality but of course reduces the amount of data available.

When the goal is to obtain only a selected protein or cDNA sequence from MPS data sets, a similar approach has to be used, i.e., reference data is required, and the reads are compared to it. In addition to the challenges mentioned above, the reads are derived from genomic DNA and therefore include intronic sequences. Protein or cDNA sequences, in contrast, only represent exons. Furthermore, when comparing protein sequences to DNA data, the latter has to be translated into all six reading frames for the comparison. While technically not difficult to implement, this results in longer computing runtimes, compared to a DNA-DNA comparison. Additional factors are that single-copy genes are much easier to assemble than members of gene families, where the short ancient DNA reads may fit equally well to different members of the gene family. Similarly, proteins containing conserved, shared functional domains or repetitive sequence motifs are also problematic.

Recently, approaches for the assembly of only specific genes of interest have been developed. The SAT-Assembler (Zhang, Sun, & Cole, 2014), for example, first assigns MPS reads to profile Hidden Markov Models of selected gene families and then uses a graph-theoretic solution to assemble all reads for a given family into contigs and scaffolds. With data from RNA-Seq and metagenomic experiments in mind, this approach could also be applied to whole genome shotgun data from ancient material. The recently developed SRAssembler (McCarthy, Chou, & Brendel, 2019) similarly aims to assemble only specific genes; it was developed for eukaryotic whole genome data and for broader applications that also include the exploration of assembly parameters or assembly gap-filling, for example. The SRAssembler implements an in-silico chromosome walking strategy by recursively searching for and assembling reads that are homologous to a query protein or cDNA sequence. Assembled contigs of one round are used as query for the next. The final contigs are then compared to the protein or cDNA query using a splice-aware aligner. We have used the SRAssembler for the protocol described below.

3. Assembling a Denisovan gene sequence based on a human homolog

As an example, we used a relatively simple system, i.e., a relatively small and single-exon protein, a high-coverage paleogenome (very few of these are available), and a closely related reference. Specifically, we downloaded the publicly available data for the Denisovan genome (http://cdna.eva.mpg.de/denisova/, Meyer et al., 2012), from which we aimed to assemble a sequence homologous to the (modern) human olfactory receptor 51J1 (NCBI Acc.No. NP_001335153) using the SRAssembler software.

3.1 Equipment

The only equipment that is required is access to computing power, i.e., a computing server on which the required software can be installed. Multiple CPUs are highly recommended. Note that high-coverage paleogenome data are very large to begin with, and that the SRAssembler outputs considerable additional amounts of data. In the example shown in this chapter, the data exceeded 1.5 TB during the analyses. Therefore, sufficient data storage volume is required.

4. Protocol

1. Install the SRAssembler (McCarthy et al., 2019) software on any Linux server. For our example it was installed on a computing server running under Scientific Linux after all pre-requisites were installed according to the software's GitHub page at https://github.com/BrendelGroup/SRAssembler.
2. Prepare the sequence reads in fastq or fasta format. As input for our example served the adaptor-trimmed and merged paired-end reads from five Denisovan libraries in fasta format; these data represent ~30× coverage of the genome.
3. Prepare the query sequence as protein or cDNA sequence. Our query was the human olfactory receptor 51J1 protein sequence with a length of 316 amino acids.
4. Select parameter options of the SRAssembler according to the software's manual. Note that some parameters are set directly within the program call, while other are listed in a *SRAssembler.conf* that is found in the *demo*

folder and can be copied and modified. For our example, we used default parameters of the SRAssembler, for single-end sequence data, although a multitude of parameters allow the fine-tuning of results for more difficult templates or lower-coverage MPS data.

5. Analyses of the data obtained

The output files include *hit_contigs.fasta* and *output.aln*, which contain the final contig(s) and its/their spliced alignment to the query sequence, respectively.

For our example, we were able to recover a protein of 310 amino acids from the Denisovan data set; only its first six amino acids were missing. When aligning the modern query sequence to the newly assembled Denisovan homolog, we observed a single non-synonymous substitution: codon 69 of the Denisovan sequence is a CAT that codes for histidine, whereas the corresponding position in the modern human sequence contains an CGT codon coding for arginine. The protein alignment of the two sequences is shown in Fig. 1, with NP_001335153 the query sequence and Den_contig16 the translated sequence of the assembled Denisovan homolog.

```
NP_001335153      1 MKISNNSLGFLPTTFILVGIPGLESEHLWISVPFSLIYIIIFLGNGIILH    50
                          ||||||||||||||||||||||||||||||||||||||||||||
Den_contig16      1 ------SLGFLPTTFILVGIPGLESEHLWISVPFSLIYIIIFLGNGIILH    44

NP_001335153     51 VIRTDIALHQPMYLFLAMLALAEVRVSASTLPTVLGIFLFGNTEISLEAC   100
                    ||||||||||||||||||||||||.|||||||||||||||||||||||||
Den_contig16     45 VIRTDIALHQPMYLFLAMLALAEVHVSASTLPTVLGIFLFGNTEISLEAC    94

NP_001335153    101 LFPDVLHPFFIHDGASCAAGHVFGPLYSHLQPTELHSYPDTAQGLWHRSY   150
                    ||||||||||||||||||||||||||||||||||||||||||||||||||
Den_contig16     95 LFPDVLHPFFIHDGASCAAGHVFGPLYSHLQPTELHSYPDTAQGLWHRSY   144

NP_001335153    151 YRTEKHYAHGSVAHSLMASALLWPQCPLTFLLSAPQSYLSCGNISVNNIY   200
                    ||||||||||||||||||||||||||||||||||||||||||||||||||
Den_contig16    145 YRTEKHYAHGSVAHSLMASALLWPQCPLTFLLSAPQSYLSCGNISVNNIY   194

NP_001335153    201 GIFIVTSTFGLDSLLIVISYGLILHTVLGIATGEGRKKALNTCGSHVCAV   250
                    ||||||||||||||||||||||||||||||||||||||||||||||||||
Den_contig16    195 GIFIVTSTFGLDSLLIVISYGLILHTVLGIATGEGRKKALNTCGSHVCAV   244

NP_001335153    251 LAYYVPMIGLSIVHRLGHRVSPLLQAMMANAYLFFPPVVNPIVYSIKTKE   300
                    ||||||||||||||||||||||||||||||||||||||||||||||||||
Den_contig16    245 LAYYVPMIGLSIVHRLGHRVSPLLQAMMANAYLFFPPVVNPIVYSIKTKE   294

NP_001335153    301 IHGAIVRMLLEKRRRV    316
                    ||||||||||||||||
Den_contig16    295 IHGAIVRMLLEKRRRV    310
```

Fig. 1 Alignment of the human olfactory receptor 51J1 protein sequence (NP_001335153) with the protein sequence (Den_contig16) reconstructed from the Denisova paleogenome data (Meyer et al., 2012).

6. Related techniques

As mentioned above, prior to the introduction and widespread use of MPS techniques, nuclear gene sequences from ancient DNA were obtained using multiplex PCR (e.g., Lalueza-Fox et al., 2007; Römpler, Dear, et al., 2006; Römpler, Rohland, et al., 2006). This approach is still a viable alternative, but only if genomic sequences, including intron sequences, that allow designing primers across the complete gene are available from a closely related species. Moreover, since the fragmented nature of ancient DNA only allows amplification of short fragments via PCR, even with multiplex PCR, the analyses of larger genes becomes challenging. However, it remains a possible approach, especially for samples with reasonable DNA preservation but low endogenous content, since obtaining palaeogenomic data of sufficient coverage may be prohibitively expensive for such samples.

7. Troubleshooting and optimization

Prior to further analyses (like protein expression or in-silico functional comparisons) of the obtained DNA sequences, it is highly recommended to scrutinize the results for possible errors. Since the correct sequence is by default unknown, this has to be done in an indirect way. There are several parameters that can be investigated. First, it should be tested whether the reconstructed sequence contains any stop codons, which would be an obvious sign for an incorrect assembly, at least for an essential gene. A subtler signal of sequence errors are unusual amino acids at one or more positions in the amino acid sequence of the reconstructed protein. These need to be evaluated with care, as in phylogenetically more distant species, amino acid differences are of course expected.

Different types of problems may arise with genes with multiple and long introns or when genomic coverage of the original data is low. We did not test all these possibilities, as this would be beyond the scope of this chapter. However, the SRAssembler allows adjusting multiple parameters to address these problems as much as possible. For details, we refer the reader to the relevant publication (McCarthy et al., 2019) as well as the detailed github page of the program (https://github.com/BrendelGroup/SRAssembler).

8. Summary

Although paleogenomics has become a research field in its own rights in recent years, researchers who want to analyze complete gene sequences of specific genes may encounter the fact that paleogenome assemblies are often incomplete. This is generally the case when the reference genome used for mapping is of insufficient quality or too distant to allow mapping of all parts of the genome. In such cases, it will become necessary to go back to the primary sequence data and attempt assembly of the gene of interest as described in this chapter. We chose a relatively simple example as proof-of-principle, but more challenging genes can be analyzed using the same approach.

References

Alberti, F., Gonzalez, J., Paijmans, J. L. A., Basler, N., Preick, M., Henneberger, K., et al. (2018). Optimized DNA sampling of ancient bones using computed tomography scans. *Molecular Ecology Resources*, *18*, 1196–1208.

Barlow, A., Cahill, J. A., Hartmann, S., Theunert, C., Xenikoudakis, G., Fortes, G. G., et al. (2018). Partial genomic survival of cave bears in living brown bears. *Nature Ecology & Evolution*, *2*, 1563–1570.

Campbell, K. L., Roberts, J. E., Watson, L. N., Stetefeld, J., Sloan, A. M., Signore, A. V., et al. (2010). Substitutions in woolly mammoth hemoglobin confer biochemical properties adaptive for cold tolerance. *Nature Genetics*, *42*, 536–540.

Dabney, J., Knapp, M., Glocke, I., Gansauge, M. T., Weihmann, A., Nickel, B., et al. (2013). Complete mitochondrial genome sequence of a middle Pleistocene cave bear reconstructed from ultrashort DNA fragments. *Proceedings of the National Academy of Sciences of the USA*, *110*, 15758–15763.

Damgaard, P. B., Margaryan, A., Schroeder, H., Orlando, L., Willerslev, E., & Allentoft, M. E. (2015). Improving access to endogenous DNA in ancient bones and teeth. *Scientific Reports*, *5*, 11184.

Dannemann, M., & Kelso, J. (2017). The contribution of Neanderthals to phenotypic variation in modern humans. *American Journal of Human Genetics*, *101*, 578–589.

Dannemann, M., Prüfer, K., & Kelso, J. (2017). Functional implications of Neandertal introgression in modern humans. *Genome Biology*, *18*, 61.

Feigin, C. Y., Newton, A. H., Doronina, L., Schmitz, J., Hipsley, C. A., Mitchell, K. J., et al. (2018). Genome of the Tasmanian tiger provides insights into the evolution and demography of an extinct marsupial carnivore. *Nature Ecology & Evolution*, *2*, 182–192.

Ghurye, J., & Pop, M. (2019). Modern technologies and algorithms for scaffolding assembled genomes. *PLoS Computational Biology*, *15*, e1006994.

Glocke, I., & Meyer, M. (2017). Extending the spectrum of DNA sequences retrieved from ancient bones and teeth. *Genome Research*, *27*, 1230–1237.

Green, R. E., Krause, J., Briggs, A. W., Maricic, T., Stenzel, U., Kircher, M., et al. (2010). A draft sequence of the Neandertal genome. *Science*, *328*, 710–722.

Greenwood, A. D., Capelli, C., Possnert, G., & Pääbo, S. (1999). Nuclear DNA sequences from late Pleistocene megafauna. *Molecular Biology & Evolution*, *16*, 1466–1473.

Henneberger, K., Barlow, A., & Paijmans, J. L. A. (2019). Double-stranded library preparation for ancient and other degraded samples. *Methods in Molecular Biology, 1963*, 65–73.

Huerta-Sánchez, E., Jin, X., Asan, Bianba, Z., Peter, B. M., Vinckenbosch, N., et al. (2014). Altitude adaptation in Tibetans caused by introgression of Denisovan-like DNA. *Nature, 512*, 194–197.

Huynen, L., Suzuki, T., Ogura, T., Watanabe, Y., Millar, C. D., Hofreiter, M., et al. (2014). Reconstruction and in vivo analysis of the extinct tbx5 gene from ancient wingless moa (Aves: Dinornithiformes). *BMC Evolutionary Biology, 14*, 75.

Jablonski, D. (2004). Extinction: Past and present. *Nature, 427*, 589.

Jaworski, C. C., Allan, C. W., & Matzkin, L. M. (2020). Chromosome-level hybrid de novo genome assemblies as an attainable option for non-model insects. *Molecular Ecology Resources*. https://doi.org/10.1111/1755-0998.13176, (published online 23 April 2020., Epub ahead of print).

Lalueza-Fox, C., Römpler, H., Caramelli, D., Stäubert, C., Catalano, G., Hughes, D., et al. (2007). A melanocortin 1 receptor allele suggests varying pigmentation among Neanderthals. *Science, 318*, 1453–1455.

Li, R., Fan, W., Tian, G., Zhu, H., He, L., Cai, J., et al. (2010). The sequence and de novo assembly of the giant panda genome. *Nature, 463*, 311–317.

McCarthy, T. W., Chou, H.-C., & Brendel, V. P. (2019). SRAssembler: Selective recursive local assembly of homologous genomic regions. *BMC Bioinformatics, 20*, 371.

McHugo, G. P., Dover, M. J., & MacHugh, D. E. (2019). Unlocking the origins and biology of domestic animals using ancient DNA and paleogenomics. *BMC Biology, 17*, 98.

Meyer, M., Kircher, M., Gansauge, M. T., Li, H., Racimo, F., Mallick, S., et al. (2012). A high-coverage genome sequence from an archaic Denisovan individual. *Science, 338*, 222–226.

Michael, T. P., & VanBuren, R. (2020). Building near-complete plant genomes. *Current Opinion in Plant Biology, 54*, 26–33.

Noonan, J. P., Hofreiter, M., Smith, D., Priest, J. R., Rohland, N., Rabeder, G., et al. (2005). Genomic sequencing of Pleistocene cave bears. *Science, 309*, 597–599.

Paijmans, J. L. A., Baleka, S., Henneberger, K., Taron, U. H., Trinks, A., Westbury, M. W., et al. (2017). Sequencing single-stranded libraries on the Illumina NextSeq 500 platform. *ArXiv, 1711*, 11004.

Paijmans, J. L., Gilbert, M. T., & Hofreiter, M. (2013). Mitogenomic analyses from ancient DNA. *Molecular Phylogenetics and Evolution, 69*, 404–416.

Palkopoulou, E., Lipson, M., Mallick, S., Nielsen, S., Rohland, N., Baleka, S., et al. (2018). A comprehensive genomic history of extinct and living elephants. *Proceedings of the National Academy of Sciences of the USA*, 115, E2566–E2574.

Park, S. D., Magee, D. A., McGettigan, P. A., Teasdale, M. D., Edwards, C. J., & Lohan, A. J. (2015). Genome sequencing of the extinct Eurasian wild aurochs, Bos primigenius, illuminates the phylogeography and evolution of cattle. *Genome Biology, 16*, 234.

Pinhasi, R., Fernandes, D., Sirak, K., Novak, M., Connell, S., Alpaslan-Roodenberg, S., et al. (2015). Optimal ancient DNA yields from the inner ear part of the human petrous bone. *PLoS One, 10*, e0129102.

Reich, D., Green, R. E., Kircher, M., Krause, J., Patterson, N., Durand, E. Y., et al. (2010). Genetic history of an archaic hominin group from Denisova Cave in Siberia. *Nature, 468*, 1053–1060.

Römpler, H., Dear, P. H., Krause, J., Meyer, M., Rohland, N., Schöneberg, T., et al. (2006). Multiplex amplification of ancient DNA. *Nature Protocols, 1*, 720–728.

Römpler, H., Rohland, N., Lalueza-Fox, C., Willerslev, E., Kuznetsova, T., Rabeder, G., et al. (2006). Nuclear gene indicates coat-color polymorphism in mammoths. *Science, 313*, 62.

Shapiro, B., & Hofreiter, M. (2014). A paleogenomic perspective on evolution and gene function: New insights from ancient DNA. *Science*, *343*, 1236573.
Simpson, J. T., & Pop, M. (2015). The theory and practice of genome sequence assembly. *Annual Review of Genomics and Human Genetics*, *16*, 153–172.
Skoglund, P., & Mathieson, I. (2018). Ancient genomics of modern humans: The first decade. *Annual Review of Genomics and Human Genetics*, *19*, 381–404.
Smith, O., & Gilbert, M. T. P. (2018). Ancient RNA. In C. Lindqvist & O. P. Rajora (Eds.), *Paleogenomics* (pp. 53–74). Cham: Springer Nature Switzerland.
Zhang, Y., Sun, Y., & Cole, J. R. (2014). A scalable and accurate targeted gene assembly tool (sat-assembler) for next-generation sequencing data. *PLoS Computational Biology*, *10*, e1003737.

CHAPTER THREE

Bioinformatic, genomic and evolutionary analysis of genes: A case study in dipteran *CSPs*

Jean-François Picimbon*
School of Bioengineering, QILU University of Technology, Jinan, Shandong, China
*Corresponding author: e-mail address: jfpicimbon@gmail.com

Contents

Abstract

Understanding physiological tasks may start by a good understanding of evolutionary traits in a given protein gene family. The successful completion of various genome projects is a key step forward for comparative analysis of similar and/or orthologous genes between species, measuring genetic relatednesses, studying evolutionary changes among different behaviors, helping to identify specifically conserved genes or genes that are variable enough to determine a new strain or organism. How evolutionary data can improve the understanding of a protein gene family is exploited here in the case of the insect Chemosensory Proteins (CSPs) as a guidance model. In this chapter, Diptera are presented not only as a huge diversity of appearances (phenotypes), behaviors and

Methods in Enzymology, Volume 642
ISSN 0076-6879
https://doi.org/10.1016/bs.mie.2020.05.012

lifestyles, but also as major differences that are gene copy numbers, gene family members, polymorphisms and evolutionary rate variations, without any *a priori* assumption about the origin and function of the sampled gene family.

1. Introduction

A CSP is a protein characterized by four cysteine residues located in two adjacent disulfide bridges at conserved positions (Picimbon, 2003). It folds into a flexible triangular pyramidal structure generally made of six α-helices. The polygon base is highly hydrophobic, while the apex is free (Jansen et al., 2007; Jansen, Zídek, Löfstedt, Picimbon, & Sklenar, 2006; Lartigue et al., 2002; Tomaselli et al., 2006).

The naturally occurring distribution of CSPs is throughout various parts of the insect body, not only in the antennae, palps and legs, but also in many various non-sensory organs, lymphs and tissues such as hemolymph, epidermis, gut and fat body (Angeli et al., 1999; Ozaki et al., 2005; Picimbon, Dietrich, Angeli, et al., 2000; Picimbon, Dietrich, Breer, & Krieger, 2000; Picimbon, Dietrich, Krieger, & Breer, 2001; Picimbon & Leal, 1999; Xuan et al., 2015).

In moths, the synthesis of CSP starts during very early phases of the adult development, much earlier than the appearance of chemosensory neurons, and continues in the adult stage at a high level of activity, particularly in response to insecticide stress (Picimbon et al., 2001; Xuan et al., 2015). The sex pheromone gland of the female silkworm contains 14 CSPs and expresses an even higher number of amino acid variants in CSP proteins through RNA editing (Picimbon, 2019; Xuan et al., 2014, 2015; Xuan, Rajashekar, Kasvandik, & Picimbon, 2016; Xuan, Rajashekar, & Picimbon, 2019). In honeybees, specific *CSP* gene knock out leads to archaic development of the head (Maleszka, Forêt, Saint, & Maleszka, 2007), strongly suggesting a function in tissue development, growth and/or regeneration for this gene family, as proposed in the early report by Nomura, Kawasaki, Kubo, and Natori (1992).

These development and tissue lipids data integrate not only with gene expression profiles, but also with binding data, *i.e.*, the characterization of the functional ligand binding properties of the protein. CSPs are known to interact with lipid compounds of different chain lengths (C12-C18, *i.e.*, 12–18 carbons chain length) with eventually some particular or specific discrimination property (Briand et al., 2002; Dani et al., 2010). Accordingly, they have been shown to retain propensity for conformational change and interaction with multiple lipids or lipid chains (Campanacci et al., 2001, 2003; Lartigue et al., 2002). However, not all CSPs interact with lipids. In the sweet potato or silverleaf whitefly *Bemisia tabaci*, it has been shown that CSP1s bind linoleic acid (C18: 2, *i.e.*, 18 carbon chain length with two double bonds), while CSP2s and CSP3s interact more specifically with small cyclic compounds such as cinnamaldehyde, a main constituent of plant oil that is recognized to be rather repellent and/or very toxic for the insect (Liu, Ma, Xie, Xuan, & Picimbon, 2016; Liu, Ma, Xie, Xuan, Xia, Fan, et al., 2016; Liu, Arnaud, Offmann, & Picimbon, 2017; Liu et al., 2020), strongly supporting a role for CSPs in lipid metabolism and/or xenobiotic degradation (Einhorn & Imler, 2019; Liu et al., 2020).

In this chapter, we examine the molecular evolution of *CSP* genes in dipteran species with particular focus given to flies and mosquitoes to show the set of data required for a complete and relevant evolutionary analysis of a specific protein gene family. We describe how to utilize the complete genome database from the dipterans *Drosophila ananassae*, *D. erecta*, *D. grimshawi*, *D. melanogaster*, *D. mojavensis*, *D. persimilis*, *D. pseudoobscura*, *D. sechellia*, *D. simulans*, *D. virilis*, *D. willistoni*, *D. yakuba*, *Aedes aegypti*, *Anopheles gambiae* and *Culex pipiens*, to identify 127 copies for the *CSP* gene and report about gene structural properties and reconstructed phylogenies of all the sequences (amino acid, nucleotide, exon and intron). We also seek to explain about the need of knowledge about retrotranspositions (*Feilai-Twin*) and their associations with expansion and evolution of a specific gene family. Then, we provide guidance for the whole analysis (genomic organization, shift of intron boundary and phylogenetic distribution) and interpretation in Diptera as a classical model system not only for retrotransposition, but also for many various chromosomal rearrangements (deletions, duplications, insertions, inversions, mutations and translocations) that clearly influence gene and genome evolution.

2. Analysis *in silico* of genomes

The method and guidance provided here cannot substitute for functional studies. It can help analyze peptide, enzyme, receptor, chaperone and binding protein genes, annotate them properly on a genome involving mapping features such as exons, introns, intron boundaries and retroposons, build parsimony or likelihood tree and/or study evolution in a given metabolic gene family. For example, it can help survey the evolutionary patterns of genomes within a specific gene family, focusing on Diptera (most diverse insect orders, >110 genome assemblies and several hundreds of population-level genomes sequenced; Wiegmann & Richards, 2018) and small soluble proteins (CSPs) known to carry long lipid chains and xenobiotic compounds for insect defense (Liu et al., 2020). This is for a unified protocol for all in biology who are interested in genome-wide investigations of evolutionary questions, such as studies of gene families that are important for development, reproduction, immunity and physiology. This is an essential part of comparative genomics in a way similar to Vieira and Rozas (2011), but we concentrate here on genomic details in dipterans without any links to origin and evolutionary history of olfaction as a model system for coevolution of binding proteins and receptors.

2.1 Preparation of templates (sequences) for bioinformatic analysis

The first step of research is often linked with "physiology." (1) Searching for olfactory and/or developmental genes can generate a clone or a newly discovered protein subjected to N-terminal sequencing. (2) This allows us to adjust primers specifically tuned to the N-terminal region in order to amplify by polymerase chain reaction or PCR (when used with oligo-dT) the complete nucleotide sequence and full-length cDNA clone of the protein. (3) Subsequently, the newly identified clone can be used as a probe to screen for other isoforms in a tissue-specific cDNA library (Picimbon & Leal, 1999; Picimbon, Dietrich, Breer, et al., 2000; Picimbon, Dietrich, Angeli, et al., 2000; Picimbon et al., 2001). This molecular approach generates a set of sequences that can be used to extract templates from genomic databases.

Genomic databases such as AnoBase, FlyBase, VectorBase, Ensembl Metazoa, Drosophila Genome Browser Gateway and Baylor College Medicine Human Genome Sequencing Center are useful to retrieve genomic sequences by blasting the nucleotide sequence of cDNA (complementary DNA = DNA copy synthesized from RNA by *in vitro* reverse transcription) against the assembly of scaffolds reconstructed from gaps and contiguous genomic clones or overlapping DNA sequences generated by genome sequencing (contigs). Using Blastn search selects scaffolds on the basis of hit rate (100%) and e-value ($\leq$0.001).

Like in the case study presented here, we can use a COBBLER consensus sequence to BLAST the genomes as "traditionally" done for genomic analyses of specific gene families. Using bio.tools (GenBank and European Molecular Biology Laboratory-European Bioinformatics Institute; Stoesser et al., 2002) helps establish the repertoire of genes in specific organisms: locus name, gene size and accession numbers (see GenBank and EMBL-EBI data for genomic CSP sequences in Diptera; Tables 1 and 2). Exon and intron nucleotide sequences are identified by plotting each cDNA sequence against the scaffold (or contig) open reading frame in a search launcher nucleotide algorithm, including sequence translation and alignment interface (BCM Search Launcher—Nucleic Acid Research Tool, pitt.edu).

ClustalW at *www.genome.jp* allows the precise determination of the intron-exon structure of the gene by directly comparing genomic DNA sequence (query) with cDNA template. Such alignment also helps to identify the position of initiation codon (ATG encoding Met-1) and stop codon (TAA, TAG or TGA). If intron implies retroposon, simple sequence repeats (microsatellites) and/or other genetic markers can be assessed using Blastn from the National Center for Biotechnology Information (NCBI) using intron/scaffold sequence as template (Figs. 1 and 2).

2.2 Phylogenetic analysis

The complete DNA sequences extracted from genomic databases can be used as templates (or taxa) in a phylogenetic analysis aimed to infer orthology groups and evolutionary relatednesses between sequences, but never to establish something in regard to function. An "orthologous" group is defined as a group of genes with orthologous relations (result from speciation event) to every other member in the group. Orthologous groups are defined by points (or nodes) at which the resulting tree or phylodendogram branches into species-specific clades. It is a part of determining homology—same nature—but not necessarily—same speech. "Homologous" means "corresponding in structure and therefore having common ancestry", which does not necessarily mean "have the same function." The phylogenetic study as that presented here shows homologs (orthologs or paralogs-same species-) that arose by gene duplication and code for proteins that may or may not have the same function.

In an evolutionary analysis of genes, phylogenetic trees are established using amino acid and nucleotide sequence data (see Tables 1–2 and Figs. 3–5). DNA or protein data are processed *via* multiple sequence alignment algorithms such as MUSCLE, MAFFT, Mauve, MegAlign Pro or Clustal. CLUSTAL X 1.8. is usually used for Phylogenetic Analysis Using PAUP (PAUP*, PAUP4.0b10 software, Altivec, Swofford, 2002).

Table 1 Repertoire of *CSP* genes identified by *in silico* analysis of fly genome databases (FlyBase, http://www.fruitfly.org, Drosophila Genome Browser Gateway, http://genome.ucsc.edu and HGSC database, http://www.hgsc.bcm.tmc.edu) in 12 species of *Drosophila*: *D. ananassae, Dana; D. erecta, Dere; D. grimshawi, Dgrim; D. melanogaster, Dmel; D. mojavensis, Dmoj; D. persimilis, Dper; D. pseudoobscura, Dpse; D. sechellia, Dsec; D. simulans, Dsim; D. virilis, Dvir; D. willistoni, Dwil; D. yakuba, Dyak* (Adams et al., 2000; Celniker et al., 2002; Celniker & Rubin, 2003; Delcher, Pop, & Salzberg, 2004; Drosophila 12 Genomes Consortium, 2007; Hoskins et al., 2015; McKay et al., 2012; Misra et al., 2002; Muzny et al., 2002; Richards et al., 2005; Stapelton et al., 2002; Stapleton, Carlson, et al., 2002; Stapleton, Liao, et al., 2002; Tweedie et al., 2009).

Name	Genome size (Mb)	Gene size (bps)	Accession
Drosophila ananassae	215.16 ± 8.80		
DanaGF11516		384	*CH902619*, XM_001976491, XP_001960430, EDV37252
DanaGF12039		391	*CH902619*, XM_001959435, XP_001959471, EDV36293
DanaGF12892		381	*CH902619*, XM_001959432, XP_001959468, EDV36290
DanaGF20186*		186	*CH902621*, XM_001963599, XP_001963635, EDV44711
Drosophila erecta	136.92 ± 3.91		
DereGG15834		520	*CH954178*, XM_001972942, XP_001972978, EDW52004
DereGG19868		366	*CH954179*, XM_001976673, XP_001976709, EDV57109
DereGG19963		449	*CH954179*, XM_001976494, XP_001976530, EDW56930
DereGG22923		381	*CH954179*, XM_001976491, XP_001976527, EDV56927
Drosophila grimshawi	234.72 ± 4.89		
DgriGH12519		454	*CH916370*, XM_001991134, XP_001991170, EDW99794
DgriGH20166		635	*CH916367*, XM_001987047, XP_001987083, EDW01950
DgriGH21706		384	*CH916367*, XM_001987025, XP_001987061, EDW01928
DgriGH22042		369	*CH916367*, XM_001987624, XP_001987660, EDW02527

Drosophila melanogaster	180		
DmelOS-D, A10		463	*AE014296*, AE002602, AE003525, AJ973480, BT025814, NM_079397, AC010069, AC10070, NP_524121, U02546, U05244, AAF49381, ABF85714, Q27377
CG9358		*366*	*AE013599*, AE002575, AE003465, AJ697736, AJ973479, AY071015, BT120076, NM_138146, AC007884, AY23914, AAF47307, AAL48637, AAR09937, ADA70413, CAG26926, CAJ01526
CG30172		465	*AE013599*, AE002575, AE003462, AJ697735, AJ973477, NM_166649, AC099018, BT001865, NP_726402, AAN71635, AAM68292, CAG26928, CAJ01524
PebIII		381	*AE013599*, AJ973478, AY113640, BT122148, NM_080227, AC099018, AC004642, U08281, AAA87058, AAF47140, AAM29645, AAR10129, ADE21091, CAJ01525
Drosophila mojavensis	130		
DmojGI14872		605	*CH933810*, XM_002010073, XP_002010109, EDW07426
DmojGI18859		381	*CH933808*, XM_002005810, XP_002005846, EDW09781
DmojGI19129		381	*CH933808*, XM_002005307, XP_002005343, EDW09278
DmojGI20684		768	*CH933808*, XM_002005793, XP_002005829, EDW09764
Drosophila persimilis	195.6 ± 4.89		
DperGL10431		428	*CH479181*, XM_002016532, XP_002016568, EDW32458
DperGL10889		378	*CH479181*, XM_002015650, XP_002015686, EDW31576
DperGL11652		381	*CH479181*, XM_002016529, XP_002016565, EDW32455
DperGL15312		473	*CH479228*, XM_002027420, XM_002027419, XP_002027456, EDW35359

Continued

Table 1 Repertoire of *CSP* genes identified by *in silico* analysis of fly genome databases (FlyBase, http://www.fruitfly.org, Drosophila Genome Browser Gateway, http://genome.ucsc.edu and HGSC database, http://www.hgsc.bcm.tmc.edu) in 12 species of *Drosophila*: *D. ananassae*, *Dana*; *D. erecta*, *Dere*; *D. grimshawi*, *Dgrim*; *D. melanogaster*, *Dmel*; *D. mojavensis*, *Dmoj*; *D. persimilis*, *Dper*; *D. pseudoobscura*, *Dpse*; *D. sechellia*, *Dsec*; *D. simulans*, *Dsim*; *D. virilis*, *Dvir*; *D. willistoni*, *Dwil*; *D. yakuba*, *Dyak* (Adams et al., 2000; Celniker et al., 2002; Celniker & Rubin, 2003; Delcher, Pop, & Salzberg, 2004; Drosophila 12 Genomes Consortium, 2007; Hoskins et al., 2015; McKay et al., 2012; Misra et al., 2002; Muzny et al., 2002; Richards et al., 2005; Stapelton et al., 2002; Stapleton, Carlson, et al., 2002; Stapleton, Liao, et al., 2002; Tweedie et al., 2009).—cont'd

Name	Genome size (Mb)	Gene size (bps)	Accession
Drosophila virilis	361.86 ± 12.7		
DvirGJ19536		534	*CH940651*, XM_002055724, XP_002055760, EDW65961
DvirGJ20437		716	*CH940648*, XM_002049958, XP_002049994, EDW61187
DvirGJ20447		381	*CH940648*, XM_002049945, XP_002049981, EDW61174
DvirGJ22262		378	*CH940648*, XM_002050592, XP_002050628, EDW61821
Drosophila sechellia	166.26 ± 3.91		
DsecGM11711		366	*CH480838*, XM_002043128, XP_002043164, EDW49295
DsecGM18283		458	*CH480820*, XM_002037453, XP_002037489, EDW53907
DsecGM18290		381	*CH480820*, XM_002037450, XP_002037486, EDW53904
DsecGM24353		520	*CH480815*, XM_002030820, XP_002030856, EDW41842
Drosophila simulans	166.26 ± 1.9		
DsimGD11826		381	*CM000362*, XM_002082876, XP_002082912, EDX08497
DsimGD12430		520	*CM000363*, XM_002085189, XP_002085225, EDX10810
DsimGD24902		366	*CM000362*, XM_002083028, XP_002083064, EDX08649
DsimGD24980		450	*CM000362*, XM_002082867, XP_002092903, EDX08488

Drosophila willistoni	224.94 ± 7.82		
DwilGK19383		775	*CH963849*, XM_002063438, XP_002063474, EDW74460
DwilGK19506		384	*CH963848*, XM_002062763, XP_002062799, EDW73785
DwilGK21931		384	*CH963849*, XM_002063445, XP_002063481, EDW74467
DwilGK24936		582	*CH964239*, XM_002071812, XP_002071848, EDW82834
Drosophila yakuba	161		
DyakGE11392		366	*CM000158*, XM_002092892, AY231914, AJ973446, XP_002092928, EDW92640
DyakGE11495		400	*CM000158*, XM_002092712, XP_002092748, EDW2460
DyakGE14360		381	*CM000158*, XM_002092709, AJ973447, AY232106, XP_002092745, EDW2457, CAJ01494, AAR10129
DyakGE22174		521	*CM000159*, XM_002095014, XP_002095050, EDW94762

Genome sizes are from Bosco, Campbell, Leiva-Neto, and Markow (2007). Access numbers in italic refer to genomic DNA clones. * Pseudogene.

Table 2 Repertoire of *CSP* genes identified by *in silico* analysis of mosquito genome databases (AnoBase: http://www.genoscope.cns.fr, http://www.ensembl.org/Anopheles_gambiae; *Aedes* and *Culex*: FlyBase, http://www.fruitfly.org, HGSC database, http://www.hgsc.bcm.tmc.edu and VectorBase, http://www.vectorbase.org).

Name	Genome size (Mb)	Gene size (bps)	Accession
Aedes aegypti	1376		
AAEL001957		60,389	*CH477232*, XM_001660702, AY431926, XP_001660752, EAT46830
AAEL001963		11,140	*CH477232*, XM_001660699, AY432716, XP_001660759, EAT46827
AAEL001964		381	*CH477232*, XM_001660736, AY432096, AY432096, XP_001660786, EAT46864
AAEL001967		387	*CH477232*, XM_001660718, XP_001660768, EAT46846
AAEL001969		381	*CH477232*, XM_001660704, AY432713, XP_001660754, EAT46832
AAEL001975		387	*CH477232*, XM_001660710, XP_001660760, EAT46838
AAEL001980		387	*CH477232*, XM_001660727, XP_001660777, EAT46855
AAEL001981		480	*CH477232*, XM_001660712, XP_001660762, EAT46840
AAEL001985		562	*CH477232*, XM_001660703, XP_001660753, EAT46831
AAEL001987		387	*CH477232*, XM_001660726, XP_001660776, EAT46854
AAEL001988		381	*CH477232*, XM_001660725, XP_001660775, EAT46853
AAEL001989		378	*CH477232*, XM_001660708, XP_001660758, EAT46836
AAEL001991		381	*CH477232*, XM_001660722, XP_001660772, EAT46850
AAEL001993		387	*CH477232*, XM_001660721, XP_001660771, EAT46849
AAEL001994		378	*CH477232*, XM_001660707, XP_001660757, EAT46835

Table 2 Repertoire of *CSP* genes identified by *in silico* analysis of mosquito genome databases (AnoBase: http://www.genoscope.cns.fr, http://www.ensembl.org/Anopheles_gambiae; *Aedes* and *Culex*: FlyBase, http://www.fruitfly.org, HGSC database, http://www.hgsc.bcm.tmc.edu and VectorBase, http://www.vectorbase.org).—cont'd

Name	Genome size (Mb)	Gene size (bps)	Accession
AAEL001995		387	*CH477232*, XM_001660715, XP_001660765, EAT46843
AAEL001999		387	*CH477232*, XM_001660713, XP_001660763, XP_001660769, EAT46841, EAT46847
AAEL002001		381	*CH477232*, XM_001660730, XP_001660780, EAT46858
AAEL002003		378	*CH477232*, XM_001660732, AY433711, XP_001660782, EAT46860
AAEL002004		381	*CH477232*, XM_001660724, XP_001660774, EAT46852
AAEL002007		387	*CH477232*, XM_001660731, XP_001660781, EAT46859
AAEL002015		381	*CH477232*, XM_001660723, XP_001660773, EAT46851
AAEL002017		387	*CH477232*, XM_001660729, XP_001660779, XP_001660781, EAT46857, EAT46859
AAEL002021		387	*CH477232*, XM_001660719, XP_001660763, XP_001660769, EAT46841, EAT46847
AAEL002022		384	*CH477232*, XM_001660735, XP_001660785, AY432031, EAT46683
AAEL002024		387	*CH477232*, XM_001660714, XP_001660764, EAT46842
AAEL002026		364	*CH477232*, XM_001660706, XP_001660756, EAT46834
AAEL002027		378	*CH477232*, XM_001660733, XP_001660783, EAT46861
AAEL002028		378	*CH477232*, XM_001660734, XP_001660784, EAT46862

Continued

Table 2 Repertoire of *CSP* genes identified by *in silico* analysis of mosquito genome databases (AnoBase: http://www.genoscope.cns.fr, http://www.ensembl.org/Anopheles_gambiae; *Aedes* and *Culex*: FlyBase, http://www.fruitfly.org, HGSC database, http://www.hgsc.bcm.tmc.edu and VectorBase, http://www.vectorbase.org).—cont'd

Name	Genome size (Mb)	Gene size (bps)	Accession
AAEL002029		387	*CH477232*, XM_001660716, AY432755, XP_001660766, EAT46844
AAEL002034		384	*CH477232*, XM_001660709, XP_001660759, EAT46837
AAEL002035		381	*CH477232*, XM_001660720, XP_001660770, EAT46848
AAEL002038		387	*CH477232*, XM_001660728, XP_001660778, XP_001660779, EAT46856, EAT46857
AAEL002040		384	*CH477232*, XM_001660711, XP_001660761, EAT46839
AAEL002041		387	*CH477232*, XM_001660717, XP_001660767, EAT46845
AAEL005682		375	*CH477355*, XM_001651291, XP_001651341, EAT42813
AAEL005687		375	*CH477355*, XM_001651295, XP_001651345, EAT42817
AAEL005691		375	*CH477355*, XM_001651294, AY431832, XP_001651344, EAT42816
AAEL005698		375	*CH477355*, XM_001651292, XP_001651342, EAT42814
AAEL005710		375	*CH477355*, XM_001651290, XP_001651340, EAT42812
AAEL005711		375	*CH477355*, XM_001651293, XP_001651343, EAT42815
AAEL012383		42,749	*CH477872*, XM_001662474, XP_001662524, EAT35447
Anopheles gambiae	260		
AgCG50174		604	*AAAB1008964*, AJ697732, XM_317404, XP_317404, EAA12702, CAG26925, AGAP008055

Table 2 Repertoire of *CSP* genes identified by *in silico* analysis of mosquito genome databases (AnoBase: http://www.genoscope.cns.fr, http://www.ensembl.org/Anopheles_gambiae; *Aedes* and *Culex*: FlyBase, http://www.fruitfly.org, HGSC database, http://www.hgsc.bcm.tmc.edu and VectorBase, http://www.vectorbase.org).—cont'd

Name	Genome size (Mb)	Gene size (bps)	Accession
AgCG50175		11,052	*AAB01008964*, AJ697734, XM_317401, XP_317401, EAA12703, CAG26927, AGAP029127
AgCG50200		684	*AAAB01008964*, AJ697731, AJ697730, XM_3170400, XP_317400, EAA12591, CAG26924, AGAP008059
AgCG50201		384	*AAAB01008964*, AF437891, BX_048417, XM_317408, XP_317408, EAA12322, AAL84186, AGAP008051
AgCG50208		20,307	*AAAB01008964*, XM_317397, AJ697733, XP_317397, EAA12601, CAG26926, AGAP008062
AgCG50210		384	*AAAB01008964*, BX062711, XM_317405, AJ697729, XP_317405, EAA12338, CAG26922, AGAP008054
AgCG50220		384	*AAAB01008964*, AJ697728, XM_317407, XP_317407, EAA12353, AGAP008052
AgAP001303		7317	*AAB01008987*, CM000357, AGF68546, KC189156, AGAP001303-RA, AGAP001303-RB
Culex pipiens	579		
CPIJ002601		141	*DS231854*, XM_001844618, XP_001844670, EDS38021
CPIJ002605		15,904	*DS231854*, XM_001844622, XP_001844674, EDS38025
CPIJ002607		17,241	*DS231854*, XM_001844624, XP_001844676, EDS38027
CPIJ002608		587	*DS231854*, XM_001844625, XP_001844677, EDS38028
CPIJ002609		381	*DS231854*, XM_001844626, XP_001844678, EDS38029

Continued

Table 2 Repertoire of *CSP* genes identified by *in silico* analysis of mosquito genome databases (AnoBase: http://www.genoscope.cns.fr, http://www.ensembl.org/Anopheles_gambiae; *Aedes* and *Culex*: FlyBase, http://www.fruitfly.org, HGSC database, http://www.hgsc.bcm.tmc.edu and VectorBase, http://www.vectorbase.org).—cont'd

Name	Genome size (Mb)	Gene size (bps)	Accession
CPIJ002611		390	*DS234497*, *DS231854*, XM_001844628, XP_001844680, XP_001870510, EDS34221, EDS38031
CPIJ002612		390	*DS231854*, XM_001844629, XP_001844681, EDS38032
CPIJ002613		390	*DS231854*, XM_001844630, XP_001844682, EDS38033
CPIJ002614		390	*DS231854*, XM_001844631, XP_001844683, EDS38034
CPIJ002615		390	*DS231854*, XM_001844632, XP_001844684, EDS38035
CPIJ002616		411	*DS231854*, XM_001844633, XP_001844685, EDS38036
CPIJ002617		375	*DS231854*, XM_001844634, XP_001844686, EDS38037
CPIJ002618		378	*DS231854*, XM_001844635, XP_001844687, EDS38038
CPIJ002619		378	*DS231854*, XM_001844636, XP_001844688, EDS38039
CPIJ002620		378	*DS231854*, XM_001844637, XP_001844690, XP_001844690, EDS38040, EDS38041
CPIJ002621		378	*DS231854*, XM_001844638, XP_001844690, XP_001844690, EDS38040, EDS38041
CPIJ002622		378	*DS231854*, XM_001844639, XP_001844691, EDS38042
CPIJ002623		378	*DS231854*, XM_001844640, XP_001844692, EDS38043
CPIJ002624		378	*DS231854*, XM_001844641, XP_001844693, EDS38044, ABH88165

Table 2 Repertoire of *CSP* genes identified by *in silico* analysis of mosquito genome databases (AnoBase: http://www.genoscope.cns.fr, http://www.ensembl.org/Anopheles_gambiae; *Aedes* and *Culex*: FlyBase, http://www.fruitfly.org, HGSC database, http://www.hgsc.bcm.tmc.edu and VectorBase, http://www.vectorbase.org).—cont'd

Name	Genome size (Mb)	Gene size (bps)	Accession
CPIJ002625		378	*DS231854*, XM_001844642, XP_001844694, EDS38045
CPIJ002626		381	*DS231854*, XM_001844643, XP_001844695, EDS38046
CPIJ002627		387	*DS231854*, XM_001844644, XP_001844696, EDS38047
CPIJ002628		384	*DS231854*, XM_001844645, XP_001844697, EDS38048
CPIJ002629		378	*DS231854*, XM_001844646, XP_001844698, EDS38049
CPIJ017094		27,992	*DS232856*, XM_001868095, XP_001868130, EDS26167
CPIJ019985		390	*DS234497*, *DS231854*, XM_001870475, XP_001844680, XP_001870510, EDS34221, EDS38031
CPIJ019986		324	*DS234497*, XM_001870476, XP_001870511, EDS34222

Access numbers in italic refer to genomic DNA clones (Arensburger, Hice, Wright, Craig, & Atkinson, 2011; Arensburger et al., 2010; Atkinson, 2007; Giraldo-Calderón et al., 2015; Gomez et al., 2005; Holt et al., 2002; Lawson et al., 2007, 2009; Loftus et al., 2005, 2006; Matthews et al., 2018; Mongin, Louis, Holt, Birney, & Collins, 2004; Nene et al., 2007; Reddy, Labbe, & Corbel, 2012; Severson et al., 2004; Topalis et al., 2005; Wiegmann & Richards, 2018; Worley et al., 2001).

Similar to IQ-TREE, RAxML, PhyML and R-CRAN, PAUP is used to test the preferred hypothesis, best account for rate variation in the group of sequences (or clades) and weight branching process depending on multiple variable characters. PAUP analysis provides branch support of character-based trees in maximum likelihood (ML) or maximum parsimony (MP). In the analysis of a gene family such as *CSPs*, which includes remote and recent paralogous copies, the ML-based approaches are probably the most appropriate. The ML-based approach draws a tree with the maximum likelihood between a long set of genetic data, estimating total variance, parallel evolution and phylogenetic heritability, which is extremely relevant to analyze the high degree of variance among species-specific sequences. In the presence

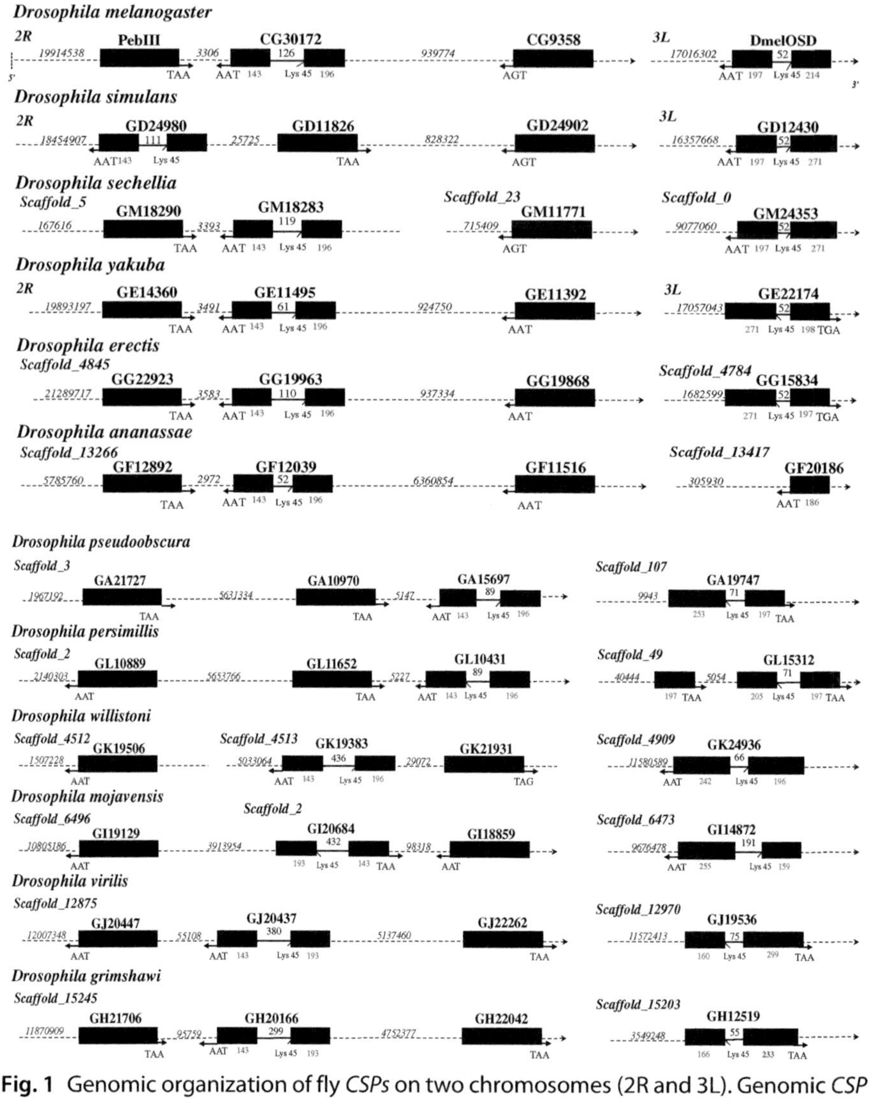

Fig. 1 Genomic organization of fly *CSPs* on two chromosomes (2R and 3L). Genomic *CSP* repertoire in 12 *Drosophilan* species: *Drosophila ananassae, D. erecta, D. grimshawi, D. melanogaster, D. mojavensis, D. persimilis, D. pseudoobscura, D. sechellia, D. simulans, D. virilis, D. yakuba* and *D. willistoni* (see Table 1). Exons are shown as black boxes and introns as plain lines. Dotted lines indicate the distances between genes. The numbers above indicate the size of each portion of gene or intergenic distances (italics) in base pairs. The numbers in red indicate the size of exon1 and exon 2 in single intron genes. The orientation of the arrows indicates the orientation of the gene: 5′-3′(right) or 3′-5′ (left). The residues at position 46 (after conserved Lysine45) are intron insertion sites. Stop codons are indicated, showing that fly *CSP* genes have TAA or TGA stop codons.

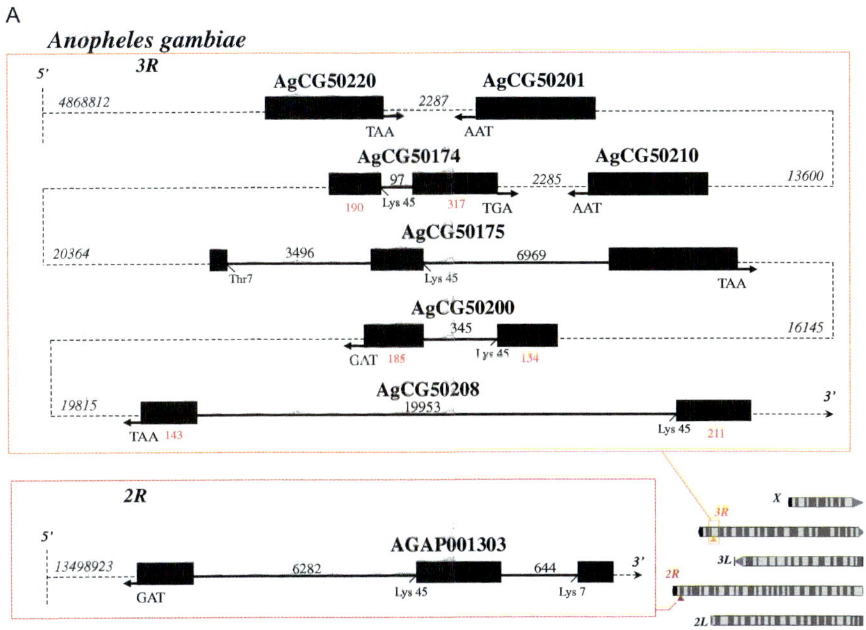

Fig. 2 Genomic organization of mosquito *CSPs* on two chromosomes (2R and 3R, 2 and 3). Genomic *CSP* repertoire in three mosquito species: *Anopheles gambiae* (AgCG) (A), *Aedes aegypti* (AAEL) (B) and *C. pipiens* (*quinquefasciatus*) (CPIJ) (C) (see Table 2). Exons are shown as black boxes and introns as plain lines. Dotted lines indicate the distances between genes. The numbers above indicate the size of each portion of gene or intergenic distances in base pairs. The numbers in red indicate the size of exon1 and exon2 in single intron genes. The orientation of the arrows indicates the orientation of the gene: 5′-3′(right) or 3′-5′ (left). The basic shape shows the position of retroposon Feilai (cross), Wujin-Aa4/Wuneng (circle), Twin (parallelogram) and APE (scalene) in the intron of *Aedes* or *Culex* CSP genes. The orientation of the diagonals indicates the orientation of the retroposon: 5′-3′ (downward) or 3′-5′ (upward). The residues at position 46 (after conserved Lysine45) and 8 (Arg) are intron insertion sites in *Anopheles*. The residues at position 46 (after conserved Lysine45), signal peptide, 8 (Arg) and 179 (Pro) are intron insertion sites in *Aedes*. The residues at position 46 (after conserved Lysine45), signal peptide, 8 (Arg), 164 (Ala) and 190 (Ser) are intron insertion sites in *Culex*. Stop codons are indicated, showing that mosquito *CSP* genes have TAA, TAG or TGA stop codons.

B

Aedes aegypti

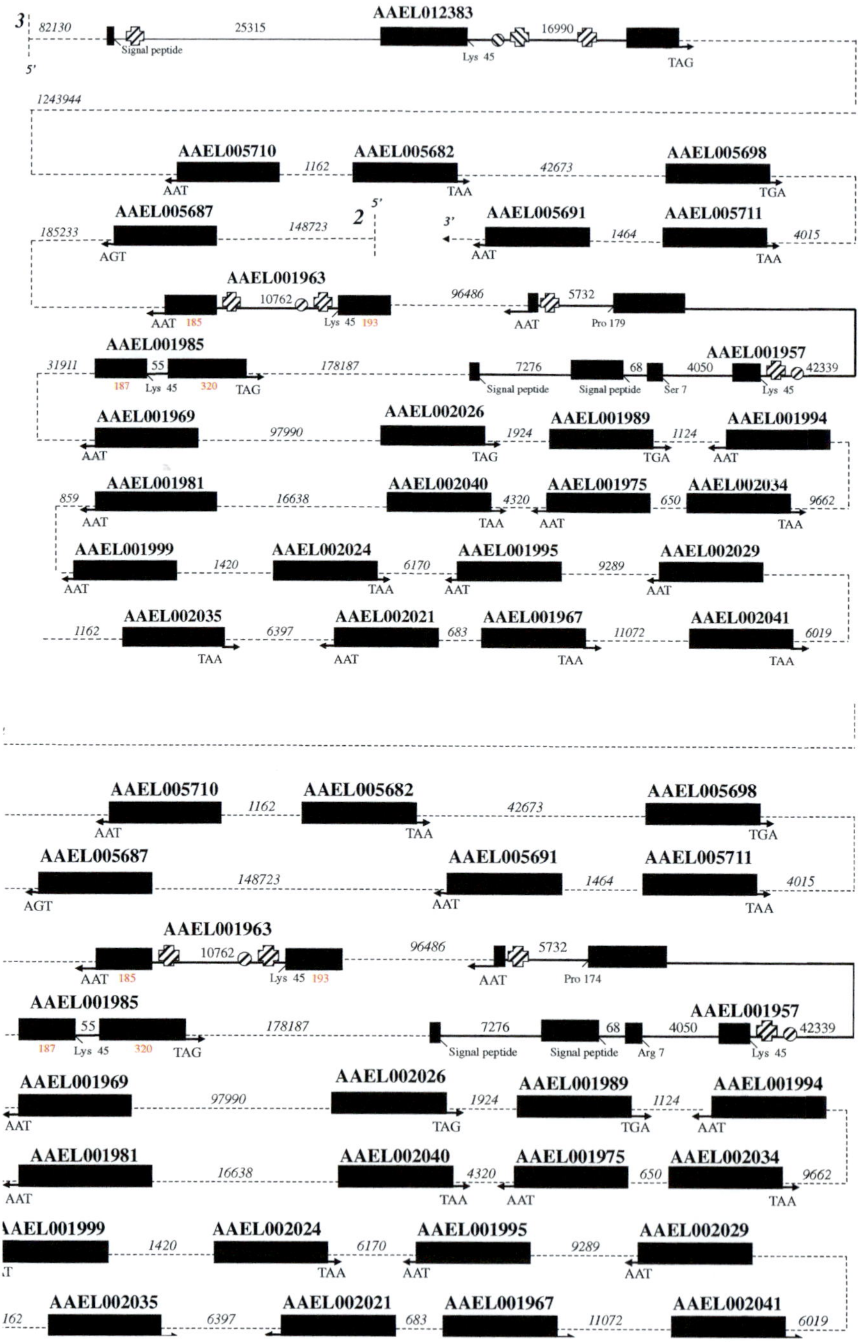

Fig. 2—Cont'd

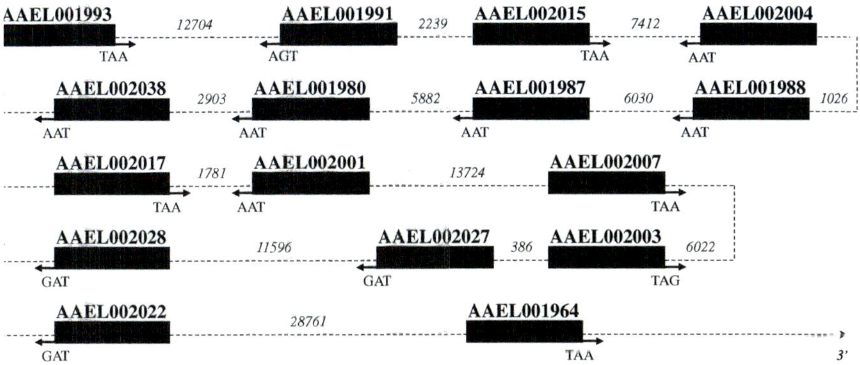
AAEL001993
12704
AAEL001991
2239
AAEL002015
7412
AAEL002004
TAA
AGT
TAA
AAT
AAEL002038
2903
AAEL001980
5882
AAEL001987
6030
AAEL001988
1026
AAT
AAT
AAT
AAT
AAEL002017
1781
AAEL002001
13724
AAEL002007
TAA
AAT
TAA
AAEL002028
11596
AAEL002027
386
AAEL002003
6022
GAT
GAT
TAG
AAEL002022
28761
AAEL001964
GAT
TAA
3'

C

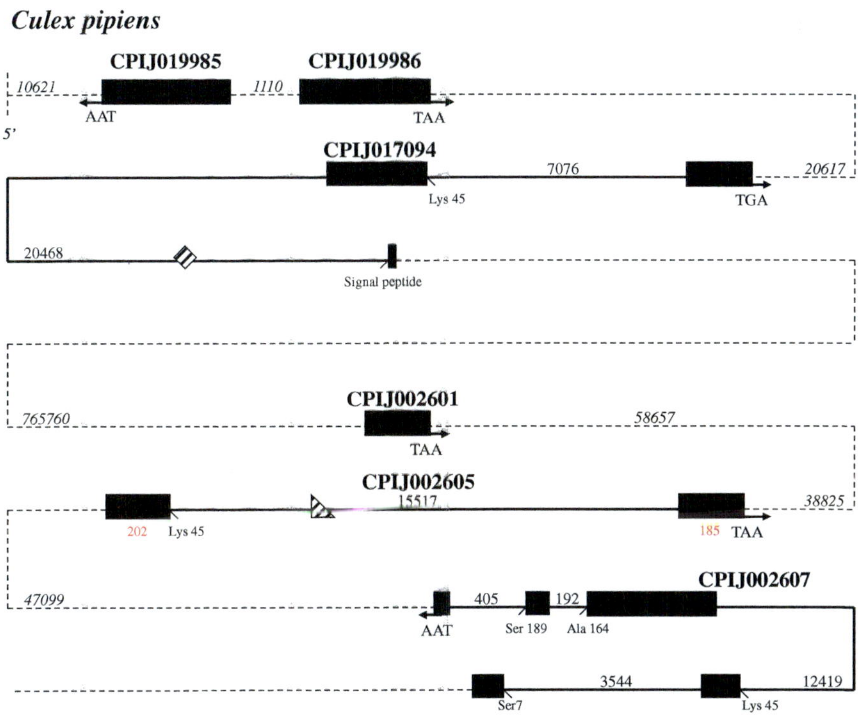
Culex pipiens
CPIJ019985
CPIJ019986
10621
1110
AAT
TAA
5'
CPIJ017094
7076
20617
Lys 45
TGA
20468
Signal peptide
CPIJ002601
765760
58657
TAA
CPIJ002605
15517
38825
202
Lys 45
185
TAA
CPIJ002607
47099
405
192
AAT
Ser 189
Ala 164
3544
12419
Ser7
Lys 45

Fig. 2—Cont'd

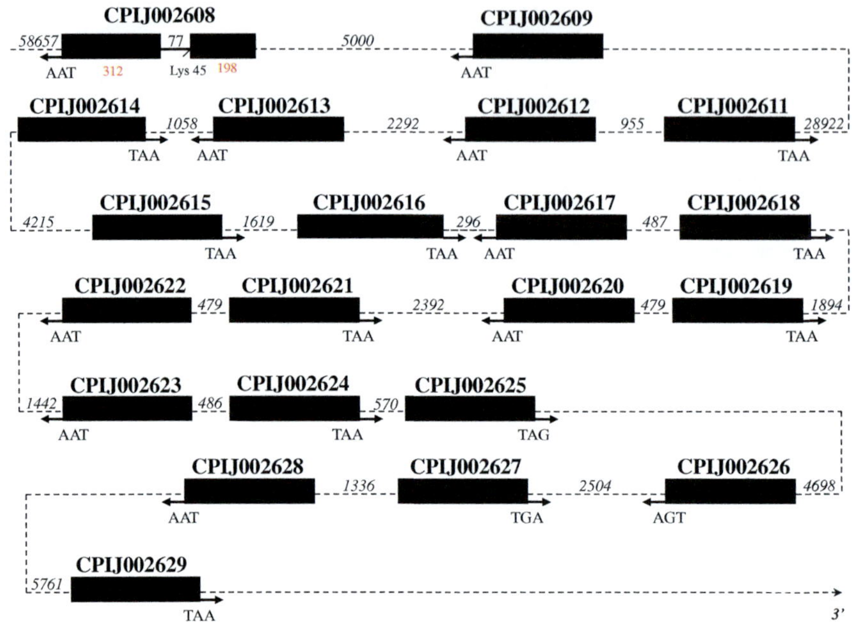

Fig. 2—Cont'd

of a complex data set, with no hypothesis on the common origin of genes, without any assumption about mutation rates or conserved sites, the MP-based approach should also make phylogeny and reproducibility of the results, although for remote homologous, MP-based approaches are clearly powerless.

In both MP and ML methods, gene structure should be plotted against the tree topology. This will aim to check for divergence, intron gain/loss, duplication with loss of function (pseudogene branch), common ancestry, common descent and evolutionary processes of gene birth/death for modeling the fates of gene duplication/gene loss on specific branch of the tree within a species or shared between species.

In a "blind" test of computational analysis for sequence comparison from phenotypes as different as those of flies and mosquitoes, one may not be interested on specific base changes. One may be more interested on pairwise evolutionary distances between sequences (one count of the number of base changes for minimum evolution). Phylogenetic methods help calculate the branch lengths based on pairwise comparison, measure bootstrap value and describe overall similarity between groups, which is mandatory to build a

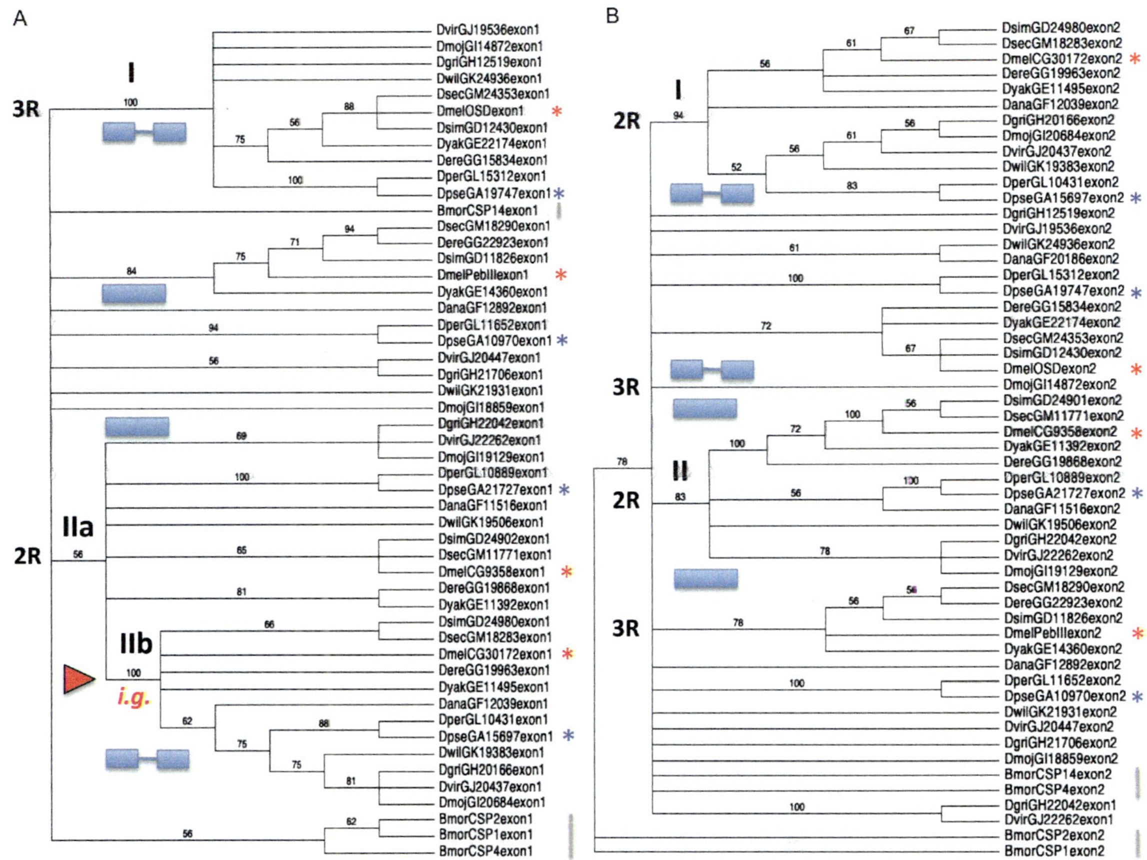

Fig. 3 See figure legend on next page

good evolutionary analysis with null-hypothesis. So, one take home message is use ML (or Bayesian Inference) to draw a good hypothesis then use other methods to confirm it, keeping in mind that every hypothesis in evolution remains a dark matter, pure theoretical science with no proof but strong evidence.

2.3 Expressed sequence tags (EST) analysis

Gene/protein functional analysis requires at least gene/protein expression data sets from Northern blot, real-time PCR, *in situ* hybridization and/or immunocytochemistry techniques. Molecular biology is usually more specific than immunoassay, which largely depends on antibody specificity. There is often a high antibody concentration required for field of view, particularly in electron microscopy (Angeli et al., 1999; Picimbon, Dietrich, Breer, et al., 2000; Picimbon, Dietrich, Angeli, et al., 2000; Picimbon et al., 2001).

Expressed Sequence Tags (EST) databases, consisting of more than 30,000 mRNA sequences from n tissue libraries, contain enough information for the association of gene expression with tissues of various functions. EST-cDNAs encoding a protein of interest are sorted from a BLASTn search of the EST database available for a specific organism using hit rate (100%) and e-value ($\leq$0.001), similarly to a BLASTn search of the genomic database. BLASTn search of the EST database available for flies and mosquitoes involves FlyBase and VectorBase (Celniker et al., 2002; Giraldo-Calderón et al., 2015; Lawson et al., 2009; Loftus et al., 2005; Nene et al., 2007; Noriega et al., 2006; Oduol, Xu, Niare, Natarajan, & Vernick, 2000;

Fig. 3—Cont'd Phylogenetic distribution of fly *CSPs*. Phylogenetic nucleotide exon1 (A) and exon2 (B) trees of *CSP* genes from 12 *Drosophila* species: *Drosophila ananassae, D. erecta, D. grimshawi, D. melanogaster, D. mojavensis, D. persimilis, D. pseudoobscura, D. sechellia, D. simulans, D. virilis, D. yakuba* and *D. willistoni* (see Table 1 and Fig. 1). *BmorCSP1, BmorCSP2, BmorCSP4* and *BmorCSP14* are used as outgroup (Picimbon, Dietrich, Angeli, et al., 2000; Xuan et al., 2014, 2015). Trees were established using nucleotide sequence data (see Table 1) aligned using CLUSTAL X 1.8 and assessed by bootstrapping in ML (Jackknifing, PAUP4.0b10 Altivec). Only bootstrap scores >50% at inner nodes are considered. The red asterisk shows the position of *D. melanogaster CSP* genes. The blue asterisk shows the position of *CSP* orthologs in *D. pseudoobscura*. The gray line in bold shows the position of *B. mori CSP* sequences compared to the two main orthologous cluster groups of *Drosophila CSPs* (I-II). Gene structure is shown for the main groups. The red arrow indicates a recent gene duplication and intron gain (i.g.) in the set of *CSP* genes on chromosome 2R (IIa-IIb).

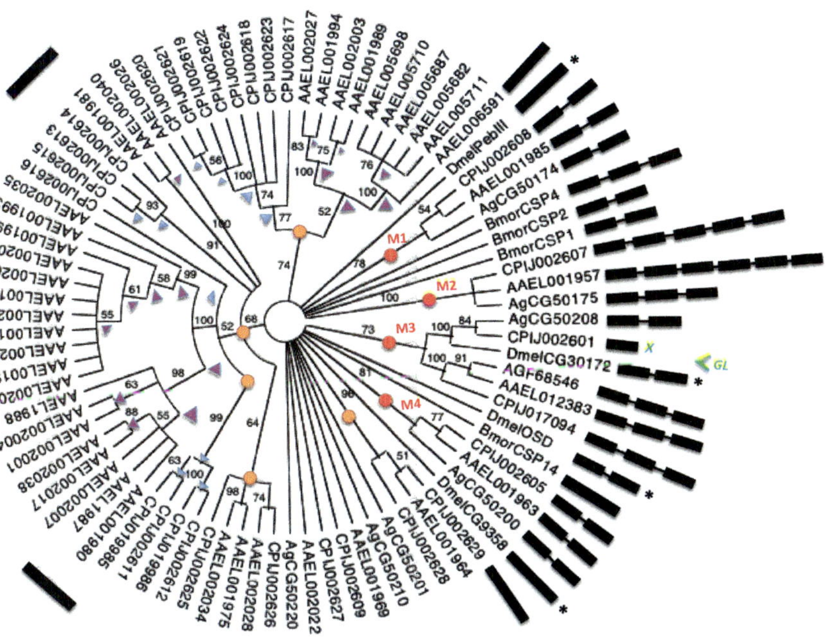

Fig. 4 Phylogenetic distribution of mosquito *CSPs*. Maximum parsimony phylogenetic analysis of amino acid sequences of *CSP* genes from *An. gambiae*, *Ae. aegypti* and *C. pipiens* with counterparts in *D. melanogaster* (see Tables 1–2 and Figs. 1–2). *BmorCSP1*, *BmorCSP2*, *BmorCSP4* and *BmorCSP14* are used as outgroup (Picimbon, Dietrich, Angeli, et al., 2000; Xuan et al., 2014, 2015). Multiple sequences were aligned using CLUSTAL X 1.8. Branch support of character-based trees in MP was assessed by bootstrapping (Jackknifing, PAUP4.0b10 Altivec). Only bootstrap scores >50% at inner nodes are considered. The black asterisk shows the position of *D. melanogaster CSP* genes. M1-M4: Mosquito-specific orthology groups of *CSPs*. The green arrow indicates gene loss (GL) and loss of gene function (X) in M3. Red dots indicate ancient gene duplication profiling in mosquitoes. Orange dots indicate recent gene duplication profiling in *Aedes* and *Culex*. Purple triangles indicate specific duplications of intronless genes in *Aedes*. Blue triangles indicate specific duplications of intronless genes in *Culex*. Gene structure is mapped to the phylogenetic tree. Exons are shown as black boxes and introns as plain lines, showing gene structural variation in M2.

Stapelton et al., 2002; Tweedie et al., 2009; Verjovski-Almeida et al., 2005; see Tables 1–2). EST data need to be interpreted with caution. For instance, an EST cloned from the antennal tissue indicates that the corresponding gene is expressed in the antennae, but it does not mean that the gene is solely or specifically expressed in the antennae. It needs to know more about tissue-distribution profiling of gene expression using a more precise quantitative method such as qRT-PCR to come out with this conclusion.

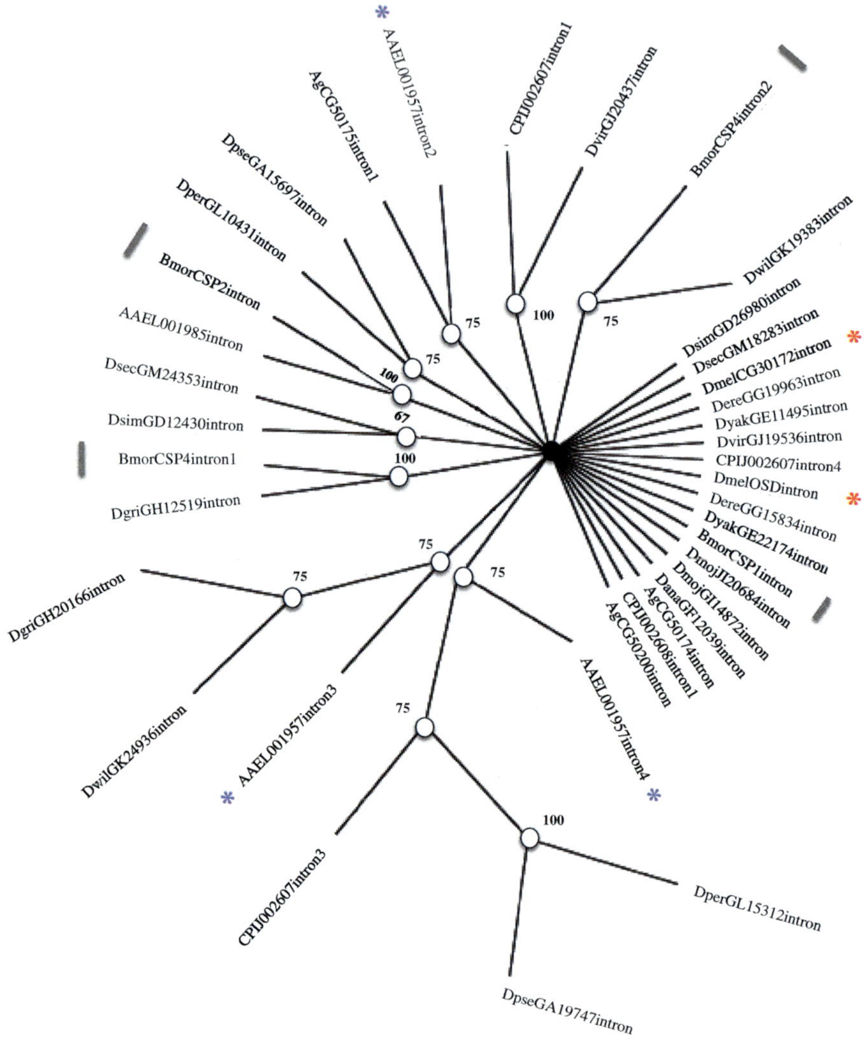

Fig. 5 Phylogenetic distribution of intron sequences in *dipteran CSP* genes. Unrooted ML analysis of nucleotide sequences of introns from fly and mosquito *CSP* genes (see Tables 1–2 and Figs. 1–2). Nucleotide sequences of introns were aligned using CLUSTAL X 1.8 (clustal aligned score 0: introns were not considered for PAUP). Branch support of character-based nucleotide trees was assessed by bootstrapping (Jackknifing, PAUP4.0b10 Altivec). Only bootstrap scores >50% at inner nodes are considered. Red asterisks show the position of intron sequences from *D. melanogaster CSP* genes (Table 1 and Fig. 1). Blue asterisks show the position of introns from *AAEL00195* (Table 2 and Fig. 2). The gray line in bold shows the position of introns from *B. mori CSPs* compared to *dipteran* intron taxa. The black dot stands for duplication node. Empty black circles stand for speciation nodes.

The function of an antennal-specific gene is certainly different than the function of a gene expressed in the antennae and other tissues. Eventually, the number of EST sequences in a specific tissue library can be used to infer transcript levels, but EST data should preferably not be used for quantitative gene expression analysis. They are informative in functional genomics analysis, identifying when, where and how specific genes are expressed to orientate research about protein function during infection, for instance, but expression of a certain transcript or protein candidate needs to be confirmed and localized to discuss further about function or physiological role.

3. Troubleshoots

In a bioinformatic approach aimed at identifying a number of genes, it must be considered that the number of ESTs or cDNA clones does not necessarily reflect the number of genes, especially in a protein gene family with RNA/peptide variance *via* editing (Deng et al., 2020).

The analysis of CSP sequences in the malaria vector mosquito, *An. gambiae*, is a good example to remind about the difference between number of EST-cDNA sequences and number of genes. *An. gambiae CSPs* has been estimated to be about seven to nine on the basis of bioinformatics-based sequence analysis using genome or EST database. However, many sequences turn out to be synonymous (AgamIR7-AgCG50210-EAA12338) or do not code for CSP (EAA12703). The sequence XP_317401 (ENSANGP 00000028374) derived from EAA12703 does not display the characteristic profiling of four cysteine residues. The sequence XP_555405 (91 amino acid residues) is mRNA encoding a part of CSP5 (#CAG26927). A shortened EST or mRNA sequence could represent a truncated RNA variant of a specific gene. In contrast, a longer RNA sequence could represent a precursor molecule. The number of genes can only be attested by identifying the number of genomic DNA loci or sequences on specific chromosomes. Genome analysis reports about eight genomic structures encoding CSP in the malaria vector mosquito, *An. gambiae*: *AgCG50220*, *AgCG50201*, *AgCG50210*, *AgCG50174*, *AgCG50175*, *AgCG50200* and *AgCG502008* loci organized in this order from the first to the seventh on the chromosome 3R, and one bachelor long *CSP* gene, *AGAP001303*, on chromosome 2R of the African malaria mosquito genome (Fig. 2A and Table 2; Mongin et al., 2004; Sharakhova et al., 2002; Zdobnov et al., 2002).

Using COBBLER for an iterative (multiple pass) search to find more isoforms can generate "false-positive" or misinterpretation when identifying a new genomic sequence. Specific elements of the protein sequence need to be recognized before assigning homology to two genes. For instance, the presence of diagnostic elements such as a pattern of four Cysteines at specific positions is fundamental to assign a new clone to *CSP* gene family (Picimbon, 2019). However, the problem of specifying how noncompliance can exist between two sequences need to be analyzed and discussed with caution. It may bring unknown links between different protein families. Using a COBBLER *Pediculus* CSP sequence in Ensembl Metazoa extracted a sequence (AGAP029265) related to dynein (six exons on chromosome 3L) with some similarity to *CSPs* on exon 2 (exclusive threshold below 1e-1). This is a situation similar to a hypothetical protein from *Lysobacter capsici* (WP_096417339) that is an uncharacterized membrane protein attached to a CSP fragment (Liu, Yue, Rajashekar, & Picimbon, 2019).

Then, it is always possible to run the sequence screen one more time, use the same known genes to estimate how many additional genes are found in another species, and perhaps find more genes with the wrong inaccurate incorrect assumption that all chromosomes or species have the same number of genes despite different evolutionary paths. BLASTing the *Aedes* or the *Culex* (subfamily Culicinae) sequences against the *Anopheles* (subfamily Anophelinae) genome does not allow us to find new *CSP* genes in *A. gambiae*, which simply shows that, in contrast to flies, mosquitoes (which diverged 95–120 Mya) do not have the same number of *CSPs* (nb = 8–42 in mosquito species; Fig. 2 and Table 2).

4. Jargon of bioinformatics and evolutionary analysis

Algorithm: The logical sequence of steps by which a *computational task* can be performed.

Amino acid sequence: The order of *amino acids* in a protein molecule.

Base sequence: The order of *nucleotide* bases in a DNA molecule.

Bayesian Inference (*BI*): The process of deducing properties or relatednesses about a population or probability distribution from data using *Bayes's Theorem*: describe the probability of an event, based on prior knowledge of conditions that might be related to the event, *i.e.*, find a probability when other probabilities are known.

Bioinformatics and Genomics: The field of the application of computer, evolutionary softwares and statistical techniques to the management of DNA/genetic/genomic information. In *genome* projects, bioinformatics includes the development of methods to search databases quickly, analyze

DNA sequence information, identify new protein coding genes and draw evolutionary models from DNA/protein sequence data.

Birth and death model of evolution: The evolutionary scenario in which new genes are built by successive *gene duplication* events, some duplicate genes are maintained in the genome (*gene gain*), other duplicates are deleted or truncated (*gene loss*).

Clade: Group of organisms or genes that consists of a *common ancestor* and all its lineal descendants or offsprings.

Clustal: Series of programs used for multiple (*clustered*) sequence alignment.

Cobbler: Gap-filling; identify conserved motif sequence and build consensus *blocks*.

Consensus (*block*): Calculated order of most frequent amino acid residues in a *protein block*.

Conserved sequence: A *base sequence* in a DNA molecule (or an *amino acid* sequence in a *protein*) that has remained essentially unchanged throughout evolution.

Domain (*or block*): A discrete *portion of a gene* with its own function. The combination of domains in a single protein determines its overall function.

Duplication: The mechanism through which new genetic material (*gene*) is generated during molecular and organismal evolution.

Evolution: The change in the heritable characteristics (*mutations*) of organisms or species over successive generations, including genes that are passed on from parents to offsprings.

Exon: Part of gene or pre-mature RNA that encodes for protein (*coding region*).

Expressed Sequence Tag (*EST*): Fragment of RNA *transcripts*, short sequencing products using randomly selected cDNA clones from a given tissue library.

Gene families: Groups of closely related *genes that make similar products*.

Genetic data: Raw material of bioinformatics and evolution study. Entire *DNA properties* of an organism, heritable and inheritable genes and traits, and their related expression.

Genome: Organism's complete set of *DNA information*.

Homologous sequences: Sequences that are related by *divergence from a common ancestor*. Homology is not a synonym for similarity.

Horizontal gene transfer: The movement of genetic material (genes) between cells or organisms other than by the transmission from parents to offspring (*vertical gene transfer*).

In silico: Biological experiment carried out *via computer simulation* or algorithm.

Intron: Part of gene or pre-mature RNA removed by alternative splicing (*non-coding region*).

Lineage: Temporal series of organisms or genes linked by a continuous *branch of descent* from parents to offsprings or ancestors to descendants, resulting in diversity or tree of life.

Maximum likelihood: Method to infer relationships between genes or species on the basis of all possible *characters*, parameters and number of changes, *i.e.*, ML seeks for all variable sites.

Maximum parsimony: Method to infer relationships between genes or species on the basis of the smallest (most parsimonious) number of *changes*, *i.e.*, MP seeks for conserved sites.

Microsatellite DNA: Typical tract of *repetitive sequence* in which some certain DNA motifs can be repeated over 5 to 50 times.

Motif: A consecutive string of nucleotides in a gene whose general character is repeated, or conserved, in all sequences in a multiple alignment at a particular position. Motifs are of interest because they may correspond to *structural or functional elements* within the gene sequences they characterize.

Multiplexing: A *sequencing* approach that uses several pooled samples simultaneously, greatly increasing sequencing speed.

Mutation: A *change*, not necessarily an alteration, in the DNA, RNA or protein sequence that helps produce a new gene, RNA or protein *variant*, prelude to new function and adaptation in a given protein gene family in responses to environmental changes.

Outgroup: Lineage that falls outside the clade (*ingroup*) being studied but is close enough to that clade, related enough to be used as a reference to study the relatednesses in the ingroup.

PAUP: Phylogenetic Analysis Using Parsimony resource to build *phylogenetic trees* using distance, likelihood and/or parsimony methods in advanced sequence statistics (Swofford).

Parallel evolution: The *similar development* of a trait (or gene) in two distinct species that share a similar original trait in response to similar evolutionary challenge or pressure.

Phenotype: All composite observable characteristics (or *traits*) of an organism or species corresponding to the result of its interaction and adaptation to the natural environment.

Phylogeny (*or Phylogenetic Tree*): The evolutionary history of a genetically related group of organisms or species in reference to *lines of descent* and relationships *via* the study of genes and the comparison of homologous sequences.

Phylogenetics: Part of systematics in biology that addresses the *evolutionary relationships* among or within groups of organisms or species.

RAxML: Randomized Axelerated Maximum Likelihood to build fast phylogenetic trees of large data sets under *maximum likelihood* (Stamatakis, 2014).

Retrogenes: Processed (stripped down) *copies of genes* inserted into another genomic region than the parental or progenitor gene and producing an identical or nearly identical product.

Retroposons: *Mobile DNA fragments* that undergo reverse transcription from any RNA and (re)integrate into genome on a different location.

Retrotransposons: *Autonomous retroposons* that can encode reverse transcriptase and therefore mediate their own retrotranscription, *i.e.*, can copy and paste themselves into genome (on different locations).

Scaffold: The first complete *sequence assembly* reconstructed from end-sequenced whole-genome clones (contigs and gaps) to which new reads can be aligned.

Sequence alignment: A linear comparison of amino (or nucleic) acid sequences in which insertions are made in order to bring equivalent positions in adjacent sequences into the correct register. Alignments are the basis of sequence analysis methods, and are used to pinpoint the occurrence of *conserved motifs*.

Speciation: The evolutionary process by which new living species emerge from a common ancestor (formation of new clades, cladogenesis or *splitting of lineages*).

5. Results and interpretation in an evolutionary context

5.1 Number of genes

Analyzing Drosophila genome databases, we found that *Drosophila ananassae*, *D. erecta*, *D. grimshawi*, *D. melanogaster*, *D. mojavensis*, *D. persimilis*, *D. pseudoobscura*, D. sechellia, *D. simulans*, *D. virilis*, *D. willistoni* and D. yakuba have strictly the same number of *CSP* genes: four (Fig. 1 and Table 1). There is a low number of *CSPs* also in the mosquito *Anopheles gambiae sensu* lato: eight (Picimbon, 2003; Vieira & Rozas, 2011; Fig. 2A). However, we identify 42 *CSP* genes as attested by the number of genomic DNA sequences in the yellow fever (dengue, chikungunya, Zika and Mayaro) mosquito, *Ae. aegypti* (Fig. 2B). We find that the common house Culex mosquito, *C. pipiens* (quinquefasciatus), retains about 29 *CSP* genes (Fig. 2C).

Therefore, RNA editing and retrotransposition mechanisms may have led to higher genetic diversity in this particular family not only in beetles and moths, but also in mosquitoes (Liu, Arnaud, Offmann, & Picimbon, 2017; Liu et al., 2020; Picimbon, 2019; Xuan et al., 2014, 2015; Xuan, Rajashekar, Kasvandik, & Picimbon, 2016; Xuan, Rajashekar, & Picimbon, 2019; Fig. 2 and Tables 3–4). Another possibility to explain such genetic divergence between species is the chromosomal evolution. In contrast to culicids, it is known that *An. gambiae* evolved to highly differentiated and heteromorphic sex chromosomes X and Y *via* an excess of movement off X after the split with *Aedes* and *Culex*, a phenomenon of genome evolution largely mediated by retrogenes or retrotransposed gene duplicates (Hall et al., 2016; Toups & Hahn, 2010). Therefore, there may be some trends between retrogenes, linkage with sex chromosomes, heteromorphism, recombination rates, genome size and genetic diversity in a group of organisms, as illustrated here for the case of *CSP* genes (see Tables 3–4).

5.2 Common ancestral gene structure and intron gain

Using genomic analysis, we can observe a site-specific intron insertion in dipteran *CSPs*. This simple observation has fundamental consequences for

Table 3 Relationships between ploidy (number of homologous chromosomes), genome size (in Megabases), number (nb) of chromosomes, total number of protein-coding genes, percentage of transposable elements (retroposons) and the number of *CSPs* in different insect genera.

Genus	Ploidy	Genome size (Mb)	Nb chromosomes	Nb genes	% Retroposons	Nb *CSPs*
Anopheles	Diploid	280	6	12,800	16	8
Aedes	Diploid	1250	6	15,419	47	42
Apis	Haploid	236	16	10,157	1	6
Bombyx	Polyploid	530	28	14,473	16	20
Culex	Diploid	1380	6	18,883	30	29
Drosophila	Diploid	180	4	15,500	1–10	4
Nasonia	Haploid	295	5	24,388	10–16	8
Pediculus	Diploid	108	6	11,000	1	6
Tribolium	Diploid	160	20	16,561	6	19

Table 4 Genetic diversity in insect *CSPs*.

Genus	pseudo	intronless	2e-1i	3e-2i	4e-3i	5e-3i	6e-5i
Anopheles	0	3	2	2	0	0	0
Aedes	0	38	2	1	0	0	1
Apis	0	0	5	1	0	0	0
Bombyx	3	1	13	3	0	0	0
Culex	1	20	2	1	0	1	0
Drosophila	0	2	2	0	0	0	0
D. ananassae	1*	2	1	0	0	0	0
Nasonia	0	0	7	1	0	0	0
Pediculus	0	0	5	1	0	0	0
Tribolium	1′	1	17	1	0	0	0

Pseudo: pseudogene (truncated gene), 2e-1i: two exons/one intron, 3e-2i: three exons/two introns, 4e-3i: four exons/three introns, 5e-4i: five exons/four introns, 6e-5i: six exons/five introns. 1*: one truncated gene only in *D. ananassae*. 1′: exon1 duplication (Liu et al., 2017, 2020, 2019; Xuan et al., 2015).

our understanding of the evolution of the gene family. Similarly to other insect orders, the intron of *CSPs* in flies and mosquitoes is always inserted in the codon for amino acid 46, after Lys45 (Figs. 1–2; Liu et al., 2017, 2020, 2019; Xuan et al., 2015). This indicates a very ancient common ancestor to these genes and strongly suggests that the insertion of intron occurred prior to the divergence of all major insect groups (>400 Mya).

The understanding of intron insertion site extends to highly conserved splicing regulation. In the Drosophilan flies, the length of the intron of *CSPs* is strictly maintained (52–436 bps). There are no genes that have a prominent intron (see Fig. 1). This may suggest that co-transcription and fast splicing rates are strongly required for whole gene family expression (Liu, Arnaud, Offmann, & Picimbon, 2017; Liu, Yue, Rajashekar, & Picimbon, 2019; Pai et al., 2017; Rigau, Juan, Valencia, & Rico, 2019; Wallace & Beggs, 2017).

The size of the gene is also important when considering evolution and regulation of gene.

It is interesting to note that, while only "small" *CSPs* are found in all fly species investigated, "big" *CSPs* can be found in *Aedes*, *Anopheles* and *Culex* mosquitoes (see Figs. 1–2). In *Aedes*, *Anopheles* and *Culex*, genes can strongly differ by addition of very prominent introns (>6000–25,000 bps; Fig. 2 and Table 2). Findings about intron size, abundance and distribution can have implications for theories of gene length, genome evolution and organismal diversity.

One may say that longer first introns are a general property of Eukaryotic gene structure (Bradnam & Korf, 2008). This is not the case if we consider mosquito *CSP* genes (Fig. 2B). In the "rain barrel" mosquito, *C. pipiens*, voluminous introns inserted right in the middle of the gene after Lys45, but also in the 5′ and 3′ends (Fig. 2C), showing that intron insertion patterns (*i.e.*, splicing site locations with respect to codon positions) across different types of organisms can be completely different (Fig. 2B–C). There are no clear rules as whether intron density, occupancy or positioning should be higher in 5′ or 3′ ends (Hong, Scofield, & Lynch, 2006; Fig. 2). All "big" genomes may have genes with numerous introns in variable sizes, but introns will insert in 5′ or 3′ end depending on species (see Fig. 2B–C). Identification of genomic structures and genetic mapping allows detection of intron insertion that can be correlated with the emergence of a particular phenotype. It can be a main concept that intron gain in specific protein gene families is associated with particular phenotypic changes or traits in the morphology, the development, the physiology and/or the behavior of a living species (Gorlova, Fedorov, Logothetis, Amos, & Gorlov, 2014; Mukherjee et al., 2018; Ruzzante, Reijnders, & Waterhouse, 2019). Like in the case of mosquitoes exposed here (Fig. 2 and Tables 2–4), different intron patterns and boundaries suggest that intron gain occurred rather recently, *i.e.*, after the divergence of the culicid lineages (about 120 Mya). Big genes bearing multiple intron and mobile element insertions in specific organisms are useful to study phenotype variation, gene regulation, splicing change and preponderant adaptation to environmental conditions.

It needs to be well understood that intron gain in coding genes is a rather rare event in many various organisms, including many insect species such as flies (Coulombe-Huntington & Majewski, 2007a, 2007b; Janice, Pande, Weiner, Lin, & Makalowski, 2012; Marais, Nouvellet, Keightley, & Charlesworth, 2005; Shepard, McCreary, & Fedorov, 2009; Tarrío, Ayala, & Rodríguez-Trelles, 2008; Yenerall, Krupa, & Zhou, 2011). However, in many other organisms including the particular case of mosquitoes, intron gain is a much more common phenomenon as more than 90–94% of coding genes is found to be with introns. This is mainly due to ancient horizontal genetic transfer between endosymbiotic bacteria and their hosts (Woolfit, Iturbe-Ormaetxe, McGraw, & O'Neil, 2009).

5.3 Presence of pseudogenes and retroposons, the signs of high gene duplication rates

Analysis of a complete gene repertoire in an entire genome is required to bring implications for the evolution of species diversity. Only repertoire-wide gene structure analyses can help draw an evolutionary model for a given protein gene family. For instance, genomic sequences in Diptera confirms that *CSP* genes in general possess one intron or contains no intron (intronless gene) (Liu et al., 2017, 2020, 2019; Xuan et al., 2015; Tables 1–4 and Figs. 1–2). Analysis of 12 *Drosophila* genomes shows that all Drosophilids retain two single-intron and two intronless *CSPs*, except for *D. ananassae*. *GF20186* gene in *D. ananassae* is found to be a truncated gene (pseudogene; Fig. 1 and Table 4). One pseudogene, also one single exon2 (*CPIJ002601*), can be found in *Culex* (Table 4 and Fig. 2C). Loss of function *via* pseudogenes needs to be demonstrated by quantitative molecular biology, *i.e.*, real-time PCR. However, this is a clear example of species-specific genes that arose by gene duplication, and after that lost the function *via* pseudogenes or gene truncations as found not only for the *CSPs* but also for the *OBP* gene family (Liu et al., 2017; Sánchez-Gracia, Vieira, & Rozas, 2009; Vieira, Sánchez-Gracia, & Rozas, 2007; Xuan et al., 2015; Fig. 1).

In addition of pseudogenes, presence of numerous transposable elements is informative of a high level of gene duplication rate in the gene family (Xuan et al., 2015; Fig. 2 and Table 3). It takes to blast the introns that are prominent in size. *Aedes* introns are associated with Feilai, MITE, Wujin-Aa4 and Wuneng (Boulesteix & Biémont, 2005; Tu, 1997, 1999, 2001). *Culex* introns are associated with APE and Twin Cp6 transposons (Crainey, Garvey, & Malcolm, 2005; Feschotte, Fourrier, Desmons, & Mouchès, 2001), as well as with microsatellite DNA and simple sequence repeats (SSRs). It is interesting to find transposons and SSRs associated with specific genes such as *CSPs* because both retroposons and SSRs have a functional importance in gene expression. They can regulate intron splicing and therefore provide a genetic basis for fast adaptation to environmental changes (Elbarbary, Lucas, & Maquat, 2016; Li, Korol, Fahima, & Nevo, 2004; Luco, 2016). Another important point is to check the orientation of the gene and its retroposons. Inverted retroposons are known to mediate genome instability and gene amplification (Lin, Lin, Lyu, & Whang-Peng, 2001). Interestingly, most *CSPs* identified in insects are in inverted position (Liu et al., 2017, 2020, 2019; Xuan et al., 2015; Figs. 1 and 2), which is an argument to say retrotransposition has played a key role in the diversification of the gene family (Fig. 2 and Tables 3–4).

Gene duplication events and gene family expansion are also suggested by the existence of tandem genes. Then, it is crucial to locate them on a specific chromosome to form a reliable evolutionary hypothesis. Diptera show a separation of *CSP* genes on two different chromosomes (*Drosophila*: 2R/3L, *Anopheles*: 2R/3R, *Aedes*: 2/3; Figs. 1–2), indicating that *CSP* gene duplication (amplification) occurred at a very early time in the history of lineages and species (Liu et al., 2017, 2020, 2019; Xuan et al., 2015), perhaps accordingly to diet-dependent life history and/or feeding preferences as postulated for human genes. Like us, insects probably are what they eat (Hunter, 2008).

Gene duplication as a major mechanism through which new gene is generated (*de novo* gene birth) at a specific point in an organism's evolutionary history is main general textbook. Genome sequence comparison has been an important approach to demonstrate that gene duplication occurs in an extremely wide variety of organisms. However, analyzing the genes of one particular organism and comparing them to those of another organism, it remains important to measure the rate at which these duplications occur and if gene duplication occurs at high rates in all evolutionary lineages. Analyzing *CSPs*, multiple gene duplication events have occurred at least in various lineages of three major groups of insects: moths, beetles and mosquitoes. In addition to the size and the structure, the organization of genes and gene duplicates is extremely relevant to the evolutionary origins of *de novo* genes and expansion of the protein family. Like *CSPs* in *A. aegypti* and *C. pipiens*, *de novo* gene duplicates are small units linked together, one after the other, like pearls in a jade necklace. Like pearls strung along a necklace, genes can be of different lengths, orientations and structures (Fig. 2). The genomic organization of genes such as *CSPs* in *Aedes* and *Anopheles* strongly contrasts with that observed in *Culex* (Fig. 2 and Tables 3–4). Such observation suggests that specific evolutionary forces can shape an organism's genome, enabling profiling and expression of specific gene repertoires (see the example of *CSPs* in insects; Table 4).

5.4 Variations of the lengths of exons and introns

Another interesting aspect of genome analysis and evolution study is to mark variations in exons *versus* introns. Having both the exon and intron lengths can help not only examining genetic variation but also in refining a hypothesis on the evolutionary history of the protein gene family. Similarly to flies, exons 1 and exons 2 of *CSP* genes in bees, beetles, body lice, mosquitoes, moths and wasps are about all of the same size (from 155 to 226 nucleotides) (Liu, Arnaud, Offmann, & Picimbon, 2017; Liu, Yue, Rajashekar, & Picimbon, 2019; Liu et al., 2020; Xuan et al., 2015; Figs. 1–2). This suggests that exon1 and exon2 in the *CSP* gene family originates from exon duplication that has probably occurred before the separation or split of the different insect species (*i.e.*, around 412–479 Mya; Misof et al., 2014).

There is usually a co-variation principle between exon and flanking intron in their length in eukaryotic genomes, which means that one could assume that exons and introns have very similar evolutionary rates (Zhu et al., 2009). It is not the case for hymenopteran *CSPs* (Liu et al., 2020). Analyzing flies and mosquito *CSPs* as used here as an example of species genetic models of gene duplication and diversification, an increase in exon length is also not necessarily correlated to an increase in the length of the flanking intron (Figs. 1–2). *Drosophila* genes such as *DmelOSD*, *GD12430*, *GM24353*, *GE22174*, *GG15834*, *GA19747*, *GL15312*, *GK24936*, *GI14872*, *GJ19536* and *GH12519* have an exon2 of 214–299 bps and an average intron length of only about 52–191 bps (Fig. 1). The *Culex* gene *CPIJ002608* (an exon2 of 312 nucleotides) and the *Anopheles* gene *AgCG50174* (an exon2 of 317 nucleotides) have an intron of only 77 and 97 bps, respectively (Fig. 2). These are examples of the importance of intron on exon formation during the evolution of genomes (Roy, Kim, Xing, & Lee, 2008).

5.5 Phylogenetic distribution

More than gene structure, phylogenetic distribution is essential to see differences in mutational rates, lineage-specificity and origin, as well as fate of a group of genes, defining orthologous groups and determining the presence or absence of specific clades in each group. This may help understand the contribution of specific genes in acquiring and developing a particular phenotype. The phylogenetic distribution of the gene family at the exon, intron and protein level may help propose or confirm an hypothesis for modeling within-species or within-population evolution. If performing phylogeny among species of the same order, genus or closely related, ML is preferably used to analyze exon and intron nucleotide sequences that show much higher rate variation. ML tree is best suited when studying and interpreting whole protein gene family evolution.

Here is an example (not fictional) evolutionary/phylogenetic analysis (ML and MP) involving intron, exon and complete amino acid sequences as it should be presented in a research work. The phylogenetic trees below represent relationships between specific genes at the level of coding region (exon), expressed sequence (protein) and noncoding silent DNA region (intron) (Figs. 3–5). Main data analysis and interpretation using fly and mosquito *CSPs* are as follows:

- There are two main orthology groups (I and II) supported by high bootstrap values (56–100%) at the level of the exon1. *DmelOSD*exon1 is highly conserved across *Drosophila* species (Group I, 100% bootstrap; Fig. 3A). Exon1s from *DmelCG9358* and *DmelCG30172* are descended from a common ancestral gene on 2R (Group II, a and b, 56% bootstrap; Fig. 3A). The pattern of branching of exons1 *via* ML is such that these *Drosophila CSPs* on 2R evolved *via* gene duplication and intron gain (see internal branch support for IIb, 100% bootstrap).
- Using ML, exon1s and exon2s result in distinct topologies with very strong differences. All members of the groups of single intron (I) appear to be more closely related to each other than they are on exon1 tree. Similarly, intronless genes fall close to each other and appear to be more closely related to a single gene that diverged long ago before the separation of chromosomes 2R and 3R (Fig. 3B). *DmelCG30172*exon2 is most highly conserved across *Drosophila* species (Group I, 100% bootstrap; Fig. 3B). Exon2s are much more divergent (Fig. 3B), strongly suggesting that exons in Diptera *CSP* did not arise by parallel evolution. Exon1 and exon2 in *CSPs* have been subject to different evolutionary pressure, assuming therefore that N- and C-tails of CSPs have evolved to completely different functions.
- Four major orthology groups of *CSPs* (M1-M4) are found in mosquitoes, showing members in each clade investigated (*Aedes*, *Anopheles* and *Culex*). A high number of mutations and gene duplication events specifically took place in Neoculex and Aedini species, respectively (Fig. 4). M2 and M4 group with high bootstrap value (81–100% identity; Fig. 4), they might derive from a *CSP* that duplicated before *Aedes*, *Anopheles* and *Culex* diverged in evolution of insects about 150 Mya (Arensburger et al., 2010; Holt et al., 2002; Nene et al., 2007; Rai & Black, 1999). M3 group that includes *DmelCG30172* might derive from

the last common ancestor of the fruit fly and mosquitoes. Mosquito/fly divergence is estimated to have occurred about 260 Mya (Chen et al., 2015; Fig. 4).

- Pairs of orthologous introns are found in good agreement with the phylogenetic distances between dipteran insect species (see multiple speciation nodes; Fig. 5). However, introns of *CSPs* from phylogenetically more distant insect species such as moths and flies are more surprisingly related. In MLs, intron1 from *BmorCSP4* and *D. grimshawi GH12519* intron form an orthology group with significative bootstrap value (100%; Fig. 5). A similar result is found comparing *BmorCSP2*intron and *AAEL001985*intron, which may suggest a common ancestral single intron gene of moth, mosquito and fly in a distant past. The maximum age of the divergence of Diptera and Lepidoptera is estimated at about 190 Mya, which is based on the oldest lepidopteran fossil (Grimaldi & Engel, 2005). Molecular phylogenetics and evolution of insects estimate the origins of Lepidoptera around 245 Mya (Mitter, Davis, & Cummings, 2017).

5.6 EST-cDNA tissue expression: Species similarities and functional significance

Gene structure and phylogenetic tree alone are insufficient to debate about function. Eventually, the determination of structure-evolutionary relations inside the gene family requires combining EST data (gene expression) in the phylogenetic tree. Genetic mapping, phylogeny and gene expression have become the *sine qua non* for a complete genomic study.

EST data are particularly useful to confront *in situ* experiments that usually focus on the expression levels of a specific gene in the cells of a given tissue at a given time. For instance, *in situ* hybridization to RNA in antennal tissue sections shows that *DmelOS-D* expression associates with the sacculus and patches of sensilla coeloconica distributed on different parts of the adult antennae (McKenna, Hekmat-Scafe, Gaines, & Carlson, 1994; Pikielny, Hasan, Rouyer, & Rosbach, 1994; Shanbhag, Singh, & Singh, 1995). However, *Drosophila* EST database reports *CSP* gene expression across the whole of the insect body (Celniker et al., 2002; Stapelton et al., 2002). *CSP*-EST gene expression profiling in flies is related not only to the antennae, but also to the adult head, the fat body and embryonic gonads, similarly to *CSPs* from the mosquito *A. gambiae* (Fig. 6; Lawson et al., 2009; Tweedie et al., 2009). The *in situ* approach has sub-cellular resolution, but EST profilings are untargeted transcriptome wide all tissues purposes.

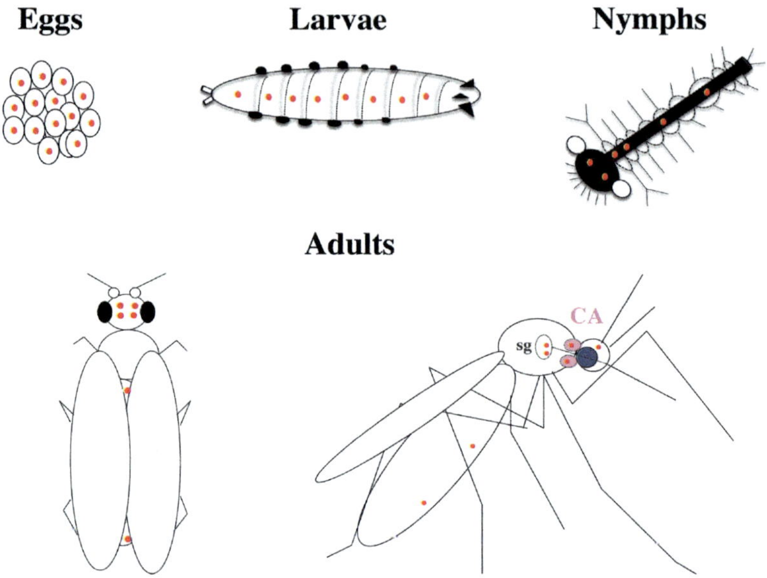

Fig. 6 Developmental regulation and tissue expression of dipteran *CSPs* based on EST data. EST-cDNAs encoding CSP are sorted from a BLASTn search of the fly and mosquito EST database through FlyBase and VectorBase (Tables 1–2). Main expression sites for *CSPs* are head, abdomen (fat body), salivary glands (sg) and *corpora allata* (CA, sites of JH biosynthesis).

Alike ML or MP phylogenetic trees, EST data work with the chance that any hypothesis is true. They carry questioning to the point of confidence in the original hypothesis, refrain from assuming a unique gene function and bring support to a new, more suitable hypothesis.

Suppose that we analyze now *CSP* expression in mosquitoes using broad scale EST data. In *A. gambiae*, *AgCG50210* is highly expressed in an immune responsive library of adults (Oduol et al., 2000). Using EST data shows that many other *CSP* genes in mosquitoes are highly expressed in various tissues and particularly in immune responsive libraries. *CSPs* are expressed in bacteria-infected *Culex* and in *Aedes* libraries infected with *Plasmodium gallinaceum*, *Brugia malayi* or Dengue-virus, in particular in transcriptomes of immune-activated hemocytes (Bartholomay et al., 2004; Loftus et al., 2005; Nene et al., 2007; Noriega et al., 2006; Verjovski-Almeida et al., 2005). This is in agreement with the overexpression of hemolymph *CSPs* in response to inoculation of antigenes of viral or bacterial origins as described in *D. melanogaster* (Einhorn & Imler, 2019; Sabatier et al., 2003).

Also similar to the" ectopic" expression of *CSPs* in non-chemosensory systems of flies, EST database analysis shows that *CSPs* in *Ae. aegypti* express in many non-sensory fluids and tissues in both larvae and adults: (1) the salivary glands, (2) *corpora allata* and (3) fat body (Tables 1–2 and Fig. 6). This is in agreement with the findings in moths where *CSPs* are found abundantly expressed not only in mandibular glands at the roof of the mouth in larvae, but also in the gut-fat body metabolic tract of adults after immune challenge (Celorio-Mancera et al., 2012; Xuan et al., 2015). Such tissue distribution from saliva to *corpora allara* (or CA, endocrine glands with function tuned to juvenile hormone-JH production) strongly suggests a function for *CSPs* in relation with lipid metabolism, immune function and JH biosynthesis.

6. Summary

Not only physiology, but also comparative genomics and gene family evolution are necessary to understand appearance and development of organisms fully adapted to their environment or ecosystem. More knowledge about species or population phenotypic changes and behavioral adaptation probably requires a full-description of the genetic basis of organismal formation, *i.e.*, reproduction, growth, development, defense and sensing for trends of interaction in order to settle in some new ecological niches. When we strive to predict organismal evolution, we may try depict gene evolution and relatednesses among species. This is a theoretical formula to be applied on the analysis of the widest sample of protein and sequence families. This is much beyond sampling a gene family in a small set of fly species, as exemplified here by the focus on *CSP* evolution. This chapter is a guidance to use genomes for proper protein gene evolution analysis from gene identification to phylogenetic reliability, with an adopted motto: understanding evolution helps tell when and how two species came to the same function (adaptation) and thus remains the only clarification for the diversity of life.

Acknowledgments

Shandong Province Overseas High-Level Talent and Taishan Scholar (NO.tshw20091015), I am very much grateful to all the international consortiums of researchers that have made insect genomes available (National Institute of Agricultural Science, Tsukuba, Japan; University of California, Lawrence Berkeley National Laboratory, Berkeley, USA; Carnegie Institution of Washington, Baltimore, USA; Baylor College of Medicine, Houston, USA; The National Human Genome Research Institute at NIH, Rockville Pike, USA; The Pasteur Institute, Paris, France; Celera Genomics, San Francisco, USA; Bioinformatics Resource Center (BRC), University of Notre Dame, Indiana, USA;

Wellcome Trust and European Bioinformatics Institutes, United Kingdom; The European Molecular Biology Laboratory, Heidelberg, Germany; The Institute for Genomic Research TIGR, Rockville, USA; The Institute of Molecular Biology and Biotechnology, Crete, Greece).

References

Adams, M. D., et al. (2000). The genome sequence of *Drosophila melanogaster*. *Science*, *287*, 2185–2195.

Angeli, S., Ceron, F., Scaloni, A., Monti, M., Monteforti, G., Minnocci, A., et al. (1999). Purification, structural characterization, cloning and immunocytochemical localization of chemoreception proteins from *Schistocerca gregaria*. *European Journal of Biochemistry*, *262*, 745–754.

Arensburger, P., Hice, R. H., Wright, J. A., Craig, N. L., & Atkinson, P. W. (2011). The mosquito *Aedes aegypti* has a large genome size and high transposable element load but contains a low proportion of transposon-specific piRNAs. *BMC Genomics*, *12*, 606.

Arensburger, P., et al. (2010). Sequencing of *Culex quinquefasciatus* establishes a platform for mosquito comparative genomics. *Science*, *330*, 86–88.

Atkinson, P.W., et al. (2007). *Culex pipiens quinquefasciatus* EST sequencing. #EV332855, EV358553, EV317590, EV330446, EV330633, EV332538, FF150040, FF150041, FF154211, FF154212, FF169164, FF169165, FF173833, FF177015, FF177016, FF184371, FF184372, "..."

Bartholomay, L. C., Cho, W. L., Rocheleau, T. A., Boyle, J. P., Beck, E. T., Fuchs, J. F., et al. (2004). Description of the transcriptomes of immune response-activated hemocytes from the mosquito vectors *Aedes aegypti* and *Armigeres subalbatus*. *Infection and Immunity*, *72*, 4114–4126.

Bosco, G., Campbell, P., Leiva-Neto, J. T., & Markow, T. A. (2007). Analysis of *Drosophila* species genome size and satellite DNA content reveals significant differences among strains as well as between species. *Genetics*, *177*, 1277–1290.

Boulesteix, M., & Biémont, C. (2005). Transposable elements in mosquitoes. *Cytogenetic and Genome Research*, *110*, 500–509.

Bradnam, K. R., & Korf, I. (2008). Longer first introns are a general property of eukaryotic gene structure. *PLoS One*, *3*, e3093.

Briand, L., Swasdipan, N., Nespoulos, C., Bézirard, V., Blon, F., Huet, J. C., et al. (2002). Characterization of a chemosensory protein (ASP3c) from honeybee (*Apis mellifera* L.) as a brood pheromone carrier. *European Journal of Biochemistry*, *269*, 4586–4596.

Campanacci, V., Lartigue, A., Hällberg, B. M., Jones, T. A., Giuici-Orticoni, M. T., Tegoni, M., et al. (2003). Moth chemosensory protein exhibits drastic conformational changes and cooperativity on ligand binding. *Proceedings of the National Academy of Sciences of the United States of America*, *100*, 5069–5074.

Campanacci, V., Mosbah, A., Bornet, O., Wechselberger, R., Jacquin-Joly, E., Cambillau, C., et al. (2001). Chemosensory protein (CSP2) from the moth *Mamestra brassicae*: Expression and secondary structure from ^{1}H and ^{15}N NMR. *European Journal of Biochemistry*, *286*, 4731–4739.

Celniker, S. E., & Rubin, G. M. (2003). The *Drosophila melanogaster* genome. *Annual Review of Genomics and Human Genetics*, *4*, 89–117.

Celniker, S. E., et al. (2002). Finishing a whole-genome shotgun: Release 3 of the *Drosophila melanogaster* euchromatic genome sequence. *Genome Biology*, *3*, research0079.

Celorio-Mancera, M. L., Sundmalm, S. M., Vogel, H., Rutishauser, D., Ytterberg, A. J., Zubarev, R. A., et al. (2012). Chemosensory proteins, major salivary factors in caterpillar mandibular glands. *Insect Biochemistry and Molecular Biology*, *42*, 796–805.

Chen, X. G., et al. (2015). Genome sequence of the Asian tiger mosquito, *Aedes albopictus*, reveals insights into its biology, genetics, and evolution. *Proceedings of the National Academy of Sciences of the United States of America, 112*, E5907–E5915.

Coulombe-Huntington, J., & Majewski, J. (2007a). Intron loss and gain in *Drosophila. Molecular Biology and Evolution, 24*, 2842–2850.

Coulombe-Huntington, J., & Majewski, J. (2007b). Characterization of intron loss events in mammals. *Genome Research, 17*, 23–32.

Crainey, J. L., Garvey, C. F., & Malcolm, C. A. (2005). The origin and evolution of mosquito APE retroposons. *Molecular Biology and Evolution, 22*, 2190–2197.

Dani, F. R., Iovinella, I., Felicioli, A., Niccolini, A., Calvello, M. A., Carucci, M. G., et al. (2010). Mapping the expression of soluble olfactory proteins in the honeybee. *Journal of Proteome Research, 9*, 1822–1833.

Delcher, A., Pop, M., & Salzberg, S. (2004). Drosophila pseudoobscura *strain MV2–25 whole genome shotgun sequencing project*. (Direct GenBank submission #AAFS01000001-AAFS01012430).

Deng, P., Khan, A., Jacobson, D., Sambrani, N., McGurk, L., Li, X., et al. (2020). Adar RNA editing-dependent and -independent effects are required for brain and innate immune functions in *Drosophila. Nature Communications, 11*, 1580.

Drosophila 12 Genomes Consortium. (2007). Evolution of genes and genomes on the *Drosophila* phylogeny. *Nature, 450*, 203–218.

Einhorn, E., & Imler, J. L. (2019). Insect immunity; from systemic to chemosensory organs protection. In J. F. Picimbon (Ed.), *Olfactory concepts of insect control-alternative to insecticides* (pp. 205–229). Basel: Springer Nature Switzerland AG.

Elbarbary, R. A., Lucas, B. A., & Maquat, L. E. (2016). Retrotransposons as regulators of gene expression. *Science, 351*, aac7247.

Feschotte, C., Fourrier, N., Desmons, I., & Mouchès, C. (2001). Birth of a retroposon: The twin SINE family from the vector mosquito *Culex pipiens* may have originated from a dimeric tRNA precursor. *Molecular Biology and Evolution, 18*, 74–84.

Giraldo-Calderón, G. I., Emrich, S. J., MacCallum, R. M., Maslen, G., Dialynas, E., Topalis, P., et al. (2015). VectorBase: An updates bioinformatics resource for invertebrate vectors and other organisms related with human diseases. *Nucleic Acid Research, 43*, D707–D713.

Gomez, S. M., Eiglmeier, K., Segurens, B., Dehoux, P., Couloux, A., Scarpelli, C., et al. (2005). Pilot *Anopheles gambiae* full-length cDNA study: Sequencing and initial characterization of 35,575 clones. *Genome Biology, 6*, R39.

Gorlova, O., Fedorov, A., Logothetis, C., Amos, C., & Gorlov, I. (2014). Genes with a large intronic burden show greater evolutionary conservation on the protein level. *BMC Evolutionary Biology, 14*, 50.

Grimaldi, D., & Engel, M. S. (2005). *Evolution of the insects*. Cambridge, New York, Melbourne: Cambridge University Press.

Hall, A. B., et al. (2016). Radical remodeling of the Y chromosome in a recent radiation of malaria mosquitoes. *Proceedings of the National Academy of Sciences of the United States of America, 113*, E2114–E2123.

Holt, R. A., et al. (2002). The genome sequence of the malaria mosquito *Anopheles gambiae. Science, 298*, 129–149.

Hong, X., Scofield, D. G., & Lynch, M. (2006). Intron size, abundance, and distribution within untranslated regions of genes. *Molecular Biology and Evolution, 23*, 2392–2404.

Hoskins, R. A., et al. (2015). The release 6 reference sequence of the *Drosophila melanogaster* genome. *Genome Research, 25*, 445–458.

Hunter, P. (2008). We are what we eat. The link between diet, evolution and non-genetic inheritance. *EMBO Reports, 9*, 413–415.

Janice, J., Pande, M., Weiner, J., Lin, C. F., & Makalowski, W. (2012). U12-type spliceosomal introns of insecta. *International Journal of Biological Sciences*, *8*, 344–352.

Jansen, S., Chmelik, J., Zídek, L., Padrta, P., Novak, P., Zdrahal, Z., et al. (2007). Structure of *Bombyx mori* chemosensory protein 1 in solution. *Archives of Insect Biochemistry and Physiology*, *66*, 135–145.

Jansen, S., Zídek, L., Löfstedt, C., Picimbon, J. F., & Sklenar, V. (2006). 1H, 13C, and 15N resonance assignment of *Bombyx mori* chemosensory protein 1 (BmorCSP1). *Journal of Biomolecular NMR*, *36*, 47.

Lartigue, A., Campanacci, V., Roussel, A., Larsson, A. M., Jones, T. A., Tegoni, M., et al. (2002). X-ray structure and ligand binding study of a moth chemosensory protein. *Journal of Biological Chemistry*, *277*, 32,094–32,098.

Lawson, D., et al. (2007). VectorBase: A home for invertebrate vectors of human pathogens. *Nucleic Acids Research*, *35*, D503–D505.

Lawson, D., et al. (2009). VectorBase: A data resource for invertebrate vector genomics. *Nucleic Acids Research*, *37*, D583–D587.

Li, Y. C., Korol, A. B., Fahima, T., & Nevo, E. (2004). Microsatellites within genes: Structure, function and evolution. *Molecular Biology and Evolution*, *21*, 991–1007.

Lin, C. T., Lin, W. H., Lyu, Y. L., & Whang-Peng, J. (2001). Inverted repeats as genetic elements for promoting DNA inverted duplication: Implications in gene amplification. *Nucleic Acids Research*, *29*, 3529–3538.

Liu, G. X., Arnaud, P., Offmann, B., & Picimbon, J. F. (2017). Genotyping and bio-sensing chemosensory proteins in insects. *Sensors*, *17*, 1801.

Liu, G. X., Ma, H. M., Xie, H. Y., Xuan, N., & Picimbon, J. F. (2016). Sequence variation of *Bemisia tabaci* chemosensory protein 2 in cryptic species B and Q: New DNA markers for whitefly recognition. *Gene*, *576*, 284–291.

Liu, G. X., Ma, H. M., Xie, H. Y., Xuan, N., Xia, G., Fan, Z. X., et al. (2016). Biotype characterization, developmental profiling, insecticide response and binding property of *Bemisia tabaci* chemosensory proteins: role of CSP in insect defense. *PLoS One*, *11*, e0154706.

Liu, G. X., Xuan, N., Rajashekar, B., Arnaud, P., Offmann, B., & Picimbon, J. F. (2020). Comprehensive history of CSP genes: evolution, phylogenetic distribution, and functions. *Genes*, *11*, 413.

Liu, G. X., Yue, S., Rajashekar, B., & Picimbon, J. F. (2019). Expression of chemosensory protein (CSP) structures in *Pediculus humanus corporis* and *Acinetobacter baumannii*. *SOJ Microbiology and Infectious Diseases*, 7, 1–17.

Loftus, B.J., Utterback, T., Pertea, G., Koo, H., Mori, A., Schneider, J., et al. (2005). *Aedes aegypti* cDNA sequencing. #DV263125, DV263127, DV289920, DV289921, DV297938, DV314711, DV314712, DV316589, DV316619, DV334734, DV334735, DV335057, DV335058, DV344048, DV347318, DV347319, DV365747, DV365766, DV357763, DV368339, DV368340, DV393559, DV400785, DV400787, "...".

Loftus, B. J., et al. (2006). *The genome sequence of* Aedes aegypti *(strain Liverpool)*. The Broad Institute Genome Sequencing Platform Consortium.

Luco, R. F. (2016). Retrotransposons jump into alternative-splicing regulation via a long noncoding RNA. *Nature Structural & Molecular Biology*, *23*, 952–954.

Maleszka, J., Forêt, S., Saint, R., & Maleszka, R. (2007). RNAi-induced phenotypes suggest a novel role for a chemosensory protein CSP5 in the development of embryonic integument in the honeybee (*Apis mellifera*). *Development Genes and Evolution*, *217*, 189–196.

Marais, G., Nouvellet, P., Keightley, P. D., & Charlesworth, B. (2005). Intron size and exon evolution in *Drosophila*. *Genetics*, *170*, 481–485.

Matthews, B. J., et al. (2018). Improved reference genome of *Aedes aegypti* informs arbovirus vector control. *Nature*, *563*, 501–507.

McKay, T. F. C., et al. (2012). The *Drosophila melanogaster* genetic reference panel. *Nature*, *482*, 173–178.

McKenna, M. P., Hekmat-Scafe, D. S., Gaines, P., & Carlson, J. R. (1994). Putative *Drosophila pheromone*-binding-proteins expressed in a subregion of the olfactory system. *Journal of Biological Chemistry, 269*, 16,340–16,347.

Misof, B., et al. (2014). Phylogenomics resolves the timing and pattern of insect evolution. *Science, 46*, 763–767.

Misra, S., et al. (2002). Annotation of the *Drosophila melanogaster* euchromatic genome: A systematic review. *Genome Biology, 3,* Research0083.

Mitter, C., Davis, D. R., & Cummings, M. P. (2017). Phylogeny and evolution of Lepidoptera. *Annual Review of Entomology, 62*, 265–283.

Mongin, E., Louis, C., Holt, R. A., Birney, E., & Collins, F. H. (2004). The *Anopheles gambiae* genome: An update. *Trends in Parasitology, 20*, 49–52.

Mukherjee, D., Saha, D., Acharya, D., Mukherjee, A., Chakraborty, S., & Gosh, T. C. (2018). The role of introns in the conservation of the metabolic genes of *Arabidopsis thaliana*. *Genomics, 110*, 310 317.

Muzny, D., et al. (2002). Drosophila melanogaster *3L BAC RP98-18E21 (Roswell Park Cancer Institute* Drosophila *BAC Library) complete sequence.* Direct GenBank submission #AC010070.

Nene, V., et al. (2007). Genome sequence of *Aedes aegypti*, a major arbovirus vector. *Science, 22*, 1718–1723.

Nomura, A., Kawasaki, K., Kubo, T., & Natori, S. (1992). Purification and localization of p10, a novel protein that increases in nymphal regenerating legs of *Periplaneta americana* (American cockroach). *International Journal of Developmental Biology, 36*, 391–398.

Noriega, F. G., Ribeiro, J. M. C., Koener, J. F., Valenzuela, J. G., Hernandez-Martinez, S., Pham, V. M., et al. (2006). Comparative genomics of insect juvenile hormone biosynthesis. *New Phytologist, 36*, 366–374.

Oduol, F., Xu, J., Niare, O., Natarajan, R., & Vernick, K. D. (2000). Genes identified by an expression screen of the vector mosquito *Anopheles gambiae* display differential molecular immune response to malaria parasite and bacteria. *Proceedings of the National Academy of Sciences of the United States of America, 97*, 11,397–11,402.

Ozaki, M., Wada-Katsumata, A., Fujikawa, K., Iwasaki, M., Yokohari, F., Satoji, Y., et al. (2005). Ant nestmate and non-nestmate discrimination by a chemosensory sensillum. *Science, 309*, 311–314.

Pai, A. A., Henriques, T., Paggi, J., Burkholder, A., Adelman, K., & Burge, C. B. (2017). Intron length and recursive sites are major determinants of splicing efficiency in flies. *eLife, 6*, e323537.

Picimbon, J. F. (2003). Biochemistry and evolution of CSP and OBP proteins. In G. J. Blomquist & R. G. Vogt (Eds.), *Insect pheromone biochemistry and molecular biology, the biosynthesis and detection of pheromones and plant volatiles* (pp. 539–566). London, San Diego: Elsevier Academic Press.

Picimbon, J. F. (2019). Evolution of protein physical structures in insect chemosensory systems. In J. F. Picimbon (Ed.), *Olfactory concepts of insect control-alternative to insecticides* (pp. 231–263). Basel: Springer Nature Switzerland AG.

Picimbon, J. F., Dietrich, K., Angeli, S., Scaloni, A., Krieger, J., Breer, H., et al. (2000). Purification and molecular cloning of chemosensory proteins from *Bombyx mori*. *Archives of Insect Biochemistry and Physiology, 44*, 120–129.

Picimbon, J. F., Dietrich, K., Breer, H., & Krieger, J. (2000). Chemosensory proteins of *Locusta migratoria* (Orthoptera: Acrididae). *Insect Biochemistry and Molecular Biology, 30*, 233–241.

Picimbon, J. F., Dietrich, K., Krieger, J., & Breer, H. (2001). Identity and expression pattern of chemosensory proteins in *Heliothis virescens* (Lepidoptera, Noctuidae). *Insect Biochemistry and Molecular Biology, 31*, 1173–1181.

Picimbon, J. F., & Leal, W. S. (1999). Olfactory soluble proteins proteins of cockroaches. *Insect Biochemistry and Molecular Biology, 30*, 973–978.

Pikielny, C. W., Hasan, G., Rouyer, F., & Rosbach, M. (1994). Members of a family of *Drosophila* putative odorant-binding proteins are expressed in different subsets of olfactory hairs. *Neuron, 12*, 35–49.

Rai, K. S., & Black, W. C. (1999). Mosquito genomes: Structure, organization and evolution. *Advances in Genetics, 41*, 1–32.

Reddy, B. P., Labbé, P., & Corbel, V. (2012). Culex genome is not just another genome for comparatibe genomics. *Parasites & Vectors, 5*, 63.

Richards, S., et al. (2005). Comparative genome sequencing of *Drosophila pseudoobscura*: Chromosomal, gene, and cis-element evolution. *Genome Research, 15*, 1–18.

Rigau, M., Juan, D., Valencia, A., & Rico, D. (2019). Intronic CNVs and gene expression variation in human populations. *PLoS Genetics, 15*, e1007902.

Roy, M., Kim, N., Xing, Y., & Lee, C. (2008). The effect of intron length on exon creation ratios during the evolution of mammalian genomes. *RNA, 14*, 2261–2273.

Ruzzante, L., Reijnders, M. J. M. F., & Waterhouse, R. M. (2019). Of genes and genomes: Mosquito evolution and diversity. *Trends in Parasitology, 35*, 32–51.

Sabatier, L., Jouanguy, E., Dostert, C., Zachary, D., Dimarcq, J. L., Bulet, P., et al. (2003). Pherokine-2 and -3: two Drosophila molecules related to pheromone/odor-binding proteins induced by viral and bacterial infections. *European Journal of Biochemistry, 270*, 3398–33,407.

Sánchez-Gracia, A., Vieira, F. G., & Rozas, J. (2009). Molecular evolution of the major chemosensory gene families in insects. *Heredity (Edinb), 103*, 208–216.

Severson, D. W., deBruyn, B., Lovin, D. D., Brown, S. E., Knudson, D. L., & Morlais, I. (2004). Comparative genome analysis of the yellow fever mosquito *Aedes aegypti* with *Drosophila melanogaster* and the malaria vector mosquito *Anopheles gambiae*. *Journal of Heredity, 95*, 103–113.

Shanbhag, S. R., Singh, K., & Singh, R. N. (1995). Fine structure and primary sensory projections of sensilla located in the sacculus of the antenna of *Drosophila melanogaster*. *Cell Tissue Research, 282*, 237–249.

Sharakhova, M. V., Hammond, M. P., Lobo, N. F., Krzywinski, J., Unger, M. F., Hillenmeyer, M. E., et al. (2002). *Update of the* Anopheles gambiae *PEST genome assembly, AgamP3. The Anopheles Genome Consortium.* Direct GenBank submission #EAA12703, ENSANGP00000011551.

Shepard, S., McCreary, M., & Fedorov, A. (2009). The peculiarities of large intron splicing in animals. *PLoS One, 4*, e7853.

Stamatakis, A. (2014). RAxML version 8: A tool for phylogenetic analysis and post-analysis of large phylogenies. *Bioinformatics, 30*, 1312–1313.

Stapelton, M., et al. (2002). *Berkeley* Drosophila *genome project.* Direct GenBank submission #AAM29645.

Stapleton, M., Carlson, J., Brokstein, P., Yu, C., Champe, M., George, R., et al. (2002). A *Drosophila* full-length cDNA resource. *Genome Biology, 3*, research0080.1.

Stapleton, M., Liao, G., Brokstein, P., Hong, L., Carninci, P., Shiraki, T., et al. (2002). The *Drosophila* gene collection: identification of putative full-length cDNAs for 70% of *D. melanogaster* genes. *Genome Research, 12*, 1294–1300.

Stoesser, G., Baker, W., van den Broek, A., Camon, E., Garcia-Pastor, M., Kanz, C., et al. (2002). The EMBL nucleotide sequence database. *Nucleic Acids Research, 30*, 21–26.

Swofford, D. L. (2002). PAUP* version 4.0 b10. In *Phylogenetic analysis using parsimony (* and other methods).* Sunderland: Sinauer.

Tarrío, R., Ayala, F. J., & Rodríguez-Trelles, F. (2008). Alternative splicing: A missing piece in the puzzle of intron gain. *Proceedings of the National Academy of Sciences of the United States of America, 105*, 7223–7228.

Tomaselli, S., Crescenzi, O., Sanfelice, D., Ab, E., Wechselberger, R., Angeli, S., et al. (2006). Solution structure of a chemosensory protein from the desert locust *Schistocerca gregaria*. *Biochemistry, 45*, 1606–1613.

Topalis, P., Koutsos, A., Dialynas, E., Kiamos, C., Hope, L. K., Strode, C., et al. (2005). AnoBase: A genetic and biological database of anophelines. *Insect Molecular Biology, 14*, 591–597.

Toups, M. A., & Hahn, M. W. (2010). Retrogenes reveal the direction of sex-chromosome evolution in mosquitoes. *Genetics, 186*, 763–766.

Tu, Z. (1997). Three novel families of miniature inverted-repeat transposable elements are associated with genes of the yellow fever mosquito, *Aedes aegypti. Proceedings of the National Academy of Sciences of the United States of America, 94*, 7475–7480.

Tu, Z. (1999). Genomic and evolutionary analysis of Feilai, a diverse family of highly related SINEs in the yellow fever mosquito, *Aedes aegypti. Molecular Biology and Evolution, 16*, 760–772.

Tu, Z. (2001). Eight novel families of miniature inverted repeat transposable elements in the African malaria mosquito, *Anopheles gambiae. Proceedings of the National Academy of Sciences of the United States of America, 98*, 1699–1704.

Tweedie, S., et al. (2009). FlyBase: Enhancing *Drosophila* gene ontology annotations. *Nucleic Acids Research, 37*, D555–D559.

Verjovski-Almeida, S., Eiglmeier, K., El-Dorry, H., Gomes, S. L., Menck, S. F. M., Nascimento, A. L., et al. (2005). *FAPESP and Institut Pasteur/AMSUD Network* Aedes aegypti *cDNA sequencing project*. #EG001037.

Vieira, F. G., & Rozas, J. (2011). Comparative genomics of the odorant-binding and chemosensory protein gene families across the Arthropoda: Origin and evolutionary history of the chemosensory system. *Genome Biology and Evolution, 3*, 476–490.

Vieira, F. G., Sánchez-Gracia, A., & Rozas, J. (2007). Comparative genomic analysis of the odorant-binding protein family in 12 *Drosophila* genomes: Purifying selection and Birth-and-death evolution. *Genome Biology, 8*, R235.

Wallace, E. W. J., & Beggs, J. D. (2017). Extremely fast and incredibly close: Cotranscriptional splicing in budding yeast. *RNA, 23*, 601–610.

Wiegmann, B. M., & Richards, S. (2018). Genomes of Diptera. *Current Opinion in Insect Science, 25*, 116–124.

Woolfit, M., Iturbe-Ormaetxe, I., McGraw, E. A., & O'Neil, S. C. (2009). An ancient horizontal gene transfer between mosquito and the endosymbiotic bacterium *Wolbachia pipientis. Molecular Biology and Evolution, 26*, 367–374.

Worley, K. C., et al. (2001). *Baylor Genome Sequencing Center*. Direct GenBank submission #AC010070.

Xuan, N., Bu, X., Liu, Y. Y., Yang, X., Liu, G. X., Fan, Z. X., et al. (2014). Molecular evidence of RNA editing in Bombyx chemosensory protein family. *PLoS One, 9*, e86932.

Xuan, N., Guo, X., Xie, H. Y., Lou, Q. N., Bo, L. X., Liu, G. X., et al. (2015). Increased expression of CSP and CYP genes in adult silkworm females exposed to avermectins. *Insect Science, 22*, 203–219.

Xuan, N., Rajashekar, B., Kasvandik, S., & Picimbon, J. F. (2016). Structural components of chemosensory protein mutations in the silkworm moth, *Bombyx mori. Agri Gene, 2*, 53–58.

Xuan, N., Rajashekar, B., & Picimbon, J. F. (2019). DNA and RNA-dependent polymerization in editing of Bombyx chemosensory protein (CSP) gene family. *Agri Gene, 12*, 100,087.

Yenerall, P., Krupa, B., & Zhou, L. (2011). Mechanisms of intron gain and loss in *Drosophila. BMC Evolutionary Biology, 11*, 364.

Zdobnov, E. M., et al. (2002). Comparative genome and proteome analysis of *Anopheles gambiae* and *Drosophila melanogaster. Science, 298*, 149–159.

Zhu, L., Zhang, Y., Zhang, W., Yang, S., Chen, J. Q., & Tian, D. (2009). Patterns of exon-intron architecture variation of genes in eukaryotic genomes. *BMC Genomics, 10*, 47.

CHAPTER FOUR

Proteomics of arthropod soluble olfactory proteins

Francesca R. Dani[a],*, Giuseppe Pieraccini[b]

[a]Department of Biology, University of Firenze, Firenze, Italy
[b]CISM, Mass Spectrometry Centre, Department of Health Sciences, University of Firenze, Firenze, Italy
*Corresponding author: e-mail address: francescaromana.dani@unifi.it

Contents

Abstract

Shotgun proteomics is a very sensitive bottom-up approach used to study complex mixtures of proteins through a combination of high-performance liquid chromatography (HPLC) and high-resolution mass spectrometry (HRMS). This technique has

Methods in Enzymology, Volume 642
ISSN 0076-6879
https://doi.org/10.1016/bs.mie.2020.04.069

been successfully applied to the study of olfactory appendixes of insects and other arthropods. Since extracting soluble proteins from tissues and processing them for shotgun proteomics is quite straightforward, the technique has proved to be very suitable for studying soluble olfactory proteins, such as odorant binding proteins (OBPs), chemosensory proteins (CSPs) and Niemann-Pick type 2 proteins. Here, we describe the main principles and methods at the basis of shotgun proteomics, including its use to quantify proteins through label-free quantification. We also provide operational protocols for the experimental workflow steps, i.e., insect dissection, protein extraction, protein enzymatic digestion, sample purification, HPLC-MS analysis, search of mass spectral data against polypeptide databases and first steps in the analysis of the search results.

1. Introduction

Proteomics studies on olfactory organs of several insects and a few other arthropods have been reported in the literature in the last 15 years. The increased sensitivity of the latest generation mass spectrometers, combined with a higher resolving power of the liquid chromatography systems, has made it possible to successfully analyze the protein extracts prepared from a limited number of antennas or other olfactory organs dissected from small-size specimens. Besides, the accuracy of these mass spectrometers has allowed a high number of proteins (up to a few thousand) to be identified with high confidence. This proteomic approach is particularly suitable for soluble proteins, including carriers of odorants involved in perireceptor olfactory processes, while very few proteomic studies have identified olfactory receptors. The same microscale proteomic approach has also been successfully applied to other organs or tissues, such as exocrine pheromonal glands, where olfactory soluble proteins can play a regulatory function in the release of small hydrophobic molecules with a semiochemical function.

Proteins involved in perireceptor events in insects and other arthropods belong to different families (Leal, 2013; Pelosi, Iovinella, Felicioli, & Dani, 2014; Pelosi, Iovinella, Zhu, Wang, & Dani, 2018), namely olfactory receptors (ORs), carriers of hydrophobic olfactants and pheromones (odorant binding proteins—OBPs, chemosensory proteins—CSPs, and Niemann-Pick type 2 proteins—NPC2), and odorant degrading enzymes (ODEs). In some insect orders, OBP, CSP and OR gene families may contain up to several dozen members, which are differentially expressed in the two sexes, and/or in individuals of different ages or physiological conditions (Pelosi et al., 2014, 2018). Expression patterns of olfactory soluble proteins and ODEs can quite easily be studied using bottom-up proteomics, while this method is less suitable for membrane hydrophobic proteins, such as ORs. In bottom-up proteomics, proteins are identified and eventually characterized for post-translational modifications through mass spectrometry (MS) analysis of the peptides obtained after their proteolytic digestion. Mass spectra obtained for above-mentioned peptides are compared with those predicted by the *in-silico* digestion of protein databases; since peptide fragmentation pattern is predictable (Wang, Kish, & Wedemiotis, 2005), the experimental spectra obtained through MS/MS analysis of the peptide ions can also be compared with the theoretical ones predicted for the above-mentioned *in-silico*-generated peptides. Shotgun proteomics is a bottom-up approach allowing to identify proteins in complex mixtures using high-performance liquid chromatography (HPLC) combined with mass spectrometry.

The protocols reported in shotgun proteomics studies of arthropod olfactory proteins differ for several aspects which go from the sample preparation to the analytical methods and, most importantly, to the analytical equipment. Since increasing performances in chromatographic resolving power and mass spectrometry sensitivity and accuracy allow to increase the number of measured peptides and consequently to identify more proteins, the output of these experiments is strongly dependent on the available pieces of equipment.

In the subsequent sections, we report general protocols for the extraction of soluble olfactory proteins from insect olfactory organs and softer tissues (such as exocrine glands) as well as for their digestion and purification before HPLC-MS analysis, based on the most widespread pieces of equipment. Variation and/or alternatives to this general method will also be mentioned. The whole experimental workflow and the steps detailed in the protocols are graphically represented in Fig. 1.

2. Extraction of olfactory proteins from arthropod tissues

2.1 Material, equipment and reagents for the extraction of proteins from the biological tissues

A stereoscope;
fine-tipped tweezers;
1.5–2 mL microcentrifuge tubes (possibly low protein binding)
ice or cooler for the 1.5–2 mL tubes;
liquid nitrogen;

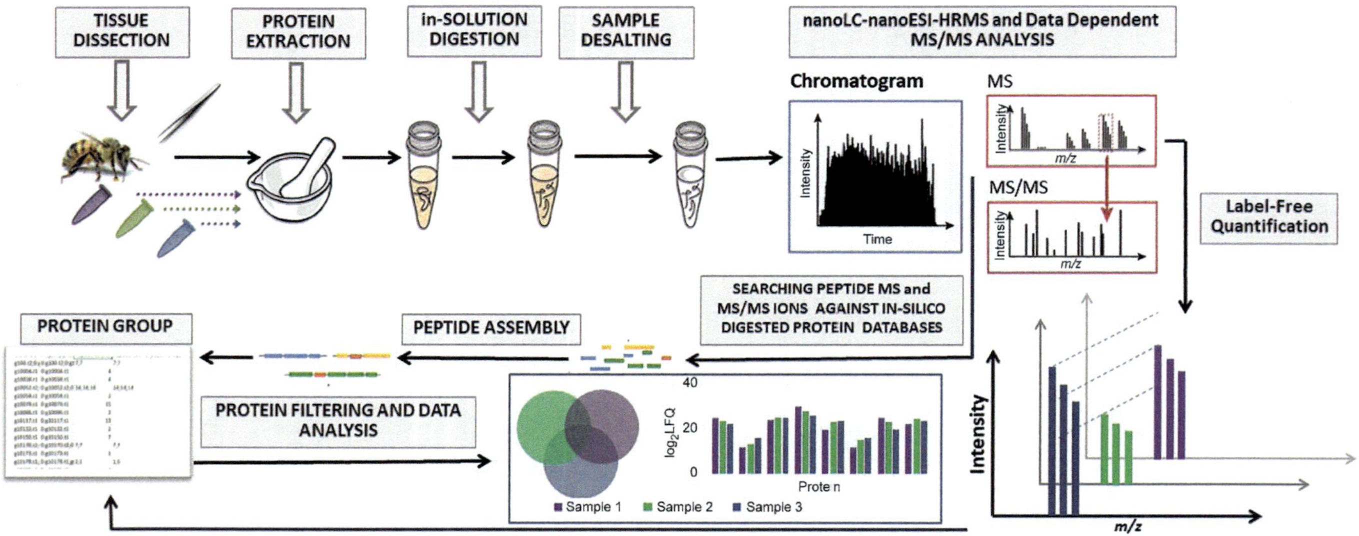

Fig. 1 Graphical representation of the experimental workflow of a shotgun proteomic experiment.

mortars and pestles;
Tris-HCl;
urea (and thiourea);
protease inhibitor (optional);
a refrigerated centrifuge for 1.5–2 mL tubes;
acetone, or chloroform-methanol solution (optional).

2.2 Protocol for the extraction of proteins from the biological tissues

1. Arthropods are better to be anesthetized by freezing and need to be maintained frozen (possibly at −80 °C) until dissection.
2. Dissection of arthropod olfactory organs (antennae, other sensory appendages such as the first pair of legs in ticks, or pheromonal glands) must be quickly performed, possibly on a cool surface. A stereoscope is needed for small-sized specimens. Since insect olfactory proteins are mainly located within the chemosensilla found on the flagellomeres, the flagellum can be separated from the basal antennal segments (scape and pedicel) to reduce the extraction of non-target proteins. Pooling the tissue of several specimens is normally needed for small-size arthropods, to obtain enough material and whenever technical and biological replicates are needed. Based on the protocol described in the dedicated section of this manuscript, we found that the extraction of the first legs of 35 females from the honeybee mite *Varroa destructor* (where chemosensory organs are located, Häußermann, Ziegelmann, Bergmann, & Rosenkranz, 2015) allowed us recovering about 200 μg of total proteins (Iovinella et al., 2018). Given the small amount of tissue, contamination with exogenous proteins, like human keratins, should be minimized, since they could compromise the success of the experiment (Table 2). In the case of tiny organs difficult to handle, dissected parts can be placed directly in the mortar where they will be extracted.
3. Dissected tissue is transferred into a (ceramic or alabaster) small-size mortar, added with a small volume (40–100 μL, depending on the amount of tissue) of 8 M urea or 6 M urea, 2 M thiourea in Tris-HCl (50 mM, pH 7.4), and manually grounded very finely using a pestle, carefully adding very small volumes of liquid nitrogen. Great care must be taken in this process, to avoid dispersion of the sample outside the mortar with the nitrogen vapor. Commercial homogenizers can be used instead of manual grounding in liquid nitrogen (Parker et al., 2012). Dithiothreitol (DDT) as well as protease inhibitors can also be added to the urea (or urea/thiourea) solution (Parker et al., 2012). The extraction of antennal proteins in

the absence of urea and thiourea has also been reported. As an example, Rund et al. (2013) extracted *Anopheles gambiae* antennae using a 50:50 mixture of trifluoroethanol and ammonium bicarbonate (50 mM), supplemented with ethylenediaminetetraacetic acid (EDTA) and phenylmethylsulfonyl fluoride (PMSF). Studies comparing the efficacy of protein extraction methods from insect antennal samples have never been reported. Since antennal structures and sizes are very diverse among different insects, the same protein extraction method may produce different protein recovery results on different organism species.

4. Extracts are transferred into 1.5–2 mL tubes, centrifuged at 14,000 rpm, for 30 min, at 4 °C, followed by supernatant recovery. Pellets can be washed with a small volume of extraction buffer and further centrifuged (14,000 rpm, 4 °C for 15 min); the supernatant can be added to the first one. A step of protein precipitation, generally in ice-cold acetone, can also be performed (Rund et al., 2013; Zheng, Xia, & Keyhani, 2020). In this case, proteins must be resuspended before proceeding to *in-solution* enzymatic digestion.

3. In-solution enzymatic digestion of protein extracts

The following protocol is mainly based on protocols reported by Foster, De Hoog, and Mann (2003), and it is adapted to enzymatically digest 15 μg of protein to be further purified according to the protocol by Rappsilber, Mann, and Ishihama (2007), as reported below. Fractionation of the extracted proteins before in-solution digestion or of the digested proteins (e.g., through ion-exchange chromatography and reverse-phase HPLC, see Smoluch, Mielczarek, Drabik, & Silberring, 2012) can be performed, but it has rarely been employed in insect proteomic studies, possibly because it increases the risk of sample loss.

Trypsin is the proteolytic enzyme most commonly used in proteomic bottom-up experiments. Sequential enzymatic digestion firstly with endoprotease Lys-C and then with trypsin helps to overcome the resistance of non-denaturated proteins to proteolysis and leads to an increasing number of identified proteins (Glatter et al., 2012).

3.1 Material, equipment, and reagents for in-solution digestion

1,4-Dithiothreitol (DTT);
iodoacetamide (IAM);
ammonium bicarbonate (AMBIC);

HPLC-grade water;
MS-grade endoprotease Lys-C (optional);
sequencing-grade modified trypsin;
trifluoroacetic acid (TFA);
a UV-absorbance spectrophotometer and the reagents for Protein Quantification Assay (e.g., the Bio-Rad Protein Assay based on Bradford's method) or a microvolume fluorometer and the kit for protein quantification (e.g., Qubit Protein Assay kit);
A thermomixer for 1.5–2 mL microcentrifuge tubes.

Solutions to be prepared

50 mM AMBIC solution in water, to be used as digestion buffer; store it at room temperature;
10% v/v TFA solution in water;
an aqueous solution of DTT (0.5 μg/μL), to be used as a stock solution; store it at −20 °C;
an aqueous solution of IAM (2.5 μg/μL), to be used as a stock solution; separate it into 100 μL aliquots to be stored at −20 °C in the dark,
an aqueous solution of endoprotease Lys-C (MS-grade) (0.4 μg/μL); separate it into small aliquots (e.g., 5 μL) and store at −20 °C;
resuspend sequencing-grade modified trypsin according to the manufacturer's instructions, normally to a concentration of 1 ng/μL; separate it into small aliquots (e.g., 4 μL) to be stored at −20 °C.

3.2 Protocol for in-solution digestion

1. Protein quantification can be performed on a UV-absorbance spectrophotometer using a colorimetric assay, such as Bradford's, or using a microvolume fluorometer, such as Nanodrop (ThermoFisher Scientific), following the manufacturer's instructions for the instrument and the reagents.
2. For each sample to be analyzed, transfer a volume containing 15 μg of protein in a clear 1.5- or 2-mL tube (better low protein binding tubes).
3. Add 1 μL of DTT stock solution and incubate for 30 min, at room temperature.
4. Dilute sample with four volumes of digestion buffer. This step is necessary to get an adequate working solution for both enzymes, i.e., high pH for Lys-C and low urea concentration for trypsin.
5. Add 1 μL of Lys-C solution and incubate for 3 h by gently shaking in the thermomixer.

6. Add 1 μg of trypsin solution and incubate overnight at 37 °C by gently shaking in the thermomixer.
7. Stop the reaction by lowering the pH at 2.5 or below by adding TFA solution (1–2 μL). The digested peptides mixture may be stored at −20 °C.

4. Sample purification

Different methods to desalt, purify and concentrate the peptides resulting from protein digestion for further HPLC-MS (or matrix assisted laser desorption ionization-MS; MALDI-MS) analyses have been reported. A general discussion about the interferences caused by detergents, salts and other contaminants in mass spectrometric analyses of peptides (Table 2) and about sample purification procedures has been reported by Feist and Hummon (2015).

Here, we report the stop and go extraction (STAGE) protocol described by Rappsilber et al. (2007), which allows to process up to 15 μg of digested proteins on in-house made small volume columns and to concentrate the samples. Alternative methods based on commercial columns are also shortly reported.

4.1 Material, equipment, and reagents for STAGE purification and sample concentration

Acetic acid, acetonitrile, and HPLC-grade water, to prepare the following buffers:
Buffer A: 0.5% v/v acetic acid in water;
Buffer B: 80:20 v/v acetonitrile:water solution, containing 0.5% acetic acid;
methanol;
C18 Empore™, 0.5 mm thickness extraction disks;
a 17- or 18-Gauge blunt-tipped syringe needle;
a plunger thinner than the internal diameter of the syringe needle (e.g., the plunger of a 10 μL Hamilton syringe);
200 μL micropipette tips;
low volume (100 μL), polypropylene vials for HPLC system autosamplers or autosampler plates;
a centrifugal vacuum concentrator system.
Alternatively, the digestion products can be desalted using commercial low volume C18 columns (see below).

4.2 Protocol for STAGE purification and sample concentration

1. Punch out, one after the other, 3 disks from a C18 Empore extraction disk using the syringe needle and eject them into a 200 μL micropipette tip using the plunger; ensure that the disks are securely wedged in the bottom of the tip. Each disk binds about 5 μg of peptides. Prepare as many desalting columns as necessary to process the digested samples.
2. Condition the column by placing 50 μL of methanol on the top of the Empore disks, and force the solvent through the disks with an air-filled syringe fitted to the end of the pipette tip.
3. Wash the column by placing 50 μL of buffer A on the disks, and force it through, as previously described.
4. Load a volume of the sample containing 15 μg of digested proteins on the top of the disks (check the pH to be lower than 2.5), and force the sample through the column, as previously described.
5. Wash the column by forcing 50 μL of Buffer A as previously described, and check that all the buffer is out of the tip. After this stage, the tips can be stored at 4 °C for several weeks until elution (step 6).
6. Elute the peptides by forcing 50 μL of Buffer B through the column. Elute directly into a polypropylene vial or an autosampler plate.
7. Place the samples in the centrifugal vacuum concentrator system and concentrate them (without heating) to less than 10 μL to evaporate acetonitrile. Take care not overdrying the samples.
8. Add Buffer A to reach a final volume of 20 μL.

Desalting and purification of peptide mixture can also be performed using commercial reversed phase solid phase extraction low volume columns such as ZipTipC18 (Millipore, Billerica, MA, USA), following the manufacturer's protocol (Oliveira et al., 2017) or, if a larger amount of sample is available, Strata-X C18 column (Zheng et al., 2020).

5. Nano-HPLC-nano-ESI-HRMS and MS/MS analysis

Different pieces of HPLC coupled to ESI-MS equipment have been used in shotgun proteomics studies of arthropod olfactory organs. Features of both pieces of equipment have a tremendous impact on the MS data acquired for each sample, and consequently on the number of proteins reliably identified. With regards to mass spectrometers, the availability of hybrid high-resolution, accurate-mass spectrometers has allowed enormous progress in proteomics studies. Among hybrid mass spectrometers, those based on a Linear Trap Quadrupole (LTQ)-Orbitrap™ and Quadrupole-Orbitrap analyzers (Thermo Fisher Scientific) are the most widespread (for a general description see Gross, 2017), but studies based on the use of quadrupole time-of-flight (Q-ToF) spectrometers (Parker et al., 2012; Zheng et al., 2020) have also been reported. Here, we describe a general method based on the use of a nano-HPLC system coupled to an LTQ-Orbitrap mass spectrometer through a nano-ESI source.

5.1 Material, equipment and reagents

A nano-LC system (such as a Dionex Ultimate 3000 HPLC, an Eksigent nano-LC 1D, or a Thermo Scientific™ EASY-nLC™ 1200 system) equipped with a commercial reversed phase nano-HPLC column (such as a Thermo Scientific™ Acclaim™ PepMap™ column, 75 μm ID, 150 mm length, 300 Å C18 stationary phase) linked to a Silica Tip (10.5 cm, 360 μm o.d, 20 μm i.d., 10 μm i.d. tip) nanospray emitter is generally used.

In-house prepared nano columns can also be used (Iovinella et al., 2018). Experimental information for in-house column preparation has been reported by Wilson, Vehus, Berg, and Lundanes (2015). Monolithic columns (such as a MonoCap C18 HighResolution 2000, 100 μm i.d., 200 cm length, GL Sciences) can also be used (Iovinella et al., 2018).

Trap columns can be installed before the analytical columns (Hu et al., 2016), and this setting is highly recommended when the digested samples are not desalted/purified before analysis (see above).

A LTQ-Orbitrap mass spectrometer system equipped with a nano-ESI source;

HPLC-grade water;

HPLC-grade acetonitrile;

formic acid

Prepare the following elution buffers

solvent A: 95:5 (or 97:3) v/v water:acetonitrile solution containing 0.1% formic acid;

solvent B: 20:80 v/v water:acetonitrile solution containing 0.1% formic acid;

4–5 μg of digested proteins are generally injected in the nano-HPLC-MS system.

5.2 Protocol for nano-HPLC-HRMS and MS/MS analyses

Flow rates and elution gradients depend on the columns to be used. Flow rates of 300 nL/min are generally used on most nano-HPLC columns.

Linear gradient (Iovinella et al., 2018) or combination of different gradients (Iovinella, Caputo, Michelucci, Dani, & della Torre, 2015) generally last up to 300 min whenever a LTQ-Orbitrap is used as detector; but they can be considerably shorter if more recent hybrid Orbitrap-based analyzers are used (Hu et al., 2016). Chromatographic parameters need to be optimized by injecting mixtures of digested standard proteins (such as bovin serum albumin) at femtomole levels and evaluating both the chromatographic profile and the sequence coverage based on MS and MS/MS experiments (see below).

Mass spectra are acquired in positive ion mode. Parameters for spray voltage and ion transfer capillary can vary; they need to be optimized prior sample analysis by injecting mixtures of digested standard proteins. Spray voltage is normally settled at 1.7–2.1 kV; ion transfer capillary temperature can vary between 175 and 260 °C.

MS data are acquired in data-dependent mode (DDA) with a dynamic exclusion, whose parameters need also to be optimized; survey MS scans are recorded in the Orbitrap analyzer, generally in the mass range of 300–2000 m/z. Nominal resolution is variable (15,000–70,000 at $m/z =$ 400 or 200), depending on the instrument features. Multicharged ions are then subjected to MS/MS experiments; exclusion times of 30 s are often applied, so to exclude selection of ions already subjected to two consecutive MS/MS experiments. The number of fragmentation experiments performed on multicharged ions after each survey scan depends on the instrument features; 20 MS/MS precursor ion experiments (TOP 20) can easily be performed on a Q Exactive™ Mass Spectrometer at a 15,000 resolution (Hu et al., 2016), while a lower number of MS/MS experiments is generally acquired in lower instruments. Since MS/MS experiments are performed on the most intense precursor ions (top ions), the number of MS/MS experiments impacts directly on the number of peptides that can be identified in the experiment (Table 1).

Despite dynamic exclusion, low-abundant peptides may not be selected for MS/MS experiments in complex mixtures; this issue limits the identification of proteins in samples with a high range of analyte concentration (more than three orders of magnitude). The latter condition is expected for insect olfactory tissue extracts, especially if ORs fall inside the target analytes. In this case, a DIA (data independent acquisition) approach could be preferable; this acquisition mode is often used in Q-ToF instruments. MS^E (Waters) and SWATH (Sciex) are two commercially available techniques for DIA experiments (Bond, Shliaha, Lilley, & Gatto, 2013; Ludwig et al., 2018). Using DIA acquisition, "all" the precursor ions are recorded

Table 1 Advantages and disadvantages of shotgun proteomics applied to the analysis and identification of olfactory proteins.

Pros	Cons
Sensibility	Contamination of samples with human or environmental derived proteins, especially keratins, are common and may compromise the results of the experiments, especially when working with very small amounts of biological samples
Reduced sample manipulation	Peptides derived from the digestion of high/medium mass and high-abundant proteins may dominate MS/MS experiments and impact on the identification of the low-abundant proteins. As an example, cuticular proteins and vitellogenins are often identified with a high number of peptides (Iovinella et al., 2018); depletions methods for these proteins have so far not been reported
No need for proprietary kits to process and analyze samples, allowing for a low cost of sample preparation and protein quantification (through LFQ)	Nano-HPLC systems and a high-resolution mass spectrometer are necessary for shotgun proteomics and both are quite delicate instruments whose working conditions need to be checked carefully by high-skilled personnel
Equipment for shotgun proteomics is often available in core facilities of research institutions	High-resolution mass spectrometers are expensive instruments

and fragmented and a software allows to assign the fragment ions to each precursor ion detected.

Some modern mass spectrometers also include an ion mobility separation step to allow the detection of less abundant precursor ions and better detection of their fragment ions (Distler, Kuharev, Navarro, & Tenzer, 2016).

6. Protein identification

MS and MS/MS data files (raw files) can be retrieved using proprietary software (Xcalibur software for Thermo Fisher instruments) for the first survey of the chromatographic profile, and MS and MS/MS experiments. Search against protein and/or nucleotide databases (in FASTA format) can be performed using different proprietary or freeware software, based on different search engines. A general description of the main features of some software is reported in Wojtkiewicz, Wiederin, and Ciborowski (2016). Among these, MaxQuant is a free software package (http://coxdocs.org/doku.php?id=maxquant:start, Cox & Mann, 2008) developed at the Max Planck Institute of Biochemistry and specifically tailored for analysis of high-resolution mass spectrometry data (Q-ToF and FTMS), and incorporating a search algorithm named Andromeda (Cox et al., 2011). The software also allows for quantitative proteomics based on stable isotope labeling as well as on Label-Free Quantification (LFQ). Only this latter application will be briefly considered here. LFQ is a low cost and attractive alternative to stable isotope labeling methods (Table 1), since it makes possible to compare large numbers of samples, not being limited by a pre-defined number of labels (multiplex), as it occurs in stable isotope labeling (Li et al., 2017). MaxQuant LFQ is based on the intensity of extracted ion currents measured at high-resolution, which are compared between samples (Cox et al., 2014). It is very important to point out that LFQ measurements require samples to be handled and analyzed under strictly uniform conditions, with minimal fractionation and in tight temporal sequence (Cox et al., 2014). Further hints to the use of MaxQuant for LFQ analysis can be found in Beer, Liu, Ky, Barnhart, and Speicher (2017).

The output of the MS and MS/MS data search is obviously strongly dependent on the quality and accuracy of the available databases. Although the first proteomic studies on arthropod olfactory proteins were conducted on a few model species (mainly *Anopheles gambiae* and *Apis mellifera*), for which complete theoretical protein databases translated from genomic data were available, databases resulting from transcriptomic studies of olfactory organs can also be used. Databases must also contain sequences of common contaminants, such as keratins, trypsin and some proteins commonly used as standards, such as BSA. Confident identification of proteins is based on a false discovery rate threshold, calculated any time by searching MS data against the reversed database; this Decoy database is automatically created during data processing. Protein assembly is generated based on the peptides identified in the in-silico database digestion. The main result of the research is the protein group file, i.e., a list reporting the groups of proteins identified based on at least one common peptide, either because of sequence redundancy in the database or because of the identity of part of the amino acid sequences among some proteins in the database. Protein group files generated by MaxQuant can be analyzed using Perseus software (https://maxquant.net/perseus/), which is also a part of the software platform for proteomics developed at the Max Plank Institute of Biochemistry.

6.1 Main parameters for mass spectral data search

Here, we report commonly used search parameters for protein identification in in-silico digested databases and for label-free quantification (LFQ) using MaxQuant. A detailed explanation of these and other parameters can be found in the MaxQuant documentation (http://www.coxdocs.org/doku.php?id=:maxquant:start), which also reports step by step instructions for beginner users.

Precursor ion tolerance: 4.5–10 ppm, depending on the instrument features and the calibration check output, including the presence/absence of an internal lock mass; checking ppm deviance in precursor ions on a standard sample such BSA peptides, can help in deciding the most appropriate value to be settled.

MS/MS tolerances: 0.5 Da, but lower values can be settled in the case of HR MS/MS spectra (Hu et al., 2016). However, since MS/MS fragmentation patterns are fairly specific to each peptide sequence, search results are relatively insensitive to small changes in the fragment tolerance setting (Eng, Searle, Clauser, & Tabb, 2011).

Enzyme: trypsin and endoprotease Lys-C.

Maximum missed cleavages: 2.

Fixed modification: carbamidomethylation of cysteines; alternatively, this modification could also be settled as variable, but this would considerably slow down the research.

Variable modification: oxidation of methionine, N-terminal acetylation.

Peptide spectrum match false discovery rate: (PSM FDR) 0.01.

Protein FDR: 0.01.

Match between run: enabled. This means that peptides present in several samples, but not identified via MS/MS in all of them are identified by matching them between runs. If this feature is enabled, it is possible to set a difference between the retention times of the ions to be matched. To set this interval correctly, it is advisable to check the reproducibility of the HPLC separation by inspecting the raw files with Xcalibur software.

Whenever LFQ of proteins has to be performed, *Multiplicity* must be settled at 1 in the group-specific parameters, while *LFQ minimum ratio count* is generally settled at 2 in the label-free quantification panel. This means that at least two peptides must be present for each protein to be compared pairwise between samples. Alternatively, *LFQ minimum ratio count* can be settled to 1 and proteins identified based on one single peptide can be eliminated from the results once the search is completed, to limit low-confidence identifications.

7. Precursor techniques

Two-dimensional electrophoresis on polyacrylamide gels (2D-PAGE) followed by *in-gel* digestion of protein spots and analysis of the resulting peptides in mass spectrometry can be considered a precursor and complementary technique. In this case, spot-protein digests are analyzed through matrix assisted laser desorption ionization-time of flight (MALDI-ToF) MS or HPLC-MS experiments. Protein identification can be obtained with peptide mass fingerprinting (PMF) analysis, i.e., by searching all measured molecular ion mass values in the resulting spectra against an *in-silico* digested protein database for proteins having the maximal number of peptide hits as well as the lowest peptide mass differences; results are reported according to an identification probability output. According to proteomics journals (e.g., Journal of Proteomics), identification of proteins based solely on mass fingerprinting are acceptable only if the protein database is derived from the genome of the organism under study.

2D-PAGE followed by identification through MALDI-ToF MS and MS/MS and HPLC-MS and MS/MS are further discussed in the paragraph Alternative methods/procedures.

8. Safety considerations and standards

Material safety data sheets (MSDS) of reagents and solvents used for sample preparation and analyses must be carefully considered. High voltages are applied at the mass spectrometer ion source; therefore, extreme caution must be taken in switching off voltages when manually operating on the mass spectrometer.

9. Analysis and statistics

Similar to other -omics approaches, proteomic experiments generate a massive amount of data. Analyzing these data can be relatively simple if the experiment aims at the mere identification of the sample proteins or at targeting some specific proteins, while it requires a quite complex statistical workflow in untargeted experiments or if protein profiles have to be compared between biological samples. Although well-established statistical workflow for data analysis in proteomics is still lacking, papers reporting general guidelines are available (Lualdi & Fasano, 2019; Yu, Qiu, & Meza, 2016); in this context, software providing helpful support to users have also been developed. Since the design of proteomic experiments can be very diverse, reporting protocols for statistical analyses is unfeasible and only a few practical hints about the first steps to be followed when analyzing protein groups will be given.

Biological and technical replicates. As most experimental procedures in biological research, proteomic experiments often require biological replicates; however, analyzing also technical replicates is necessary or advisable under some circumstances. In fact, processing aliquots of the same sample or checking the reproducibility of the analytical method through the comparison of different aliquots of a fully processed sample is recommended when changes in the sample processing workflow or in the analysis setting are made. Moreover, randomization in the order of the analyses of samples belonging to different biological groups is advisable to avoid systematic errors, especially if experimental measurements cannot be completed over a short period. The number of biological replicates depends on the aim of the study; this is obviously higher in proteomic experiments aiming at comparing the abundance of all the identified proteins between biological groups, and especially if several groups (e.g., age classes, different fertility status, behavioral tasks) have to be subjected to multiple comparisons. The latter condition is also true whenever a multivariate statistic approach is required. Therefore, establishing a priori the number of biological replicates is challenging (Yu et al., 2016).

Filtering protein groups. As already reported, Perseus software has been developed to analyze the search results obtained with MaxQuant. By using Perseus, protein groups can be filtered firstly for contaminants and proteins only identified based on modified peptides (only identified by side). Analysis of LFQ values is often performed by filtering for proteins that have been identified based on at least two "Unique + Razor" peptides, i.e., those peptides exclusively shared by the proteins of the same group. Moreover, the number of proteins considered in the following analyses can be further reduced by selecting those that have been identified at least in a certain fraction (e.g., 1/3) of biological replicates. Such filters aim at reducing type I error when testing differences in low-abundance proteins between biological groups. The final set of proteins can be subjected to further processing such as imputation of LFQ values (generally after Log2 transformation), for proteins that have not been quantified in some biological replicates (Robles, Cox, & Mann, 2014).

Presenting data and results for scientific publications. Some basic but important guidelines about how to report experimental data and results of proteomic studies are available in specialized journals, such as Journal of Proteome Research, Proteomics, Molecular & Cellular Proteomics, and Journal of Proteomics. These guidelines include the details to be reported about Peptide and Protein Identifications, about data treating in quantitative proteomic experiments and the documentation to be provided. Supplying at least the files reporting the protein groups and the main associated probability information (FDR thresholds), as they result from the search against databases, is required by most journals. Moreover, several journals also request to deposit original MS files (raw files) in a dedicated repository for proteomic mass spectrometry data such as PRIDE (https://www.ebi.ac.uk/pride/archive/) or other archives members of ProtemeExchange consortium (http://www.proteomexchange.org).

10. Related techniques

The success of proteomic experiments depends on the accuracy and completeness of protein or nucleotide sequence databases. In the case of non-model organisms, whose genomes have not been sequenced, nucleotide databases for the same organ/tissue to be analyzed in proteomics experiments can be obtained through a de novo transcriptome assembly (Yin et al., 2013). The obtained transcript sequences may contain variants of the same protein, which may eventually be identified at the protein level. Transcriptomic data also entail quantitative information allowing to study differential gene expression between specimens, organs/tissues or physiological conditions (Rinker et al., 2013). Transcriptomic and proteomic results can be compared both at a qualitative and quantitative level. Although transcriptomes based on Next Generation Sequencing reach a deeper level than shotgun proteomic experiments, studies have reported that up to 60% of the transcripts can find a correspondence in the identified proteins (Nagaraj et al., 2011).

11. Pros and cons in shotgun proteomics applied to olfactory proteins

The main advantages and disadvantages of shotgun proteomics applied to olfactory proteins are concisely reported in Table 1.

12. Alternative methods and procedures

An alternative to shotgun proteomics is proteomics based on two-dimensional electrophoresis on polyacrylamide gels (2D-PAGE) (Drabik, Bodzoń-Kułakowska, & Silberring, 2016; Friedman, 2005). In this case, *in-gel* digestion of protein spots is performed and peptides are submitted to MS (and MS/MS) analysis either through HPLC-MS (and MS/MS) or through matrix assisted laser desorption ionization-time of flight (MALDI-ToF and ToF/ToF) for final protein identification. This kind of approach can be very demanding and time-consuming if all the spots revealed in the gel must be identified. However, this technique is more often applied in experiments targeting some specific proteins or aiming at identifying proteins whose abundance, evaluated based on the spot intensities or through two-dimensional difference gel electrophoresis (2D DIGE), differs between biological groups (Drabik et al., 2016).

Table 2 Main problems and troubleshooting of shotgun proteomics of insect olfactory proteins.

Problem	Solution
Contamination of samples with keratins can be deleterious when processing microscale biological samples	Always use laboratory gloves and keep plastic material, mortars, pestles and solvents protected from dust
Detergents may contaminate HPLC columns, and tubing and can clog the transfer capillary. Contamination with detergents and polymers (such as polyethylene glycol, PEG) may suppress MS/MS experiments on the sample peptides	Finding the origin of contamination can be very challenging. Detergents are sometimes necessary for protein extraction or protein solubilization, especially if the experiment also targets membrane proteins. Protein precipitation in chloroform-methanol solutions or in acetone (this latter being more suitable for microproteomics) can be the first step for protein purification before sample digestion (Feist & Hummon, 2015). Salts and most detergents can be removed using C18 microcolumns or STAGE tips (see above). Some Mass Spec friendly detergents, such as N-octyl-β-glucopyranoside, have been reported in the literature; they can be used in the extraction if strictly needed (Speers & Wu, 2007)

In 2D-PAGE separation, arthropod soluble olfactory proteins are mainly found in the interval 3–7 for the isoelectric point and 10–20 kDa for the apparent molecular weight (Dani et al., 2010; Iovinella et al., 2011). A few dozen of MS analyses may be necessary to identify all the proteins present in this area for olfactory tissue extracts (Dani et al., 2010; Iovinella et al., 2011). With respect to shotgun proteomics, this technique is less sensitive, and this can be a severe limit when the biological material is limited (Mastrobuoni et al., 2013). On the other hand, separation on 2D-PAGE may help to identify protein isoforms (occurring as spot trains), whenever these species are sufficiently separated on the gel.

13. Troubleshooting and optimization

Main problems and troubleshooting are concisely reported in Table 2.

14. Summary

Shotgun proteomics is a bottom-up approach to characterize complex mixtures of proteins present in biological samples; it is based on the combination of HPLC-MS and MS/MS analysis of the peptides resulting from the enzymatic digestion of such mixtures. Nano-HPLC systems and high-resolution mass spectrometers (HRMS) are necessary to efficiently separate and measure these peptides. Protein identification is based on the comparison of the acquired MS and MS/MS data with those expected for peptides predicted based on the in-silico digestion of polypeptide (or translated nucleotide) databases, and on the assembly of peptides attributable to the same protein or group of proteins. Being suitable for microscale analysis, shotgun proteomics can be used to analyze small amounts of tissues, such as arthropod olfactory organs and pheromonal glands. Moreover, since soluble proteins can be quite easily extracted from biological tissues, the method is particularly suited for soluble olfactory proteins. Beside qualitative information, different methods can be used to compare the abundance of proteins between samples, and eventually the whole protein profile by using multivariate statistics. Among these methods, label-free quantification (LFQ) can allow comparing many samples at the same time; therefore, it is particularly suited for studies where different biological conditions (e.g., age classes, reproductive status) are compared. The workflow for shotgun proteomics requires: (i) the extraction of proteins by grinding very carefully tissues in liquid nitrogen in a denaturant buffer, avoiding as much as possible the use of surfactants;

(ii) the enzymatic digestion of the sample supernatant after reduction and derivatization of cysteine residues; (iii) desalting and purification of the sample using commercial or homemade small volume C18 columns; (iv) the analysis of an aliquot (2–5 μg) of the enzymatically digested sample on a nano-HPLC system coupled to a HRMS, by acquiring peptide MS and MS/MS spectra in data-dependent mode, to favor the acquisition of data over a wide dynamic range of protein concentration; (v) checking for the quality of the acquired data and search against polypeptide or translated nucleotide databases, using parameters consistent with the sample preparation method and the MS data acquisition setting; (vi) filtering the identified protein groups to reduce the effects of stochastic detection of low-abundance proteins, due to instrument sensitivity. Proteins to be considered in the following analyses can be filtered based on the number of peptides on which protein identification is based (often >1) and/or by setting a threshold in the percentage of biological replicates where each protein has been identified.

References

Beer, L. A., Liu, P., Ky, B., Barnhart, K. T., & Speicher, D. W. (2017). Efficient quantitative comparisons of plasma proteomes using label-free analysis with MaxQuant. *Methods in Molecular Biology*, *1619*, 339–352.

Bond, N. J., Shliaha, P. V., Lilley, K. S., & Gatto, L. (2013). Improving qualitative and quantitative performance for MSE-based label-free proteomics. *Journal of Proteome Research*, *12*, 2340–2353.

Cox, J., Hein, M. Y., Luber, C. A., Paron, I., Nagaraj, N., & Mann, M. (2014). Accurate proteome-wide label-free quantification by delayed normalization and maximal peptide ratio extraction, termed MaxLFQ. *Molecular & Cellular Proteomics*, *13*, 2513–2526.

Cox, J., & Mann, M. (2008). MaxQuant enables high peptide identification rates, individualized p.p.b.-range mass accuracies and proteome-wide protein quantification. *Nature Biotechnology*, *26*, 1367–1372.

Cox, J., Neuhauser, N., Michalski, A., Scheltema, R. A., Olsen, J. V., & Mann, M. (2011). Andromeda: A peptide search engine integrated into the MaxQuant environment. *Journal of Proteome Research*, *10*, 1794–1805.

Dani, F. R., Iovinella, I., Felicioli, A., Niccolini, A., Calvello, M. A., & Carucci, et al. (2010). Mapping the expression of soluble olfactory proteins in the honeybee. *Journal of Proteome Research*, *9*, 1822–1833.

Distler, U., Kuharev, J., Navarro, P., & Tenzer, S. (2016). Label-free quantification in ion mobility-enhanced data-independent acquisition proteomics. *Nature Protocols*, *11*, 795–812.

Drabik, A., Bodzoń-Kułakowska, A., & Silberring, J. (2016). Gel electrophoresis. In P. Ciborowski & J. Silberring (Eds.), *Proteomic profiling and analytical chemistry* (2nd ed., pp. 115–143). Elsevier B.V.

Eng, J. K., Searle, B. C., Clauser, K. R., & Tabb, D. L. (2011). A face in the crowd: Recognizing peptides through database search. *Molecular & Cellular Proteomics, 10*. https://doi.org/10.1074/mcp.R111.009522.

Feist, P., & Hummon, A. B. (2015). Proteomic challenges: Sample preparation techniques for microgram-quantity protein analysis from biological samples. *International Journal of Molecular Sciences*, *16*, 3537–3563.

Foster, L. J., De Hoog, C. L., & Mann, M. (2003). Unbiased quantitative proteomics of lipid rafts reveals high specificity for signaling factors. *Proceedings of the National Academy of Sciences of the United States of America, 100*, 5813–5818.

Friedman, D. B. (2005). Identification of proteins from two-dimensional gels. In R. M. Carpioli & M. L. Gross (Eds.), *Vol. 2. Biological applications. Part A: Peptides and proteins. The encyclopedia of mass spectrometry* (pp. 47–54), Oxford: Ed. Elsevier.

Glatter, T., Ludwig, C., Ahrné, E., Aebersold, R., Heck, A. J., & Schmidt, A. (2012). Large-scale quantitative assessment of different in-solution protein digestion protocols reveals superior cleavage efficiency of tandem Lys-C/trypsin proteolysis over trypsin digestion. *Journal of Proteome Research, 11*, 5145–5156.

Gross, J. H. (2017). *Mass spectrometry: A textbook* (3rd ed.). Springer, (chapter 4).

Häußermann, C. K., Ziegelmann, B., Bergmann, P., & Rosenkranz, P. (2015). Male mites (*Varroa destructor*) perceive the female sex pheromone with the sensory pit organ on the front leg tarsi. *Apidologie, 46*, 771–778.

Hu, H., Bienefeld, K., Wegener, J., Zautke, F., Hao, Y., Feng, M., et al. (2016). Proteome analysis of the hemolymph, mushroom body, and antenna provides novel insight into honeybee resistance against *Varroa* infestation. *Journal of Proteome Research, 15*, 2841–2854.

Iovinella, I., Cappa, F., Cini, A., Petrocelli, I., Cervo, R., Turillazzi, S., et al. (2018). Antennal protein profile in honeybees: Caste and task matter more than age. *Frontiers in Physiology, 9*, 748. https://doi.org/10.3389/fphys.2018.00748.

Iovinella, I., Caputo, B., Michelucci, E., Dani, F. R., & della Torre, A. (2015). Candidate biomarkers for mosquito age-grading identified by label-free quantitative analysis of protein expression in *Aedes albopictus* females. *Journal of Proteomics, 128*, 272–279.

Iovinella, I., Dani, F. R., Niccolini, A., Sagona, S., Michelucci, E., Gazzano, A., et al. (2011). Differential expression of odorant-binding proteins in the mandibular glands of the honey bee according to caste and age. *Journal of Proteome Research, 10*, 3439–3449.

Leal, W. S. (2013). Odorant reception in insects: Roles of receptors, binding proteins, and degrading enzymes. *Annual Review of Entomology, 58*, 373–391.

Li, H., Han, J., Pan, J., Liu, T., Parker, C. E., & Borchers, C. H. (2017). Current trends in quantitative proteomics—An update. *Journal of Mass Spectrometry, 52*, 319–341.

Lualdi, M., & Fasano, M. (2019). Statistical analysis of proteomics data: A review on feature selection. *Journal of Proteomics, 198*, 18–26.

Ludwig, C., Gillet, L., Rosenberger, G., Amon, S., Collins, B. C., & Aebersold, R. (2018). Data-independent acquisition-based SWATH-MS for quantitative proteomics: A tutorial. *Molecular Systems Biology, 14*e8126.

Mastrobuoni, G., Qiao, H., Iovinella, I., Sagona, S., Niccolini, A., Boscaro, F., et al. (2013). A proteomic investigation of soluble olfactory proteins in *Anopheles gambiae*. *PLoS One, 8*e75162.

Nagaraj, N., Wisniewski, J. R., Geiger, T., Cox, J., Kircher, M., Kelso, J., et al. (2011). Deep proteome and transcriptome mapping of a human cancer cell line. *Molecular Systems Biology, 7*, 548.

Oliveira, D. S., Brito, N. F., Nogueira, F. C. S., Moreira, M. F., Leal, W. S., Soares, M. R., et al. (2017). Proteomic analysis of the kissing bug *Rhodnius prolixus* antenna. *Journal of Insect Physiology, 100*, 108–118.

Parker, R., Guarna, M. M., Melathopoulos, A. P., Moon, K. M., White, R., Huxter, E., et al. (2012). Correlation of proteome-wide changes with social immunity behaviors provides insight into resistance to the parasitic mite, *Varroa destructor*, in the honey bee (*Apis mellifera*). *Genome Biology, 13*, R81. https://doi.org/10.1186/gb-2012-13-9-r81.

Pelosi, P., Iovinella, I., Felicioli, A., & Dani, F. R. (2014). Soluble proteins of chemical communication: An overview across arthropods. *Frontiers in Physiology, 5*, 320.

Pelosi, P., Iovinella, I., Zhu, J., Wang, G., & Dani, F. R. (2018). Beyond chemoreception: Diverse tasks of soluble olfactory proteins in insects. *Biological Reviews, 93*, 184–200.

Rappsilber, J., Mann, M., & Ishihama, Y. (2007). Protocol for micro-purification, enrichment, pre-fractionation and storage of peptides for proteomics using StageTips. *Nature Protocols, 2*, 1896–1906.

Rinker, D. C., Pitts, R. J., Zhou, X., Suh, E., Rokas, A., & Zwiebel, L. J. (2013). Blood meal-induced changes to antennal transcriptome profiles reveal shifts in odor sensitivities in *Anopheles gambiae*. *Proceedings of the National Academy of Sciences of the United States of America, 110*, 8260–8265.

Robles, M. S., Cox, J., & Mann, M. (2014). In-vivo quantitative proteomics reveals a key contribution of post-transcriptional mechanisms to the circadian regulation of liver metabolism. *PLoS Genetics, 10,* e1004047.

Rund, S. S., Bonar, N. A., Champion, M. M., Ghazi, J. P., Houk, C. M., & Leming, et al. (2013). Daily rhythms in antennal protein and olfactory sensitivity in the malaria mosquito *Anopheles gambiae*. *Scientific Reports, 3*, 2494.

Smoluch, M., Mielczarek, P., Drabik, A., & Silberring, J. (2012). Online and offline sample fractionation. In P. Ciborowski & J. Silberring (Eds.), *Proteomic profiling and analytical chemistry* (2nd ed., pp. 65–99). Elsevier B.V.

Speers, A. E., & Wu, C. C. (2007). Proteomics of integral membrane proteins—Theory and application. *Chemical Reviews, 107*, 3687–3714.

Wang, P., Kish, M. M., & Wedemiotis, C. (2005). Fragmentation mechanisms of peptide ions. In R. M. Carpioli & M. L. Gross (Eds.), *Vol. 2. Biological applications. Part A: Peptides and proteins. The encyclopedia of mass spectrometry* (pp. 139–151). Oxford, Ed. Elsevier.

Wilson, S. R., Vehus, T., Berg, H. S., & Lundanes, E. (2015). Nano-LC in proteomics: Recent advances and approaches. *Bioanalysis, 7*, 1799–1815.

Wojtkiewicz, M., Wiederin, J., & Ciborowski, P. (2016). Proteomic database search and analytical quantification for mass spectrometry. In P. Ciborowski & J. Silberring (Eds.), *Proteomic profiling and analytical chemistry* (2nd ed., pp. 193–210). Elsevier B.V.

Yin, X. W., Iovinella, I., Marangoni, R., Cattonaro, F., Flamini, G., Sagona, S., et al. (2013). Odorant-binding proteins and olfactory coding in the solitary bee *Osmia cornuta*. *Cellular and Molecular Life Science, 70*, 3029–3039.

Yu, F., Qiu, F., & Meza, J. (2016). Design and statistical analysis of mass-spectrometry based quantitative proteomics data. In P. Ciborowski & J. Silberring (Eds.), *Proteomic profiling and analytical chemistry* (2nd ed., pp. 211–237). Elsevier B.V.

Zheng, R., Xia, Y., & Keyhani, N. O. (2020). Sex-specific variation in the antennal proteome of the migratory locust. *Journal of Proteomics, 216*, 103681.

CHAPTER FIVE

Analysis of post-translational modifications in soluble proteins involved in chemical communication from mammals and insects

Andrea Scaloni*
Proteomics and Mass Spectrometry Laboratory, ISPAAM, National Research Council, Naples, Italy
*Corresponding author: e-mail address: andrea.scaloni@ispaam.cnr.it

Contents

Abstract

Different soluble, small polypeptide molecules isolated from biological tissues/body fluids of mammals and insects have been reported to bind odorant compounds and pheromones, and have been accordingly named odorant-binding proteins, pheromone-binding proteins and chemosensory proteins. Mammal and insect odorant-binding and pheromone-binding proteins show evident differences in their primary structure; in insects, the same condition occurs for odorant- and

Methods in Enzymology, Volume 642
ISSN 0076-6879
https://doi.org/10.1016/bs.mie.2020.04.062

pheromone-binding proteins, which belong to the same protein family, and chemosensory proteins. In order to understand their molecular characteristics, these proteins have been subjected to an extensive analysis for their post-translational modifications through mass spectrometry approaches. Depending on proteins, they have been characterized as having specific disulfide bridge arrangements, which were consistent among molecular sequence homologs. Only selected odorant-binding proteins from mammals have been demonstrated bearing N-linked glycan structures and other modifications. In this review article, we discuss the application of mass spectrometry procedures for the analysis of post-translational modifications in odorant-binding proteins, pheromone-binding proteins and chemosensory proteins. Different example proteins are reported, and protocols are presented for obtaining definitive information in this context.

1. Introduction

In the last two decades, the molecular mechanisms related to the perception of pheromones and odorant compounds in animals were the object of a number of dedicated studies (Block, 2018). This interest was due to the attempt of the scientific community to elucidate molecules and reactions involved in semiochemical perception in animals, and the consequent hope possibly to regulate odor-driven behaviors, such as food search, reproduction and other intra-/inter-specific interactions. In this context, early studies identified abundant soluble proteins present in the aqueous perireceptor space of olfactory neurons, which are able to bind hydrophobic pheromones/odorant molecules and allow their delivery to corresponding receptors. Depending on their nature, these proteins were named odorant-binding proteins (OBPs), pheromone-binding proteins (PBPs) and chemosensory proteins (CSPs). In mammals, these proteins are generally synthesized by glands of the nasal epithelium and/or the oral cavity, and are secreted in the corresponding mucus and saliva (Paolini et al., 1998; Loebel et al., 2000; Scaloni et al., 2001; D'Innocenzo et al., 2006; Mastrogiacomo et al., 2014; Zhu et al., 2017). They also occur at high concentration in other secreted biological fluids related to animal communication, such as urine, seminal fluid and sweat, being produced by kidney, liver and/or skin/anal sebaceous glands, respectively (D'Innocenzo et al., 2006; Mastrogiacomo et al., 2014). In insects, they are abundant components of the sensillar lymph, *i.e.*, the aqueous medium filling the space between the dendrites of chemosensory neurons and the cuticular wall of specialized sensilla, which are generally present in antennae, tarsi and labrum (Scaloni, Monti, Angeli, & Pelosi, 1999;

Angeli et al., 1999; Picimbon et al., 2000; Marchese et al., 2000; Ban, Scaloni, Brandazza, et al., 2003; Ban, Scaloni, D'Ambrosio, et al., 2003; Calvello et al., 2003).

Molecular biology studies allowed deducing the amino acid sequence of these proteins, demonstrating that insect and mammalian OBPs and PBPs do not have a similar primary structure (Ban, Scaloni, D'Ambrosio, et al., 2003; Calvello et al., 2003; Loebel et al., 2000; Marchese et al., 2000; Scaloni et al., 2001). Similarly, CSPs were shown to occur only in insects and other arthropods, and to be different from the above-mentioned proteins (Angeli et al., 1999; Ban, Scaloni, Brandazza, et al., 2003; Calvello et al., 2003; Picimbon et al., 2000). All these studies evidenced the need for additional, dedicated biochemical investigations aimed at elucidating protein primary, tertiary and quaternary structure, with the ultimate goal of understanding the molecular basis regulating the interaction of pheromones/odorant compounds with the corresponding protein carriers. In this context, our laboratory has used different proteomic approaches to identify post-translational modifications (PTMs) present in OBPs, PBPs and CSPs.

In the last two decades, the development of matrix-assisted laser desorption/ionization (MALDI) and electrospray ionization (ESI) as soft ionization methods for mass spectrometry (MS) of biomolecules allowed investigating PTMs directly at the protein level, assigning their nature and adduction site(s) (Bachi, Dalle-Donne, & Scaloni, 2013; D'Ambrosio, Salzano, Arena, Renzone, & Scaloni, 2007). Direct protein mass measurement, peptide mass fingerprinting (PMF) and fragment fingerprinting (MS/MS) upon collision induced fragmentation (CID) experiments were widely used to this purpose. A high sequence coverage output is a crucial prerequisite for the comprehensive assignment of PTMs in proteins; this result is dependent on the amount of the protein(s) under investigation. In bottom-up experiments, this is generally obtained by combining protein proteolytic digestion with enzymes having a high sequence specificity with micro/nano-reverse phase high-performance liquid chromatography (RP-HPLC) of resulting digests followed by *on-line* MALDI-time of flight (TOF)-MS or ESI-quadrupole (Q), -ion trap (IT), -TOF, -linear ion trap (LIT), -Orbitrap MS and/or MS/MS analysis of eluted fractions. Hybrid mass spectrometers were also used for this purpose. Since peptides bearing PTMs may possibly occur at a low concentration with respect to non-modified counterparts, their selective enrichment/trapping by affinity techniques was also used to facilitate their detection (Arena, Salzano, Renzone, D'Ambrosio, & Scaloni, 2014; Bachi et al., 2013; D'Ambrosio et al., 2007). Similarly, optimized CID and electron transfer dissociation

(ETD) processes were developed to fragment peptides bearing labile PTMs, generating sequence-informative MS/MS spectra allowing for modification assignment (Arena et al., 2014; Bachi et al., 2013). In all cases reported above, bioinformatic analysis of resulting mass spectrometric data was essential for interpreting instrument outputs toward a final protein structural characterization.

This review summarizes how combined protein chemistry and mass spectrometry procedures were used to detect disulfide bridges, N-linked glycan structures and phosphate groups present in OBP, PBPs and CSPs, to assign the nature of the adducted moieties as well as the site(s) of the modification(s). Specific data on selected proteins are described here as paradigmatical examples. Elucidation of the present cystines in OBPs, PBPs and CSPs provided important information to determine common structural motifs and structure-function relationships in these proteins, before the resolution of their three-dimensional structure, whereas identification of additional PTMs in these components highlighted amino acids present on their molecular surface.

2. Assignment of disulfide bridges in OBPs, PBPs and CSPs

In order to assign PTMs in proteins, the primary structure of interest (as deriving from cloning of the corresponding gene or as deduced from sequence database) has to be verified. This is generally obtained by combining MS measurements of the purified intact, native species, with that of corresponding peptide digests after component alkylation under denaturing, reducing conditions, according to a typical peptide mapping experimental approach. The general experimental conditions used in our laboratory for OBPs, PBPs and CSPs are here reported below in dedicated sections, and were the same we utilized for other proteins (Amiconi et al., 2000; Cecconi et al., 2002; Orrù et al., 1997; Raimondo et al., 2005).

In the case of *Bombyx mori* PBP and GOBP2, preliminary ESI-Q-MS analysis of the purified, intact, native species allowed detecting the occurrence of one or multiple components in each sample, while corresponding measured mass value(s) permitted ascertaining its perfect agreement with theoretical one(s), and evaluating the possible occurrence of PTMs and/or proteolytic processing at polypeptide N- and C-termini. A good match with the predicted protein sequence was verified for both native species only assuming all Cys residues (6 in number) being involved in disulfide bridges (exp. $M_{Av\text{-}BPB} = 15{,}878.9 \pm 1.3$ Da; theor. $M_{Av\text{-}BPB} = 15{,}878.2$ Da; exp. $M_{Av\text{-}GOBP2} = 16{,}157.1 \pm 1.1$ Da; theor. $M_{Av\text{-}GOBP2} = 16{,}157.5$ Da) (Scaloni et al., 1999). These experiments also highlighted the occurrence of equimolar truncated protein species lacking a single amino acid at protein C-terminus (measured $\Delta M = -99.2$ and -162.6 Da for BPB and GOBP2, corresponding to missing Val and Tyr, respectively). No other PTMs were hypothesized. Additional ESI-Q-MS experiments on purified, intact PBP and GOBP2 alkylated with iodoacetamide in denaturing, non-reducing conditions allowed verifying the absence of free thiol groups in the natural proteins, as deduced by the measured experimental mass values very similar to those reported above, while experiments performed on samples alkylated in denaturing, reducing conditions confirmed alkylation of all molecular cysteines (exp. $M_{Av\text{-}BPB\text{-}RedCAM} = 16{,}225.9 \pm 1.6$ Da; theor. $M_{Av\text{-}BPB\text{-}RedCAM} = 16{,}226.5$ Da; exp. $M_{Av\text{-}GOBP2\text{-}RedCAM} = 16{,}505.1 \pm 1.8$ Da; theor. $M_{Av\text{-}GOBP2\text{-}RedCAM} = 16{,}505.8$ Da). A comparison of the measured molecular mass values in above-mentioned experiments always showed for species alkylated under denaturing, reducing conditions a $\Delta M \approx 348.4\,\text{Da} \approx 6 \times (1 + 57.1)$ Da, with respect to counterparts modified in denaturing, non-reducing conditions, definitively proving the occurrence of three disulfide bridges in both PBP and GOBP2. Analogous experiments on intact native and recombinant OBPs from *Locusta migratoria* and *Polistes dominula* similarly demonstrated the unique occurrence of cystine-containing species (Ban, Scaloni, D'Ambrosio, et al., 2003; Calvello et al., 2003).

Disulfide bridge assignment in *B. mori* PBP and GOBP2 was obtained by chemical and/or enzymatic hydrolysis reactions on native proteins followed by MALDI-TOF-MS analysis of resulting peptide mixtures (Scaloni et al., 1999). In this case, structural assignment was achieved by parallel experiments on unfractionated peptide mixtures treated or not with dithiothreitol (DTT), which allowed easily recognizing the disappearance of signals associated with disulfide-linked peptides and the concomitant appearance of signals corresponding to reduced counterparts. Often, an independent experimental confirmation was obtained by sequence analysis of RP-HPLC-purified Cys-containing peptides from native protein digests, and identification of phenylthiohydantoin (PTH)-cystine at the expected Edman degradation cycle.

In particular, native PBP was hydrolyzed with cyanogen bromide (CNBr) and subjected to MALDI-TOF-MS analysis, which generated the spectrum reported in Fig. 1. Different signals were associated with

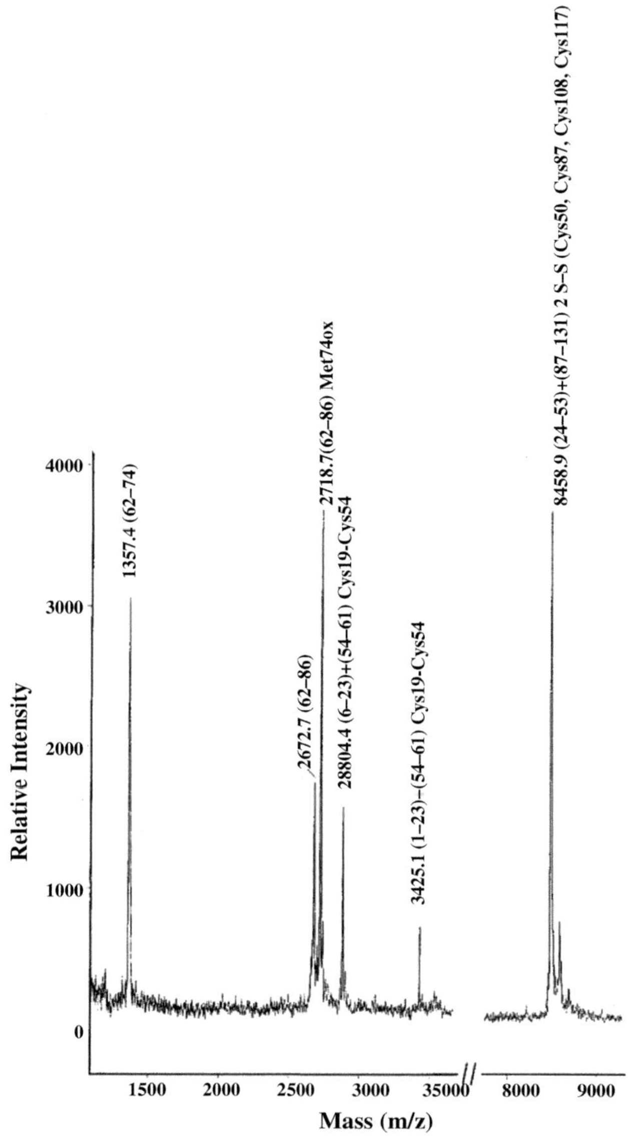

Fig. 1 MALDI-TOF mass spectrum of native PBP after hydrolysis with CNBr. Signals assigned to linear and disulfide-bridged peptides are reported. In the latter case, each signal was associated with the corresponding peptide pair, and the Cys residues involved in the disulfide bond are indicated. Signals associated with disulfide-bridged peptides disappeared following sample treatment with DTT.

disulfide-linked peptides based on their unique mass values and their disappearance following incubation with DTT. Indeed, the MH^+ signals at *m/z* 2880.4 and 3425.1 were assigned to cross-linked peptides (6–23)S-S(54–61) and (6–23)S-S(54–61), respectively, both containing the disulfide bridge Cys19-Cys54. On the other hand, the MH^+ signal at *m/z* 8458.9 was associated with a two-peptide cluster involving fragments (24–53) and (87–131) linked by two disulfide bonds between Cys50, Cys97, Cys108 and Cys117. An independent digestion of native PBP with trypsin, at pH 6.5, allowed assigning the remaining unknown cystines. In this case, the spectrum showed two evident signals at *m/z* 3134.2 and 4257.1. The former was associated with cross-linked peptides (80–99)S-S(111–119) containing the disulfide bond Cys97-Cys117, which originated from aspecific hydrolysis at Trp110. The latter was assigned to a three-molecular cluster involving peptides (47–53), (80–99) and (108–119) linked by two S-S bridges that, according to above-mentioned data, were identified as Cys50-Cys108 and Cys97-Cys117. This peptide was generated as result of aspecific proteolysis at Met53. Data on aspecific hydrolysis at Trp110 and Met53 were already recorded during peptide mapping experiments on the carboxamidomethylated protein. In conclusion, the combination of the above-reported experiments demonstrated the arrangement of the PBP cysteine pairing as Cys19-Cys54, Cys50-Cys108 and Cys97-Cys117.

The disulfide bridge pairing in native *B. mori* GOBP-2 was similarly established by MALDI-TOF-MS analysis of its tryptic digest, as obtained at pH 6.5 (Scaloni et al., 1999); in particular, evident spectrum signals were observed at *m/z* 1493.1, 3029.1 and 3358.4 that disappeared following treatment with DTT. The first one corresponded to the peptide species (93–99)S-S(114–119) linked by the disulfide Cys97-Cys117. The latter signals were assigned to three-peptide clusters (15–20)S-S(47–58)S-S(100–110) and (15–20)S-S(47–55)S-S(100–110), respectively, involving Cys19, Cys54, Cys50 and Cys108; in the case of the second molecule, an aspecific hydrolysis at Met55 was also observed. Attempts to subdigest these peptide clusters by an optimal reaction failed because of the impracticability of hydrolyzing the polypeptide chain between Cys50 and Cys54. Nevertheless, Edman degradation of the RP-HPLC-purified peptides provided definitive information on Cys residues pairing; in fact, no PTH-Cys was detected at the degradation cycles corresponding to Cys19 and Cys50, while PTH-cystine was revealed at the ones corresponding to Cys54 and Cys108. These results definitively proved remaining protein disulfides as Cys19-Cys54 and Cys50-Cys108.

Very similar procedures were used to assign the S-S bridges arrangement in native and recombinant OBP1 from *L. migratoria* and *P. dominula*, affording definitive information in this context (Ban, Scaloni, D'Ambrosio, et al., 2003; Calvello et al., 2003). A final comparison of the disulfide patterns ascertained in *B. mori* PBP and GOBP2, *L. migratoria* OBP1 and *P. dominula* OBP1 demonstrated a common conservative arrangement among homologous protein members from lepidoptera, orthopteran and hymenoptera orders, namely C1-C3, C2-C5 and C4-C6, notwithstanding their limited sequence identity (Fig. 2A). These results allowed predicting an identical S-S bridge arrangement also in PBPs and OBPs from other orders, when a certain sequence homology was observed.

An identical approach was also used to unveil disulfide arrangement in insect CSPs. In this context, original information was obtained following the analysis of purified *Schistocerca gregaria* CSP isoforms (Angeli et al., 1999). In particular, determination of the accurate molecular mass value of the native proteins before and after alkylation of molecules with iodoacetamide under denaturing, reducing conditions demonstrated that all Cys residues present in the polypeptide chains (4 in number) were involved in disulfide bridges [$\Delta M \approx 232.4\,Da \approx 4 \times (1+57.1)$ Da]. Based on their good agreement with theoretical data as deduced after cloning of corresponding genes, these experimental measurements also excluded the occurrence of additional protein PTMs.

The corresponding arrangement of the two intramolecular S-S bonds was established in four native protein isoforms following their digestion with endoprotease Lys-C, at pH 6.5 (Angeli et al., 1999). Corresponding peptide mixtures were in parallel analyzed by MALDI-TOF-MS and resolved by RP-HPLC, collecting eluted fractions. All MH^+ signals were interpreted according to corresponding protein sequences and protease specificity. For the CSP4 digest, two chromatographic peaks contained peptides with a MH^+ signal at *m/z* 1819.6, 1837.6 and 2001.6, which were assigned to molecular fragments (47–63)S-S, (47–59)S-S(60–63) and (26–43)S-S, respectively, thus demonstrating the occurrence of the disulfide bridges Cys57-Cys60 and Cys29-Cys38. These assignments were confirmed by MALDI-TOF-MS analysis of the corresponding products from DTT treatment, which showed the occurrence of the reduced counterparts. When these peptides were subjected to Edman degradation, they gave the expected amino acid sequence; no PTH-Cys was observed at the cycles corresponding to Cys29 and Cys57, while PTH-cystine was detected at the ones corresponding to Cys38 and Cys60. Very similar results were obtained for three other native CSP isoforms.

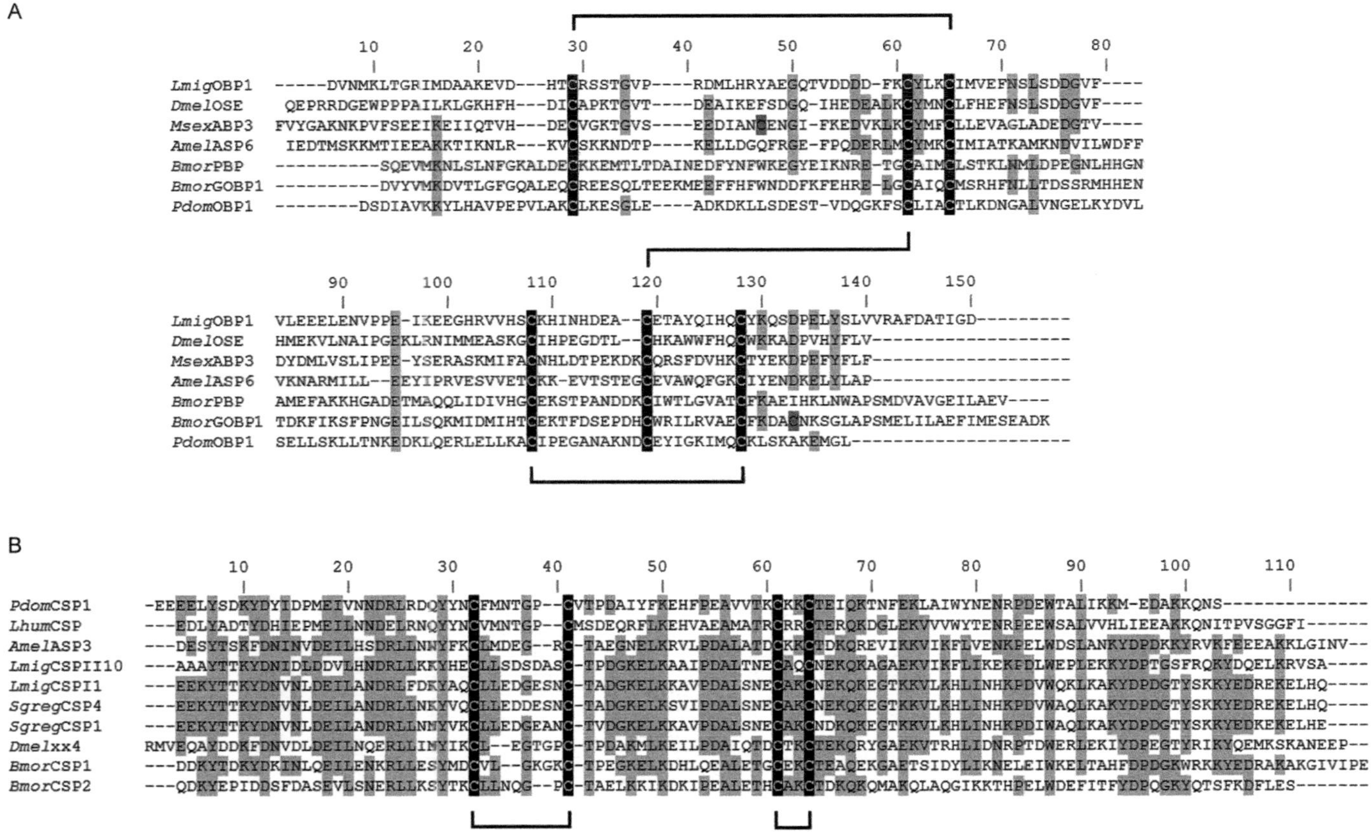

Fig. 2 Sequence analysis of insect OBPs, PBPs and CSPs as related to corresponding ascertained PTMs. Amino acid sequence alignment of OBPs and PBPs (A) and CSPs (B) from different insect orders. Cysteines that were ascertained as involved in disulfide bonds are highlighted in black, with corresponding S-S bridges indicated; cysteines that occurred in reduced form are evidenced in dark gray; amino acids that were conserved at a specific position (frequency $\geq$ 0.5) are reported in light gray. *Lmig*, *Locusta migratoria*; *Dmel*, *Drosophila melanogaster*; *Msex*, *Manduca sexta*; *Amel*, *Apis melliphera*; *Bmor*, *Bombyx mori*; *Pdom*, *Polistes dominula*; *Lhum*, *Linepithema humile*; *Sgreg*, *Schistocerca gregaria*.

A similar procedure was also used to assign protein disulfides in native CSP1 from *B. mori* and *Eurycantha calcarata*, as well as in recombinant CSP-II-10 from *L. migratoria*, CSP1 from *P. dominula*, and CSP4 from *S. gregaria* (Ban, Scaloni, Brandazza, et al., 2003; Calvello et al., 2003; Marchese et al., 2000; Picimbon et al., 2000; Picone et al., 2001; Tomaselli et al., 2006). A comparison of the ascertained S-S bridge pairings of the above-reported CSPs demonstrated the common arrangement C1-C2 and C3-C4 in protein members from lepidoptera, orthoptera, hymenoptera and phasmatodea orders, also highlighting a significant sequence homology of the corresponding sequences (Fig. 2B). When a conserved localization of corresponding cysteines was observed, these results allowed predicting an identical S-S bridges arrangement in additional CSPs from other orders. Therefore, the disulfide arrangement present in insect CSPs totally differed from that observed in OBPs counterparts (Fig. 2A).

Due to the frequent occurrence of an odd number of Cys residues in the corresponding protein sequence, particular care was used during the assignment of protein disulfides in native mammalian OBPs and related lipocalins. In fact, the concomitant occurrence of reduced and S-S-linked cysteines should have determined the occurrence of disulfide scrambling phenomena during protein digestion at neutral-alkaline pH values, ultimately affecting the final structural determination. To this purpose, all proteins were preventively subjected in parallel to alkylation with iodoacetamide in denaturing non-reducing conditions as well as in reducing conditions; this allowed always blocking thiol groups not involved in disulfide bonds. Resulting purified products were then directly subjected to comparative peptide mapping experiments. In fact, preliminary ESI-Q-MS measurements on intact proteins as obtained in both conditions often evidenced very complex spectral profiles, suggesting the occurrence of multiple PTMs in original species. As reported above, peptide mapping experiments on protein samples alkylated under denaturing, reducing conditions allowed verifying the primary structure of interest as well as evaluating the possible occurrence of protein isoforms, proteolytic processing at polypeptide N- and C-termini, and/or additional PTMs (see below). On the other hand, a comparison of these results with that on protein samples alkylated under denaturing, non-reducing conditions allowed to easily recognize in the MALDI-TOF-MS spectrum disulfide-linked peptides, which disappeared after component treatment with DTT, together with the concomitant appearance of signals corresponding to reduced counterparts.

For example, the tryptic digest of porcine OBPII alkylated under denaturing, reducing conditions showed in MALDI-TOF-MS evident MH^+ signals at *m/z* 2096.5, 2126.2, 2323.6 and 2947.4, which were associated with carboxamidomethylated (CAM) peptides (95–111), (138–157), (93–111) and (43–69), respectively (Scaloni et al., 2001). The protein counterpart alkylated under denaturing, non-reducing conditions maintained the MH^+ signals corresponding to peptides (95–111)CAM and (93–111)CAM, but showed the occurrence of a component at *m/z* 4955.7 corresponding to the disulfide-linked peptide (43–69)S-S(138–157), which disappeared after treatment with DTT. These experiments definitively demonstrated the occurrence in porcine OBPII of the disulfide Cys61-Cys152, while Cys101 is present as a reduced species (Scaloni et al., 2001).

Similar experiments were performed on porcine OBPI demonstrating the occurrence of a disulfide bond in the protein linking together Cys63 and Cys155 (Paolini et al., 1998). A combination of above-mentioned experiments with Edman degradation of purified Cys-containing peptides from protein samples alkylated under different conditions allowed establishing that the boar salivary lipocalin (SAL) binding steroidal sex pheromones contains the S-S bridge Cys68-Cys160, while its Cys75 and Cys141 are both present in reduced form (Loebel et al., 2000). On the other hand, parallel peptide mapping experiments on protein samples treated with iodoacetamide as reported above identified the disulfide Cys59-Cys152 in rabbit OBP3 from nasal and seminal fluid, also demonstrating the occurrence therein of reduced Cys66 and Cys133 (Mastrogiacomo et al., 2014).

A comparison of the identified cystines and reduced cysteines in the mammalian OBPs and PBPs described above demonstrated the conservation of the ascertained disulfide bond at specific amino acid positions, as shown in Fig. 3, whereas eventual thiols occurred at various non-conserved sites. The significant sequence identity of above-mentioned proteins with other mammalian OBPs as well as the conserved occurrence of Cys residues at the positions where cystine were already identified suggested the occurrence of a S-S bridge also in the former cases, with the exception of bovine OBP.

3. Assignment of additional PTMs in OBPs

Above-reported studies excluded the occurrence of additional PTMs in insect OBPs, PBPs and CSPs, while provided evidences of eventual other modifications in mammalian OBPs and related lipocalins. Generally, the

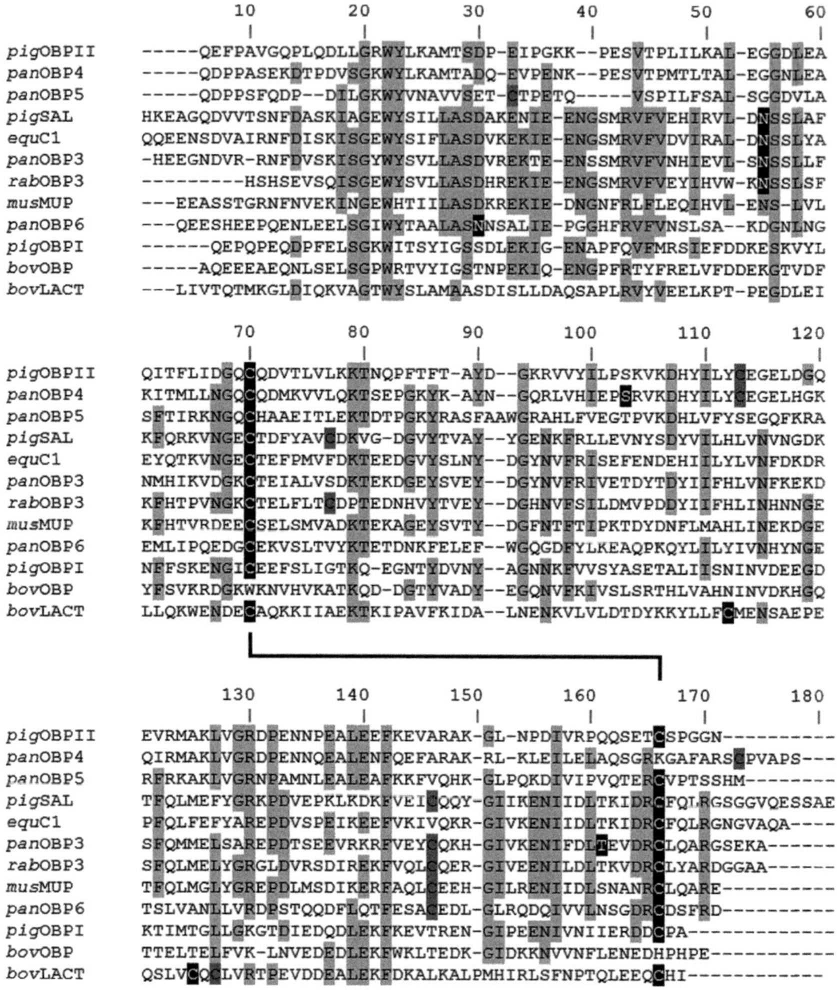

Fig. 3 Sequence analysis of mammalian OBPs, PBPs and additional lipocalins involved in chemoreception as related to corresponding ascertained PTMs. Amino acid sequence alignment of various OBPs, PBPs and lipocalins from different mammals. Cysteines that were ascertained as involved in disulfide bonds are highlighted in black, with the corresponding S-S bridge indicated; cysteines that occurred in reduced form are evidenced in dark gray. Asparagines that were ascertained as N-linked to oligosaccharide chains are highlighted in black; serines and threonines that were ascertained as phosphorylated are highlighted in black. Amino acids that were conserved at a specific position (frequency $\geq$ 0.5) are reported in light gray. *Pig, Sus scrofa; pan, Ailuropoda melanoleuca; equ, Equus caballus; rab, Oryctolagus cuniculus; mus, Mus musculus; bov, Bos taurus.*

latter phenomenon occurred in the case of proteins that migrated in SDS-PAGE as broad electrophoretic bands, showing an apparent molecular mass much higher than the theoretical one calculated according to corresponding protein sequence. In this case, proteins also showed a sensitivity of their mass value to *N*-glycosidase but not to *O*-glycosidase treatment. These findings were integrated by ESI-Q-MS measurements on intact, native species that showed, similarly to corresponding peptide digests from protein derivatives alkylated with iodoacetamide under denaturing, reducing conditions, the occurrence of a significant molecular heterogeneity, as already detected in proteins bearing N-linked oligosaccharide chains. Glycosylation was already described in other mammalian lipocalins. Its occurrence was associated with a positive effect in increasing the solubility of these proteins, which are generally present at a high concentration in body fluids.

In order to identify the N-linked glycosylated amino acids and to characterize corresponding polysaccharide moieties, proteins of interest alkylated with iodoacetamide under denaturing, reducing conditions were subjected to proteolytic digestion with a specific enzyme, and resulting digests were resolved by RP-HPLC; collected fractions were analyzed by MALDI-TOF-MS. In this case, the presence in the spectra of adjacent peaks differing by 162, 203 and 291 mass units made the fraction(s) containing glycopeptides immediately recognizable. Alternatively, protein digests were directly enriched for glycopeptides with HILIC columns, whose eluates were finally analyzed by MALDI-TOF-MS. In both cases, recorded spectrum signals were associated with glycopeptides based on measured mass values, protein sequence and known pathways of glycoprotein biosynthesis. A final confirmation of glycopeptides assignment was obtained by direct peptide sequencing, and/or after components treatment with PNGase F and further MALDI-TOF-MS or MALDI-TOF-TOF-MS analysis of resulting products, which demonstrated glycopeptides transformation into deglycosylated species containing Asn > Asp conversion. Subsequent introduction of novel nanoRP-HPLC-ESI-Q-Orbitrap-MS/MS instruments allowed us to avoid above-mentioned experiments and to analyze directly protein digests, obtaining simultaneous detection of glycopeptides as well as their structural characterization (Zhu et al., 2017). In all above-mentioned cases, bioinformatic analysis of MALDI-TOF-MS or nanoRP-HPLC-ESI-Q-Orbitrap-MS/MS data by dedicated software including modified amino acid and glycan structure unit options was essential to interpret instrumental outputs toward a final protein structural characterization.

When a purified sample of EquC1 reduced and alkylated with iodoacetamide under denaturing conditions was subjected to trypsin digestion, the resulting digest allowed resolving by RP-HPLC corresponding peptides that were then subjected to MALDI-TOF-MS analysis (D'Innocenzo et al., 2006). The occurrence of a specific chromatographic fraction presenting adjacent signals differing by above-mentioned mass units, which disappeared after PNGase F treatment, made glycopeptides immediately recognizable. In fact, signals in the *m/z* range of 1807.0–4831.7 were associated with glycosylated forms of peptide (50–63), as also confirmed by Edman degradation analysis. Based on the criteria reported above, these signals were assigned to the above-mentioned peptide bearing complex-type N-linked glycans bound to Asn53. A mixture of bi-, tri-, and tetra-antennary structures was ascertained, together with partially degraded oligosaccharide species. Aliquots of the non-glycosylated peptide (50–63) (MH^+ signal at *m/z* 1603.7) were also detected in another chromatographic fraction. These results, together with the ascertained occurrence of peptides (35–42) and (33–42) as non-glycosylated species (MH^+ signals at *m/z* 936.2 and 1193.4, respectively), demonstrated that a partial protein glycosylation occurred on Asn53, whereas no modification was present at Asn38. Before this study, these Asn residues were predicted being putatively *N*-glycosylated by bioinformatic searching for putative conserved modification motifs, but no experimental data were recorded. This information was essential since EquC1 is also an allergen, whose glycosylation was already reported affecting corresponding allergenic properties.

Similar MALDI-TOF-MS experiments were performed on porcine SAL, proving the occurrence of an N-linked protein glycosylation site at Asn53 bearing complex-type mono-, bi-, tri-, and tetra-antennary structures, which varied in their nature depending on the sample biological source, whenever male saliva or male/female nasal mucus (Loebel et al., 2000; Scaloni et al., 2001). In this case, this modification was verified as being exhaustive at this site. Additional putative modified amino acids bearing N-linked glycans (Asn38 and Asn99), which were predicted *in silico* according to bioinformatics, were not ascertained in the protein; in this case, only non-modified peptides were characterized, thus proving the absence of *N*-glycosylation therein. On the other hand, analogous MALDI-TOF-MS-based peptide mapping experiments on rabbit OBP3 tryptic digest samples further enriched for glycopeptides through HILIC identified Asn44 as the site of N-linked protein glycosylation bearing mono-, bi- and tri-antennary complex glycan structures (Mastrogiacomo et al., 2014). Extensive MALDI-TOF-TOF-MS analysis of the fractions recovered from a parallel nanoRP-HPLC purification confirmed above-mentioned results (also identifying Asn > Asp conversion in glycopeptide products after its deglycosylation with PNGase F), proved the completeness of the modification at Asn44, and demonstrated the absence of modification at the other predicted N-linked glycosylation site (Asn29).

Recently, nanoRP-HPLC-ESI-Q-Orbitrap-MS/MS analysis of panda OBPs digests, as obtained by trypsinolysis of corresponding SDS-PAGE bands, allowed us to identify directly various PTMs (including N-linked glycosylation) present in these proteins. In particular, sequence analysis using NetNGlyc 1.0 software predicted the presence of two N-linked glycosylation sites in AimelOBP3 (at Asn36 and Asn51), the second of which was actually found to be modified with a complex-type N-linked glycan moiety. Fig. 4 shows the MS and MS/MS spectra of some coeluting glycopeptides detected in the AimelOBP3 digest; the complete list of the N-linked glycopeptides detected in AimelOBP3 is also reported. No glycopeptides containing Asn36 were detected in the protein digest. On the other hand, no glycosylation was predicted for AimelOBP4, AimelOBP5, and AimelOBP6. Nevertheless, a modified peptide bearing a N-linked glycan was detected in AimelOBP6, which resulted from the modification at Asn27. N-linked glycosylation of AimelOBP3 and AimelOBP6 was also evidenced by digestion of these proteins with *N*-glycosidase and analysis of the resulting products by SDS/PAGE and Western blot.

On the other hand, NetPhos 3.1 software predicted several sites of potential phosphorylation in AimelOBPs. nanoRP-HPLC-ESI-Q-Orbitrap-MS/MS analysis of panda OBPs digests showed actual phosphorylation of AimelOBP3 and AimelOBP4 at Thr154 and Ser91, respectively. Non-phosphorylated peptide counterparts were also detected. Only preliminary evidences were reported in the past for phosphorylation of mammalian OBPs; it was suggested to be a way of modifying the binding specificity of these proteins. However, the function of this modification on OBPs still remains to be experimentally shown.

4. Protocols

4.1 Protein purification

OBPs and PBPs from mammals and insects as well as CSPs from insects were purified by gel filtration chromatography followed by anion exchange chromatography, as previously reported (Angeli et al., 1999; Ban, Scaloni, Brandazza, et al., 2003; Ban, Scaloni, D'Ambrosio, et al., 2003; Calvello et al., 2003; D'Innocenzo et al., 2006; Loebel et al., 2000;

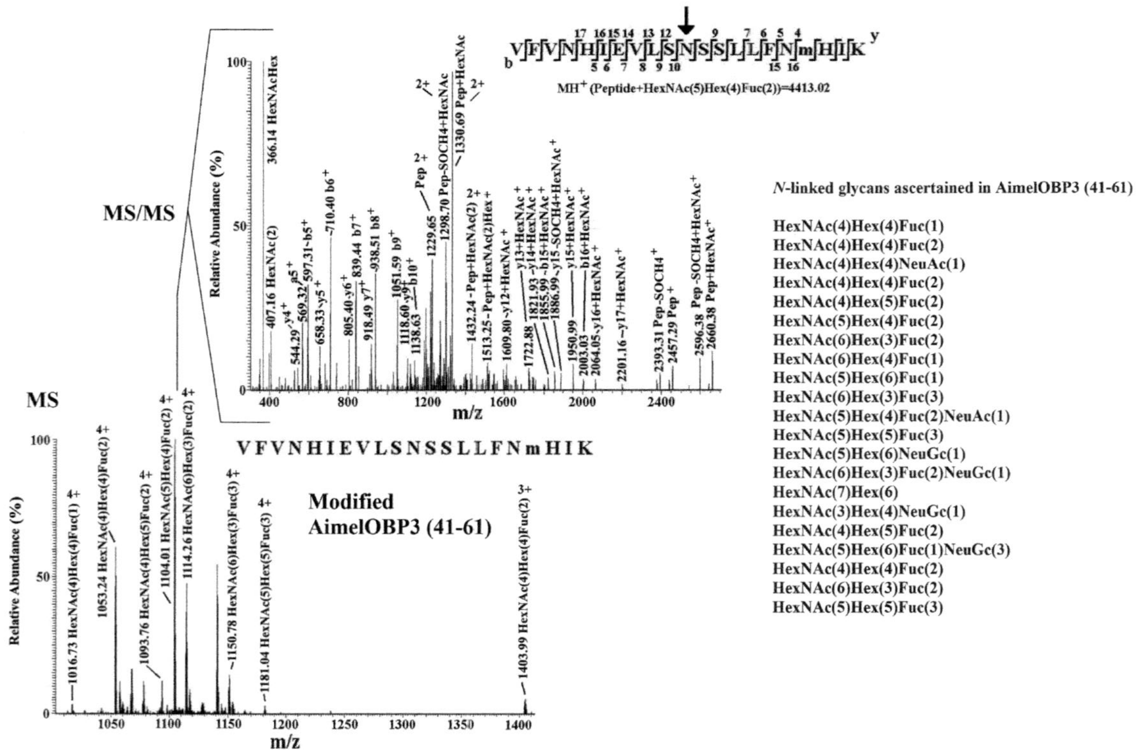

Fig. 4 Mass spectrometric analysis of the N-linked glycopeptides detected in the PanOBP3 digest. Lower spectrum. ESI-Q-Orbitrap-MS spectrum of some N-linked glycated forms of the oxidized PanOBP3 peptide (41–61). Assignment to specific glycation structures was based on fragmentation data. Upper spectrum. ESI-Q-Orbitrap-MS/MS spectrum of the ion at m/z 1104.01 reported in the lower panel. Assigned peptide fragments are highlighted in the peptide sequence reported within the panel. The assigned N-linked glycan structure is also shown. The fragmentation spectrum also shows fragments due to the characteristic loss of methanesulfenic acid from the side chain of oxidized Met derivative (−64 and −32 for singly and doubly protonated ions, respectively) (Bachi et al., 2013). The latter derivative is reported as m; the peptide moiety is indicated with Pep. On the right side, reported is also the complete set of N-linked glycan structures associated with the PanOBP3 peptide (41–61), as detected by nanoRP-HPLC-ESI-Q-Orbitrap-MS/MS experiments.

Marchese et al., 2000; Mastrogiacomo et al., 2014; Paolini et al., 1998; Picimbon et al., 2000; Scaloni et al., 1999, 2001); alternatively, they were directly excised from SDS-PAGE bands of enriched fraction of animal body fluids or tissues, as already described (Zhu et al., 2017). In the first case, proteins were subjected to a final chromatographic step on a Vydac C_4 column 214TP54 (250 × 4.6 mm, 5 μm, 300 Å pore size) (The Separation Group, USA). Proteins were dissolved in 0.1% trifluoroacetic acid (TFA), loaded onto the column and eluted by a linear gradient from 5% to 70% of acetonitrile (ACN) in 0.1% TFA over 30 min, at a flow rate of 1 mL/min. In both cases, protein fractions were collected, lyophilized, and stored at −20 °C, until their use.

4.2 Protein alkylation with iodoacetamide under different experimental conditions

In order to obtain proteins alkylated in denaturing, reducing conditions, samples of interest (30–100 μg) were firstly solved in 0.25 M Tris-HCl, 1.25 mM EDTA, 6 M guanidinium chloride, pH 7.0, containing a 10:1 M excess of DTT over the thiol groups, at 37 °C, for 2 h; then, they were added with an equal volume of 2.2 M iodoacetamide in the same buffer, not containing the reducing agent, at 25 °C, allowing reaction for 1 min, in the dark. In order to obtain proteins alkylated in denaturing, non-reducing conditions, samples of interest (30–100 μg) were directly alkylated with 1.1 M iodoacetamide in 0.25 M Tris-HCl, 1.25 mM EDTA, containing 6 M guanidinium chloride pH 7.0, at 25 °C, for 1 min, in the dark.

In both cases, alkylated proteins were freed from salt and reagent excess by passing the reaction mixture through an analytical Vydac C_4 column, as previously reported (Amiconi et al., 2000; Cecconi et al., 2002; Orrù et al., 1997). Protein samples were collected and lyophilized.

4.3 Mass spectrometric analysis of intact proteins

Intact proteins were subjected to ESI-Q-MS analysis, using a Platform single quadrupole mass spectrometer (Micromass, UK). Samples were dissolved in 1% (v/v) acetic acid and 2–10 μL were injected into the mass spectrometer at a flow rate of 10 μL/min. The quadrupole was scanned from m/z 500 to 1800 at 10 s/scan and the spectra were acquired and elaborated using the MassLynx and Maximum Entropy software (Micromass, UK). Mass calibration was performed by means of the multiply charged ions from a separate injection of horse heart myoglobin (molecular mass 16,951.5 Da).

4.4 Protein digestion

Carboxamidomethylated proteins (20–50 μg) obtained as reported above were digested with trypsin and/or endoprotease LysC (Boehringer, Germany) in 0.4% NH_4HCO_3, pH 6.5 to avoid scrambling phenomena, at 37 °C for 18 h, by using an enzyme/substrate ratio of 1:30 (w/w). CNBr hydrolysis was performed by adding a reagent granule (Sigma) to the protein solved in 70% TFA under N_2 atmosphere, overnight, in the dark.

Alternatively, protein bands (5–20 μg) were excised from the SDS-PAGE gel, triturated and washed with water. Proteins were *in-gel* reduced with 10 mM DTT in 100 mM NH_4HCO_3 (45 min, at 55 °C) and S-alkylated with 55 mM iodoacetamide in 100 mM NH_4HCO_3 (30 min, at 25 °C, in the dark). Gel particles were washed with 50 mM NH_4HCO_3 and acetonitrile, dried and rehydrated with the digestion solution (12.5 ng/μL of trypsin in 50 mM NH_4HCO_3, 5 mM $CaCl_2$). After incubation for 1 h at 5 °C, the supernatant was replaced by 50 mM NH_4HCO_3, 5 mM $CaCl_2$, and gel particles were incubated at 37 °C, overnight. Gel particles were extracted with 25 mM NH_4HCO_3/acetonitrile (1:1 v/v) by sonication and peptide mixtures were concentrated. Samples were desalted using μZipTipC18 pipette tips (Millipore, Bedford MA, USA) before MALDI-TOF-MS or nanoRP-HPLC-ESI-Q-Orbitrap-MS/MS analysis.

4.5 Peptide separation

Peptide digests were fractionated by RP-HPLC on a Vydac C_{18} column 214TP52 (250 × 2.1 mm, 5 μm, 300 Å pore size) (The Separation Group, USA) by using a linear gradient from 5% to 60% of ACN in 0.1% TFA over 60 min, at a flow rate of 0.2 mL/min. Individual components were manually collected and concentrated in a SpeedVac centrifuge (Savant), or lyophilized.

4.6 Mass spectrometric analysis of isolated peptides

MALDI-TOF mass spectra were recorded using a Voyager DE-PRO mass spectrometer (Applied Biosystems, USA). Peptide mixtures were loaded on the instrument target, using the dried droplet technique and α-cyano-4-hydroxycinnamic acid as matrix. In the case of whole peptide digests, internal mass calibration was performed with peptides deriving from protease autoproteolysis. In the case of purified peptide fractions, internal mass calibration was performed with molecular ions from bovine insulin and the matrix; the latter species were preventively mixed and loaded on the instrument target together with samples. Spectra were acquired in linear and/or reflectron mode, and further elaborated using the DataExplorer 5.1 software (Applied Biosystems, USA).

4.7 Peptide treatment with PNGase F

Glycopeptides were deglycosylated by reacting with 0.2 unit of PNGase F (Roche) in 50 mM NH_4HCO_3, pH 8, at 37 °C for 12 h. Then they were desalted on μZipTipC18 pipette tips (Millipore) using 50% (v/v) ACN, 5% (v/v) formic acid (FA) as eluent, before MALDI-TOF-MS analysis.

4.8 Peptide treatment with dithiothreitol

Peptides containing disulfide bonds were reduced by reacting with 10 mM DTT in 50 mM NH_4HCO_3, pH 8, at 37 °C for 3 h. Then, they were desalted on μZipTipC18 pipette tips (Millipore) using 50% (v/v) ACN, 5% (v/v) FA as eluent before MALDI-TOF-MS analysis.

4.9 Edman degradation of isolated peptides

Amino acid sequence was determined by direct Edman degradation using a Perkin Elmer-Applied Biosystems 477A pulsed-liquid protein sequencer equipped with a Perkin Elmer-Applied Biosystems 120A RP-HPLC apparatus for phenylthiohydantoin-amino acid identification.

4.10 nanoRP-HPLC-ESI-Q-Orbitrap-MS/MS analysis of protein digests

Digest samples from SDS-PAGE were desalted with μZip-TipC_{18} pipette tips (Millipore) using 50% (v/v) ACN, 5% (v/v) FA as eluent. Resulting peptide mixtures were analyzed with a nanoRP-HPLC-ESI-Q-Orbitrap-MS/MS instrument based on an UltiMate 3000 HPLC RSLC nano system (Dionex) coupled to a Q-Exactive^{Plus} mass spectrometer through a Nanoflex ion source (Thermo Fisher Scientific). Peptides were loaded on an Acclaim PepMap™ RSLC C18 column (150 mm × 75 μm ID, 2 μm, 100 Å pore size) (Thermo Fisher Scientific), and eluted with a gradient of solvent B [19.92/80/0.08 (v/v/v) water/ACN/FA] in solvent A [99.9/0.1 (v/v) water/FA], at a flow rate of 300 nL/min. The gradient of solvent B started at 3%, increased to 40% over 40 min, raised to 80% over 5 min, remained at this percentage for 4 min, and finally returned to 3% in 1 min, at which remained for additional 20 min. The mass spectrometer operated in data-dependent mode, using a full scan (*m/z* range

375–1500, nominal resolution of 70,000), followed by MS/MS scans of the 10 most abundant ions. MS/MS spectra were acquired in a scan *m/z* range 200–2000, using a normalized collision energy of 32%, an automatic gain control target of 100,000, a maximum ion target of 100 ms, and a resolution of 17,500. A dynamic exclusion value of 20 s was used.

4.11 Bioinformatic analysis of mass spectrometric data

Observed ESI-Q and MALDI-TOF mass values were assigned to proteins, peptides, glycopeptides or disulfide-linked peptides using the GPMAW software (Lighthouse Data, Odense, Denmark). This software generated a mass/fragment database output based on protein sequence, protease specificity, nature of the amino acids susceptible to eventual PTMs and the molecular mass of the modifying groups (including peptide cross-linking). Mass values were matched to protein regions using a 0.02% mass tolerance value. PTMs assignments were always confirmed by additional MS experiments on modified peptides as treated above. Peptide numbering refers to mature proteins.

MS and MS/MS raw data files generated following nanoRP-HPLC-ESI-Q-Orbitrap-MS/MS analysis were loaded into Proteome Discoverer software (Thermo Scientific) and searched with Mascot software (Matrix Science, UK) against a homemade organism protein database containing UniProtKB and NCBI sequence entries. For PTMs discovery, Bionics™ (Protein Metrics) and PEAKS Studio (Bioinformatics Solutions) software were used. In all cases, the following search parameters were used: carbamidomethylation of Cys as a fixed modification; oxidation of Met, deamidation of Asn and Gln, pyroglutamate formation of Gln, phosphorylation of Ser/Thr/Tyr, and glycation of Asn with common mammalian N-linked glycans as variable modifications. Peptide mass tolerance was set to ±20 ppm and the fragment mass tolerance to ±0.05 Da. Proteolytic enzyme and maximal number of missed cleavages were set to trypsin and 3, respectively. Protein candidates assigned on the basis of at least 2 sequenced peptides with an individual peptide expectation value <0.05 (corresponding to a confidence level for peptide identification $>95\%$) were considered confidently identified. Definitive peptide assignment was always associated with manual spectra visualization and verification. Results were filtered to 1% false discovery rate.

5. Summary

Scientists who seek to use protein chemistry and mass spectrometry to investigate how post-translational modifications may affect the biochemical properties of odorant-binding proteins, pheromone-binding proteins and chemosensory proteins may encounter various challenges, which are generally related to the number of technological tools to be simultaneously utilized and the complexity of them. This chapter reports on it, attempting to highlight the specific, punctual procedures and to describe the combination of analytical techniques for overwhelming them. It also describes the results of the positive application of these approaches in originally elucidating specific post-translational modifications in above-mentioned protein families, which unveiled conserved disulfide bridge arrangements as well as other specific modifications therein.

References

Amiconi, G., Amoresano, A., Boumis, G., Brancaccio, A., De Cristofaro, R., De Pascalis, A., et al. (2000). A novel venombin B from *Agkistrodon contortrix contortrix*: Evidence for recognition properties in the surface around the primary specificity pocket different from thrombin. *Biochemistry*, *39*, 10294–10308.

Angeli, S., Ceron, F., Scaloni, A., Monti, M., Monteforti, G., Minnocci, A., et al. (1999). Purification, structural characterization, cloning and immunocytochemical localization of chemoreception proteins from *Schistocerca gregaria*. *European Journal of Biochemistry*, *262*, 745–754.

Arena, S., Salzano, A. M., Renzone, G., D'Ambrosio, C., & Scaloni, A. (2014). Non-enzymatic glycation and glycoxidation protein products in foods and diseases: An interconnected, complex scenario fully open to innovative proteomic studies. *Mass Spectrometry Reviews*, *33*, 49–77.

Bachi, A., Dalle-Donne, I., & Scaloni, A. (2013). Redox proteomics: Chemical principles, methodological approaches and biological/biomedical promises. *Chemical Reviews*, *113*, 596–698.

Ban, L., Scaloni, A., Brandazza, A., Angeli, S., Zhang, L., Yan, Y., et al. (2003). Chemosensory proteins of *Locusta migratoria*. *Insect Molecular Biology*, *12*, 125–134.

Ban, L., Scaloni, A., D'Ambrosio, C., Zhang, L., Yan, Y., & Pelosi, P. (2003). Biochemical characterization and bacterial expression of an odorant-binding protein from *Locusta migratoria*. *Cellular and Molecular Life Science*, *60*, 390–400.

Block, E. (2018). Molecular basis of mammalian odor discrimination: A status report. *Journal of Agricultural and Food Chemistry*, *66*, 13346–13366.

Calvello, M., Guerra, N., Brandazza, A., D'Ambrosio, C., Scaloni, A., Dani, F. R., et al. (2003). Soluble proteins of chemical communication in the social wasp *Polistes dominulus*. *Cellular and Molecular Life Science*, *60*, 1933–1943.

Cecconi, I., Scaloni, A., Rastelli, G., Moroni, M., Vilardo, P. G., Costantino, L., et al. (2002). Oxidative modification of aldose reductase induced by copper ion. Definition of the metal-protein interaction mechanism. *Journal of Biological Chemistry*, *277*, 42017–42027.

D'Ambrosio, C., Salzano, A. M., Arena, S., Renzone, G., & Scaloni, A. (2007). Analytical methodologies for the detection and structural characterization of phosphorylated proteins. *Journal of Chromatography B, Analytical Technologies in the Biomedical and Life Sciences*, *849*, 163–180.

D'Innocenzo, B., Salzano, A. M., D'Ambrosio, C., Gazzano, A., Niccolini, A., Sorce, C., et al. (2006). Secretory proteins as potential semiochemical carriers in the horse. *Biochemistry*, *45*, 13418–13428.

Loebel, D., Scaloni, A., Paolini, S., Fini, C., Ferrara, L., Breer, H., et al. (2000). Cloning, post-translational modifications, heterologous expression and ligand-binding of boar salivary lipocalin. *Biochemical Journal*, *350*, 369–379.

Marchese, S., Angeli, S., Andolfo, A., Scaloni, A., Brandazza, A., Mazza, M., et al. (2000). Soluble proteins from chemosensory organs of *Eurycantha calcarata* (insects, Phasmatodea). *Insect Biochemistry and Molecular Biology*, *30*, 1091–1098.

Mastrogiacomo, R., D'Ambrosio, C., Niccolini, A., Serra, A., Gazzano, A., Scaloni, A., et al. (2014). An odorant-binding protein is abundantly expressed in the nose and in the seminal fluid of the rabbit. *PLoS One*, *9*, e111932.

Orrù, S., Scaloni, A., Giannattasio, M., Urech, K., Pucci, P., & Schaller, G. (1997). Amino acid sequence, S-S bridge arrangement and distribution in plant tissues of thionins from *Viscum album*. *Biological Chemistry*, *378*, 989–996.

Paolini, S., Scaloni, A., Amoresano, A., Marchese, S., Napolitano, E., & Pelosi, P. (1998). Amino acid sequence, post-translational modifications, binding and labelling of porcine odorant-binding protein. *Chemical Senses*, *23*, 689–698.

Picimbon, J. F., Dietrich, K., Angeli, S., Scaloni, A., Krieger, J., Breer, H., et al. (2000). Purification and molecular cloning of chemosensory proteins from *Bombyx mori*. *Archives of Insect Biochemistry and Physiology*, *44*, 120–129.

Picone, D., Crescenzi, O., Angeli, S., Marchese, S., Brandazza, A., Ferrara, L., et al. (2001). Bacterial expression and conformational analysis of a chemosensory protein from *Schistocerca gregaria*. *European Journal of Biochemistry*, *268*, 4794–4801.

Raimondo, D., Andreotti, G., Saint, N., Amodeo, P., Renzone, G., Sanseverino, M., et al. (2005). A folding-dependent mechanism of antimicrobial peptide resistance to degradation unveiled by solution structure of distinctin. *Proceeding of National Academy of Sciences of the United States of America*, *102*, 6309–6314.

Scaloni, A., Monti, M., Angeli, S., & Pelosi, P. (1999). Structural analysis and disulfide-bridge pairing of two odorant-binding proteins from *Bombyx mori*. *Biochemical and Biophysical Research Communications*, *266*, 386–391.

Scaloni, A., Paolini, S., Brandazza, A., Fantacci, M., Bottiglieri, C., Marchese, S., et al. (2001). Purification, cloning and characterisation of odorant- and pheromone-binding proteins from pig nasal epithelium. *Cellular and Molecular Life Sciences*, *58*, 823–834.

Tomaselli, S., Crescenzi, O., Sanfelice, D., Ab, E., Wechselberger, R., Angeli, S., et al. (2006). Solution structure of a chemosensory protein from the desert locust *Schistocerca gregaria*. *Biochemistry*, *45*, 10606–10613.

Zhu, J., Arena, S., Spinelli, S., Liu, D., Zhang, G., Wei, R., et al. (2017). Reverse chemical ecology: Olfactory proteins from the giant panda and their interactions with putative pheromones and bamboo volatiles. *Proceeding of National Academy of Sciences of the United States of America*, *114*, E9802–E9810.

CHAPTER SIX

Bacterial expression and purification of vertebrate odorant-binding proteins

Marine Brulé, Margot Glaz, Christine Belloir, Nicolas Poirier, Lucie Moitrier, Fabrice Neiers, Loïc Briand*

Centre des Sciences du Goût et de l'Alimentation, AgroSup Dijon, CNRS, INRAE, Université Bourgogne Franche-Comté, Dijon, France

*Corresponding author: e-mail address: loic.briand@inrae.fr

Contents

Abstract

Vertebrate odorant-binding proteins (OBPs) are small soluble proteins abundantly secreted in the olfactory mucus of many animal species, including humans. Vertebrate OBPs reversibly bind odorant molecules with micromolar range affinities.

Methods in Enzymology, Volume 642
ISSN 0076-6879
https://doi.org/10.1016/bs.mie.2020.05.002

Although their physiological role is not clearly understood, OBPs are proposed to carry airborne odorants toward membrane olfactory receptors through the nasal mucus. Measurements of odorant-OBP interactions and structural studies require a large amount of pure OBPs devoid of ligands. The bacterial expression system is the first choice for expressing vertebrate OBPs used in our laboratory and others. This system generally produces OBPs in large amounts without major problems. In this chapter, we describe the milligram-scale production of recombinant pig OBP1 (pOBP1) in *E. coli*. The different steps of expression and purification are presented and discussed. Protocols for secondary structures investigation by circular dichroism and binding properties of the recombinant protein are also provided. More generally, these approaches can be used to produce and characterize any vertebrate OBPs for use in functional and structural studies.

1. Introduction

Vertebrate odorant-binding proteins (OBPs) are small carrier proteins (~20 kDa) secreted by the nasal epithelium at high concentrations. They have been shown to reversibly bind odorant molecules with dissociation constants in the micromolar range. Vertebrate OBPs have been observed in a variety of mammalian species (Pelosi, 2001; Tegoni et al., 2000), such as cows, pigs, mice, rats, rabbits, giant pandas and humans (Briand et al., 2002). Different OBP subtypes have been reported to occur simultaneously in the same animal species. For instance, three OBPs have been described in rat olfactory mucus with complementary odorant binding properties (Löbel, Jacob, Völkner, & Breer, 2002; Löbel, Marchese, Krieger, Pelosi, & Breer, 1998; Löbel, Strotmann, Jacob, & Breer, 2001).

Vertebrate OBPs belong to the superfamily of transport proteins named lipocalins. They share a conserved scaffold made of eight antiparallel beta-sheets flanked by a short alpha-helix located close to the C-terminus. The beta-barrel defines a central apolar cavity, whose role is to bind hydrophobic compounds such as odorant molecules (Tegoni et al., 2000). Vertebrate OBPs do not typically contain post-translational modifications. However, different OBP isoforms have been observed in pig and giant panda, resulting from glycosylation and phosphorylation (Nagnan-Le Meillour, Vercoutter-Edouart, Hilliou, Le Danvic, & Lévy, 2014; Scaloni et al., 2001; Zhu et al., 2017). In pigs, native and recombinant OBP isoforms differing in these post-translational modifications have been shown to display different binding affinities for specific ligands (Brimau et al., 2010; Nagnan-Le Meillour et al., 2019).

Although various hypotheses have been proposed for their physiological function, vertebrate OBPs are suspected to transport airborne hydrophobic odorants through the aqueous olfactory mucus toward the olfactory receptors (Heydel et al., 2013; Pelosi, 2001). Recently, vertebrate OBPs have been suggested to show antimicrobial activity against pathogenic bacteria and fungi (Bianchi et al., 2019).

The compact folding of vertebrate OBPs makes them resistant to organic solvents, thermal denaturation and proteolytic degradation (see Chapter "Solution structure of insect CSP and OBPs by NMR" by Delia Picone). They can be easily expressed in bacteria, and their binding properties can be modified using mutagenesis to design mutants with desired specific binding properties (Zhu et al., 2017; Zaremska, Tan, Lim, Knoll, & Pelosi, 2020; see also Chapter "Effect of OBPs on the response of olfactory receptors" by Guirong Wang). OBPs have thus great potential to be used for biotechnological applications (Pelosi, Mastrogiacomo, Iovinella, Tuccori, & Persaud, 2014) and as sensing elements in biosensors (Barou, Sigoillot, Bouvet, Briand, & Meunier-Prest, 2015; Hurot et al., 2019; Larisika et al., 2015; Mulla et al., 2015).

Large-scale production methods are required to produce milligram quantities of purified and homogenous protein for X-ray crystallographic (Zhu et al., 2017) and NMR structural investigations (Portman et al., 2014). Bacterial systems are the first choice for expressing vertebrate OBPs, which can be generally produced in large amounts and without major problems. Vertebrate OBPs can be also successfully expressed in mammalian CHO (Chinese Hamster Ovary) cells. However, the yields are too low to perform ligand binding studies (Brimau et al., 2010). Alternatively, our laboratory and others have shown that vertebrate OBPs can be expressed using the methylotrophic yeast *Pichia pastoris* (Briand, Huet, et al., 2000; Briand et al., 2002; Ferrari et al., 1997; Guiraudie-Capraz et al., 2005; Lücke et al., 1999). This cost-effective eukaryotic expression system has been demonstrated to secrete vertebrate OBPs with post-translational modifications, such as glycosylation and phosphorylation (Briand, Nespoulous, et al., 2000; Brimau et al., 2010). The advantage of *Pichia pastoris* is the use of integrative plasmids amplified in bacteria to incorporate an expression cassette encoding the protein sequence of interest. Homologous recombination is thus used to stably integrate cassettes in the yeast genome. The selection of stably producing clones is time consuming. In addition, the production of proteins is a long process and generally requires 4 to 6 days. However, the *P. pastoris* expression system represents a good choice, particularly when bacterial expression fails or post-translational modifications are necessary for the recombinant protein to be fully functional. Compared to *P. pastoris*, we found that the bacterial expression system is easier to handle, and OBPs can be expressed quicker (2 days) with similar yields. In this chapter, we provide detailed protocols for the production of pig OBP1 as a hexahistidine (His_6)-tagged protein in *Escherichia coli* for functional characterizations, structure-function relationship studies and structural investigations.

2. Expression of vertebrate OBPs in bacteria

Vertebrate OBPs have been expressed as soluble proteins in the *E. coli* BL21 (DE3) strain (Zhu et al., 2017) using the T7 promoter present in the pET vectors (Novagen, Merk Millipore). Expression of His_6-tagged vertebrate OBPs has been preferred in our hands to facilitate protein purification. Alternatively, the pQE system developed by Qiagen (see The QIAexpressionist™ manual from Qiagen) has been used by the group of Heinz Breer (University of Hohenheim, Stuttgart, Germany) to express the three rat OBPs, rOBP1, rOBP2 and rOBP3 (Löbel et al., 1998, 2001, 2002). The low copy number pQE vector utilizes two *lac* operator sequences to increase the control of the T5 promoter and ensure sufficient repression of proteins. Protein expression can be induced by lactose non-hydrolysable analog isopropyl β-D-1-thiogalactopyranoside (IPTG). For protein expression, the pQE system requires *E. coli* host strains, such as M15 [pREP4] or SG13009 [pREP4], which contain the repressor pREP4 plasmid. This plasmid encodes a *lac* repressor that ensures tight regulation of expression. This protein expression system was proven, in our experience, to be especially successful for the production of various vertebrate OBPs. Notably, we showed that a rat OBP (rOBP3) could not be expressed using the classical pET system, whereas it was expressed very well using the pQE system (unpublished data). Although we have no evidence, we suspect that rOBP3 exhibits toxicity to host cell bacteria. At low levels, due to leaky expression under the control of the T7 promoter, we hypothesize that rOBP3 binds *E. coli* metabolites and interferes with the normal proliferation and homeostasis of the microorganism. The tighter control of expression of the pQE vector, the lower potency of the T5 promoter and its low copy number should probably explain why rOPB3 is less detrimental for the host cell. Fig. 1 summarizes our expression construct that efficiently expresses pig OBP1 (pOBP1). This construct incorporates an N-terminal His_6 metal tag

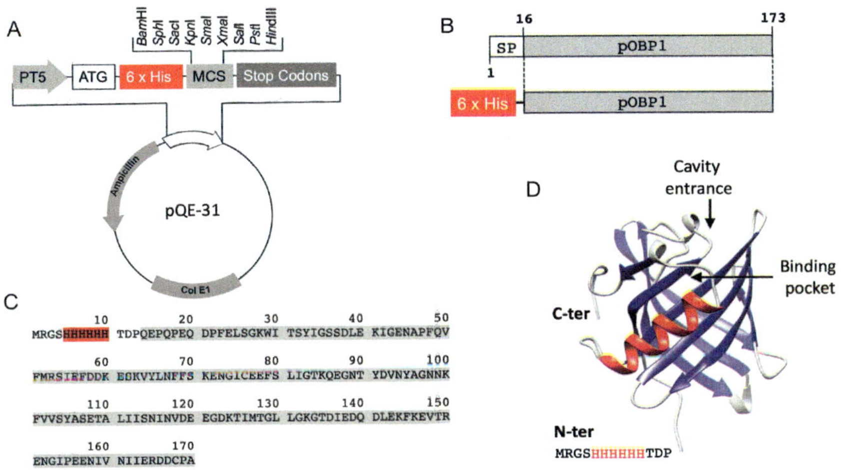

Fig. 1 Strategy used to express pig OBP1 in *E. coli*. (A) Pig OBP1 (pOBP1) was expressed as a mature protein minus its signal peptide (SP). The expressed protein incorporates an N-terminal 6 x His tag to aid purification and detection of the protein. The numerical positions of the amino acid residues of pOBP1 are indicated. (B) The pQE-31 plasmid used for pOBP1 expression. The *E. coli* optimized cDNA sequence of pOBP1 was cloned between the *BamHI* and *HindIII* restriction sites of pQE-31. The vector contains a His_6-tag sequence upstream of the OBP encoding sequence, PT5: T5 promoter; ATG: start codon; MCS: multicloning site, Col E1: Col E1 origin of replication, Ampicillin: ampicillin resistance gene. (C) Amino acid sequence of expressed pOBP1. His_6-Tag and mature amino acid sequences of pOBP1 are shown in black and gray, respectively. Numbers refer to amino acid residues of expressed pOBP1. (D) Three-dimensional structure of pig OBP1, folded in a typical β-barrel motif with eight strands (blue) flanked by a single α-helix (red). The position of the N-terminal 6x His tag is indicated.

with affinity to pOBP1 for purification and immunodetection. The cDNA encoding the mature pOBP1 minus its signal peptide was subcloned into the *Bam*HI and *Hind*III restriction sites of the pQE-31 vector according to the manufacturer's instructions. Alternatively, a synthetic cDNA sequence encoding OBPs could be optimized for expression in *E. coli*. We observed that *E. coli* codon optimization approximately doubles the level of expression of the three rat OBPs (rOBP1, rOBP2 and rOBP3). Note that the pQE-31 vector should be propagated and amplified using *E. coli* strains harboring the $lacI^q$ mutation, such as XL1 Blue, JM109 and TG1 cells. In our laboratory, we used the JM109 strain for plasmid amplification and *E. coli* strain M15 [pREP4] for the expression of vertebrate OBPs.

The protocol below is designed for a one-liter culture, though we often grow up multiple liter cultures simultaneously since we found that bacterial pellets can be stored at −20 °C for 6 months without appreciable modifications of the recombinant OBPs. SDS-PAGE analysis was used to assess the target OBP production level. Typical growth and purification from a 1-L culture yields approximately 50 mg of purified pOBP1 after 3 h of IPTG induction at 37 °C (Fig. 2A).

2.1 Equipment

- Water bath
- Incubator and shaking incubator
- Microbiological safety cabinet
- Visible/UV spectrophotometer
- Refrigerated centrifuge
- 14-mL sterile ventilated culture tubes (e.g., Falcon 352049)
- 50-mL sterile ventilated culture tubes (e.g., Corning 431720)
- 4 × 2-L sterile baffled Erlenmeyer shaker flasks
- Standard SDS-PAGE setup
- SDS-PAGE gels (12%)

2.2 *E. coli* strain and plasmid

- M15 [pREP4] competent cells (pQE expression system, QIAGEN)
- pQE-31 plasmid containing the cDNA encoding the mature OBP

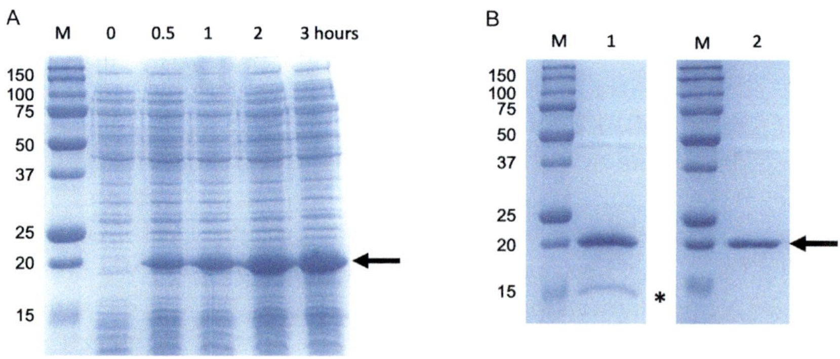

Fig. 2 SDS-PAGE analysis of pOBP1 expressed in *E. coli*. The proteins were separated by 12% SDS-PAGE and stained with Coomassie blue. The molecular mass markers (kDa) are indicated. (A) *E. coli* M15 [pREP4] cells were transformed with the pQE-31-pOBP1 plasmid before and after IPTG induction for 0.5, 1, 2 and 3 h. pOBP1 is indicated with an arrow. (B) SDS-PAGE analysis of purified pOBP1 by IMAC (1) followed by gel filtration chromatography (2). The star indicates a band probably corresponding to an N-terminal fragment of pOBP1.

2.3 Buffers and reagents

- Sterile ampicillin solution (100 mg/mL) (Sigma-Aldrich)
- Sterile kanamycin solution (50 mg/mL) (Sigma-Aldrich)
- LB medium, autoclaved, with 100 μg/mL ampicillin and 50 μg/mL kanamycin
- LB agar plates with 100 μg/mL ampicillin and 50 μg/mL kanamycin
- Super optimal broth with catabolite repression (SOC) media, autoclaved
- Freshly prepared sterile IPTG (isopropyl-β-thiogalactoside, 1 M stock) (Sigma-Aldrich)
- Lysis buffer (50 mL): 50 mM NaH_2PO_4, 300 mM NaCl, 20 mM imidazole, pH 7.5

2.4 Protocol

2.4.1 Transformation of M15 [pREP4] competent cells

1. Mix 1 μL of purified plasmid (20 ng) with 20 μL of M15 [pREP4] competent cells placed in 14-mL culture tubes.
2. Evenly mix the cells with the pipette. Leave the tube on ice for 20 min.
3. Heat shock for 45 s in a water bath at 42 °C and place the tube on ice for 5 min.
4. Add 250 μL of SOC medium and shake smoothly.
5. Place the tube in the incubator at 200 rpm and 37 °C for 60 min.
6. Plate (50 μL of culture) onto LB/ampicillin/kanamycin agar plates and leave overnight at 37 °C.

2.4.2 Growth phase

1. The day before the expression, pick a single transformed colony and inoculate two 50-mL Corning Mini Bioreactor tubes with ventilated caps, each containing 20 mL of LB medium (100 μg/mL ampicillin, 25 μg/mL kanamycin).
2. Let the colonies grow overnight at 37 °C (20 h, 200 rpm).

2.4.3 Induction phase

1. Use 20 mL of starter culture to inoculate two 2 L baffled Erlenmeyer flasks each containing 500 mL of LB/100 μg/mL ampicillin/50 μg/mL kanamycin.
2. Incubate at 37 °C with 200 rpm shaking to grow the culture to an OD_{600} (optical density at 600 nm) value between 0.5 and 0.7. This typically takes 3 h.

3. Add freshly prepared IPTG (1 M) to reach a final concentration of 0.2 mM to induce protein expression.
4. Induce expression for 3 h at 37 °C with 200 rpm shaking.
5. Centrifuge cells for 20 min at 4500 *g* and 4 °C.
6. Remove the supernatant and resuspend the cell pellet with 30 mL of lysis buffer. The cell suspension can be stored for weeks at −20 °C.

2.5 Notes

1. The M15 strain was derived from *E. coli* K12 and has the phenotype NaI^S, Str^S, Rif^S, Thi^-, Lac^-, Ara^+, Gal^+, Mtl^-, F^-, $RecA^+$, Uvr^+, Lon^+. *E. coli* strain M15 does not harbor a chromosomal copy of the $lacI^q$ mutation, so pREP4 must be maintained by selection for kanamycin resistance.
2. *E. coli* strains such as XL1 Blue and TG1, which harbor the lacIq mutation, produce enough lac repressor to efficiently block transcription. We use JM109 cells to store and propagate the pQE-31 plasmid.
3. Plates containing transformed colonies can be kept at 4 °C for up to 2 weeks. The transformed cells can also be stored as glycerol or DMSO stocks.
4. This is a generic protocol for vertebrate OBP production, used for instance for pOBP1, rOBP1 and rOBP3 production. Depending on the expressed OBP, it can be necessary to optimize protein expression by modifying the expression conditions (i.e., IPTG concentration, induction time and temperature). For instance, we found that some vertebrate OBPs should be expressed for 6 h at 37 °C or overnight at 20 °C.
5. We typically use LB medium for expression. However, we observed that a nutritionally rich medium (Terrific Broth, TB) can be used, leading to higher yields of some OBPs.

3. Purification of OBP using immobilized metal affinity chromatography (IMAC)

Vertebrate OBPs free of endogenous ligands are difficult to obtain because of their binding affinities, with dissociation constants in the micromolar range and their generally broad specificity to a range of ligands. For instance, natural bovine OBP (bOBP) has been copurified and co-crystallized with the naturally occurring 1-octen-3-ol ligand (Ramoni et al., 2001). The recombinant bOBP heterologously produced in bacteria was similarly copurified with a serendipitous ligand (Ramoni et al., 2008).

This putative metabolite odorant, probably originating from *E. coli,* was not identified. As expected, we observed that rOBP1 is able to bind airborne odorants in conditions close to those of peri-receptor physiology (Briand, Huet, et al., 2000). To avoid the presence of odorants or volatiles present in the laboratory atmosphere, we strongly recommend purifying OBPs under a fume cupboard. Indeed, vertebrate and insect OBPs are able to accommodate a large range of organic compounds. For instance, the 3D structure of the honey bee pheromone-binding protein ASP1 was solved with a serendipitous ligand, n-butyl-benzene-sulfonamide, an additive used in plastics (Lartigue et al., 2004). For all these reasons, we strongly recommend avoiding plastic materials during all the steps of purification and washing the glassware with solvents. We suggest using reagents and solvents of the highest possible degree of purity and stocking the purified OBPs in glass vials.

We found that the expression of OBPs as His_6-tagged proteins is useful for purification because OBPs can be immobilized using a nickel affinity column. Finally, and most importantly, we observed that solvents such as methanol, ethanol or acetonitrile at concentrations higher than 5% were able to efficiently displace the fluorescent probe 1-AMA bound to rat OBP1 (Briand, Huet, et al., 2000). Thus, because His_6-OBP can be immobilized using IMAC, we developed this protocol to purify rOBP3 free of endogenous ligands. This purification combines 7 washing steps using solvents such as acetonitrile and ethanol used at concentrations sufficient to release putative endogenous ligands from OBP but low enough to avoid protein denaturation. Using isothermal-titration calorimetry (ITC), we found that the binding stoichiometry of purified rOBP3 is close to 1, as expected for vertebrate OBPs, indicating that the OBP is free of ligands after purification. We also observed, using fluorescent binding assays, that the presence of a unknown ligand led to significant misevaluations of the binding constants (Pelosi, Zhu, & Knoll, 2018; see also Chapter "Modeling of OBPs and docking simulations" by Chris Oostenbrink) (data not shown). We observed that this purification protocol can be successfully applied to rOBP1 and pOBP1 and is likely to prove applicable to other OBPs.

3.1 Equipment

- Fast Protein Liquid Chromatography (FPLC) system or low-pressure chromatographic system with UV detection capabilities (e.g., AKTA Prime Plus FPLC, GE Healthcare Life Science)

- Centrifuge tubes (e.g., 50 mL; Corning)
- Dialysis bottles (e.g., 2 L; Schott Duran)
- Dialysis membranes (e.g., 6–8 kDa; Spectrum Laboratories)
- Standard SDS-PAGE setup
- SDS-PAGE (12%)
- IMAC column (e.g., HisTrap™ HP 5 mL; GE Healthcare Life science)
- Peristaltic pump (e.g., Minipuls II; Gilson)
- Refrigerated centrifuge (e.g., Multifuge X3 FR; Thermo Scientific)
- Syringe filters (e.g., 1.2 and 0.45 μm; Minisart® Sartorius)
- Sonic cell disruptor (e.g., Bioblock Scientific Vibracell)
- Microvolume UV/visible spectrophotometer (e.g., NanoDrop™ One/One^{C}; Thermo Scientific)
- 4-mL Glass vials (e.g., Wheaton)
- 5-mL Glass collection tubes (e.g., NAFVSM)
- Glass pipettes

3.2 Chemicals

- Acetonitrile, LC-MS grade (Carlo Erba Reagents)
- Ammonium acetate (Merck Millipore)
- Ethanol, LC-MS grade (Carlo Erba Reagents; Absolute anhydrous)
- Methanol, LC-MS grade (Carlo Erba Reagents)
- Water, HPLC grade (Carlo Erba Reagents, Water RS For HPLC PLUS)
- Imidazole (Reagent Plus®; Sigma-Aldrich)
- Sodium phosphate dibasic (Sigma-Aldrich)
- Sodium phosphate monobasic (Sigma-Aldrich)
- Sodium chloride (Sigma-Aldrich)
- Sodium hydroxide (Sigma-Aldrich)

3.3 Buffers

The glassware should be washed 3 times with 100% methanol and rinsed 3 times with ultrapure HPLC grade water under a fume hood. The buffers can be prepared the day before purification with HPLC grade water and stored in glass bottles at 4 °C. Compared to ultrapure grade water, we found that the Millipore Milli-Q™ water is not pure enough to carry out OBP purification. Note that the ammonium acetate buffer is substituted with phosphate buffer to avoid phosphate precipitation caused by ethanol and acetonitrile. The buffers are not filtered to avoid pollution by plastic compounds. All the materials, especially the glass tubes, are rinsed with methanol and HPLC water and ventilated until they are dried under the fume hood.

- Buffer A (500 mL): 50 mM sodium phosphate buffer, 300 mM NaCl, 20 mM imidazole, pH 7.5
- Buffer B (300 mL): 50 mM sodium phosphate, 300 mM NaCl, 250 mM imidazole, pH 7.5
- Buffer C (500 mL): 50 mM ammonium acetate, pH 7.5
- Buffer D (100 mL): 50 mM ammonium acetate, ethanol 20% (*v*/v), pH 7.5
- Buffer E (100 mL): 50 mM ammonium acetate, acetonitrile 30% (v/v), pH 7.5
- Dialysis buffer (1.5 L): 100 mM sodium phosphate, pH 7.5

3.4 Protocol

3.4.1 *Cell lysis*

1. Prepare buffers A to E, and dialysis buffer, and store them at −4 °C until needed.
2. Defrost the bacterial cell suspension on ice (in 50-mL Falcon tube).
3. Using a tip sonicator, lyse cells on ice. We typically use two cycles of 2 min each, using a power amplitude of 30%, with a pulser on time of 5 s and pulser off time of 3 s. Let cells cool down for a few minutes between sonication cycles to avoid overheating and protein denaturation.
4. Clarify the supernatant by centrifugation at 20,000 *g* for 45 min at 4 °C. All of the following steps should be carried out under a fume hood using a peristaltic pump.
5. Filter the supernatant by using 1.2- and 0.45-μm syringe filters and place it on ice.

3.4.2 *OBP purification using IMAC*

1. Connect the IMAC column (HisTrap™ HP 5 mL) to the peristaltic pump.
2. Wash the column with 5 column volumes (CV) of ultrapure water at a flow rate of 5 mL/min.
3. Equilibrate the column with 5 CV of buffer A.
4. Wash the column with 5 CV of buffer B.
5. Run 10 CV of buffer A.

6. Load filtrated supernatant onto the column (flowrate of 2.5 mL/min) to allow His-tagged-OBP to bind to the IMAC resin.
7. Wash the column with 5 CV of buffer A.
8. Run 4 CV of buffer C.
9. Wash the column with 8 CV of buffer D.
10. Run 4 CV of buffer C.
11. Run 5 CV of buffer E.
12. Wash with 4 CV of buffer C.
13. Wash with 5 CV of buffer A.
14. Connect the column to the FPLC system.
15. Elute the protein with a linear gradient of buffer B (100% buffer A to 100% buffer B) with a flow rate of 2.5 mL/min over 50 mL. Collect 2 mL fractions in washed glass tubes and put them immediately on ice under a fume hood.
16. Save 10 μL of each fraction for SDS-PAGE analysis.
17. Pool the fractions containing OBP.
18. Dialyze purified his-tagged OBP at 4 °C in a closed glass bottle with stirring (usually approximately 10 mL sample into 0.5 L dialysis buffer). Perform two dialysis baths with dialysis buffer supplemented with 5% acetonitrile (*v*/v) followed by two dialysis baths with dialysis buffer. Change dialysis baths every 2 h.
19. The following day, after dialysis, save a 10 μL sample for gel electrophoresis. SDS-PAGE analysis should reveal a major band corresponding to purified OBP (Fig. 2B).
20. Quantify the concentration of the OBP spectrophotometrically at 280 nm using its molar extinction coefficient. We typically use a microvolume spectrophotometer to avoid contamination of the purified OBP using a cuvette.
21. Aliquot the protein in glass vials and store it at −20 °C.

4. Purification of OBP via size-exclusion chromatography

To further purify OBPs for structural studies, it is advisable to use gel filtration. Care must be taken to clean the chromatographic system and the column to avoid contamination of the IMAC purified OBP with ligands during this separation phase. We recommend cleaning the column with 3 CVs of 20% ethanol and 30% acetonitrile to remove organic compounds potentially bound onto the resin. Do not concentrate the protein with plastic concentrators to avoid OBP contamination.

4.1 Equipment

- Fast Protein Liquid Chromatography (FPLC) system or low-pressure chromatographic system with UV detection capabilities (e.g., AKTA Prime Plus FPLC; GE Healthcare Life Science)
- Preparative size-exclusion column (e.g., HiPrep™, Sephacryl® S-100 HR 26 × 600 mm; GE Healthcare Life Science)
- Clear glass laboratory bottles (500 mL) with screw caps (e.g., Schott Duran)
- Dialysis membranes (e.g., Spectrum Laboratories, 6–8 kDa)
- Standard SDS-PAGE protein gel setup
- SDS-PAGE gels (12%)
- Syringe filters (0.45 and 0.22 μm; Pall)
- Microvolume UV/visible spectrophotometer (e.g., NanoDrop™ One/One^{C}; Thermo Scientific)
- Glass vials (4 mL) (e.g., Wheaton)
- 5-mL glass collection tubes (e.g., NAFVSM)
- Glass pipettes

4.2 Buffer

Gel filtration buffer (1.5 L): 50 mM sodium phosphate, 150 mM NaCl, pH 7.5 prepared with HPLC grade water.

4.3 Protocol

1. Equilibrate the column with at least one CV of water and two CVs of 50 mM sodium phosphate and 150 mM NaCl, pH 7.5.
2. Load 12 mL of pOBP1 solution on the gel filtration column and elute at a flow rate of 1 mL/min.
3. The chromatogram should reveal a large peak, along with a smaller peak corresponding to aggregates or proteolyzed OBP fragments. We collect approximately 2-mL fractions in washed glass tubes. Immediately put the fractions on ice under a fume hood.
4. Save 10 μL of each fraction for SDS-PAGE analysis and pool the fractions containing OBP. The SDS-PAGE should show a pure OBP with a single band (Fig. 2B).
5. Estimate the protein concentration using a microvolume spectrophotometer as previously described.
6. Aliquot the protein in glass vials and store it at −20 °C.

Note: To remove co-purified ligands, Löbel and collaborators dialyzed the recombinant rOBP3 against 1 mM of a low affinity ligand, 1-octanol, and then against buffer (Löbel et al., 2001).

5. OBP characterization

5.1 Circular dichroism analysis

Circular dichroism (CD) spectroscopy is an excellent technique for rapid estimation of the secondary structure and folding properties of purified protein. The most widely used applications of OBP CD are to determine whether a purified OBP is folded (Briand et al., 2002; Briand, Huet, et al., 2000; Burova, Choiset, Jankowski, & Haertlé, 1999; Löbel et al., 1998) or if an amino acid substitution affects its conformation (Tcatchoff, Nespoulous, Pernollet, & Briand, 2006). In addition, CD has also been used to study OBP-odorant interactions (Nespoulous, Briand, Delage, Tran, & Pernollet, 2004).

5.1.1 Equipment

- CD spectrophotometer (e.g., JASCO J-815 spectrophotometer equipped with a Peltier temperature controller)
- Demountable quartz cuvette with short pathlength (0.1 mm) (e.g., Hellma analytics)
- Cuvette mount
- Nitrogen gas
- UV spectrophotometer
- Dialysis cassettes or tubing (MW cutoff of 10 kDa)

5.1.2 Chemicals

Many buffers can be used, such as Tris-HCl or phosphate buffer. The choice of the buffer depends on the selected pH. Some buffers, such as sodium fluoride (NaF), present the advantage of not absorbing the range of wavelengths used for dichroic spectra records. In the selected example, we used 50 mM NaF buffer at pH 7.5.

5.1.3 Protocol

1. The purified OBP is placed in semipermeable membrane dialysis tubing with a 10 kDa cutoff. The dialysis tubing was sealed and placed into a 2 L container containing buffer (e.g., 50 mM NaF buffer, pH 7.5). After 2 h, the dialysis tubing was placed into a second 2 L container

containing a new preparation of the same buffer. After 2h, the OBP was removed from the dialysis tubing and placed into a 1.5mL tube.

2. The 1.5mL tube containing the OBP is centrifuged for 20min at 20,000 *g*. The supernatant is placed in a new clean tube, and the $OD_{280\ nm}$ is measured. Using the extinction molar coefficient, the OBP concentration is calculated and adjusted to 40μM.
3. The nitrogen gas is turned on with a flow rate of 3L/min. Then, the light source of the CD spectrophotometer is turned on, followed by the computer.
4. The Peltier temperature control is set at the desired temperature (e.g., 20°C)
5. A demountable 0.1mm CD cuvette is filled with 50μL buffer (from the second 2-L container used for dialysis), and the spectrum corresponding to the baseline is acquired from 240 to 180nm using a scan speed of 50nm/min and a data pitch of 0.5nm with 3scans.
6. The demountable 0.1mm CD cuvette is filled with 50μL of the solution containing the OBP, and the spectrum is acquired from 240 to 180nm using a scan speed of 50nm/min and a data pitch of 0.5nm. The OBP concentration is then adjusted by trial to observe a spectrum with the highest intensity and the lowest noise (spikes on the spectrum). If the intensity of the spectrum is too low, the protein concentration can be increased or a 1mm CD cuvette can be used. If the noise is too high, the protein can be diluted, or a 0.01mm CD cuvette can be used.
7. When the best concentration is determined, spectra are acquired from 240 to 180nm with 15 scans using a scan speed of 50nm/min and a data pitch of 0.5nm. If the intensity of the spectra is low and consequently the resolution is low, the scan numbers and the scan speed can be increased to improve the resolution.
8. The data are analyzed by subtracting the baseline, correcting the baseline (adjusting to zero the highest wavelengths) and adapting the units as desired, usually molar ellipticity, after normalization to the protein concentration, as shown in Fig. 3 for pOBP1.

5.2 Isothermal titration calorimetry

Isothermal titration calorimetry (ITC) is a sensitive, accurate and label-free technique that can directly characterize the thermodynamics of ligand-protein interactions. ITC has been successfully used to characterize the binding properties of vertebrate OBPs (Barou et al., 2015;

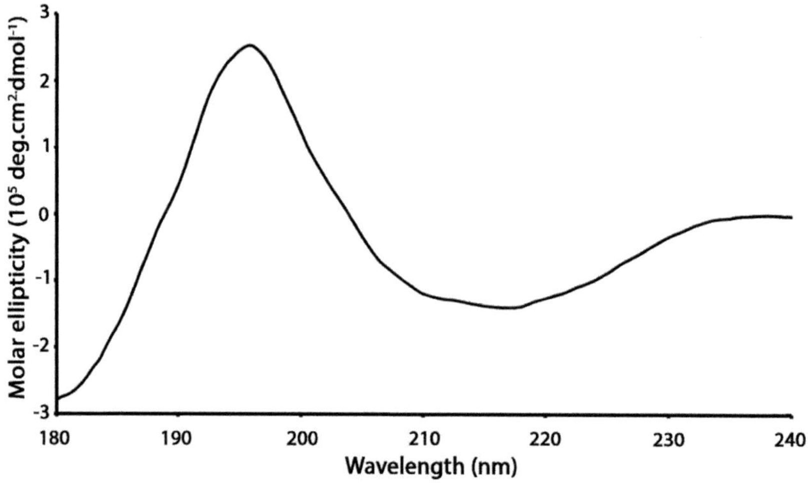

Fig. 3 Far-UV circular dichroism spectrum of pOBP1. The spectrum was recorded using a JASCO J-815 spectrophotometer equipped with a Peltier temperature control set at 20 °C. The spectra were recorded in NaF 50 mM buffer at pH 7.5 with pOBP1 at 40 μM in a 0.1 mm cuvette. CD measurements were normalized to protein concentration and presented as molar ellipticity in deg.cm^2.dmol^{-1} using Jasco Spectra Analysis software.

Borysik, Briand, Taylor, & Scott, 2010; Burova et al., 1999; Nespoulous et al., 2004; Yabuki, Scott, Briand, & Taylor, 2011). This method directly measures the heat absorbed or generated during the binding process between OBPs and odorants, allowing the determination of the binding stoichiometry. The ITC instrument is made of three essential parts: a sample cell loaded with a solution of the protein of interest, a reference cell filled with the control buffer used for protein solubilization, and an injection syringe loaded with ligand solution. In practice, the syringe injects precise amounts of the ligand into the sample cell, leading to the formation of a certain amount of protein/ligand complex, which is accompanied by the generation (exothermic reaction) or absorption (endothermic reaction) of heat within the sample cell. The feedback system of the ITC instrument adjusts the thermal power applied to compensate for the temperature imbalance between the reference and the sample cells. The recorded signal (amount of power in μcal/s) shows a peak at each injection. Progressively, the titrated solution becomes saturated, and the heat signal decreases until reaching the baseline level. Integration of the peak areas by Origin software allows determination of the following thermodynamic parameters: enthalpy (ΔH), entropy (ΔS), affinity constant (K_a) and stoichiometry (n).

5.2.1 Equipment

- VP-ITC microcalorimeter (e.g., Microcal LLC or equivalent)
- ThermoVac station for degassing samples (e.g., Microcal LLC or equivalent)
- Origin Scientific Graphing and Analysis Software (OriginLab)
- Two 2.5 ml long-needle syringes (e.g., Hamilton SYN161714)
- 5 mL test tubes, plastic vials with caps for ThermoVac (MALVERN PANALYTICAL INC)

5.2.2 Chemicals

- Solution to titrate: purified OBP (20–30 μM)
- Titrant solution: odorant compound (e.g., 2-isobutyl-3-methoxypyrazine (IBMP), 250 μM)
- Methanol HPLC grade (50 mL)
- Phosphate buffer (200 mL) used for OBP dialysis during the final purification step (50 mM K_2HPO_4 pH 7.4, prepared in HPLC water)
- Water HPLC grade (Carlo Erba Reagents, Water RS For HPLC PLUS) (50 mL)
- 20% solution (50 mL) of Decon® 90 (Fisher Scientific)

5.2.3 Protocol

This protocol is adapted for the OBP-odorant binding reaction with VP-ITC. The time required for a complete ITC measurement at 25 °C is approximately 2 h.

5.2.3.1 Select concentration of reactants

The heat variation associated with the binding reaction depends on the concentration of interacting molecules and must be in the calorimetric determination range of the ITC instrument. In practice, considering stoichiometry (1:1) and the apparent K_D values in the micromolar range for OBP-odorant interactions, we recommend using a ligand concentration of 250 μM, which is 10 times more concentrated than the OBP solution (25 μM), to guarantee a thermogram with a sigmoidal inflection point that finally reaches the baseline point. If ΔH is too low, the concentration of both reactants or the amount of the ligand should be increased. However, if the isotherm curve is too high, the ligand concentration could be decreased. It is also possible to shift the ITC measurement temperature by 5 to 10 degrees to obtain a better signal.

5.2.3.2 Prepare samples

During all the steps, the OBP sample should be protected from odorant exposure by working under a chemical fume hood as much as possible. Rinse all the material, especially plastic tubes, with methanol and HPLC water and let them ventilate under the fume hood until they are dry.

Since ITC measurements are very sensitive to the buffer solution, concentration, pH, ionic strength and additives, it is critical to use the same buffer for OBP and odorant preparation. Thus, we recommend to use the OBP dialysis buffer.

The minimum OBP solution required is 2.1 mL to properly fill the 1.4 mL sample cell, without producing air bubbles. The same should be applied for the syringe, which holds 240 μL but requires 400 μL of ligand solution.

Degas the OBP solution for 5 min with the ThermoVac system, but do not degas the odorant solutions.

5.2.3.3 Load solutions

Sample cell: rinse the sample cell two times with dialysis buffer using a 2.5 mL long-needle syringe and then remove all the liquid. Next, load 2.1 mL of OBP solution. During loading, take care to prevent air bubbles in the cell. Finally, pipet the overflow from the cell.

Injection syringe: rinse with dialysis buffer and then fill the syringe with ligand solution according to the manufacturer's instructions. After closing the fill port, perform a purge-refill cycle to avoid air bubbles inside the syringe. Carefully wipe the outside of the needle with lint-free paper and then gently insert the syringe into the sample cell.

5.2.3.4 Set instrument settings and start measurements

Set the VP-ITC microcalorimeter to perform twenty-six injections of 10 μL, with the injection duration set to 20 s (except for the first injection of only 3 μL with a duration of 6 s) with 210 s intervals between injections, and the filter period to 2 s. Set the measurement temperature to 25 °C. Set the reference power to 10 μcal/s and the initial delay between the start and first injection to 60 s. A stirring speed of 300 rpm is selected to ensure good mixing. Select high for feedback mode and fast equilibration for ITC equilibrium. After baseline equilibration, the injection sequence starts, and the titration is performed until saturation is observed.

5.2.3.5 Clean the ITC

Once the titration is finished, the injection syringe should be washed with water (20 mL), then methanol (20 mL) and water again. Let it dry or rinse it with buffer to start a new measurement. The sample cell should be emptied manually with the long-needle syringe. Using the microcalorimeter ThermoVac system, wash the sample cell with water (100 mL) and rinse it with methanol (50 mL). Then, perform a final wash with water (50 mL), and keep the cell in water until the next ITC measurement.

5.2.3.6 Analysis of the data

Data analysis is performed using the sigmoidal fitting function with OriginLab software. First, adjust the baseline and integrate the peaks. Remove the first data point from the 3 μL injection. Subtract the heat mean of the last three injections to remove the heat dilution of the ligand in the buffer. To allow calculation, enter concentration values for the OBP and odorant. Choose the binding model fitting "One set of sites". Do as many iterations as necessary until a stable "chi ˆ2" value is obtained and then select the "Done" box. The thermodynamic parameters, including n (number of binding sites), K_a, ΔH and ΔS should appear. A typical titration calorimetric thermogram obtained with pOBP1 and IBMP clearly shows that IBPM binding is saturable with approximately one site per monomer, as indicated by the abscissa of the inflection point (Fig. 4). Fig. 5 shows titration calorimetric thermograms of rOBP3 purified without or with solvent cleaning. In the absence of solvent cleaning (Fig. 5A), the binding stoichiometry (n-value) is 0.25, meaning that the rOBP3 binding pocket is partially occupied by an endogenous ligand. In contrast, after the cleaning procedure described here, the stoichiometry should be as expected (n value = 0.99), indicating that the purification strategy has efficiently removed all ligand (Fig. 5B).

5.3 Fluorescent binding assay

The competitive fluorescent binding assay is the method of choice to study OBP binding properties because this technique is simple, rapid and requires small amounts of OBP proteins (Pelosi et al., 2018).

The principle of this method is based on the ability of OBP to bind a fluorescent probe. The affinity of nonfluorescent odorants can be evaluated by their ability to compete with the fluorescent probe. A variety of different probes have been used, but 1-AMA (1-aminoanthracene) and 1-NPN (*N*-phenyl naphthylamine) are the most commonly used (please refer to

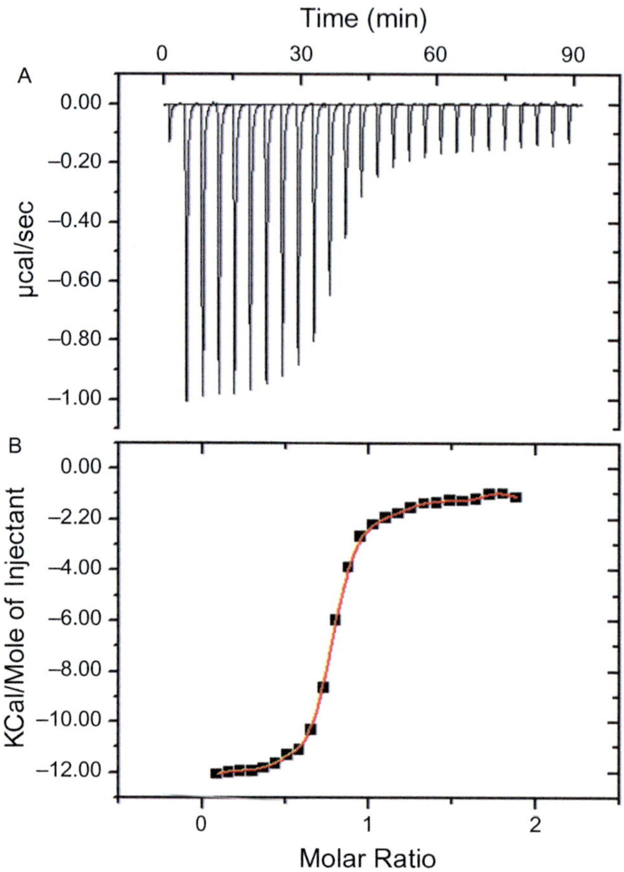

Fig. 4 Isothermal titration calorimetry thermogram of IBMP binding to pOBP1. (A) Raw data of representative ITC experiments; upper curve, control with buffer in the cell and odorant in the syringe; lower curve, protein solution in the cell and odorant in the syringe. (B) Fitted binding isotherms corrected for heat of ligand dilution. pOBP1 concentration, 22 µM in 50 mM phosphate buffer at pH 7.4. Controlled temperature, 25 °C. The negative peaks indicate that the interaction was exothermic ($\Delta H < 0$). The fitted binding stoichiometry (IBMP:pOBP1 ratio) was 0.91 ± 0.02, close to the value of 1.0 ± 0.0.

Chapter 10. Ligand binding assays of this volume). We present here typical results of binding of 1-AMA onto pOBP1, which was completed by IBMP with high affinity. To determine the binding affinities of 1-AMA, titration experiments were performed. Titration of pOBP1 with 1-AMA was saturable, leading to a K_D value of 1.02 ± 0.14 µM (Fig. 6A). As shown in Fig. 6B, the fluorescence intensity of the pOBP1/1-AMA complex was severely reduced by the addition of increasing concentrations of IBMP with a K_D value of 1.44 ± 0.15 µM.

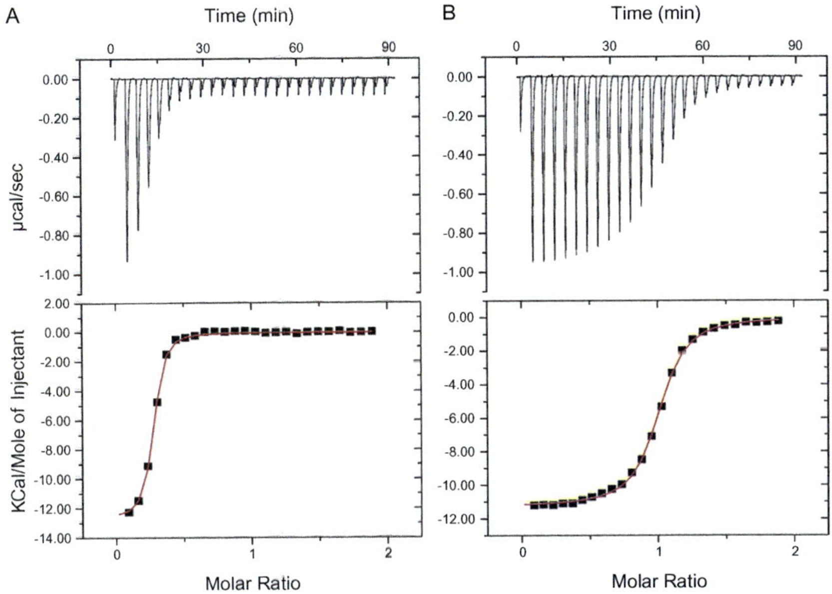

Fig. 5 Isothermal titration calorimetry thermogram of rOBP3 with β-ionone (A) without or (B) with ethanol and acetonitrile washing steps. The experiment was performed in 50 mM phosphate buffer, pH 7.4, at 25 °C. The concentrations of the reactants were 25 μM rOBP3 and 250 mM β-ionone. The fitted binding stoichiometries (β-ionone: rOBP3 ratio) were 0.25 (before washing) and 0.99 (after washing), close to the value of 1.0 ± 0.0. The parameters obtained for the "One set of Sites" binding model were as follows: for (A): $n = 0.25$, $Ka = 6.04 \times 10^6\,M^{-1}$, $\Delta H = -1.27 \times 10^4$ cal/mol, $\Delta S = -11.7$ cal/mol/deg., and for (B): $n = 0.99$, $Ka = 3.68 \times 10^6\,M^{-1}$, $\Delta H = -1.13 \times 10^4$ cal/mol, $\Delta S = -7.7$ cal/mol/deg.

5.3.1 Equipment

- Spectrofluorometer equipped with electronic magnetic stirrer and Peltier temperature control (e.g., Cary Eclipse, Agilent Technologies)
- 4 X Quartz cells for magnetic stirrers with stopper (Hellma, Fisher Scientific)
- PTFE micro stirring bar, 6 × 3 mm (e.g., Fisher Scientific)
- 2 mL glass vials (Wheaton)
- Hamilton 0.5 μL syringe

5.3.2 Chemicals

- 1-Aminoanthracene fluorescent probe (Sigma-Aldrich)
- 50 mL of Hellmanex® III diluted at 2% (Sigma-Aldrich)
- 50 mL of methanol RS for 2.5-L LC/MS (Carlo Erba Reagents)
- 50 mL of Water RS for 2.5-L LC/MS (Carlo Erba Reagents)

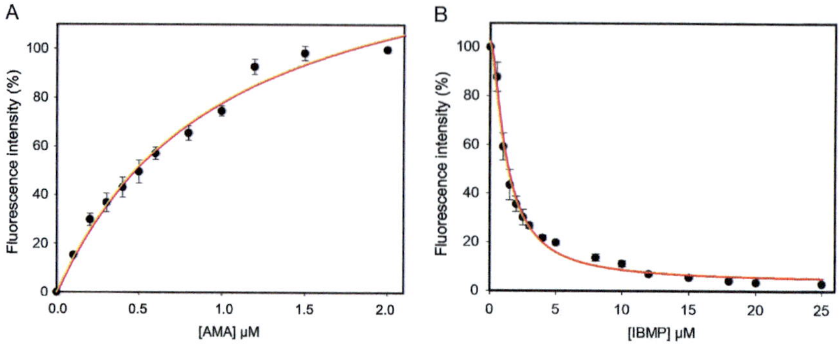

Fig. 6 Binding of ligands to pOBP1 assessed by competitive fluorescence assay. (A) Titration curve of pOBP1 with 1-AMA. Fluorescence emission spectra were recorded at 25 °C at 485 nm for 1-AMA with a 255 nm excitation wavelength. The protein concentration was 2 µM. (B) Competitive binding assays of 1-AMA with 2-isobutyl-3-methoxypyrazine (IBMP) leading to a half-maximal value (IC_{50}) of 1.26 ± 0.09 µM.

- 250 mL of 50 mM sodium phosphate buffer pH 7.5 prepared with LC/MS water
- Odorant, 2-isobutyl-3-methoxypyrazine (Sigma-Aldrich)

5.3.3 Protocol

The fluorimeter cuvettes should be washed with 100% methanol and ultrapure grade water and dried at ambient temperature. We recommend performing these manipulations under a fume hood.

1. Preparation of the 1-AMA stock solution

Prepare an amber glass vial containing 1 mL of 100 mM 1-AMA prepared in 100% MeOH. Prepare the 10 × serial dilutions: 10 mM 1-AMA (100% MeOH LC/MS), 1 mM 1-AMA, and 0.1 mM 1-AMA (50% MeOH LC/MS and 50% 50 mM sodium phosphate buffer pH 7.5 *v*/v). Store at −20 °C and use on ice and protected from light.

2. Preparation of odorant stock solution

  1. Prepare a glass vial of 1 mL of 100 mM odorant (IBMP) in 100% MeOH.

  2. Prepare serial dilutions of 10 × 10 mM odorant (100% MeOH LC/MS), 1 mM odorant (100% MeOH LC/MS) and 0.1 mM odorant (50% MeOH LC/MS and 50% buffer v/v).

  3. Store vials at −20 °C and use them on ice.

3. Cleaning of fluorometer cells and beaker

  1. Store fluorimeter cells in Hellmanex® III solution diluted at 2% (Sigma-Aldrich).

2. Remove the Hellmanex® solution by rinsing twice with LC/MS water.
3. Add MeOH LC/MS (100%) to the cells and beaker to eliminate this last reagent.
4. Rinse fluorometer cells with LC/MS water and then rinse with phosphate buffer.

4. Titration of pOBP1 with 1-AMA
In a clean washed beaker, prepare enough pOBP1 solution at 2 μM in 50 mM sodium phosphate buffer for four 1-mL cells. Use three cells for measurement of the fluorescence, and one as a control (buffer only). Add successive 1-μL aliquots of 1-aminoanthracene (1-AMA) probe to 1 mL of 2 μM pOBP1 solution. After a 3 min incubation time, record the spectra at 25 °C using an excitation wavelength of 255 nm. Fluorescence spectra should be recorded from 400 to 550 nm.

Calculation is performed from a plot of fluorescence intensity vs. probe concentration obtained with a nonlinear regression method using SigmaPlot software.

5. Competition of 1-AMA with IBMP
Perform the competitive binding assays with 2 μM pOBP1 solution using 2 μM 1-AMA. In a clean washed beaker, prepare enough pOBP1/1-AMA complex solution in 50 mM sodium phosphate buffer for four 1-mL cells. Use three cells for measurement of the fluorescence, and one as a control (MeOH/buffer solution in absence of odorant). The methanol concentration in the binding assay never exceeded 0.2% v/v, leading to a maximum relative fluorescence decay of 10%. Add successive 0.1 μL IBMP aliquots to 1 mL of pOBP1 solution using a Hamilton syringe. After the addition of odorant, the fluorescence of the pOBP1:1-AMA complex should be recorded after stabilization of the signal (2–3 min). The concentration of odorant causing fluorescence decay to half-maximal intensity can be taken as the IC_{50} value. Calculation of the IC_{50} value is performed using SigmaPlot software. The apparent K_D value for IBMP was calculated as $K_D = [IC_{50}]/(1 + [L]/K_{D\ 1\text{-}AMA})$ with [L] being the free fluorophore concentration and K_D being the OBP-fluorophore complex dissociation constant.

6. Summary

Investigators who seek to express recombinant vertebrate OBPs may encounter a number of specific difficulties. This chapter describes the difficulties and attempts to explain techniques and approaches to overcome them. In general, bacteria are able to produce large quantities of his-tagged vertebrate OBPs that are easy to purify using IMAC. Importantly, careful attention to avoid plastic and odorant contamination will minimize problems and render structure-function relationship studies and structural analyses feasible.

Acknowledgments

This work was supported by a grant from the Agence Nationale de la Recherche ANR-18-CE42-0012-01.

References

Barou, E., Sigoillot, M., Bouvet, M., Briand, L., & Meunier-Prest, R. (2015). Electrochemical detection of the 2-isobutyl-3-methoxypyrazine model odorant based on odorant-binding proteins: The proof of concept. *Bioelectrochemistry*, *101*, 28–34.

Bianchi, F. A.-O., Flisi, S., Careri, M., Riboni, N., Resimini, S., Sala, A., et al. (2019). Vertebrate odorant binding proteins as antimicrobial humoral components of innate immunity for pathogenic microorganisms. *PLoS One*, *14*(3), e0213545.

Borysik, A. J., Briand, L., Taylor, A. J., & Scott, D. J. (2010). Rapid odorant release in mammalian odour binding proteins facilitates their temporal coupling to odorant signals. *Journal of Molecular Biology*, *404*(3), 372–380.

Briand, L., Eloit, C., Nespoulous, C., Bézirard, V., Huet, J.-C., Henry, C., et al. (2002). Evidence of an odorant-binding protein in the human olfactory mucus: Location, structural characterization, and odorant-binding properties. *Biochemistry*, *41*(23), 7241–7252.

Briand, L., Huet, J.-C., Perez, V., Lenoir, G., Nespoulous, C., Boucher, Y., et al. (2000). Odorant and pheromone binding by aphrodisin, a hamster aphrodisiac protein. *FEBS Letters*, *476*, 179–185.

Briand, L., Nespoulous, C., Perez, V., Rémy, J.-J., Huet, J.-C., & Pernollet, J.-C. (2000). Ligand-binding properties and structural characterization of a novel rat odorant-binding protein variant. *European Journal of Biochemistry*, *267*(10), 3079–3089.

Brimau, F., Cornard, J.-P., Le Danvic, C., Lagant, P., Vergoten, G., Grebert, D., et al. (2010). Binding specificity of recombinant odorant-binding protein isoforms is driven by phosphorylation. *Journal of Chemical Ecology*, *36*(8), 801–813.

Burova, T. V., Choiset, Y., Jankowski, C. K., & Haertlé, T. (1999). Conformational stability and binding properties of porcine odorant binding protein. *Biochemistry*, *38*(45), 15043–15051.

Ferrari, E., Lodi, T., Sorbi, R. T., Tirindelli, R., Cavaggioni, A., & Spisni, A. (1997). Expression of a lipocalin in *Pichia pastoris*: secretion, purification and binding activity of a recombinant mouse major urinary protein. *FEBS Letters*, *401*(1), 73–77.

Guiraudie-Capraz, G., Clot-Faybesse, O., Pageat, P., Malosse, C., Cain, A.-H., Ronin, C., et al. (2005). Heterologous expression of piglet odorant-binding protein in *Pichia pastoris*: A comparative structural and functional characterization with native forms. *Journal of Biotechnology*, *117*(1), 11–19.

Heydel, J.-M., Coelho, A., Thiebaud, N., Legendre, A., Bon, A.-M. L., Faure, P., et al. (2013). Odorant-binding proteins and xenobiotic metabolizing enzymes: Implications in olfactory perireceptor events. *The Anatomical Record*, *296*(9), 1333–1345.

Hurot, C., Brenet, S., Buhot, A., Barou, E., Belloir, C., Briand, L., et al. (2019). Highly sensitive olfactory biosensors for the detection of volatile organic compounds by surface plasmon resonance imaging. *Biosensors and Bioelectronics*, *123*, 230–236.

Larisika, M., Kotlowski, C., Steininger, C., Mastrogiacomo, R., Pelosi, P., Schütz, S., et al. (2015). Electronic olfactory sensor based on *A. mellifera* odorant-binding protein 14 on a reduced graphene oxide field-effect transistor. *Angewandte Chemie International Edition, 54*, 13245–13248.

Lartigue, A., Gruez, A., Briand, L., Blon, F., Bézirard, V., Walsh, M., et al. (2004). Sulfur single-wavelength anomalous diffraction crystal structure of a pheromone-binding protein from the honeybee *Apis mellifera* L. *The Journal of Biological Chemistry, 279*(6), 4459–4464.

Löbel, D., Jacob, M., Völkner, M., & Breer, H. (2002). Odorants of different chemical classes interact with distinct odorant binding protein subtypes. *Chemical Senses, 27*(1), 39–44.

Löbel, D., Marchese, S., Krieger, J., Pelosi, P., & Breer, H. (1998). Subtypes of odorant-binding proteins. *European Journal of Biochemistry, 254*(2), 318–324.

Löbel, D., Strotmann, J., Jacob, M., & Breer, H. (2001). Identification of a third rat odorant-binding protein (OBP3). *Chemical Senses, 26*(6), 673–680.

Lücke, C., Franzoni, L., Abbate, F., Löhr, F., Ferrari, E., Sorbi, R. T., et al. (1999). Solution structure of a recombinant mouse major urinary protein. *European Journal of Biochemistry, 266*, 1210–1218.

Mulla, M. Y., Tuccori, E., Magliulo, M., Lattanzi, G., Palazzo, G., Persaud, K., et al. (2015). Capacitance-modulated transistor detects odorant binding protein chiral interactions. *Nature Communications, 6*, 6010.

Nagnan-Le Meillour, P., Joly, A., Le Danvic, C., Marie, A., Zirah, S., & Cornard, J.-P. (2019). Binding specificity of native odorant-binding protein isoforms is driven by phosphorylation and O-N-acetylglucosaminylation in the pig *Sus scrofa*. *Frontiers in Endocrinology, 9*(816).

Nagnan-Le Meillour, P., Vercoutter-Edouart, A.-S., Hilliou, F., Le Danvic, C., & Lévy, F. (2014). Proteomic analysis of pig (*Sus scrofa*) olfactory soluble proteome reveals O-linked-N-acetylglucosaminylation of secreted odorant-binding proteins. *Frontiers in Endocrinology, 5*, 202.

Nespoulous, C., Briand, L., Delage, M.-M., Tran, V., & Pernollet, J.-C. (2004). Odorant binding and conformational changes of a rat odorant-binding protein. *Chemical Senses, 29*(3), 189–198.

Pelosi, P. (2001). The role of perireceptor events in vertebrate olfaction. *Cellular and Molecular Life Sciences, 58*(4), 503–509.

Pelosi, P., Mastrogiacomo, R., Iovinella, I., Tuccori, E., & Persaud, K. C. (2014). Structure and biotechnological applications of odorant-binding proteins. *Applied Microbiology and Biotechnology, 98*(1), 61–70.

Pelosi, P., Zhu, J., & Knoll, W. (2018). From radioactive ligands to biosensors: Binding methods with olfactory proteins. *Applied Microbiology and Biotechnology, 102*, 8213–8227.

Portman, K. L., Long, J., Carr, S., Briand, L., Winzor, D. J., Sean Searle, M., et al. (2014). Enthalpy/entropy compensation effects from cavity desolvation underpin broad ligand binding selectivity for rat odorant binding protein 3. *Biochemistry, 53*(14), 2371–2379.

Ramoni, R., Spinelli, S., Grolli, S., Conti, V., Merli, E., Cambillau, C., et al. (2008). Deswapping bovine odorant binding protein. *Biochimica et Biophysica Acta, Proteins and Proteomics, 1784*(4), 651–657.

Ramoni, R., Vincent, F., Grolli, S., Conti, V., Malosse, C., Boyer, F.-D., et al. (2001). The insect attractant 1-Octen-3-ol is the natural ligand of bovine odorant-binding protein. *Journal of Biological Chemistry, 276*(10), 7150–7155.

Scaloni, A., Paolini, S., Brandazza, A., Fantacci, M., Bottiglieri, C., Marchese, S., et al. (2001). Purification, cloning and characterisation of odorant- and pheromone-binding proteins from pig nasal epithelium. *Cellular and Molecular Life Sciences, 58*(5), 823–834.

Tcatchoff, L., Nespoulous, C., Pernollet, J.-C., & Briand, L. (2006). A single lysyl residue defines the binding specificity of a human odorant-binding protein for aldehydes. *FEBS Letters, 580*(8), 2102–2108.

Tegoni, M., Pelosi, P., Vincent, F., Spinelli, S., Campanacci, V., Grolli, S., et al. (2000). Mammalian odorant binding proteins. *Biochimica et Biophysica Acta, Protein Structure and Molecular Enzymology, 1482*(1), 229–240.

Yabuki, M., Scott, D. J., Briand, L., & Taylor, A. J. (2011). Dynamics of odorant binding to thin aqueous films of rat-OBP3. *Chemical Senses, 36*(7), 659–671.

Zaremska, V., Tan, J., Lim, S., Knoll, W., & Pelosi, P. (2020). Isoleucine residues determine chiral discrimination of odorant-binding protein. *Chemistry: A European Journal,* (in press).

Zhu, J., Arena, S., Spinelli, S., Liu, D., Zhang, G., Wei, R., et al. (2017). Reverse chemical ecology: Olfactory proteins from the giant panda and their interactions with putative pheromones and bamboo volatiles. *Proceedings of the National Academy of Sciences of the United States of America, 114*(46), E9802–E9810.

CHAPTER SEVEN

Structure of odorant binding proteins and chemosensory proteins determined by X-ray crystallography

Anaïs Gaubert, Béatrice Amigues, Silvia Spinelli, Christian Cambillau*

Architecture et Fonction des Macromolécules Biologiques, Centre National de la Recherche Scientifique (CNRS) & Aix-Marseille Université, Marseille, France

*Corresponding author: e-mail addresses: cambillau@afmb.univ-mrs.fr; ccambillau@gmail.com

Contents

Abstract

OBPs and CSPs are small soluble proteins used by organisms as shuttle to transport odorant molecules between air and the membrane-embedded receptors. Deciphering the interactions of these proteins with their ligands at a molecular level may give clue on the function and specificity of the olfactory chain. To reach this goal, protein

Methods in Enzymology, Volume 642
ISSN 0076-6879
https://doi.org/10.1016/bs.mie.2020.04.070

crystallography is very helpful with more than hundred entries available in the protein data bank (PDB). In this chapter, we present the peculiarities of OBPs and CSPs concerning their crystallization and 3D structure determination by X-ray diffraction.

1. Introduction

Odorant Binding Proteins (OBPs) in insects and mammals share a same function: extracting protein at the air-lymph interface and carrying them to the olfactory receptors. In insects, OBPs are located in the antennal sensillary lymph, whereas they are located in the nasal mucus of mammals. Both are small molecules of ~10–20 kDa. To note, OBPs that bind insects' specific pheromones are called PBPs (Pheromone Binding Protein), and OBPs that bind general odorant molecules are sometimes called GOPBs (General OBPs). A special group of insect proteins, the Chemosensory Proteins (CSPs), have been shown to bind chemicals that play also a role in insect communication or interactions.

Despite their similar function, mammalian and insect OBPs possess totally different three-dimensional structures. Classical insects' OBPs assemble a core of six α-helices, in the center of which is located the binding cavity (Fig. 1A and B). In classical OBPs, these helices are held together by three disulfide bridges (Horst et al., 2001; Sandler et al., 2000; Tegoni, Campanacci, & Cambillau, 2004). Large variations have been observed diverging from this simple classical model and including additional α-helices, more or less disulfide bridges (Spinelli et al., 2012) and extra domains (Lagarde, Spinelli, Qiao, et al., 2011). Chemosensory Proteins (CSPs) assemble also six α-helices, as for insects' OBPs, but arranged in a totally different topology with the binding cavity located within the α-helices core (Fig. 1C) (Campanacci et al., 2003; Lartigue et al., 2002). More structural knowledge on insects' OBPs became available in recent years with more than 100 structures available today in the Protein Data Bank (PDB: https://www.rcsb.org/), and several recent reviews. In contrast with insects' OBPs, mammalian OBPs are a sub-class of lipocalins formed of an eight β-stranded β-barrel linked to an α-helix by a turn, and followed by a ninth β-strand antiparallel to the first strand (Fig. 1D). Their binding cavity is located within the β-barrel (Bianchet et al., 1996; Tegoni, Ramoni, Bignetti, Spinelli, & Cambillau, 1996; Zhu et al., 2017).

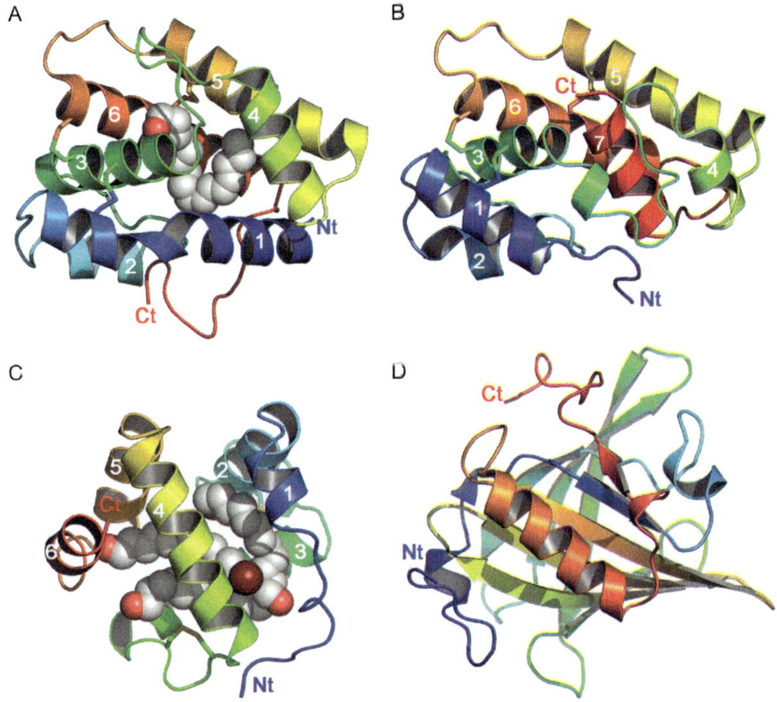

Fig. 1 The crystal structures of OBPs and CSPs. (A) Ribbon view representation of the structure of *Bombyx mori* PBP in complex with bombykol (1DQE) determined at pH 8.5 (Sandler, Nikonova, Leal, & Clardy, 2000). Helices are numbered 1–6. (B) Ribbon view representation of the structure of *Bombyx mori* apo-PBP (1LS8) (Horst et al., 2001; Lautenschlager, Leal, & Clardy, 2007). Helices are numbered 1– 7. (C) Ribbon view representation of the structure of *Mamestra brassicae* CSP A6 in complex with three molecules of bromo-decanol (1N8U) (Campanacci et al., 2003). Helices are numbered 1–6. (D) Ribbon view representation of the structure of the giant Panda *Ailuropoda melanoleuca* (5NGH), formed of a β-barrel and a unique helix (Zhu et al., 2017).

2. Crystallographic issues relevant to OBPs and CSPs structural studies

Three methods can be used to determine a protein structure at atomic resolution: X-ray diffraction, NMR and cryoEM (cryo-electron microscopy). This last method can be ruled out for OBPs and CSPs structural studies as it can be applied (to date) only to molecules larger than ~30–50 kDa. In contrast, NMR can be applied mostly to molecules smaller than ~30–50 kDa. X-ray diffraction has no size limitation as small organic molecules up to viruses can be used for structural studies with this method. To note, atomic resolution cryoEM is a method of choice to study olfactory receptors as exemplified by the recently published ORCO structure (Butterwick et al., 2018).

2.1 Crystallization of proteins and proteins–ligand complexes

A difficulty in X-ray diffraction experiments is the necessity to obtain suitable proteins crystals, *i.e.*, an ordered 3D lattice of proteins able to diffract X-rays. The main method to obtain crystals is to mix a solution of a highly purified protein with a "precipitating" agent solution that clusters water molecules. Practically, a solution of pure and concentrated protein is mixed with a solution of precipitating agent and a drop of the mixture is placed in a closed volume containing a reservoir of the precipitant solution used for the mixture. Since mixing the protein solution and the precipitant solution results in a dilution of the precipitant, a chemical gradient will be established between the drop and the reservoir, leading to water vapor diffusion from the drop to the reservoir (Fig. 2A). As a result, the concentration of protein and precipitant in the drop will increase slowly inducing formation of crystals. A large variety of precipitating agents is available, that can be classified into neutral polymers, salts, and alcohols, by decreasing order of popularity (McPherson, 2017). The most used neutral polymers are polyethylene glycols (PEG), whose general structure is $HO\text{-}(CH_2\text{-}CH_2\text{-}O\text{-})_n\text{-}OH$. The molecular weights of PEGs used for crystallization vary between ~350 and 20,000 Da, hence with 8–500 repeats. The great advantage of PEGs is that they interfere little with the proteins and therefore do not disturb their local structure. The favored salt precipitants are ammonium sulfate or sodium chloride (SO_4^{2-} $2NH_4^+$ and NaCl). They are very efficient precipitating agents, but can interfere with the proteins, providing salt bridges between charged amino acids. Finally, alcohols such as methylpentanediol (MPD), ethanol, isopropanol, are less popular due to their volatility that makes crystallization conditions more difficult to reproduce, with the exception of MPD. Considerations on the protein structure based on sequence analysis cannot be used as guides for choosing the most favorable precipitating agent. Instead, the solution is a random approach trying a variety of "ready to use" crystallization kits commercially available (McPherson & Cudney, 2014; McPherson & Gavira, 2014).

Another important crystallization parameter is the pH. With some enzymes (*e.g.*, serine proteases) the pH has to be chosen within a certain

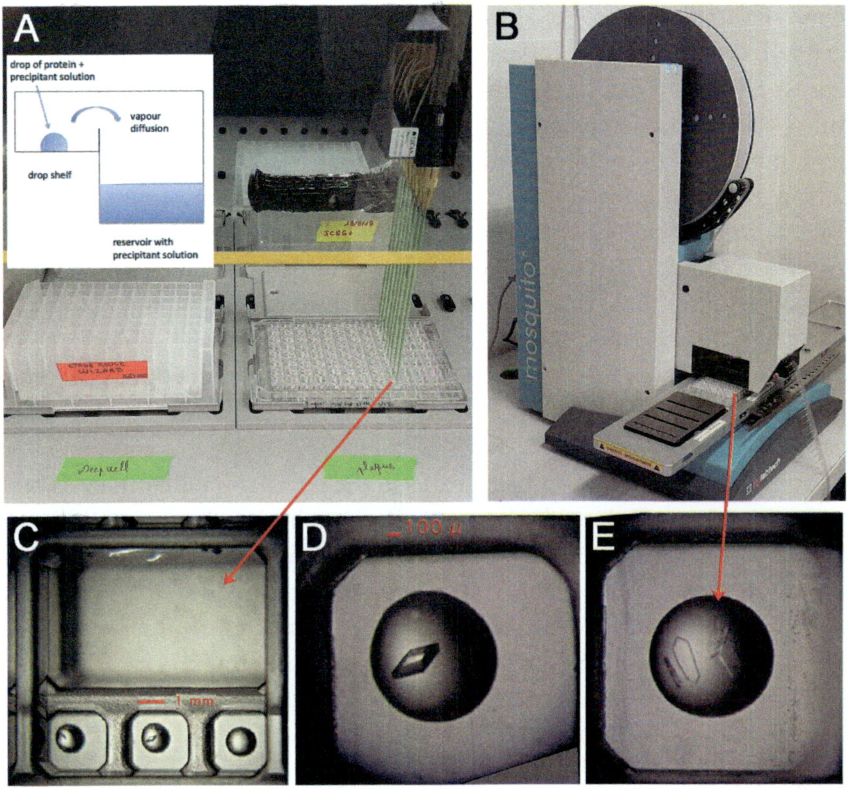

Fig. 2 Crystallization of olfactory proteins. (A) The crystallization plates and precipitant dispensing in the reservoir; inset: principle of vapor diffusion in crystallization plates. (B) a nono drops dispensing robot (Mosquito, TTP Labtech). (C) The reservoir and the three drops shelves setting in a Greiner crystallization plate. (D and E) Crystals in nanodrops.

range to keep the protein in its active conformation. With OBPs, pH can be tremendously important as, in some cases, it can modulate the structure and hence their binding capacity. The effect of pH on binding has been observed in the case of two well studied OBPs. The structure of *Bombyx mori* OBP1, in complex with the *Bombyx* pheromone bombykol was determined in 2000 by Clardy's group using X-ray crystallography (Sandler et al., 2000). The crystals were obtained with PEG 20,000 as precipitating agent, at pH 8.2. They revealed the presence of bombykol inside the hydrophobic cavity formed by the six PBP α-helices. Later on, Wüthrich's group solved the NMR structure of the same PBP at pH 4.5, revealing that at acidic pH the C-terminal segment forms a seventh α-helix that fits inside the ligand cavity (Horst et al., 2001). However, the same structure could also be obtained later by X-ray diffraction at pH 7.5 in the absence of ligand (Lautenschlager, Leal, & Clardy, 2005). The structure of *Apis melifera* PBP is also pH dependent as the C-terminus structure can be tuned both by pH and by the presence of ligand, leading to either a monomeric or a domain swapped form (Pesenti et al., 2008, 2009). Some other OBPs or PBPs do not exhibit such pH sensitivity, as revealed by the shorter *Leucophaea maderae* PBP structure (Lartigue et al., 2004). These observations tell us two things: first, that the C-terminal segment plays an important role in some OBPs, and should therefore be kept free of any tag. Second, that a large range of pH (*e.g.*, 4.5–9) should be used in crystallization assays in order to eventually capture different crystal forms of different structures (Pesenti et al., 2009).

As the concentration of OBPs/PBPs in the sensillar lymph can be as high as 10 mM (Vogt, Kohne, Dubnau, & Prestwich, 1989), the solubility of expressed proteins has to be consequently very high. Hence, high concentrations of proteins (20–40 mg/L or more) are often required in crystallization assays. However, considering the small quantities of protein solution dispensed by nano-pipetting crystallization robots, hundreds of different conditions can be used in OBPs crystallization assays, allowing for testing wide ranges of precipitant, pH, concentration and additives.

Practically an OBP/PBP/CSP solution is prepared at a protein concentration as high as possible (20–40 mg/L) and pH is adjusted with a low concentration of neutral buffer, *e.g.*, ~10 mM, in order to have the final pH determined by the precipitant solution. Three drops of protein of 100, 200 or 300 nL are dispensed by the robot in the cups of the crystallization plate (McPherson & Gavira, 2014), and the operation is repeated in the 96 × 3 cups of the plate (Fig. 2B). A second step is the deposition of a fixed volume of 150 nL of precipitant over the protein, before sealing the plate with a plastic adhesive cover (Lartigue, Gruez, Briand, et al., 2003). The plate is then stored in shelves or in a crystal plates observation robot where the plates are observed or imaged every day. Several days or weeks can pass before crystals appear. If well formed, large enough, crystals appear (20– 200 μm), the next steps will be the OBP structure determination.

2.2 Structure determination

2.2.1 Data collection

Once crystals are obtained, they should be placed in a fiber loop and deep frozen immediately. Then they are brought to a synchrotron beam line, where data are collected at a certain X-ray wavelength (see paragraph 2–3).

Very often these X-ray beam lines are micro-focused, which enables the collection of data from very small crystals (10–50 μm). Data collection is fast (a few minutes), and in a first step, diffraction quality should be assessed by screening or collecting data sets from different crystals. Once data are collected, they are integrated and reduced by various programmes that provide a complete set of data and relevant statistics. Processing X-ray diffraction images may be conveniently divided into four major stages: (i) Spot finding and auto indexing. (ii) Parameter refinement. (iii) Integration (the actual measurement of spot intensities). (iv) Scaling and merging. However, since diffraction provides only the intensities of the diffracted beams, but not their phases, a further step is necessary to obtain these missing data.

2.2.2 Phasing and model building

Once a data set of a well-diffracting crystal form is obtained, the next step, phasing, can be prepared. Two methods are used to obtain the phases of a diffraction data set: *ab-initio* phasing and molecular replacement. The theoretical aspects of phasing will not be discussed here, but only the practical implications will be presented.

Molecular replacement uses a known protein structure to obtain the phases of a dataset for a protein of unknown structure adopting an appropriate set of programmes. Both structures should be close enough, generally within 1.5 Å root mean square deviation (r.m.s.d.), to ensure a positive outcome. As the second structure is not known, the r.m.s.d. is also unknown. A rule of the thumb for guessing the success of molecular replacement is that when the two protein sequences are more than ~30% identical, the method is likely to work. This is often the case with mammalian OBPs that, being lipocalins, share a rigid β-barrel. In contrast, insects OBPs, PBPs and CSPs being formed of an α-helix bundle, are very flexible and can adopt different conformations sufficiently divergent to hamper the success of molecular replacement. A striking example is that of the *Mamestra brassicae* CSP. The structure of the protein in complex with the ligand (bromo-dodecanol) could not be determined using molecular replacement with the apo-protein, despite a 100% identity, as a large conformational change was observed between apo- and bound forms (Campanacci et al., 2003). However, due to the speed and the easy use of molecular replacement programmes, this procedure should be tried, especially if several starting models with good identity with your target are available. For example, the structure of *Anopheles gambiae* OBP7 could be solved using that of *Anopheles gambiae* OBP1 (Lagarde, Spinelli, Tegoni, et al., 2011). If molecular replacement fails, you have to embark in an *ab-initio* phasing procedure.

Ab-initio phasing is much more demanding as compared to molecular replacement. In brief, phases can be obtained from an OBP derivative bearing anomalous scattering atoms. These atoms can be selenium, iodine, caesium, mercury, *etc.* (a complete list is available at http://skuld.bmsc.washington.edu/scatter/). Selenium can replace sulfur in methionine, then called Se-methionine (Se-Met). Procedures exist for expressing proteins in a medium supplemented with Se-Met (Doublie, 1997). A rule of the thumb is that one Se-Met is enough to phase 5 kDa of protein. Classically, for an OBP, three Se-Met may be sufficient. Once your Se-Met "labeled" OBP is obtained, you have to crystallize it again, in conditions that may be not totally identical to those of the native protein. Then, data collection should be repeated, but at a specific wavelength, that of the "absorption edge" of the selenium atoms in your protein. The whole procedure is well mastered in synchrotrons. Then specialized programmes will extract the phases from your data. This procedure has been used to obtain the first structure of an insect OBP (Sandler et al., 2000).

An alternative to Se-Met labeling is to diffuse "heavy" atoms, anomalous scatterers in your native crystals. These atoms could be mercury, platinum, tungsten (Spinelli et al., 2012) which exhibit an absorption edge in the synchrotron X-ray wavelength range, or Cs and I (or CsI) that necessitate a special procedure (Lagarde, Spinelli, Qiao, et al., 2011). A more rare, but sometimes efficient, technique of phasing relies on the weak scattering signal of sulfur (Lartigue et al., 2004). Finally, ligands containing a scattering atom can also be used. For example, the structure of the complex between *Mamestra brassicae* CSP and bromo-dodecanol was solved using the Br-K absorption edge (Campanacci et al., 2003).

Once the phases have been obtained using any of the above methods, the electron density map can be readily calculated (Perrakis, Morris, & Lamzin, 1999; Terwilliger et al., 2008). Depending on the "quality" of the phases, the map will contain more or less errors. Current programmes, however, can sketch the almost complete amino acid chain of a protein in a matter of hours. Once the initial track obtained, manual reconstruction at the display is performed using, *e.g.,* COOT (Emsley, Lohkamp, Scott, & Cowtan, 2010). This step is followed by protein refinement *e.g.,* REFMAC (Winn, Murshudov, & Papiz, 2003), Phenix (Janowski et al., 2016; Terwilliger et al., 2008), Buster (Blanc et al., 2004) which consist to refine the model geometry by molecular mechanics within restraints imposed by the diffraction data. Cycles of modeling/refinement should lead to a final model with good statistics.

3. Protocols

We will describe here four protocols involving different approaches for 3D structure determination: three deal with insect OBPs phased using Se-Met, sulfur and ligand's bromine anomalous scattering. The fourth example is that of a mammal OBP, whose structure was solved by molecular replacement.

3.1 Crystallization and structure determination the apo-PBP from the cockroach *Leucophaea maderae* (LmaPBP)

(Lartigue, Gruez, Spinelli, et al., 2003; Lartigue, Riviere, Brossut, Tegoni, & Cambillau, 2003; Riviere et al., 2003)

3.1.1 Crystallization

The LmaPBP crystallization assays were performed with a nanodrops pipetting robot, using a protein concentration of 40 mg/mL in 10 mM Tris pH 8.0, 10 mM NaCl, mixed to a precipitant solution containing 22.5% PEG10000 in 0.2 M imidazole/malate at pH 8.5. Since crystals were of sufficient size, no optimization was performed.

3.1.2 Structure determination

As in the previous cases, all attempts to solve the structure with molecular replacement failed. Se-Met substituted protein was therefore used in order to perform *ab-initio* phasing using the selenium anomalous scattering power. Expression of selenomethionine substituted LmaPBP was performed using the method of methionine biosynthesis pathway inhibition (Doublie, 1997). Crystals of the Se-Met substituted LmaPBP appeared in the same conditions as those of the unlabeled protein.

Crystals were flash-frozen at 100 °K in their mother liquor with 25% of glycerol as cryo-protectant, and the data were collected at ESRF (Grenoble). The structure of apo-LmaPBP was finally solved using SAD (Single-wavelength anomalous diffraction) method at the Se-K edge ($\lambda = 0.9792$ Å). Initial phases were obtained at 2.5 Å using SOLVE (Wang et al., 2004). Phases were improved by solvent flattening using RESOLVE (Wang et al., 2004). The experimental density map after solvent flattening was clear enough to manually build two-thirds of the model and to locate the three disulfide bridges. Phase improvement was carried out using phases coming from the constructed model with ARPwARP (Perrakis et al., 1999) and the resulting map was enough to complete the model with all its missing parts. The model was refined using Refmac in CCP4 (Winn et al., 2003, 2011).

3.2 Crystallization and structure determination of *Apis melifera* PBP1 (Amel-ASP1) apo-protein (Lartigue et al., 2004)

3.2.1 Crystallization

The protein was concentrated at 40 mg/mL in a buffer solution (10 mM Tris, 25 mM NaCl pH 8.0) and crystallization experiments were performed immediately after protein purification. Crystallization screening experiments were performed with the commercial kits: Structure Screens 1 and 2 (Jancarik, Scott, Milligan, Koshland Jr., & Kim, 1991), Clear Strategy Screens 1 and 2 (Molecular Dimensions Limited), ZetaSol (Ries-Kautt & Ducruix, 1997), Stura Footprint Screen (Molecular Dimensions Limited), and Wizard Screens I and II (Emerald BioStructures).

The sitting-drop method was used with Greiner Plates. The Greiner Plates main reservoir was filled with a TECAN pipetting robot (Fig. 2A). The protein nanodrops were dispensed in the Greiner plates side cups with a nanodrop dispensing robot (Mosquito, Fig. 2B, dispensed volumes of 100, 200 and 300 nL of protein into the three lateral cups, Fig. 2C). A 150 nL volume of the reservoir precipitant solution was dispensed by the Mosquito over to the protein drops. The plates were sealed with a transparent film and stored in a cabinet at 223 °K. Thin crystals appeared after 1–3 weeks.

3.2.2 Crystal optimization

From the initial crystallization a few hits were obtained that were improved by crystallization optimization (Lartigue, Riviere, Brossut, et al., 2003). Four stock solutions were used for optimization: (i) 1.3 M Ammonium Sulfate, 0.15 M Sodium Citrate, pH 5.0; (ii) 1.7 M Ammonium Sulfate, 0.15 M Sodium Citrate, pH 5.0. (iii) 1.3 M Ammonium Sulfate, 0.15 M Sodium Citrate, pH 6.0; (iv) 1.7 M Ammonium Sulfate, 0.15 M Sodium Citrate, pH 6.0. These solutions allowed the Ammonium Sulfate concentration to vary from 1.3 M to 1.7 M along the crystallization plate x axis, while the pH varied from 5 to 6 along the y axis (Lartigue, Riviere, Brossut, et al., 2003). A TECAN Robot was used to distribute varying amounts of the four stock solutions according to a 8 × 8 matrix in the 64 wells of a Greiner Plate. A nanodrops dispensing robot dispensed 100 nL of precipitant over 200 nL of protein solution. The plate was sealed with a transparent film and stored in a cabinet at 223 °K. The plate could be briefly removed in order to examine the drops under a microscope. Thicker crystals appeared after the optimization procedure. The crystals were cryo-cooled at 100 °K in a cryo-freezing loop using their crystallization liquor with the addition of 25% glycerol as a cryoprotectant (Lartigue, Gruez, Briand, et al., 2003).

3.2.3 Structure determination

Diffraction data are indexed and integrated with DENZO (Otwinowski & Minor, 1997) and scaled with SCALA (Winn et al., 2011). SCALA scales together multiple occurrences of the same X-ray reflection and merges them leading to an averaged intensity.

All attempts to solve the structure with molecular replacement methods using homologous models like *L. maderae* PBP (1OW4, 1ORG) or *B. mori* PBP (1DQE) failed. Therefore, *ab-initio* phasing has been undertaken, using the anomalous diffraction power of sulfur atoms present in disulfide bridges and methionines.

A fluorescence spectrum was recorded, directly proportional to the X-ray absorption. For the sulfur atoms present in proteins, there is no accessible maximum absorption; hence the wavelength chosen was 1,77 Å (remote wavelength phasing). Data were integrated and scaled with HKL2000 (Kabsch, 2010). Five sulfur atom sites were found using SHELXD (Schneider & Sheldrick, 2002) or SOLVE, which allowed phasing and the calculation of a first electron density map. An initial model was built with 85% completion using ARP/wARP (Perrakis et al., 1999). The missing parts were introduced manually with an interactive graphics programme (Emsley et al., 2010). The structure was refined using REFMAC5 (Winn et al., 2003) (PDB code 1R5R).

3.3 Crystallization and structure determination of Mamestra brassicae CSP (CSP-MbraA6) in complex with bromo-dodecanol ($BrC_{12}OH$) (Campanacci et al., 2003)

3.3.1 Co-crystallization of CSPMbraA6 with BrC12OH

Crystallization experiments were performed with the commercial crystallization kit "Stura FootPrint Screen". The protein concentration was 40 mg/mL in HEPES 10 mM pH 7.0, the precipitant was PEG 2000 MME and the bromo-dodecanol concentration was 20 mM. Initial screening was followed by optimization of the precipitant concentration/pH leading to the following condition: 34% PEG 2000 MME, 0.1 M sodium-cacodylate pH 6.5. The crystals were then flash frozen at 100 K without cryoprotectant.

3.3.2 Structure determination

Attempts to use molecular replacement with this apo-CSP structure failed, and hence *ab-initio* phasing using the anomalous diffraction power of bromine was undertaken. The wavelength at which the absorption was at its maximum was determined. A fluorescence spectrum was measured, indicating an absorption maximum at the Br-K edge $\lambda=0{,}91965$ Å in the CSPMbraA6/BrC_{12}OH complex.

All data, collected at ESRF synchrotron (Grenoble) were processed and reduced by using DENZO (Otwinowski & Minor, 1997) and CCP4 suite (Wang et al., 2004). Heavy atom refinement and phasing calculations with MLPHARE and DM gave the initial maps and WARP (Perrakis et al., 1999) could build automatically the complete initial model. Refinement was performed with REFMAC5 (Winn et al., 2003) followed by manual corrections using graphics display, thus leading to the final model (1N8U).

3.4 Crystallization and structure determination the OBP3 from the giant Panda *Ailuropoda melanoleuca* (AimelOBP3) (Zhu et al., 2017)

3.4.1 Crystallization of AimelOBP3

A Mosquito robot (TTP Labtech) was used to perform crystallization experiments with the commercial kit AmSO4 suite (Qiagen) using sitting-drop vapor diffusion at 277 °K. Crystals appeared in the following condition: 2.2 M Ammonium sulfate and 0.2 M Potassium Nitrate pH 7.0. Crystals were soaked briefly in the cryo protectant TMAO (trimethylamine oxide) before flash-cooling them in a nitrogen gas stream at 100 °K.

3.4.2 Structure determination

Data were indexed, integrated and scaled using XDS package (Kabsch, 2010). Indexing is normally carried out using an automated procedure and gives the crystal orientation, an initial estimate of the unit cell dimensions with an idea of the crystallographic symmetry and an index for each reflection. Images were scaled with XSCALE in XDS (Kabsch, 2010) (The resulting scaled data set of XDS is in ASCII format. It can be converted into CCP4 style multi-recorded MTZ file using the copy failure of SCALA in CCP4 (Winn et al., 2011)). The crystal structure of AimelOBP3 was determined by molecular replacement with MOLREP (Vagin & Teplyakov, 2010) using the structure of major mouse urinary Protein IV (3KFF) as starting model (sequence identity greater than 25%). Refinement was performed with AutoBuster (Blanc et al., 2004) alternated with display modeling with COOT (Emsley et al., 2010), primarily focused on building and validation of atomic model into 3D electron density maps.

4. Structure analysis

Starting from the sequence of the OBP/CSP, Blast on PDB data base followed by Multalign (Corpet, 1988) will give an indication on the vicinity of the query to known 3D structures. If the sequence identity is too low, Jpred (Cole, Barber, & Barton, 2008) and HHpred (Zimmermann et al., 2018) may help to determine the secondary structure and the putative fold of the query. Once the OBP/CSP structure is determined, the DALI server (Holm & Laakso, 2016; Holm & Rosenstrom, 2010) will indicate which proteins resemble most to yours. The PISA server (Krissinel & Henrick, 2007) will give indications on the oligomeric state of the protein in the crystal, and provide a list of contacts between monomers. The volume of the OBP cavity can be estimated by the server CASTp (Tian, Chen, Lei, Zhao, & Liang, 2018).

5. Related techniques

A number of related techniques are important in the field of OBPs to complement the X-ray structure determination. The alternative or complementary technique of NMR structure determination is presented elsewhere in this book (Chapter "Solution structure of OBPs by NMR" by Picone). Very important techniques deal with the determination of ligand/OBP affinity. In principle, several techniques can be used: micro calorimetry, surface plasmon resonance (Chapter "Optical sensors for monitoring odorant binding" by Dostalek) and fluorescence quenching (Chapter "Ligand binding assays of OBPs and CSPs" by Pelosi). Practically, due to the cost of the instruments and the difficulties in setting up the experiments, the two first techniques are little used. Instead, fluorescence quenching, either using the fluorescence quenching of tryptophanes by ligands (Pesenti et al., 2009) of the displacement of a fluorescent probe bound in the OBP cavity by ligands (Campanacci et al., 2001), is of wide use (Chapter "Ligand binding assays of OBPs and CSPs" by Pelosi).

Another approach to avoid structure determination by X-ray diffraction is homology modeling, often followed by simulated ligand docking

(Chapter "Ligand binding assays of OBPs and CSPs" by Pelosi). However, such approaches should be used with caution considering the high flexibility of OBPs and CSPs in the context of pH and ligand binding (Campanacci et al., 2003; Sandler et al., 2000).

6. Summary

Researchers who seek to determine the structure of insects OBPs or CSPs alone or in complex with ligands will have to overcome specific challenges linked to the high flexibility of these proteins. We have described some approaches and strategies to deal with such problems. While the crystallization of OBPs and CSPs is in general relatively easy, structure determination of the insects' proteins requires in most cases *ab-initio* phasing. In contrast, structures of mammalian OBPs can be determined readily using molecular replacement. In any case, special attention should be paid to the effect of pH and ligands on the overall OBP/CSP structure in correlation with function.

Acknowledgments

Dr. Alain Roussel (AFMB, Marseille) is greatly acknowledged for his constant support. Present and past members of the group are warmly acknowledged for their constant efforts over the 25 last years.

References

Bianchet, M. A., Bains, G., Pelosi, P., Pevsner, J., Snyder, S. H., Monaco, H. L., et al. (1996). The three-dimensional structure of bovine odorant binding protein and its mechanism of odor recognition. *Nature Structural Biology*, *3*, 934–939.

Blanc, E., Roversi, P., Vonrhein, C., Flensburg, C., Lea, S. M., & Bricogne, G. (2004). Refinement of severely incomplete structures with maximum likelihood in BUSTER-TNT. *Acta Crystallographica. Section D, Biological Crystallography*, *60*, 2210–2221.

Butterwick, J. A., Del Marmol, J., Kim, K. H., Kahlson, M. A., Rogow, J. A., Walz, T., et al. (2018). Cryo-EM structure of the insect olfactory receptor Orco. *Nature*, *560*, 447–452.

Campanacci, V., Krieger, J., Bette, S., Sturgis, J. N., Lartigue, A., Cambillau, C., et al. (2001). Revisiting the specificity of *Mamestra brassicae* and *Antheraea polyphemus* pheromone-binding proteins with a fluorescence binding assay. *The Journal of Biological Chemistry*, *276*, 20078–20084.

Campanacci, V., Lartigue, A., Hallberg, B. M., Jones, T. A., Giudici-Orticoni, M. T., Tegoni, M., et al. (2003). Moth chemosensory protein exhibits drastic conformational changes and cooperativity on ligand binding. *Proceedings of the National Academy of Sciences of the United States of America*, *100*, 5069–5074.

Cole, C., Barber, J. D., & Barton, G. J. (2008). The Jpred 3 secondary structure prediction server. *Nucleic Acids Research*, *36*, W197–W201.

Corpet, F. (1988). Multiple sequence alignment with hierarchical clustering. *Nucleic Acids Research*, *16*, 10881–10890.

Doublie, S. (1997). Preparation of selenomethionyl proteins for phase determination. *Methods in Enzymology*, *276*, 523–530.

Emsley, P., Lohkamp, B., Scott, W. G., & Cowtan, K. (2010). Features and development of Coot. *Acta Crystallographica. Section D, Biological Crystallography*, *66*, 486–501.

Holm, L., & Laakso, L. M. (2016). Dali server update. *Nucleic Acids Research*, *44*, W351–W355.

Holm, L., & Rosenstrom, P. (2010). Dali server: Conservation mapping in 3D. *Nucleic Acids Research*, *38*, W545–W549.

Horst, R., Damberger, F., Luginbuhl, P., Guntert, P., Peng, G., Nikonova, L., et al. (2001). NMR structure reveals intramolecular regulation mechanism for pheromone binding and release. *Proceedings of the National Academy of Sciences of the United States of America*, *98*, 14374–14379.

Jancarik, J., Scott, W. G., Milligan, D. L., Koshland, D. E., Jr., & Kim, S. H. (1991). Crystallization and preliminary X-ray diffraction study of the ligand-binding domain of the bacterial chemotaxis-mediating aspartate receptor of *Salmonella typhimurium*. *Journal of Molecular Biology*, *221*, 31–34.

Janowski, P. A., Moriarty, N. W., Kelley, B. P., Case, D. A., York, D. M., Adams, P. D., et al. (2016). Improved ligand geometries in crystallographic refinement using AFITT in PHENIX. *Acta crystallographica. Section D, Structural biology*, *72*, 1062–1072.

Kabsch, W. (2010). Integration, scaling, space-group assignment and post-refinement. *Acta Crystallographica. Section D, Biological Crystallography*, *66*, 133–144.

Krissinel, E., & Henrick, K. (2007). Inference of macromolecular assemblies from crystalline state. *Journal of Molecular Biology*, *372*, 774–797.

Lagarde, A., Spinelli, S., Qiao, H., Tegoni, M., Pelosi, P., & Cambillau, C. (2011). Crystal structure of a novel type of odorant-binding protein from Anopheles gambiae, belonging to the C-plus class. *The Biochemical Journal*, *437*, 423–430.

Lagarde, A., Spinelli, S., Tegoni, M., He, X., Field, L., Zhou, J. J., et al. (2011). The crystal structure of odorant binding protein 7 from *Anopheles gambiae* exhibits an outstanding adaptability of its binding site. *Journal of Molecular Biology*, *414*, 401–412.

Lartigue, A., Campanacci, V., Roussel, A., Larsson, A. M., Jones, T. A., Tegoni, M., et al. (2002). X-ray structure and ligand binding study of a moth chemosensory protein. *The Journal of Biological Chemistry*, *277*, 32094–32098.

Lartigue, A., Gruez, A., Briand, L., Blon, F., Bezirard, V., Walsh, M., et al. (2004). Sulfur single-wavelength anomalous diffraction crystal structure of a pheromone-binding protein from the honeybee *Apis mellifera L*. *The Journal of Biological Chemistry*, *279*, 4459–4464.

Lartigue, A., Gruez, A., Briand, L., Pernollet, J. C., Spinelli, S., Tegoni, M., et al. (2003). Optimization of crystals from nanodrops: Crystallization and preliminary crystallographic study of a pheromone-binding protein from the honeybee *Apis mellifera L*. *Acta Crystallographica. Section D, Biological Crystallography*, *D59*, 919–921.

Lartigue, A., Gruez, A., Spinelli, S., Riviere, S., Brossut, R., Tegoni, M., et al. (2003). The crystal structure of a cockroach pheromone-binding protein suggests a new ligand binding and release mechanism. *The Journal of Biological Chemistry*, *278*, 30213–30218.

Lartigue, A., Riviere, S., Brossut, R., Tegoni, M., & Cambillau, C. (2003). Crystallization and preliminary crystallographic study of a pheromone-binding protein from the cockroach *Leucophaea maderae*. *Acta Crystallographica. Section D, Biological Crystallography*, *59*, 916–918.

Lautenschlager, C., Leal, W. S., & Clardy, J. (2005). Coil-to-helix transition and ligand release of *Bombyx mori* pheromone-binding protein. *Biochemical and Biophysical Research Communications*, *335*, 1044–1050.

Lautenschlager, C., Leal, W. S., & Clardy, J. (2007). *Bombyx mori* pheromone-binding protein binding nonpheromone ligands: Implications for pheromone recognition. *Structure, 15*, 1148–1154.

McPherson, A. (2017). Protein crystallization. *Methods in Molecular Biology, 1607*, 17–50.

McPherson, A., & Cudney, B. (2014). Optimization of crystallization conditions for biological macromolecules. *Acta crystallographica. Section F, Structural biology communications, 70*, 1445–1467.

McPherson, A., & Gavira, J. A. (2014). Introduction to protein crystallization. *Acta crystallographica. Section F, Structural biology communications, 70*, 2–20.

Otwinowski, Z., & Minor, W. (1997). Processing of X-ray diffraction data collected in oscillation mode. *Methods in Enzymology, 276*, 307–326.

Perrakis, A., Morris, R., & Lamzin, V. S. (1999). Automated protein model building combined with iterative structure refinement. *Nature Structural Biology, 6*, 458–463.

Pesenti, M. E., Spinelli, S., Bezirard, V., Briand, L., Pernollet, J. C., Campanacci, V., et al. (2009). Queen bee pheromone binding protein pH-induced domain swapping favors pheromone release. *Journal of Molecular Biology, 390*, 981–990.

Pesenti, M. E., Spinelli, S., Bezirard, V., Briand, L., Pernollet, J. C., Tegoni, M., et al. (2008). Structural basis of the honey bee PBP pheromone and pH-induced conformational change. *Journal of Molecular Biology, 380*, 158–169.

Ries-Kautt, M., & Ducruix, A. (1997). [3] Inferences drawn from physicochemical studies of crystallogenesis and precrystalline state. *Methods in Enzymology, 276*, 23–59.

Riviere, S., Lartigue, A., Quennedey, B., Campanacci, V., Farine, J. P., Tegoni, M., et al. (2003). A pheromone-binding protein from the cockroach *Leucophaea maderae:* Cloning, expression and pheromone binding. *The Biochemical Journal, 371*, 573–579.

Sandler, B. H., Nikonova, L., Leal, W. S., & Clardy, J. (2000). Sexual attraction in the silkworm moth: Structure of the pheromone-binding-protein-bombykol complex. *Chemistry & Biology*, 7, 143–151.

Schneider, T. R., & Sheldrick, G. M. (2002). Substructure solution with SHELXD. *Acta Crystallographica. Section D, Biological Crystallography, 58*, 1772–1779.

Spinelli, S., Lagarde, A., Iovinella, I., Legrand, P., Tegoni, M., Pelosi, P., et al. (2012). Crystal structure of *Apis mellifera* OBP14, a C-minus odorant-binding protein, and its complexes with odorant molecules. *Insect Biochemistry and Molecular Biology, 42*, 41–50.

Tegoni, M., Campanacci, V., & Cambillau, C. (2004). Structural aspects of sexual attraction and chemical communication in insects. *Trends in Biochemical Sciences, 29*, 257–264.

Tegoni, M., Ramoni, R., Bignetti, E., Spinelli, S., & Cambillau, C. (1996). Domain swapping creates a third putative combining site in bovine odorant binding protein dimer. *Nature Structural Biology, 3*, 863–867.

Terwilliger, T. C., Grosse-Kunstleve, R. W., Afonine, P. V., Moriarty, N. W., Zwart, P. H., Hung, L. W., et al. (2008). Iterative model building, structure refinement and density modification with the PHENIX AutoBuild wizard. *Acta Crystallographica. Section D, Biological Crystallography, 64*, 61–69.

Tian, W., Chen, C., Lei, X., Zhao, J., & Liang, J. (2018). CASTp 3.0: Computed atlas of surface topography of proteins. *Nucleic Acids Research, 46*, W363–W367.

Vagin, A., & Teplyakov, A. (2010). Molecular replacement with MOLREP. *Acta Crystallographica. Section D, Biological Crystallography, 66*, 22–25.

Vogt, R. G., Kohne, A. C., Dubnau, J. T., & Prestwich, G. D. (1989). Expression of pheromone binding proteins during antennal development in the gypsy moth *Lymantria dispar. The Journal of Neuroscience, 9*, 3332–3346.

Wang, J. W., Chen, J. R., Gu, Y. X., Zheng, C. D., Jiang, F., Fan, H. F., et al. (2004). SAD phasing by combination of direct methods with the SOLVE/RESOLVE procedure. *Acta Crystallographica. Section D, Biological Crystallography, 60*, 1244–1253.

Winn, M. D., Ballard, C. C., Cowtan, K. D., Dodson, E. J., Emsley, P., Evans, P. R., et al. (2011). Overview of the CCP4 suite and current developments. *Acta Crystallographica. Section D, Biological Crystallography*, *67*, 235–242.

Winn, M. D., Murshudov, G. N., & Papiz, M. Z. (2003). Macromolecular TLS refinement in REFMAC at moderate resolutions. *Methods in Enzymology*, *374*, 300–321.

Zhu, J., Arena, S., Spinelli, S., Liu, D., Zhang, G., Wei, R., et al. (2017). Reverse chemical ecology: Olfactory proteins from the giant panda and their interactions with putative pheromones and bamboo volatiles. *Proceedings of the National Academy of Sciences of the United States of America*, *114*, E9802–E9810.

Zimmermann, L., Stephens, A., Nam, S. Z., Rau, D., Kubler, J., Lozajic, M., et al. (2018). A completely reimplemented MPI bioinformatics toolkit with a new HHpred server at its core. *Journal of Molecular Biology*, *430*, 2237–2243.

CHAPTER EIGHT

Solution structure of insect CSP and OBPs by NMR

Serena Leone[a], Alessandro Emendato[a], Roberta Spadaccini[b], Delia Picone[a,*]

[a]Department of Chemical Sciences, University of Naples "Federico II", Complesso Universitario Monte S. Angelo, Napoli, Italy

[b]Department of Science and Technology, Università degli Studi del Sannio, Benevento, Italy

*Corresponding author: e-mail address: delia.picone@unina.it

Contents

Abstract

Insect odorant binding proteins (OBPs) and chemosensory proteins (CSPs) are proteins deputed to the solubilization, transport and stabilization of lipophilic and odorant compounds. These proteins have a conserved fold, which undergoes massive structural rearrangements in order to accommodate medium to large-sized lipophilic ligands. Solution NMR spectroscopy, due to its intrinsically dynamic nature, is the perfect technique to extrapolate structural information and dynamic parameters and to elucidate the conformational changes that occur upon ligand binding. This chapter will describe in detail the experimental protocols for the production and purification of

Methods in Enzymology, Volume 642
ISSN 0076-6879
https://doi.org/10.1016/bs.mie.2020.04.063

isotope-labeled recombinant CSPs and OBPs for NMR studies. Detailed procedures for spectra acquisition, processing and analysis will be presented, focusing on the protein CSP-sg4 from *Schistocerca gregaria* as a model. Finally, experiments aimed at providing information on protein flexibility and ligand binding modes will also be described.

1. Introduction

Insect chemosensory proteins (CSPs) and odorant binding proteins (OBPs) are two related families of small soluble proteins that act as carriers for lipophilic pheromones and odorant substances within insect sensory structures (Pelosi, Iovinella, Felicioli, & Dani, 2014; Zhu, Iovinella, Dani, Pelosi, & Wang, 2019). OBPs have been more thoroughly studied than CSPs: in fact, over 30 structures of insect OBPs are currently present in the Protein Data Bank. OBPs from different arthropods are typically made of 130–150 residues arranged in a conserved bundle of six α-helices, strengthened by three conserved disulfide bridges (Pelosi et al., 2014; Zhu et al., 2019). CSPs are structurally related to OBPs, with a few significant differences: they are on average slightly smaller, made of 100–120 amino acids arranged in a six α-helices fold as well. Nonetheless, the few CSPs structures elucidated so far display a different arrangement of the secondary structure elements compared to OBPs and present only two conserved disulfide bridges, which do not contribute as much to the protein stability as they do to the protein great conformational flexibility (Jansen et al., 2007; Lartigue et al., 2002; Tomaselli et al., 2006). In general, the sequences of CSPs are more conserved than that of OBPs (Eyun et al., 2017; Vieira & Rozas, 2011). Moreover, CSPs expression is not confined to sensory organs, suggesting more generic functions based on their ability to bind medium and large-sized lipophilic molecules. Both OBPs and CSPs possess a large hydrophobic pocket to accommodate ligands and an elevated number of exposed charged amino acids, which confers them extraordinary solubility. This property, together with their small size, makes them ideal candidates for solution NMR studies. CSPs and OBPs undergo profound conformational rearrangements in order to bind, solubilize and store large lipophilic molecules. NMR spectroscopy is the preferred technique to directly photograph these structural reorganizations. In addition, several parameters can be extracted from NMR experiments, which provide information on the plasticity and dynamics of the protein fold in solution. These parameters are directly related to binding site dynamics and specificity, and could assist the design of mutants for applications of OBPs and CSPs in the production of specific biosensors (described in detail in Chapters "Electrochemical biosensors," by Sabine Szunerits; "Electronic biosensors, based on graphene FETs," by Wolfgang Knoll; "Electronic biosensors based on EGOFETs," by Yusuf Mulla and Luisa Torsi; "Gravimetric biosensors," by Krishna Persaud; "Optical sensors for monitoring odorant binding," by Jakub Dostalek). In order to perform NMR studies on CSPs and OBPs, alone or in complex with their ligands, one should first produce and purify consistent amounts of isotopically labeled protein. This chapter will illustrate all the steps to achieve the structure characterization of OBPs and CSPs, from the recombinant production of isotopically enriched proteins and their purification to the recording, processing and analysis of two- and three-dimensional NMR experiments and consequent structure calculation.

2. Structural studies on CSPs

2.1 Recombinant production and purification of CSPs and OBPs

Before being able to acquire NMR data on CSPs and OBPs, it is necessary to produce the desired recombinant protein in isotopically enriched medium. Depending on the need and on the planned experiments, single or double isotope labeling with ^{15}N or ^{13}C will be required. Complete structure determination with full resonance assignment requires double isotope labeling, whereas, if the structure of the protein is known and only dynamic parameters or information on conformational changes are sought, ^{15}N labeling might suffice. Nonetheless, the possibility to use triple resonance experiments (i.e., double labeling), despite requiring longer and more complicated NMR measurements, should be preferred whenever possible, because it can speed up ^{15}N and ^{1}H resonance assignment. Production of high amounts of soluble recombinant proteins for NMR studies is usually achieved in *Escherichia coli* expression systems. While this system is not the first choice for the production of proteins containing disulfide bridges, production of OBPs and CSPs is usually achieved without major problems. Some OBPs may result trickier to express, and protocols to address production issues have been described in detail in Chapter "Expression and purification of OBPs," by Loic Briand. The production of disulfide containing protein in *E. coli* can be achieved in the periplasm, because the reducing state of the cytoplasm does not allow for disulfide bridge formation (Berkmen, 2012). The chemosensory protein CSPMbraA6 from *Mamestra brassicae* and the pheromone binding proteins from *Bombyx mori* and *Antheraea polyphemus*, for instance, have been produced without affinity tags or fusion constructs using the pelB leader sequence in the pET22b(+) vector to target expression to the periplasm of *E. coli* BL21(DE3) (Campanacci et al., 2001; F. Damberger et al., 2000; Wojtasek & Leal, 1999). This system has been successfully employed for the production of several OBPs and CSPs (Leal et al., 2008; Mao et al., 2010). Other OBPs, like LUSH from *Drosophila,* could be recovered from inclusion bodies upon cytoplasmic expression (Bucci, Kruse, Thode, Alvarado, & Jones, 2006). In our studies, we have achieved the recombinant production of untagged CSP-sg4 from *Schistocerca gregaria* in the cytoplasm of *E. coli* BL21(DE3) cells using the pET5b vector (Picone et al., 2001). We found that this system allowed for the production of large amounts of soluble protein and that, upon purification, the correct disulfide bridges arrangement was found in the protein, likely thanks to the correct pairing of the cysteines on the nascent protein chain, that could assume the correct oxidation state upon cell disruption (Picone et al., 2001; Tomaselli et al., 2006). Purification of untagged OBPs and CSPs can easily be achieved by ion exchange chromatography (IEX), exploiting the elevated charge density of these proteins (Campanacci et al., 2001; Picone et al., 2001; Wojtasek & Leal, 1999). An additional polishing step of size exclusion chromatography (SEC) or HPLC may be required to purify the samples to homogeneity.

2.2 NMR spectroscopy

One of the best ways to understand the function of CSPs and OBPs relies on the knowledge of their three-dimensional structures and their binding mode. Both classes of proteins are small globular proteins and this makes them very suitable for NMR studies. The three NMR structures of CSPs elucidated so far (Jansen et al., 2007; Mosbah et al., 2003; Tomaselli et al., 2006) were solved using multidimensional NMR spectroscopy techniques, based on chemical shift assignment and distance restraints extracted from NOE data (Clore & Gronenborn, 1991; Sattler, Schleucher, & Griesinger, 1999) using ^{13}C and ^{15}N labeled proteins. All three structures are characterized by four helices arranged to form a V-shaped motif, with a fifth helix running across it and a sixth one packed against the external surface of the protein. As mentioned above, each structure is characterized by two conserved disulfide bridges (Maleszka & Stange, 1997) and presents one hydrophobic cavity. Although not all putative CSPs ligands are known, these proteins bind hydrophobic ligands, and the binding modes can be characterized by NMR: for instance, the measure of chemical shift perturbation, i.e., the measure of the variation of the chemical shift of cross-peaks in the $^{1}H,^{15}N$-HSQC spectrum as a function of different ligand concentrations, is a quick and extremely efficient way to extrapolate information on protein ligand binding mode (Williamson, 2013). NMR titrations have shown that proteins undergo large conformational rearrangements in the presence of at least two putative ligands, namely 12-bromododecanol for CSPMbraA6 (Mosbah et al., 2003) and oleamide for CSP-sg4 (Tomaselli et al., 2006). These results are in perfect agreement with those obtained by crystallography on CSPMbraA6 in complex with three molecules of 12-bromododecanol and highlight the extreme conformational flexibility of CSPs, as the inner hydrophobic cavity can expand to host three molecules of ligand.

NMR spectroscopy has contributed greatly also to the study of OBPs, allowing to unveil the function of these proteins. The most extensively characterized from the structural point of view is the pheromone binding protein BmorPBP from *Bombyx mori*. Combining NMR studies with chemical shift perturbation and hydrogen exchange (HDX) experiments, it has been shown that the protein structure is affected by pH variation and ligand binding (Damberger et al., 2000; Damberger, Ishida, Leal, & Wüthrich, 2007; Damberger, Michel, Ishida, Leal, & Wüthrich, 2013; Horst et al., 2001; Lee et al., 2002). The C-terminus of BmorPBP is rich in acidic residues, that at neutral pH are ionized, repelling each other, thus making this segment very flexible and disordered (Lee et al., 2002). In acidic solutions, instead, the residues are protonated and the C-terminal segment forms a seventh α-helix, which occupies the binding pocket of the protein (Horst et al., 2001). Competition between the seventh helix and lipophilic ligands for the binding cavity enables selective uptake of bombykol, its native ligand (Damberger et al., 2013). This conformational variation, tuned by pH, selectively regulates pheromone transport (Damberger et al., 2013). NMR studies on other pheromone binding proteins, namely ApolPBP1 from *Antheraea polyphemus* (Damberger et al., 2013; Katre, Mazumder, Prusti, & Mohanty, 2009) and AtraPBP1 from *Amyelois transitella* (Xu et al., 2010), have suggested that this interplay between pH and ligand binding might be a general phenomenon within this protein family.

3. Materials, equipment and reagents

1. Incubator with shaking and, eventually, oxygenation systems.
2. Refrigerated centrifuge with appropriate rotors.
3. Ultrasonic cell homogenizer or French press.
4. SDS-PAGE (sodium dodecyl sulfate-polyacrylamide gel electrophoresis) system.
5. UV/visible spectrophotometer.
6. FPLC or similar low-pressure chromatography system with UV detection (e.g., AKTA purifier, GE Healthcare Life Sciences).
7. CM Sepharose Fast Flow (GE Healthcare Life Sciences) cation exchange and DEAE Sepharose Fast Flow (GE Healthcare Life Sciences) anion exchange columns.
8. HPLC system (e.g., Agilent 1220 or Thermo Vanquish HPLC) equipped with a C4 reverse phase column.
9. NMR spectrometer (see Section 4.2).
10. pH meter.
11. Chemically competent *Escherichia coli* BL21(DE3) cells and the appropriate pET vector containing the gene of interest. In our examples, we will use the pET-Schi10 vector, obtained from the insertion of the gene coding for CSP-sg4 in the pET5b vector (EMD).
12. Luria-Bertani/Miller (LB) and minimal microbial (M9) growth media (Elbing & Brent, 2002).
13. Ampicillin and isopropyl β-D-1-thiogalactopyranoside (IPTG) or appropriate antibiotics and induction agents if a different expression vector is used.
14. Isotopically enriched chemicals such as D_2O, $^{15}NH_4Cl$ and $^{13}C_{6-}$ glycerol or D-glucose-$^{13}C_6$ (Cambridge Isotope Laboratories).
15. cOmplete™ Protease Inhibitor Cocktail tablets (Roche) or other protease inhibitors.
16. HPLC grade solvents (acetonitrile, water, trifluoroacetic acid).
17. NMR data acquisition and processing software (see Section 4.3).

4. Protocols

4.1 Production and purification of ^{13}C,^{15}N-labeled CSP-sg4

1. Prepare a starter cell culture by inoculating a single colony of *Escherichia coli* BL21(DE3) cells freshly transformed with the pET-Schi10 vector

(Pelosi, 1994). Alternatively, start the culture by scraping a frozen glycerol stock of previously transformed cells, into a 5 mL LB growth media. Add 100 μg/mL of ampicillin and grow the culture overnight under shaking at 37 °C.

2. Add $^{15}NH_4Cl$ and D-glucose-$^{13}C_6$ to a final concentration of 1 and 4 g/L, respectively, to 1 L of M9 growth medium (Elbing & Brent, 2002). These chemicals will be the only nitrogen and carbon sources in the growth medium and they will be incorporated in the overexpressed protein. Sterilize by filtration on a 0.22 μm filter.
3. Dilute the starter culture to a cell culture flask containing the desired volume of isotopically enriched M9 medium in a ratio of 1:100 v/v. Add ampicillin to a final concentration of 50 μg/mL and grow the cells at 37 °C until they reach OD_{600} of 0.6–0.8.
4. Induce protein expression by adding isopropyl β-D-1-thiogalactopyranoside (IPTG), to the final concentration of 0.4 mM. Allow for protein expression to proceed at 37 °C for at least 2.5 h.

   Tips: Optimal induction agent concentration and protein expression time and temperature can vary significantly depending on the vector used and the protein expressed. It is advisable to perform small scale trials changing the experimental condition: for instance, try varying induction time between 4 h and overnight, induction temperature between 16 and 37 °C, IPTG concentration between 0.1 and 1 mM.
5. Harvest the cells by centrifuging them at 4 °C, 5000 × *g* for 10 min. Freeze the pellet and store it at −80 °C until protein extraction is performed.
6. Resuspend the cell pellet from 1 L culture in 10 mL of 25 mM TRIS/HCl + 50 mM EDTA buffer at pH 7.60 containing one tablet of cOmplete™ Protease Inhibitor Cocktail, and place the sample inside an ice bath. Disrupt the cells using a cell homogenizer, e.g., a Microson Ultrasonic Homogenizer XL2000 at 100 W power, applying intermittent activation (30 s on followed by 20 s off) in order to avoid sample overheating. Centrifuge the cell lysate at 15,000 × *g* for 45 min at 4 °C: soluble recombinant CSP-sg4 will be in the supernatant. Discard the cellular debris.
7. Perform a first coarse purification by cation exchange chromatography on CM Sepharose® Fast Flow (15 mL). Equilibrate and elute the column with 25 mM TRIS/HCl buffer, pH 7.60. Since the isoelectric point of CSP-sg4 is around 5, it will be negatively charged at this pH, hence it will elute in the flow through in these conditions. Check for the presence of the protein in the chromatography fractions by SDS-PAGE, and pool them.

Tips: it could be useful concentrate the sample by a centrifugal filter unit with a cutoff of 3 kDa in order to reduce the sample volume for the next purification step.

8. Perform the second purification step by anion exchange chromatography, using a stationary phase of DEAE Sepharose® Fast Flow (15 mL). Elute the protein with a linear gradient of NaCl from 0 to 250 mM in 25 mM TRIS/HCl, pH 7.60 buffer. In this way the proteins contained in the sample will be differentially eluted based on their net negative charge. Check for the presence and purity of the protein by SDS-PAGE experiments, and pool the appropriate fractions.

Tips: check for the presence and the amount of protein after every step of this protocol by SDS-PAGE. This can be useful to identify critical steps of the procedure and optimize them to improve the final protein yield.

9. Desalt the pooled fractions by dialysis or ultrafiltration against ultrapure H_2O and lyophilize.

10. Perform a final purification step by reverse phase HPLC, using a semi-preparative Vydac C4 column (250 × 4.6 mm) (Vydac, The Separations Group, Inc., Hesparia, CA, USA). Dissolve the protein in 0.1% trifluoroacetic acid and elute with a linear gradient from 5% to 70% of acetonitrile in 0.1% trifluoroacetic acid over 30 min, at a flow rate of 1 mL/min. Lyophilize the protein and store it at −20 °C.

Tips: Recombinant CSP-sg4 is readily solubilized from the powder. When possible, lyophilization is advisable because it greatly increases the storage life of proteins. Nonetheless, other proteins might be less prone to lyophilization and each case should be assessed individually.

11. Check for effective isotope incorporation by mass spectrometry. This can be assessed calculating the ratio between the protein monoisotopic mass obtained and the theoretical monoisotopic mass considering full isotope incorporation (+1 Da for each N and C atoms in the protein sequence).

Tips: It is possible to perform step 3 growing *E. coli* cells in LB medium to achieve faster growth and higher biomass. In this case, the cells should be harvested by centrifugation (5000 × *g* for 10 min at 4 °C) and resuspended in the isotopically enriched M9 medium before induction with IPTG (step 4). This will sensibly increase the final protein yield, but isotope incorporation could be reduced, leading to partial protein labeling (Marley, Lu, & Bracken, 2001).

4.2 NMR data recording

1. Perform a preliminary check of the protein folding and stability before recording multidimensional NMR spectra. This can be done by circular dichroism spectroscopy (CD), an easy and fast method to evaluate the proteins secondary structure as a function of various parameters such as temperature, pH, ionic strength and presence of protein ligands. Alternatively, an evaluation of the protein folding can be achieved by 1D NMR proton spectra and/or 2D $^{1}H,^{15}N$-HSQC spectra, where the presence of secondary structure elements increases dramatically the signals dispersion (Fig. 1).

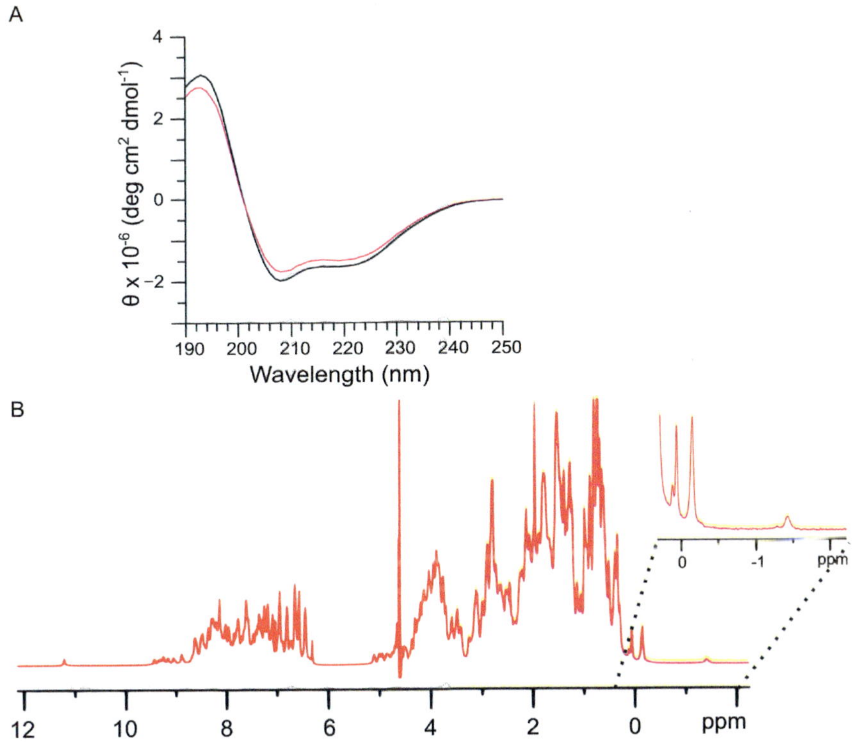

Fig. 1 (A) Comparison of the CD spectra of recombinant CSP-sg4 10 μM in 10 mM Tris/HCl, pH 7.9 (black line) with the original extracted protein (red line): the shape of the spectra confirms in both cases the elevated α-helix content. (B) 1D ^{1}H NMR spectrum of 1.2 mM purified recombinant CSP-sg4. The signal dispersion, and in particular the signal at about −1.5 ppm showed in the inset, indicates correct protein folding. *CD data reproduced with permission from Picone, D., Crescenzi, O., Angeli, S., Marchese, S., Brandazza, A., Ferrara, L., et al. (2001). Bacterial expression and conformational analysis of a chemosensory protein from Schistocerca gregaria.* European Journal of Biochemistry, *268(17), 4794–4801. https://doi.org/10.1046/j.1432-1327.2001.02408.x.*

2. Prepare the NMR sample: dissolve the recombinant CSP-sg4 at a final concentration of ~1 mM in 500 μL of 50 mM Na_2HPO_4/NaH_2PO_4 buffer, pH 6.80, containing 5% v/v D_2O. The optimal protein concentration depends on the magnitude of the magnetic field of the NMR instrument.

   Tips: If a proton-containing buffer is necessary, use the perdeuterated form. Whenever possible, use phosphate or other buffers containing only exchangeable protons.

   Tips: Record NMR spectra at different protein concentration in order to assess the possible protein oligomerization due to high concentration. The protein concentration should be as high as possible to minimize the number of scans in the 3D datasets, but it is also relevant avoid artifacts due to protein-protein interactions.

   Tips: Triple resonance equipment (instrument, probe) are needed in combination with high magnetic fields for optimal resolution and sensitivity. If a high filed instrument is not available, application for measuring time at large facilities is possible (e.g., iNEXT, http://www.inext-eu.org/). However, it's advisable to check in advance the quality of the NMR sample.
3. Record sets of 2D (^1H,^{13}C-HSQC and ^1H,^{15}N-HSQC) and 3D (HNCA, HNCACB, CBCA(CO)NH, HNCO, TOCSY- and NOESY-^1H,^{15}N-HSQC, (H)CCH-TOCSY and H(C)CH-TOCSY) heteronuclear spectra for resonances assignment of CSP-sg4. To extract NOE based distance restrains, record a set of 3D ^1H,^{15}N- and ^1H,^{13}C-NOESY-HSQC spectra with a mixing time of 100 ms. Typically, NMR manufacturers provide extensive pulse sequence libraries with the most common NMR experiments already set up; nonetheless, some NMR expertise is needed to optimize acquisition parameters on the actual samples.

   Tips: Multiple NMR samples using labeled, selectively labeled and unlabeled protein batches can be useful to resolve ambiguities in resonances assignment.

   Tips: Sometimes signals overlapping issues can be solved recording spectra at slightly different temperatures, helping the assignment procedure.
4. To perform H/D exchange experiments, lyophilize CSP-sg4 samples from 50 mM Na_2HPO_4/NaH_2PO_4 buffer, pH 6.80 and re-dissolve them in 100% D_2O. Immediately after dissolution, record a series of ^1H,^{15}N-HSQC spectra at fixed time intervals (e.g., 30 min) for at least 24 h. Depending on the protein structure, the experiment can last even longer.

4.3 Data analysis

NMR spectra processing can be performed with several free and commercial software. In particular, we used NMRPipe for processing (Delaglio et al., 1995) and NMRView for inspection, pick-picking and analysis (Johnson & Blevins, 1994). This requires homemade scripts to convert the data. Alternatively, a plethora of software performing fully automated processing of data is available, such as CcpNmr Analysis (https://www.ccpn.ac.uk/v2-software/software/analysis), TopSpin (freely available from Bruker for academic users), CARA (http://cara.nmr.ch/doku.php) or NMRFAM-SPARKY (https://nmrfam.wisc.edu/nmrfam-sparky-distribution/). The use of such software requires a bit of familiarization with the user manual and the choice is strictly dependent on the user's preferences.

4.4 Assignment procedure

The resonance assignment is performed using a combination of different multidimensional spectra.

1. Assign the resonances of the protein backbone using the HNCACB, CBCA(CO)NH, HNCA and HNCO spectra. On one hand, the HNCACB correlates each NH group with the Cα and Cβ chemical shifts of its own residue (strong cross-peak) and of the preceding residue (weak cross-peak). On the other hand, the CBCA(CO)NH correlates the NH group to the preceding Cα and Cβ chemical shifts only. From the comparative analysis of both spectra, it is possible to "walk" along the protein backbone, connecting the resonances to construct chains of subsequent residues, even though the residue type is still unknown. Moreover, the chemical shifts for the Cα and Cβ of some amino acids, such as Alanine, Serine, Threonine and Glycine (in the case of Glycine there is no Cβ), adopt peculiar values, thus, in many cases, they can be identified already at this stage of the analysis. Once a chain of connected resonances has been built, the identification of these distinctive amino acid types allows to compare this string to the whole protein sequence, facilitating the identification of other residues based on the knowledge of the primary structure. In some cases, the signal-to-noise ratio of the spectra can be not optimal, and some cross-peaks can fall below the noise level and be harder to identify.
2. Complement the initial information using a HNCA and a HNCO spectra, that correlate in the same fashion the Cα and C′ chemical shifts. These two kinds of spectra have the advantage to be more sensitive, even though they lack the Cβ information, that often turns very useful.

3. Once the backbone has been mostly reconstructed, the spin systems corresponding to the different residues can be pinpointed using TOCSY- and NOESY-[^{1}H,^{15}N]-HSQC (Fig. 2A), whereas side chain resonances can be assigned using (H)CCH-TOCSY and H(C)CH-TOCSY for carbons and protons, respectively. In particular, in the TOCSY spectra all side chain proton resonances of one residues correlate with the HN chemical shift of that residue (or with carbon chemical shift in the case of (H)CCH-TOCSY), whereas, in the NOESY spectrum, additional cross-peaks between protons within 5 Å belonging to nearby amino acids are also present.
4. At the end of the assignment process, it is possible to identify every peak in the ^{1}H,^{15}N-HSQC and link it to the corresponding amino acid residue in the protein sequence; therefore, this spectrum constitutes the actual "fingerprint" of the protein (Fig. 2B). Moreover, the HN chemical shifts often undergo sensible variations because of many different parameters, such as secondary structure elements, pH, temperature, interaction with other molecules, etc. For this reason, the ^{1}H,^{15}N-HSQC spectrum can give important structural information also before the protein 3D structure has been solved (or if the protein structure was already solved by other techniques).

5. Additional information deriving from the chemical shift assignment

5.1 Secondary structure and dihedral angles prediction

Once the protein chemical shifts are assigned, it is possible to extract further information from these data alone, before performing further structural calculation. In particular, from the deviation of measured (δ_{ex}) Cα and Cβ values, the so-called secondary chemical shifts, from typical random coil values (δ_{rc}), it is possible to predict the amount of alpha and beta secondary structure present in the protein. The chemical shifts of Hα, H^{N}, N^{H} and C′ chemical shifts are also linked to the secondary structure, though not all as strongly as the Cα and Cβ atoms. The Wishart lab has done lots of work on the topic and on its webpage many interesting servers can be found, including CSI 3.0 (Hafsa, Arndt, & Wishart, 2015), that allows the prediction of secondary structure from experimental chemical shift values (http://csi3.wishartlab.com/cgi-bin/index.php). In particular, when four or more consecutive Hα chemical shifts result upshifted ($\delta_{ex} < \delta_{rc}$) with respect to random coil value, this indicates a putative α-helix region; conversely, if they

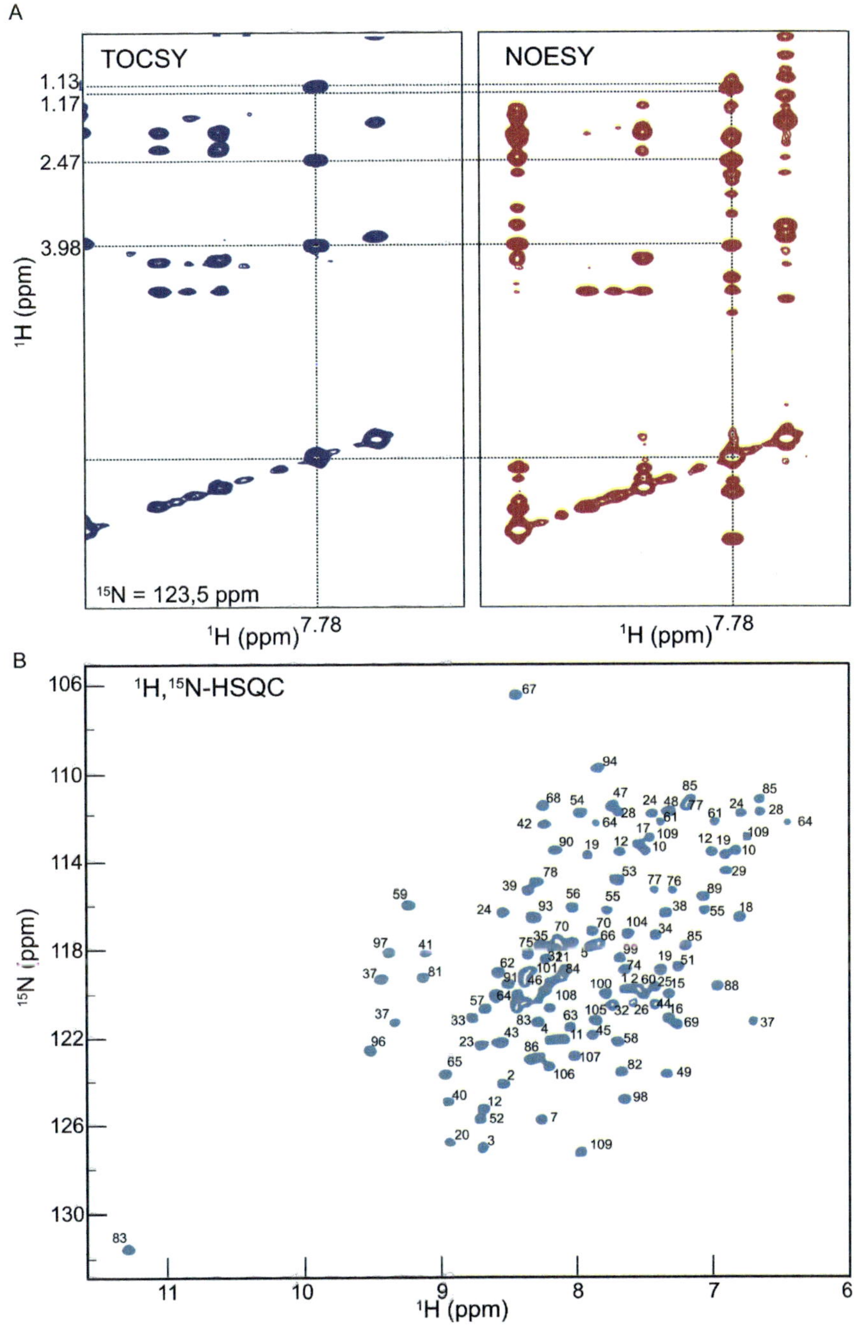

Fig. 2 (A) Representative planes from the TOCSY- and NOESY-^{1}H,^{15}N-HSQC spectra in correspondence of residue Val82 (δ_N 123.5 ppm). (B)The ^{1}H,^{15}N-HSQC spectrum of CSP-sg4 with the complete cross-peak assignment represents the fingerprint of the folded protein.

result downshifted ($\delta_{ex} > \delta_{rc}$), it indicates a putative β-sheet region. The Cα secondary chemical shifts show the opposite trend: a consensus of upshifted values indicates a β-sheet region, whereas consensus of downshifted values indicates α-helix region. These two analyses can be performed independently and can be coupled to improve the prediction accuracy. The chemical shifts data obtained for CSP-sg4 according to the instructions provided in Section 3, were analyzed using CSI 3.0 (Fig. 3, black).

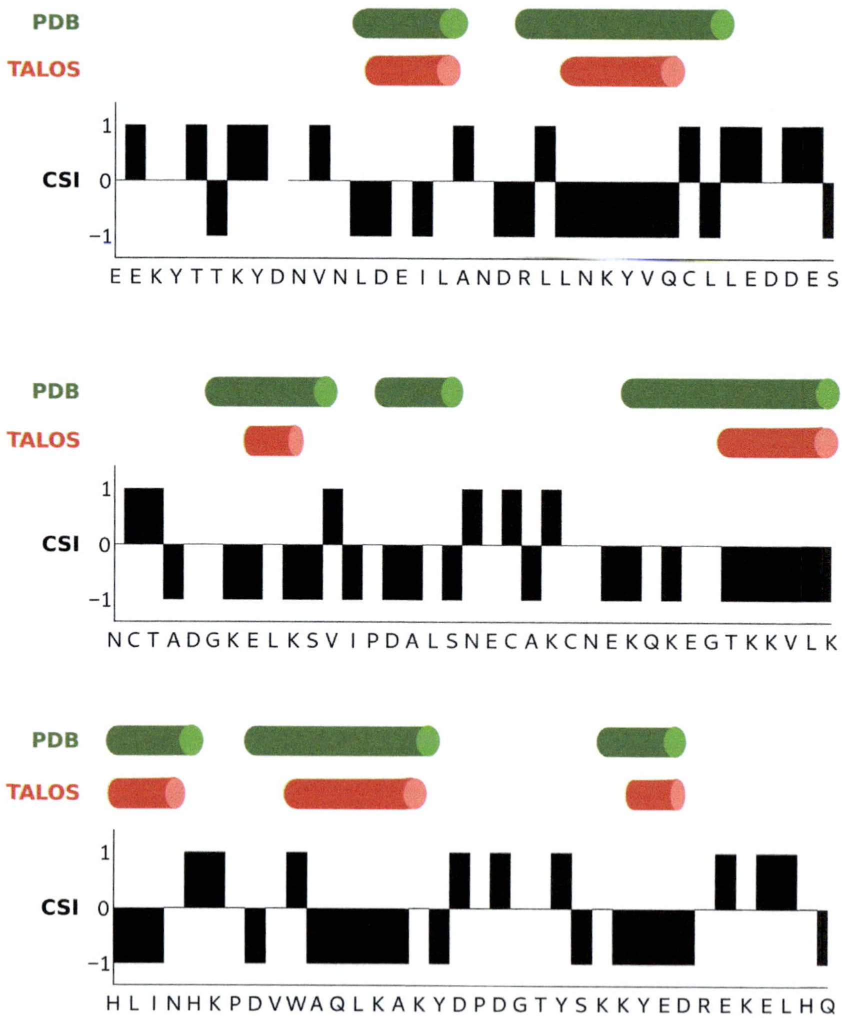

Fig. 3 Comparison between secondary structure predictions performed with CSI 3.0 (black column graph) and TALOS+ (red rods) and the final structure of CSP-sg4 solved by NMR including NOE restraints (green rods, PDB: 2gvs).

Another way of predicting the protein secondary structure from the assigned chemical shifts is to compare them with a database of chemical shift and sequence homology, to deduct protein backbone angle restraints and extract the Ramachandran plots. In fact, it is well known that regions of regular secondary structure correspond to well defined values of the dihedral angles ϕ and ψ. Consequently, it is possible to predict backbone dihedral angles, that can be used in subsequent, more refined structure calculations, from the values of secondary chemical shifts. Various software can be used for this purpose. For CSP-sg4 we have performed the prediction using the program TALOS+ (Shen, Delaglio, Cornilescu, & Bax, 2009) (Fig. 3, red), an improved version of the original program TALOS (Cornilescu, Delaglio, & Bax, 1999), that uses secondary shift and sequence information to make quantitative predictions for the protein backbone angles ϕ and ψ, providing also an estimate of the uncertainties on the predictions. TALOS+ is also available as a webserver (https://spin.niddk.nih.gov/bax/nmrserver/talos/).

The predictions for CSP-sg4 have been compared with those obtained with CSI 3.0 and with the complete structure obtained from full structure calculations based on the distance restraints from NOE spectra (Fig. 3, PDB: 2gvs).

As visible, both predictions manage to detect most of the helical regions of the fold (green rods), with the only exception of the fourth helix (residues 51–54), which was not detected by TALOS+ and only hinted by the CSI.

5.2 Amide hydrogen/deuterium exchange and calculation of the temperature coefficients

Hydrogen/deuterium exchange (HDX) is a chemical reaction in which a covalently bonded hydrogen atom is replaced by a deuterium atom from the solvent, or *vice versa*. Protons bound to strong electron-attractor atoms are easily exchanged in the presence of a suitable deuterium source, without the need of any catalyst. In proteins, exchangeable protons are typically those of the backbone amide group and the protons of the polar and charged side chains. HDX gives information about the solvent accessibility to various parts of the protein, and thus also about its tertiary structure and local flexibility. Indeed, protons that exhibit fast HDX rates are likely embedded in highly dynamic and/or unstructured regions, whereas protons characterized by slow exchange rate are likely involved in interactions (hydrogen bonds, contact surfaces in ligand-protein or protein-protein complexes, etc.). In the most basic experiment, the HDX is allowed to take place for a set time before quenching it, typically by lowering the pH. HDX can be easily followed by a series of $^{1}H,^{15}N$-HSQC spectra, as the signals intensity will decay exponentially as the hydrogen is substituted by deuterium. The data are then fitted with the exponential function:

$$I_t = I_0 e^{-k_{obs} t}$$

where I_t represents the cross-peak integral at time t and I_0 the original cross-peak integral (i.e., the intensity of the peak measured in the same experimental conditions, but for the presence of $^{1}H_2O$), to obtain the exchange constant for each protein residue simultaneously. In the case of CSP-sg4 protein, this experiment was performed recording a $^{1}H,^{15}N$-HSQC spectrum every 30 min after the protein had been dissolved in 100% D_2O. The HDX rates were calculated for every residue, and on its base the amide protons were classified as fast and slow exchanging, thus producing the solvent accessibility map (Fig. 4A).

The amide proton chemical shifts display a well-known temperature dependence, they shift upfield as increasing the temperature and this behavior is usually described as a negative temperature coefficient. This effect is

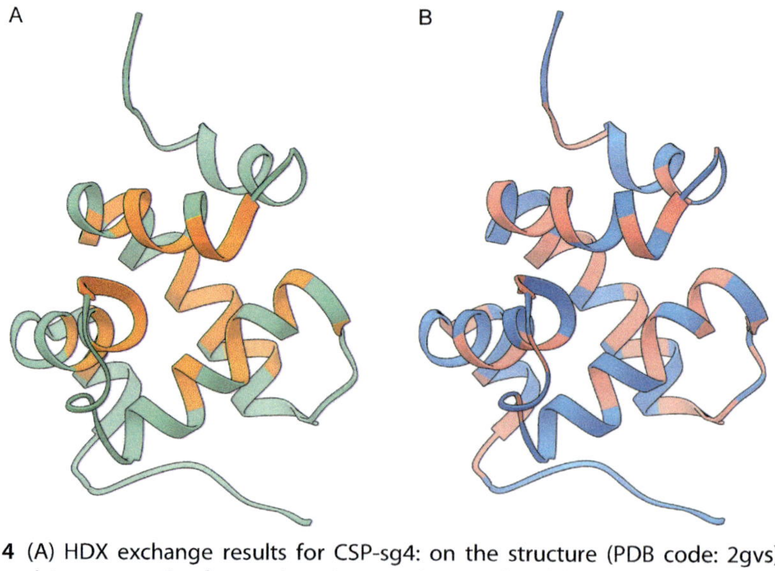

Fig. 4 (A) HDX exchange results for CSP-sg4: on the structure (PDB code: 2gvs) are mapped in orange the fast exchanging residues and in blue the slow exchangers. (B) Temperature coefficients of the amide protons: in pink are indicated the residues with the lowest temperature coefficients. The most flexible residues in both analyses roughly define the contour of the binding pocket.

due to a weakening of amide protons hydrogen bond as the temperature increase, and therefore to a minor downfield shift caused by the interaction (i.e., a relative upfield shift). The value of the amide proton temperature coefficient is still widely used to predict hydrogen bond donors, with values more negative than about −5 ppb/K being considered as an indicator of hydrogen bonding (Fig. 4B). These coefficients can be obtained taking the gradient of the best-fit line applied to the chemical shift versus temperature data. This approach is often used in combination with HDX data in order to achieve a more reliable prediction of protein secondary structure and hydrogen bonding because it compensates the pH and solvent exposure dependence of exchange data that can bias the prediction (Baxter & Williamson, 1997). Very recently a web-server tool for temperature coefficients calculation and analysis was proposed, able also to detect deviations from linearity that can be related to other structural insights such as local denaturation or fast exchange between different conformations (Trainor, Palumbo, MacKenzie, & Meiering, 2019).

5.3 Chemical shift perturbation

The measure of chemical shift perturbation, i.e., the measure of the variation of chemical shifts of the protein as a function of different ligand concentrations, is a quick and extremely efficient way to extrapolate information on protein ligand binding mode (Williamson, 2013). Usually, a series of ^{1}H,^{15}N-HSQC experiments is recorded while adding increasing amounts of a ligand and the chemical shifts are monitored. As the interaction partner is added, some chemical shifts will be perturbed. As a rule of thumb, a significant perturbation can be defined as a value higher than the average of all combined perturbations plus one standard deviation. Usually, these belong to amino acids close to the interaction surface. It is good to keep in mind that a chemical shift variation indicates a variation of the magnetic environment of a nucleus, which does not necessarily imply a direct interaction with a binding partner. Indeed, if a protein undergoes substantial conformational rearrangement upon complex formation, widespread chemical shift perturbations may be observed, even in residues far from the interaction site (which nonetheless experience a change in their structural environment).

Chemical shift perturbations are very sensitive even to subtle conformational effects and represent a suitable method to study weak interactions, better for binding constants in the mM range, such as those expected for binding and delivery of chemical stimuli. It is also possible to follow the chemical shift perturbation of the protein of interest as a function of other factors, such as pH: one can prepare different protein samples with the same concentration but at different pH and then record $^{1}H,^{15}N$-HSQC spectra and map the variation of chemical shifts as function of pH. This kind of experiment has been very useful for the understanding of the structure activity relationship in various OBPs (Damberger et al., 2013; Katre et al., 2009; Lescop, Briand, Pernollet, & Guittet, 2009; Portman et al., 2014).

6. Structure calculation and analysis

Once protein chemical shifts are assigned it is possible to move to structure calculation. The basis for a successful structural determination is the collection of a large number of structural restraints, 15/20 restraints per residue are usually needed. Among these, the most important come from NOE cross-peaks, whose intensity is related to interproton distance, but the analysis of some coupling constant, which depend on dihedral angles, can also be useful. In the first stage of structure calculation, resonance assignments (Section 4.4) are used to generate tentative assignments of NOE cross-peaks in NOESY spectra. This procedure is carried on using a computer program that converts assignments into distances and generates a first set of structures. NMR structures are always represented as a bundle of conformers, to indicate that more than one conformation in solution can fulfill experimental data and also to give a visual idea of how good is the agreement between the experimental data and the 3D structure. This first set of calculated structures is then used to discard tentative assignments that are incompatible with the majority of the calculated structures and to extend the resonance assignment list. A new set of structures is then computed and the cycle is iterated several times. Various software implements this strategy. The most used are ARIA2 (Rieping et al., 2007) and CYANA (Güntert, 2004). Automated procedures are reliable if the chemical shifts (including side chains) have been assigned above 90% completeness. The 3D structure calculation from NMR restraints is generally done using the so-called restrained molecular dynamics (rMD) simulations. The term "restrained" indicates that the simplified force field based on the protein covalent structure (van der Waals interaction and peptide plane planarity) is integrated by a *pseudo* force field that takes into account all NMR-derived conformational restraints. To overcome energy barriers and converge toward a global minimum, the rMD simulation is first carried out at higher temperature (*simulated annealing,* the atoms have high kinetic energy) and with a simplified force field to speed up the calculation. The process is repeated from several random initial structures to sample more extensively the conformational landscape. At the end of the protocol, the molecule is cooled down and a MD simulation with a complete force field (Lennard-Jones and electrostatic interactions) in a box filled with explicit water molecules is performed.

We performed structure calculations on CSP-sg4 using the CYANA 2.0 package (Güntert, 2004) 20,000 steps of torsion angle dynamics, producing 100 structures in total, and the 20 with the lowest target function, a parameter that gives a measure of how experimental restraints are fulfilled in the calculated structure, were used in the next stage.

The protocol for complete structural calculation from NMR data requires a certain degree of training and should be performed, at least at the beginning, with the help of more experienced users. Since the detailed protocol depends greatly on the software of choice, the user is advised to refer to the software manual, and the steps will be described only in general terms.

In the case of CSP-sg4, most NOE assignments were based on the automatic procedure implemented by CYANA 2.0 from the analysis of a $^{1}H,^{1}H$-NOESY and a 3D NOESY-$^{1}H,^{13}C$-HSQC spectrum and 36 NOE correlations were assigned manually, resulting in a total of 1931 NOEs derived distances. These NOE-derived distance restraints were used in conjunction with 127 dihedral restraints calculated on the basis of the chemical shift values with TALOS (Cornilescu et al., 1999) (see Section 4.3) and 47 $^{3}J_{HNH\alpha}$ coupling restraints, corresponding to an average of 19 constraints per residue.

In the final run, 200 structures were computed, and the 40 with lowest target function (i.e., the best agreement with the NMR data) were refined by 4000 steps of restrained minimization using the more realistic AMBER99 force field (Ponder & Case, 2003), using an implicit solvation model (Hawkins, Cramer, & Truhlar, 1995, 1996) to account for the influence of solvent. The AMBER-minimized structures were sorted by increasing restraint violation, and the first 20 were selected as representative protein models.

Tips: Homonuclear 2D experiments, which provide higher resolution and allow for the detection of specific signals or cross-peaks in less crowded spectral regions (i.e., Hα-Hα correlations for β-sheet refinements) can be also recorded to aid the structure determination procedure. In this case, an unlabeled protein sample can be used.

6.1 Structure quality assessment

Once structures have been calculated, it is necessary to assess their quality. Existing software packages can be used to generate validation reports for all submitted protein NMR structures (Vuister, Fogh, Hendrickx, Doreleijers, & Gutmanas, 2014). Generally, this software checks the completeness of the chemical shift data, analyzes the geometry of calculated structures and validates restraints versus model structures.

To analyze structures, it is essential a structure visualization and analysis program like Pymol (PyMOL The PyMOL Molecular Graphics System, Version 2.0 Schrödinger, LLC.) or UCSF Chimera (Pettersen et al., 2004).

One essential feature in CSPs structures is the presence of cavities able to host ligands. Once the structure has been calculated, it is possible to access the Computed Atlas of Surface Topography of proteins (CASTp) server (http://sts.bioe.uic.edu/castp/index.html?4jii) (Binkowski, Naghibzadeh, & Liang, 2003). Based on an input PDB file containing the calculated protein structure, the server locates, defines and measures concave surface regions, including pockets located on protein surfaces and voids buried in the interior of proteins. The server is also directly accessible as a plug-in from Chimera.

7. Summary

Following the protocol presented in this chapter, it is possible to perform all the steps required to achieve the full structure characterization of OBPs and CSPs by NMR spectroscopy. This approach should always be comforted by and complemented with other experimental techniques. Using very similar protocols, unlabeled or uniformly labeled proteins can be produced, which serve as substrates for different kinds of analysis. For instance, while double labeling is necessary to perform full chemical shift assignment, unlabeled proteins are needed for high resolution two-dimensional homonuclear experiments. We have described how to purify the protein to homogeneity with a simple approach based on ion exchange chromatography, followed by a finer purification on reverse phase HPLC. Since NMR spectroscopy is extremely time and resource consuming, extensive preliminary checks should be performed before the sample is submitted for a complete series of 2D and 3D experiments. These checks should include, but are not limited to, SDS-PAGE, mass spectrometry, to confirm isotope incorporation and protein purity, and CD spectroscopy, to identify diagnostic patterns of secondary structure elements. Once one or more of these checks have been performed, the protein can be used to acquire NMR data, which then need to be processed and analyzed. We have described a general procedure to obtain the full resonance assignment, and to utilize chemical shift data to extrapolate more refined structural information, such as secondary structure elements within the sequence. We have also presented the general workflow to convert NOE data into distance restraints for structure calculations with rMD. NMR spectroscopy is a very versatile technique, able to provide much more information than that presented in this chapter. We have described two examples of techniques that allow to highlight differences in flexibility and plasticity in various regions of the protein, such as HDX and determination of the temperature coefficients. Additionally, one may envision to perform studies to map the surface of the protein with paramagnetic probes, to corroborate data on the accessibility of the protein surface to larger molecules, or to perform comparative studies to measure the affinity of a given protein for different ligands. The outcomes of these experiments could easily be compared or integrate other techniques, such as those based on SPR for ligand binding detection or ITC for ligand affinity quantitation (Portman et al., 2014).

References

Baxter, N. J., & Williamson, M. P. (1997). Temperature dependence of ^{1}H chemical shifts in proteins. *Journal of Biomolecular NMR*, *9*, 359–369. https://doi.org/10.1023/A:1018334207887.

Berkmen, M. (2012). Production of disulfide-bonded proteins in Escherichia coli. *Protein Expression and Purification*, *82*(1), 240–251. https://doi.org/10.1016/j.pep.2011.10.009.

Binkowski, T. A., Naghibzadeh, S., & Liang, J. (2003). CASTp: Computed atlas of surface topography of proteins. *Nucleic Acids Research*, *31*(13), 3352–3355. https://doi.org/10.1093/nar/gkg512.

Bucci, B. K., Kruse, S. W., Thode, A. B., Alvarado, S. M., & Jones, D. N. M. (2006). Effect of *n*- alcohols on the structure and stability of the *Drosophila* odorant binding protein LUSH. *Biochemistry*, *45*(6), 1693–1701. https://doi.org/10.1021/bi0516576.

Campanacci, V., Mosbah, A., Bornet, O., Wechselberger, R., Jacquin-Joly, E., Cambillau, C., et al. (2001). Chemosensory protein from the moth Mamestra brassicae. *European Journal of Biochemistry*, *268*(17), 4731–4739. https://doi.org/10.1046/j.1432-1327.2001.02398.x.

Clore, G. M., & Gronenborn, A. M. (1991). Applications of three- and four-dimensional heteronuclear NMR spectroscopy to protein structure determination. *Progress in Nuclear Magnetic Resonance Spectroscopy*, *23*(1), 43–92. https://doi.org/10.1016/0079-6565(91)80002-J.

Cornilescu, G., Delaglio, F., & Bax, A. (1999). Protein backbone angle restraints from searching a database for chemical shift and sequence homology. *Journal of Biomolecular NMR*, *13*(3), 289–302. https://doi.org/10.1023/a:1008392405740.

Damberger, F., Horst, R., Wüthrich, K., Peng, G., Nikonova, L., & Leal, W. S. (2000). NMR characterization of a pH-dependent equilibrium between two folded solution conformations of the pheromone-binding protein from *Bombyx mori*. *Protein Science*, *9*(5), 1038–1041. https://doi.org/10.1110/ps.9.5.1038.

Damberger, F., Ishida, Y., Leal, W. S., & Wüthrich, K. (2007). Structural basis of ligand binding and release in insect pheromone-binding proteins: NMR structure of Antheraea polyphemus PBP1 at pH 4.5. *Journal of Molecular Biology*, *373*(4), 811–819. https://doi.org/10.1016/j.jmb.2007.07.078.

Damberger, F., Michel, E., Ishida, Y., Leal, W. S., & Wüthrich, K. (2013). Pheromone discrimination by a pH-tuned polymorphism of the Bombyx mori pheromone-binding protein. *Proceedings of the National Academy of Sciences of the United States of America*, *110*(46), 18680–18685. https://doi.org/10.1073/pnas.1317706110.

Delaglio, F., Grzesiek, S., Vuister, G. W., Zhu, G., Pfeifer, J., & Bax, A. (1995). NMRPipe: A multidimensional spectral processing system based on UNIX pipes. *Journal of Biomolecular NMR*, *6*(3), 277–293. https://doi.org/10.1007/BF00197809.

Elbing, K., & Brent, R. (2002). Media preparation and bacteriological tools. *Current Protocols in Molecular Biology*, *59*(1), 1.1.1–1.1.7. https://doi.org/10.1002/0471142727.mb0101s59.

Eyun, S., Soh, H. Y., Posavi, M., Munro, J. B., Hughes, D. S. T., Murali, S. C., et al. (2017). Evolutionary history of chemosensory-related gene families across the Arthropoda. *Molecular Biology and Evolution*, *34*(8), 1838–1862. https://doi.org/10.1093/molbev/msx147.

Güntert, P. (2004). Automated NMR structure calculation with CYANA. *Methods in Molecular Biology (Clifton, N.J.)*, *278*, 353–378. https://doi.org/10.1385/1-59259-809-9:353.

Hafsa, N. E., Arndt, D., & Wishart, D. S. (2015). CSI 3.0: A web server for identifying secondary and super-secondary structure in proteins using NMR chemical shifts. *Nucleic Acids Research*, *43*(W1), W370–W377. https://doi.org/10.1093/nar/gkv494.

Hawkins, G. D., Cramer, C. J., & Truhlar, D. G. (1995). Pairwise solute descreening of solute charges from a dielectric medium. *Chemical Physics Letters*, *246*(1–2), 122–129. https://doi.org/10.1016/0009-2614(95)01082-K.

Hawkins, G. D., Cramer, C. J., & Truhlar, D. G. (1996). Parametrized models of aqueous free energies of solvation based on pairwise descreening of solute atomic charges from a dielectric medium. *The Journal of Physical Chemistry*, *100*(51), 19824–19839. https://doi.org/10.1021/jp961710n.

Horst, R., Damberger, F., Luginbuhl, P., Guntert, P., Peng, G., Nikonova, L., et al. (2001). NMR structure reveals intramolecular regulation mechanism for pheromone binding and release. *Proceedings of the National Academy of Sciences of the United States of America*, *98*(25), 14374–14379. https://doi.org/10.1073/pnas.251532998.

Jansen, S., Chmelík, J., Žídek, L., Padrta, P., Novák, P., Zdráhal, Z., et al. (2007). Structure of Bombyx mori chemosensory protein 1 in solution. *Archives of Insect Biochemistry and Physiology*, *66*(3), 135–145. https://doi.org/10.1002/arch.20205.

Johnson, B. A., & Blevins, R. A. (1994). NMR view: A computer program for the visualization and analysis of NMR data. *Journal of Biomolecular NMR*, *4*(5), 603–614. https://doi.org/10.1007/BF00404272.

Katre, U. V., Mazumder, S., Prusti, R. K., & Mohanty, S. (2009). Ligand binding turns moth pheromone-binding protein into a pH sensor effect on the Antheraea polyphemus PBP1 conformation. *Journal of Biological Chemistry*, *284*(46), 32167–32177. https://doi.org/10.1074/jbc.M109.013383.

Lartigue, A., Campanacci, V., Roussel, A., Larsson, A. M., Jones, T. A., Tegoni, M., et al. (2002). X-ray structure and ligand binding study of a moth chemosensory protein. *Journal of Biological Chemistry*, *277*(35), 32094–32098. https://doi.org/10.1074/jbc.M204371200.

Leal, W. S., Barbosa, R. M. R., Xu, W., Ishida, Y., Syed, Z., Latte, N., et al. (2008). Reverse and conventional chemical ecology approaches for the development of oviposition attractants for Culex mosquitoes. *PLoS One*, *3*(8), e3045. https://doi.org/10.1371/journal.pone.0003045.

Lee, D., Damberger, F. F., Peng, G., Horst, R., Güntert, P., Nikonova, L., et al. (2002). NMR structure of the unliganded Bombyx mori pheromone-binding protein at physiological pH. *FEBS Letters*, *531*(2), 314–318. https://doi.org/10.1016/S0014-5793(02)03548-2.

Lescop, E., Briand, L., Pernollet, J.-C., & Guittet, E. (2009). Structural basis of the broad specificity of a general odorant-binding protein from honeybee. *Biochemistry*, *48*(11), 2431–2441. https://doi.org/10.1021/bi802300k.

Maleszka, R., & Stange, G. (1997). Molecular cloning, by a novel approach, of a cDNA encoding a putative olfactory protein in the labial palps of the moth Cactoblastis cactorum1GenBank accession no. U95046.1. *Gene*, *202*(1), 39–43. https://doi.org/10.1016/S0378-1119(97)00448-4.

Mao, Y., Xu, X., Xu, W., Ishida, Y., Leal, W. S., Ames, J. B., et al. (2010). Crystal and solution structures of an odorant-binding protein from the southern house mosquito complexed with an oviposition pheromone. *Proceedings of the National Academy of Sciences of the United States of America*, *107*(44), 19102–19107. https://doi.org/10.1073/pnas.1012274107.

Marley, J., Lu, M., & Bracken, C. (2001). A method for efficient isotopic labeling of recombinant proteins. *Journal of Biomolecular NMR*, *20*(1), 71–75. https://doi.org/10.1023/A:1011254402785.

Mosbah, A., Campanacci, V., Lartigue, A., Tegoni, M., Cambillau, C., & Darbon, H. (2003). Solution structure of a chemosensory protein from the moth Mamestra brassicae. *Biochemical Journal*, *369*(1), 39–44. https://doi.org/10.1042/bj20021217.

Pelosi, P. (1994). Odorant-binding proteins. *Critical Reviews in Biochemistry and Molecular Biology*, *29*(3), 199–228. https://doi.org/10.3109/10409239409086801.

Pelosi, P., Iovinella, I., Felicioli, A., & Dani, F. R. (2014). Soluble proteins of chemical communication: An overview across arthropods. *Frontiers in Physiology*, *5*, 320. https://doi.org/10.3389/fphys.2014.00320.

Pettersen, E. F., Goddard, T. D., Huang, C. C., Couch, G. S., Greenblatt, D. M., Meng, E. C., et al. (2004). UCSF chimera—A visualization system for exploratory research and analysis. *Journal of Computational Chemistry*, *25*(13), 1605–1612. https://doi.org/10.1002/jcc.20084.

Picone, D., Crescenzi, O., Angeli, S., Marchese, S., Brandazza, A., Ferrara, L., et al. (2001). Bacterial expression and conformational analysis of a chemosensory protein from Schistocerca gregaria. *European Journal of Biochemistry*, *268*(17), 4794–4801. https://doi.org/10.1046/j.1432-1327.2001.02408.x.

Ponder, J. W., & Case, D. A. (2003). Force fields for protein simulations. *Advances in Protein Chemistry*, *66*, 27–85. https://doi.org/10.1016/s0065-3233(03)66002-x.

Portman, K. L., Long, J., Carr, S., Briand, L., Winzor, D. J., Searle, M. S., et al. (2014) Enthalpy/entropy compensation effects from cavity desolvation underpin broad ligand binding selectivity for rat odorant binding protein 3. *Biochemistry*, *53*(14), 2371–2379. https://doi.org/10.1021/bi5002344.

Rieping, W., Habeck, M., Bardiaux, B., Bernard, A., Malliavin, T. E., & Nilges, M. (2007). ARIA2: Automated NOE assignment and data integration in NMR structure calculation. *Bioinformatics (Oxford, England)*, *23*(3), 381–382. https://doi.org/10.1093/bioinformatics/btl589.

Sattler, M., Schleucher, J., & Griesinger, C. (1999). Heteronuclear multidimensional NMR experiments for the structure determination of proteins in solution employing pulsed field gradients. *Progress in Nuclear Magnetic Resonance Spectroscopy*, *34*(2), 93–158. https://doi.org/10.1016/S0079-6565(98)00025-9.

Shen, Y., Delaglio, F., Cornilescu, G., & Bax, A. (2009). TALOS+: A hybrid method for predicting protein backbone torsion angles from NMR chemical shifts. *Journal of Biomolecular NMR*, *44*(4), 213–223. https://doi.org/10.1007/s10858-009-9333-z.

Tomaselli, S., Crescenzi, O., Sanfelice, D., Ab, E., Wechselberger, R., Angeli, S., et al. (2006). Solution structure of a chemosensory protein from the desert locust Schistocerca gregaria. *Biochemistry, 45*(35), 10606–10613. https://doi.org/10.1021/bi060998w.

Trainor, K., Palumbo, J. A., MacKenzie, D. W. S., & Meiering, E. M. (2019). Temperature dependence of NMR chemical shifts: Tracking and statistical analysis. *Protein Science Special Issue: Tools for Protein Science, 29*(1), 306–314. https://doi.org/10.1002/pro.3785.

Vieira, F. G., & Rozas, J. (2011). Comparative genomics of the odorant-binding and chemosensory protein gene families across the Arthropoda: Origin and evolutionary history of the chemosensory system. *Genome Biology and Evolution, 3*, 476–490. https://doi.org/10.1093/gbe/evr033.

Vuister, G. W., Fogh, R. H., Hendrickx, P. M. S., Doreleijers, J. F., & Gutmanas, A. (2014). An overview of tools for the validation of protein NMR structures. *Journal of Biomolecular NMR, 58*(4), 259–285. https://doi.org/10.1007/s10858-013-9750-x.

Williamson, M. P. (2013). Using chemical shift perturbation to characterise ligand binding. *Progress in Nuclear Magnetic Resonance Spectroscopy, 73*, 1–16. https://doi.org/10.1016/j.pnmrs.2013.02.001.

Wojtasek, H., & Leal, W. S. (1999). Conformational change in the pheromone-binding protein from Bombyx mori induced by pH and by interaction with membranes. *Journal of Biological Chemistry, 274*(43), 30950–30956. https://doi.org/10.1074/jbc.274.43.30950.

Xu, X., Xu, W., Rayo, J., Ishida, Y., Leal, W. S., & Ames, J. B. (2010). NMR structure of navel orangeworm moth pheromone-binding protein (AtraPBP1): Implications for pH-sensitive pheromone detection. *Biochemistry, 49*(7), 1469–1476. https://doi.org/10.1021/bi9020132.

Zhu, J., Iovinella, I., Dani, F. R., Pelosi, P., & Wang, G. (2019). Chemosensory proteins: A versatile binding family. In J.-F. Picimbon (Ed.), *Olfactory concepts of insect control—Alternative to insecticides* (pp. 147–169). Springer International Publishing. https://doi.org/10.1007/978-3-030-05165-5_6.

CHAPTER NINE

Stability of OBPs

Nadja Hellmann*
Department of Chemistry/Biochemistry, Johannes Gutenberg-University Mainz, Mainz, Germany
*Corresponding author: e-mail address: nhellmann@uni-mainz.de

Contents

Abstract

Odorant binding proteins (OBPs) are small proteins, some of which bind odorants with high specificity. OBPs are relatively easy to produce and show a pronounced stability toward thermal and chemical denaturation. This high stability renders OBPs attractive

Methods in Enzymology, Volume 642
ISSN 0076-6879
https://doi.org/10.1016/bs.mie.2020.05.011

candidates for the development of odorant detections systems. Unfortunately, binding of odorants is not easy to quantify due to lack of spectroscopic signals upon binding. Therefore, a possible approach to detect binding is to employ the shift in thermal or chemical stability upon ligand–protein interaction. Being a rather indirect approach, the experimental setup should be done with care. Here, the experimental results on stability of OBPs are summarized and issues which should be considered when performing stability experiments are discussed.

Abbreviations

(ATR)-FTIR (Attenuated-Total-Reflection)-Fourier-Transformed Infrared
1-AMA 1-aminoanthracene
CD circular dichroism
DHM 2,6-dimethyl-7-octen-2-ol
DMO 3,7-dimethyl-1-octanol
DSC differential scanning calorimetry
eugenol 4-Allyl-2-methoxyphenol
GdnHCl guanidine hydrochloride
geraniol 2,6-Dimethyl-*trans*-2,6-octadien-8-ol
IBMP 2-isobutyl-3-methoxy-pyrazine
ITC isothermal titration calorimetry
NMR nuclear magnetic resonance
OCT 1-octen-3-ol
OBP odorant binding proteins
PYR 2-isobuthyl-3-methoxypyrazine

1. Introduction

Odorant binding proteins (OBPs) are found in the nasal epithelium of vertebrates or in the sensillar lymph of insects, where they are supposed to function as carriers for the largely hydrophobic airborne odorants, allowing them to reach their target olfactory receptors (Steinbrecht, 1998). OBPs show a pronounced stability toward thermal and chemical denaturation. This high stability renders OBPs attractive candidates for the development of odorant detections systems (Pelosi, Zhu, & Knoll, 2018).

In the case of vertebrate OBPs, which are members of the lipocalin family, the high stability is believed to result from the compact hydrophobic core of the β-barrel. Some OBPs form dimers, and in case of bovine OBPs, these dimers are formed by domain swapping (Bianchet et al., 1996; Tegoni, Ramoni, Bignetti, Spinelli, & Cambillau, 1996). Insect OBPs are structurally completely unrelated, containing six α-helical domains, assembled in a compact structure (Sandler, Nikonova, Leal, & Clardy, 2000; Tegoni, Campanacci, & Cambillau, 2004; Ziemba, Murphy, Edlin, & Jones, 2013). For insect OBPs also dimeric variants exist, and in the case of *Anopheles gambiae* OPB48, a domain swapped dimer (Tsitsanou et al., 2013) was found. Exact values for denaturation temperatures and the unfolding mechanism can depend on species, isoform, posttranslational modifications (Nagnan-Le Meillour et al., 2019), source (purified from natural sources or heterologously expressed) (Stepanenko et al., 2014) and purification method, due to potential presence of endogenous ligands (Burova et al., 2000).

Binding of ligands to OBPs usually does not lead to changes in secondary or tertiary structure large enough to be of practical application, therefore either radioactive labelled ligands have to be used (Vincent et al., 2000) or competition experiments with a fluorescent ligand (e.g., 1-AMA, see also Chapter "Ligand-binding assays with OBPs and CSPs" by Pelosi). Alternatively, the change in stability of OBPs in the presence of ligands has been used to evaluate the binding affinity of different odorants (see Table 3).

Experimental approaches to determine OBP's stability include differential calorimetry (DSC), Nuclear Magnetic Resonance (NMR), Fourier-Transformed Infrared (FTIR), fluorescence and circular dichroism (CD) spectroscopy. In the following sections the focus is on the last three methods. The comparison of protein denaturation data obtained with different methods and under different conditions is not always straightforward due to a number of parameters determining the observed denaturation curve. In this contribution the most important aspects to be considered when evaluating chemical and thermal protein stability are discussed, based on the studies performed on OBPs.

2. Methods for studying protein stability

When studying protein stability, the native structure of the protein is disturbed by adding chaotropic agents such as Guanidine hydrochloride (GdnHCl) or urea, or increasing the temperature. Different methods address typically only certain aspects of the protein's structure (quaternary, tertiary or secondary structure changes), and therefore are often used in combination.

2.1 Fourier-transformed infrared (FTIR) spectroscopy

Of the three methods described in this chapter, FTIR spectroscopy is the one with the highest information content, but is, on the other hand, the most complex, both in terms of experimental approach, technical requirements and analysis of the data.

Infrared spectroscopy reports on the vibrational states of chemical bonds. Characteristic for proteins are the amide I band and the amide II band, originating from the C=O vibration and N—H bending modes, respectively. The amide I band (≈ 1600–$1700\,cm^{-1}$) is composed of a number of bands, which can be correlated to the secondary structure of a protein (Barth, 2007). Of importance for denaturation studies, aggregation often contributes with a characteristic band, around $1620\,cm^{-1}$, which reflects intermolecular β-sheets. Another important feature is that the position of the bands is different in H_2O and D_2O. FTIR experiments are usually performed after equilibration of the sample with buffer in D_20, in order to reduce unwanted background absorption of H_20 around $1645\,cm^{-1}$. During this procedure, the accessible amide protons of the protein are exchanged, too. If upon structural changes or partial denaturation more amide protons are exposed to the solvent, shifts in the amide I bands occur, indicating tertiary structure changes (Paolini et al., 1999). Therefore, with FTIR even small tertiary structure changes become visible due to shifts in the amide bands upon H/D exchange.

In general, the changes in absorption are not very pronounced, therefore usually FTIR spectra are presented as difference spectra in order to make the changes better visible (Schwaighofer, Kotlowski, et al., 2014). Due to the small changes involved it is also important that the baseline is very stable.

When comparing denaturation studies based on FTIR with those using CD or fluorescence, one should keep in mind that typically in FTIR experiments significantly higher protein concentrations are employed, e.g., 10 (Marabotti, Lefèvre, et al., 2008) or 36 mg/mL (Ausili et al., 2005; Marabotti, Scire, et al., 2008; Paolini et al., 1999). A high protein concentration might shift the transition between native and denatured state to higher temperatures or denaturant concentrations, if, e.g., the native state is a dimer, since at high protein concentrations the dimer is stabilized. In Attenuated Total Reflection FTIR (ATR-FTIR) spectroscopy the sample preparation is different: the protein is attached to the surface of a crystal, e.g., via a His-tag, and the remaining solution is replaced by buffer (Schwaighofer, Kotlowski, et al., 2014). Thus, the average protein concentration is rather low, but locally, at the surface, the concentration is high. Alternatively, the sample is placed on the crystal and dried to some extent (Khalil, Boubegtiten-Fezoua, Hellmann, & Hellwig, 2016). Thus, the actual protein concentration is not exactly defined, and the solution situation is probably "crowded," characterized by a low water content, which also tends to have an effect on the stability of proteins (Stepanenko et al., 2016c).

2.2 Fluorescence spectroscopy

Fluorescence spectroscopy relies on the usually pronounced dependence of a fluorophore's emission characteristics (emission maximum, quantum yield) on the local environment of the fluorophore. Upon a transfer of the fluorophore from a hydrophobic to a polar environment the emission maximum experiences a redshift. This is caused by solvent relaxation during the excited state of the fluorophore. In proteins, only tryptophan's emission depends strongly on environmental polarity. Therefore, if a protein possesses at least one tryptophan which is shielded from the solvent in the native state, denaturation processes can easily be followed by fluorescence spectroscopy. For example, tryptophans of OBPs in the native state exhibit an emission wavelength of about 335–340 nm, typical for a tryptophan buried within a hydrophobic cavity. The emission maximum is shifted to about 350 nm upon denaturation, typical for being completely solvent exposed. Usually the quantum yield of tryptophan decreases when exposed to a more polar environment, e.g., as found for bovine OBP (Stepanenko et al., 2016a). This does not seem to be the case for porcine OBP probably owing to formation of an excited complex between tryptophan and a lysine or a water molecule (Stepanenko et al., 2008).

Tryptophan fluorescence spectroscopy is sensitive solely to changes in the environment of tryptophan(s), thus reporting on local tertiary structure changes. The sensitivity of fluorescence is rather high, typically protein concentrations in the lower micromolar range are sufficient. However, the concentration should be high enough to yield a spectrum considerably more intense than the Raman-scattering peak, which inevitably overlaps with the tryptophan emission band. On the other hand, very high concentrations of protein should be avoided due to inner-filter effects. Also, addition of components which absorb strongly at the excitation or emission wavelength can lead to reduction of emitted light intensity, which could erroneously be interpreted as the result of an interaction.

Fluorescence quantum yield, and consequently the measured intensity, is intrinsically temperature dependent. At higher temperatures, more collisions with solvent molecules occur, thus the probability of radiation-less relaxation increases, and the fraction of molecules relaxing via emission of a photon decreases. Therefore, it is important to measure a spectrum only after the solution has reached the final temperature, in particular when performing thermal denaturation experiments. This effect also leads to the characteristic shape of thermal denaturation curves of proteins, with a sloping intensity before and after the transition from native to denatured. In case of chemical denaturation, e.g., GdnHCl or urea, the intensity outside the transition region is also somewhat affected by interactions between the tryptophan itself and the denaturant (see fig. 5 in (Polverini et al., 2011).

In order to monitor the denaturation process, different parameters can be employed, like emission maximum (λ_{max}), the ratio of the intensities at two wavelengths, or the intensity at a given wavelength. Not all these parameters are equally well suited for quantitative analysis, see Stepanenko et al. (2014) and Section 3.2.

More advanced parameters characterising the fluorescence emission of tryptophan, such as lifetime and anisotropy have been also employed (Staiano et al., 2007) but are not discussed here.

2.3 Circular dichroism

Circular dichroism (CD) is based on the differential extinction coefficient of chiral chromophores for left and right circularly polarized light. Different secondary structures lead to characteristic CD spectra of the peptide bond between 180 and 250 nm. An intrinsic restriction in the application of CD spectroscopy is that many substances, including buffers and denaturants, also absorb in this region and thus may decrease the signal/noise ratio. Therefore, both the composition and the concentrations of buffers have to be chosen carefully. When performing denaturation studies, inevitably the far end (up to 210 nm) of the spectrum will be lost due to high absorption of the denaturant. Since CD spectra are, unlike fluorescence, not intrinsically temperature dependent, they offer a good alternative for thermal denaturation studies. The typical protein concentration is a bit higher than in case of fluorescence spectroscopy (about 0.2 mg/mL, corresponding to about 10 μM in the case of mammalian OBPs). Since far-UV spectroscopy reports mainly on secondary structure changes, it supplements information from tryptophan fluorescence studies. If the near-UV spectral range is employed (250–300 nm), information about environment of the aromatic amino acids can be obtained (e.g., Mazzini, Polverini, Parisi, Sorbi, & Favilla, 2007; Staiano et al., 2007). However, higher concentrations of protein (100 μM) and larger volumes are required.

3. General considerations when studying protein stability

All of the analyses discussed below are based on a thermodynamic equilibrium. There is some discussion, to which extent partially irreversible processes can nevertheless be analysed based on the same approach (Brandts & Lin, 1990; Wafer, Kloczewiak, Polleck, & Luo, 2017). Fortunately, in many cases OBPs were shown to unfold reversibly, or at least they do so within a reasonable temperature range (see Tables 1 and 2), justifying the application of thermodynamic models.

3.1 Equilibrium conditions and the issue of reversibility

Strictly speaking, when investigating protein denaturation, thermodynamic equilibrium means that the fractions of protein in the various states (native, intermediate, denatured) are only determined by the conditions (e.g., denaturant concentration, temperature) and not by the path which led to this state (e.g., per unfolding or per refolding). Thus, unfolding and refolding curves should not show any hysteresis, but should overlap. Then the process is reversible, and both curves are in equilibrium.

To achieve equilibrium conditions, as indicated by a temporally stable spectroscopic signal, it might take a while: for chemical denaturation of OBPs often incubation times of 24 h were employed. The refolding process might take even longer and often displays considerable hysteresis (Stepanenko et al., 2014, 2016b). Typically, the refolding curve is the one which does not represent equilibrium conditions.

In thermal denaturation studies, different results might be obtained for different heating rates. If denaturation is slower than the change in temperature, the denatured fraction lags behind and the curves obtained do not represent equilibrium conditions.

Irreversible aggregation of the denatured protein tends to occur in thermal denaturation (Parisi et al., 2005). If the temperature at which aggregation starts (T_{agg}) is considerably higher than the denaturation temperature (T_m), it is possible to separate the two processes, and restrict the analysis to the range below T_{agg}. Not unfrequently, however, both processes take place within a narrow temperature range, and the obtained value of T_m might be distorted.

Table 1 Thermal stability of OBPs.

OBP	Method	[OBP]/μM	Pre T_m/°C	Main T_m/°C	Buffer	pH	Rev.	Ref.
Porcine								
WT (D)	FTIR (λ_{max})	2100	60…70	80…95	TRIS/HCl/D_2O	7.0		Paolini et al. (1999)
	Fluo (λ_{max})	2	–	60…85	TRIS/HCl	7.5		
WT*	DSC	≈20	–	69	K_2PO_4/EDTA	6.6	y	Burova et al. (1999)
WT	Fluo (I_{rat})	28	–	70	?	?	n	Stepanenko et al. (2008)
	CD	28	–	>90	?	?		
WT	CD (θ_{205})	10	–	74	TRIS/HCl	7.8	y	Parisi et al. (2005)
$MutC^{minus}$	CD (θ_{205})	10	–	41	TRIS/HCl	7.8	<65°	Parisi et al. (2005)
Bovine								
WT_{rec} (D)	FTIR (λ_{max})	500	75…85	85…99	TRIS/HCl/D_2O	7.4		Marabotti, Scire, et al. (2008)
WT_{rec} (D,M)	DSC	25	79.5	83.4	K_2PO_4/NaCl	7.4	(y)	Gómez-Velasco, Rojo-Domínguez, and García-Hernández (2020)
		100	79.8	85.4	K_2PO_4/NaCl	7.4	(y)	
$MutG^+$ (M)		53	–	79.6	K_2PO_4/NaCl	7.4	(y)	
MutGCC (M)	FTIR (λ_{max})	500	70…80	90…99	TRIS/HCl/D_2O	7.4		Marabotti, Scire, et al. (2008)

Insect							
Honey bee OBP14 WT	ATR-FTIR	n.a.	–	≈55	K_2PO_4/NaCl	8.0	Schwaighofer, Kotlowski, et al. (2014)
	CD (θ_{222})	42	–	56	K_2PO_4/NaCl	8.0	
Honey bee OBP14 WT	CD (θ_{222})	42	–	56	K_2PO_4/NaCl	8.0	Schwaighofer, Pechlaner, et al. (2014)
Wasp OBP2 WT	CD (θ_{222})	42	–	64.9	K_2PO_4/NaCl	8.0	Ahmed et al. (2017)
Wasp OBP2 MutC^{minus}	CD (θ_{222})	42	–	58.4	K_2PO_4/NaCl	8.0	
Wasp OBP3 WT	CD (θ_{222})	42	–	65.8	K_2PO_4/NaCl	8.0	
Wasp OBP3 MutC^{minus}	CD (θ_{222})	42	–	59.6	K_2PO_4/NaCl	8.0	

Comparison of transition temperatures or transition ranges of different OBPs and different methods. A range is given, if the data presented were not analysed to yield a T_m. The values given in the table serve as an overview; for accurate information please refer to the original publication.
OBPs used; WT: wild type extracted from tissue; WT_{rec}: recombinantly expressed wild-type; WT*: in another study, no solvent extraction step was performed, leading to sample-dependent thermograms due to co-purified ligand (Burova et al., 2000); Mutants: porcine C^{minus}: C63A/C155A; G^+: G inserted after position 121; GCC: G+ and W64C/H155C; Wasp C^{minus}: S6C; quaternary structure indicated if tested in this study: D = dimer, M = monomer. Method: measurement type, and spectral parameter employed for obtaining the denaturation curve is given: λ_{max} = wavelength of spectral maximum; θ = ellipticity at given wavelength; I = fluorescence intensity at given wavelength; I_{rat} = ratio of two wavelengths; blank = rough estimation based on general spectral changes.
[OBP] = protein concentration used, n.a. = not applicable;
Pre/Main: if only one transition is observed, this is classified as Main; if two are observed, the first is classified as Pre, the second as Main.
Buffer, pH: ? = no information available.
Rev = Reversibility; n=not reversible; y = reversible; (y) = partially reversible; <65 °C = reversible if not heated above 65 °C; blank = not addressed in the publication.

Table 2 Chemical stability of OBPs.

OBP	Method	[OBP]/μM	Pre $[C_{0.5}]$/M	Main $[C_{0.5}]$/M	Buffer	pH^e	Rev.	Ref.
Porcine								
WT	Fluo (λ_{max})	1–50	–	2.4	K_2PO_4	7.0	–	Parisi et al. (2003)
	CD (θ_{225})	5	–	2.4	K_2PO_4	7.0	y	
WT	CD (θ_{222})	28		2.3	?	?	y	Staiano et al. (2007)
	Fluo ($I_{rat,c}$)	?	< 0.4	2.3	?	?		
$MutC^{minus}$	CD (θ_{225})	4	–	0.66	TRIS/HCl	7.8	y	Parisi et al. (2005)
	Fluo (I_{rat})	4	–	0.53	TRIS/HCl	7.8		
Bovine								
WT(D)	Fluo (I^+_{345})	1.4–27	< 1	2.6–2.7	TRIS/HCl	7.8	y	Mazzini et al. (2002)
WT_{rec} (M,D)	CD (θ_{222})	≈8	<1.6	≈2.2	Na_2PO_4	7.8	y	Stepanenko et al. (2014)
	Fluo (I^+_{320})	≈8	<1.6	≈2.2	Na_2PO_4	7.8	y	
WT_{rec} (M,D)	Fluo (I^+_{320})	≈8	<1	2.1	Na_2PO_4	7.8	y	Stepanenko et al. (2016b)
WT (M)	Fluo (I^+_{345})	4	<1	1.5	Gly/OH	2.5	y	Mazzini et al. (2002)
MutGCC (M)	Fluo (I_{325})	5	–	2.9	xPO_4	7.0	y	Polverini et al. (2011)
$MutG^+$ (M)	Fluo (I^+_{320})	≈8	< 1	2.0	Na_2PO_4	7.8	y	Stepanenko et al. (2016b)
MutGCC (M)	Fluo (I^+_{320})	≈8	< 1	2.3	Na_2PO_4	7.8	y	
Insect								
Drosophila LUSH	Fluo (?)	≈100	–	2.4 (Urea)	Na_2PO_4	6.5		Bucci, Kruse, Thode, Alvarado, and Jones (2006)

Comparison of C_{50} value (or ranges) for different OBPs and different methods. The values given in the table serve as an overview; for accurate information please refer to the original publication. For details see also legend of Table 1.
Method: I^+: result for fluorescence intensity at given wavelength is presented, other parameters are also used in the publication; $I_{rat,c}$ = corrected ratio of two wavelengths; ? = parameter not provided Buffers: xPO4: salt not given.

3.2 Choice of parameter for constructing the denaturation curve

Usually, the aim is to construct a denaturation curve, which reflects the increasing fraction of denatured protein. Thus, the signal employed has to be a linear superposition of the signal from native and denatured protein, in order to faithfully represent the denaturation process.

3.2.1 *Single wavelength*

The best choice is usually to select the signal (intensity or absorbance) at one wavelength and plot this value against temperature or denaturant concentration. If the signal-to-noise ratio is not very high, this can be improved by calculating the mean over a small region of the spectrum around the selected wavelength. However, if the signal change due to denaturation is small (<10% of the signal), errors in the actual protein concentration in each measurement might lead to significant scattering in the case of chemical denaturation. Here each data point corresponds to a new sample and thus to a (possibly) slightly different protein concentration.

3.2.2 *Position of the spectral maximum*

Frequently the shift of the maximum's wavelength is employed to evaluate the denaturation of the protein. The advantage is, that this parameter does not depend on the protein concentration, but it reflects the denaturation curve well enough only under certain conditions, namely if the maxima of the two spectra corresponding to the native and denatured state are not too far apart. This presentation tends to yield steeper transitions if the spectral shift is too large (Fig. 1).

3.2.3 *Ratio at two wavelengths*

The ratio of intensities at two wavelengths as measure for the denaturation process has also the advantage of being insensitive to the concentration of the protein. However, the ratio of intensities is not proportional to the concentrations of the different species (native, denatured, etc.) and therefore could alter the shape of the denaturation curve (Fig. 1). See Staiano et al. (2007) for an example with bovine OBP.

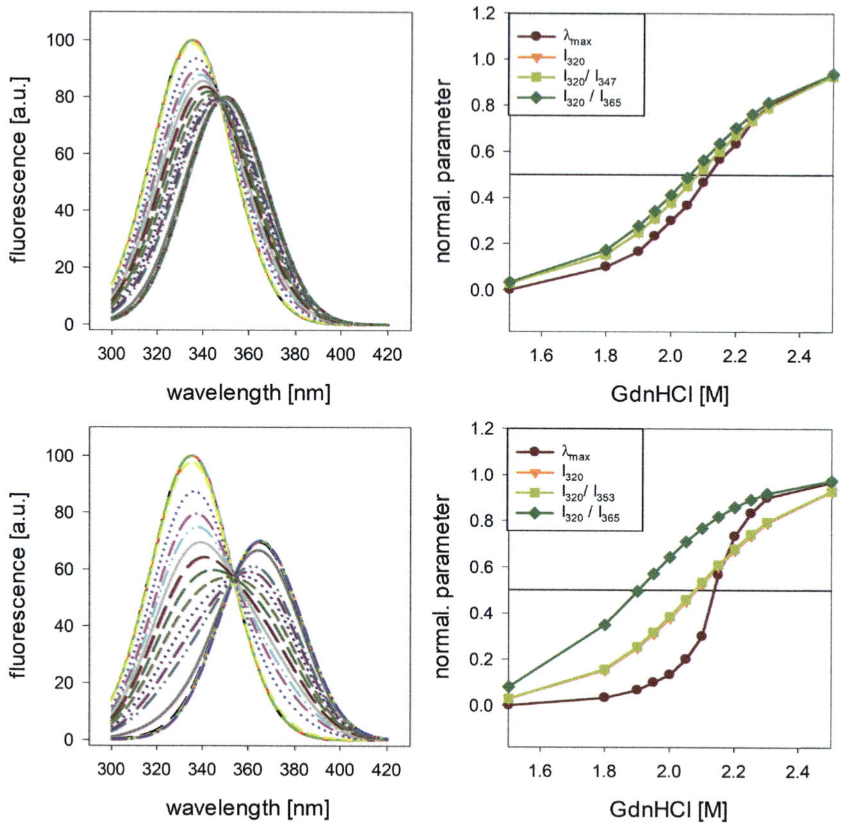

Fig. 1 Choice of spectral parameters to reconstruct a denaturation curve. A denaturation curve with a C_{50}-value of about 2.1 was simulated. Then, two different spectra were assigned to the native (λ_{max} around 335 nm) and the denatured state ($\lambda_{max} = 350$ nm (upper panel) and $\lambda_{max} = 365$ nm (lower panel)). The corresponding sum-spectra were simulated for each denaturant concentration. From the spectra, different parameters were employed to reconstruct the denaturation curve and normalised. Choosing the intensity at a given wavelength, e.g., I_{320}, faithfully represents the real binding curve. Normalisation to an isosbestic point (347 nm and 353 nm, respectively) yields the same curve. Choosing λ_{max} leads to a curve which tends to be steeper and shifted to higher concentrations. This effect is smaller when then two peaks are closer together. The ratio of the intensities at two wavelengths (e.g., I_{320}/I_{365}) yields a shallower and left-shifted curve compared to the real denaturation curve. The details of the deformation depend on the relative height, shape and λ_{max} of the spectra. However, it becomes clear that a fixed wavelength or a fixed wavelength normalised to an isosbestic point (if present) is the most appropriate parameter.

4. Stability of OBPs

4.1 Vertebrate OBPs

The following sections summarize the main results reported in the literature on the stability of porcine and bovine OBPs (Tables 1 and 2). These are the only mammalian proteins of this family to be used in such studies, apart from the rat OBP (Nespoulous, Briand, Delage, Tran, & Pernollet, 2004; Scirè et al., 2009), which is omitted here.

4.1.1 Porcine OBP

Infrared spectra obtained in the study of thermal stability of porcine OBP (Paolini et al., 1999) revealed a first transition between 60 and 70 °C, which was attributed to loosening of the tertiary structure and concomitant H_2O/D_2O exchange. The main denaturation process, the loss of secondary structure, occurred between 80 and 95 °C. Employing tryptophan fluorescence spectroscopy, instead, only one broad transition was observed between 60 and 85 °C. However, the two studies were performed at different pH values and at protein concentrations differing by a factor of 1000 (Table 1).

The DSC-experiments yielded a similar T_m (about 69 °C), despite being performed at a about 10 times higher protein concentration than the fluorescence measurements mentioned above (Burova et al., 1999). A transition temperature around 70 °C was also reported by Stepanenko et al. (2008) based on tryptophan fluorescence. Quite unexpectedly, in this study the CD spectrum remained unaltered up to 95 °C while in another study a transition at 74 °C was observed (Parisi et al., 2005). The reason for the discrepancies remains obscure.

An increase of stability with protein concentration, as observed in the FTIR-experiments, can be expected, if the protein forms oligomers, e.g., a dimer. With respect to dimerization of porcine OBP the literature is somewhat controversial. Size-exclusion chromatography (SEC) indicates dimer formation at pH 7.2 and monomers at pH 3.0, but analytical ultracentrifugation did not detect a difference in molecular weight (Burova et al., 1999). Since in this study the DSC endotherm could well be described by a two-state transition, the authors concluded that if dimers are present, their dissociation takes place before denaturation, with a low enthalpy, and therefore remains undetected. However, an alternative explanation could be suggested, hypothesising that dimer dissociation is followed immediately by denaturation. In order to distinguish between these two cases, stability at different protein concentrations needs to be performed. If the protein dissociates well before denaturation, the denaturation process involves only monomers, and then the same T_m is observed. Thus, the observation that in the FTIR experiment the T_m is much higher, could indicate that dissociation and denaturation are coupled. For a discussion of the effect of dimerization on stability bovine OBP see Mazzini et al. (2002).

In fluorescence experiments addressing chemical denaturation (Table 2), the concentration of GdnHCl needed to reach half-maximal effect (C_{50}) was about 2.4 M GdnHCl with no appreciable dependence on protein concentration between 1 and 50 μM (Parisi et al., 2003), in accordance with their data showing that the protein is monomeric in absence of denaturants. Essentially the same value for C_{50} was obtained employing CD spectroscopy. These results were confirmed by Staiano et al. (2007), yielding a C_{50} of 2.3 M for GmdHCl, both from CD and fluorescence experiments. For the latter the ratio of intensities measured at 320 and 365 nm was employed to follow denaturation. Only after proper correction the same value for $C_{0.5}$ was obtained for both methods (Staiano et al., 2007, see also Fig. 1).

Substitution of the two cysteines, which are conserved in most vertebrate OBPs, by alanin led to a strong decrease in chemical stability, (C_{50} about 0.6 M GdnHCl) and thermal stability (T_m = 41 °C) (Parisi et al., 2005). Chemical denaturation of this C^{minus}-mutant was largely reversible; thermal denaturation was reversible if not heated to temperatures higher than 65 °C. CD and fluorescence data also indicate some structural changes in the mutant already at room temperature.

4.1.2 Bovine OBPs

The quaternary structure of bovine OBPs is a special case within OBPs, since here the C-terminal domains are swapped to form a stable dimer (Tegoni et al., 1996). Dimer formation is kinetically hindered, as indicated by refolding studies (Mazzini et al., 2002) and also by the observation that heterologously expressed OBP is not completely folded in the native state, but acquires native structure after treatment with GdnHCl at low concentrations (Stepanenko et al., 2014).

Thermal stability, as monitored by FTIR, showed a minor transition between 75 °C and 85 °C, and a main transition between 85 and 99 °C (Mazzini et al., 2002). The first transition was, as for porcine OBP (Paolini et al., 1999), ascribed to increased H/D exchange, and the second corresponds to the loss of secondary structure. Data for thermostability based on fluorescence or CD are not available, but DSC data showed two transitions, one with a T_m of about 80 °C, and one at about 84 °C, with a clearly larger contribution of the second transition at higher protein concentrations (25 μM compared to 100 μM). This indicates that dissociation and denaturation events are coupled in this case (Gómez-Velasco et al., 2020). In contrast, and similarly to porcine OBP, chemical stability of bovine OBP does not depend on protein concentration in the range between 1.4 and 27 μM (Mazzini et al., 2002), with C_{50}-values of 2.6–2.7 M GdnHCl, indicating that dissociation and denaturation are uncoupled. However, SEC experiments showed that only above 2.6 M GdnHCl a second peak, corresponding to a monomer, became visible. This observation excludes the hypothesis that denaturation and dissociation are uncoupled events. At present an explanation for these seemingly contradicting results is missing.

Also similar to porcine OBP (Burova et al., 1999), lowering the pH to 2.5 leads to formation of predominantly monomers (Mazzini et al., 2007). The main transition is shifted to 1.5 M GdnHCl, compared to 2.6 M at neutral pH. Of note, denaturation and renaturation curves are much more similar than those at neutral pH, which seems to support the idea that formation of the swapped dimer is the rate limiting step for renaturation at neutral pH. However, studies on mutants which are monomeric at neutral pH do not support this interpretation. In the "G+"-mutant an extra glycine is introduced after position 121, increasing flexibility and inhibiting formation of the swapped dimer (Marabotti, Scire, et al., 2008). This mutant has basically the same stability toward GdnHCl as the recombinant WT OPB, with C_{50} = 2.0 and 2.1 or 2.2 M, respectively (Table 2). In the GCC-mutant two additional cysteines form a disulphide-bridge, present in the porcine OBP and many other members of this protein family. The chemical stability of this monomeric mutant was even higher than that of the wild-type (WT) Also, hysteresis between denaturation and refolding curves was still observed for the GCC monomer, thus domain swapping cannot be the only cause for the slow refolding (Polverini et al., 2011). A DSC study showed only one thermal transition for the GCC-mutant, corresponding to the first peak of WT OPB, and this could be fitted well by a two-state model (Gómez-Velasco et al., 2020).

FTIR-data also showed an increased stability toward thermal denaturation for the GCC mutant, since the transition reflecting loss of secondary structure is shifted to higher temperatures compared to the WT protein (Marabotti, Scire, et al., 2008). However, the onset of tertiary structure changes was shifted to lower temperatures (Marabotti, Scire, et al., 2008).

Thus, the intermediate state is more populated in case of this monomeric mutant. Aggregation occurs in both variants, starting at about 85–90 °C.

4.2 Insect OBPs

Much less information is available on the stability of insect OBPs, as compared to their mammal counterparts. The honey bee odorant binding protein 14 undergoes a conformational change around 55 °C as investigated by circular dichroism spectroscopy (Schwaighofer, Kotlowski, et al., 2014; Schwaighofer, Pechlaner, et al., 2014). However, this transition does not reflect complete loss of secondary structure, since the CD-value of the plateau value above 70 °C is still about 50% of the initial value, indicating significant residual secondary structure. ATR-FTIR spectra also showed a transition in the temperature range between 40 and 70 °C, and further, smaller spectroscopic changes in the range between 75 and 90 °C. This behaviour is not altered in a mutant, which contains a third disulfide bridge, present in most OBPs, but missing in the honeybee OBP14 (Schwaighofer, Pechlaner, et al., 2014). However, the relative ellipticity remaining at high temperatures increases to about 70%. Similarly, based on circular dichroism, the stability of two isoforms of a parasitoid wasp was investigated (Ahmed et al., 2017). For the WT protein a change in secondary structure with T_m-values of 64.9 and 65.8 °C was observed; mutants which lack one of the three disulfide bridges were less stable (T_m = 58.5 °C). Again, at high temperatures (85 °C) a significant amount of secondary structure was still present. Thus, the transition around 65 °C is not the one leading to complete loss of secondary structure. Chemical stability of the drosophila odorant binding protein LUSH was investigated employing tryptophan fluorescence, revealing a C_{50} of 2.3 M for urea (Bucci et al., 2006). At 8 M urea the tryptophans seem to be fully solvent exposed. Since usually higher concentrations of urea are required to induce denaturation, as compared to GdnHCl, we can conclude that insect OBPs are less resistant to chemical denaturation than mammalian OBPs, while thermal stability is similar or even higher, and involves stable intermediates.

5. Analysis of denaturation data

A first level of analysis is to determine the concentration of denaturant (C_{50}) or temperature (T_m), where half of the protein is denatured. This can often be done by a "model-free" approach, where a Boltzmann-like function is fitted to the denaturation curve measured via a signal S (e.g., fluorescence intensity), obtained in chemical or thermal denaturation studies, respectively:

$$S(C) = a^*C + \frac{b^*C}{1 + \exp\left[\frac{C_{50}-C}{d}\right]} \quad (1a)$$

$$S(T) = a^*T + \frac{b^*T}{1 + \exp\left[\frac{T_m-T}{d}\right]} \quad (1b)$$

Since often the signal exhibits a pre- and post-transition slope (in particular if tryptophan fluorescence is monitored), linearly changing amplitudes are included (a*C, b*C or a*T, b*T).

If the aim of the analysis is only to determine C_{50} or T_m, then Eqs. (1a) and (1b) can be used. However, if more detailed information is required, thermodynamic models have to be applied. In fact, Eq. (1a) is derived from the commonly employed model for chemical denaturation (5.1), and Eq. (1b) is a simplification of the general temperature dependence of thermodynamic variables (5.2).

5.1 Chemical denaturation

Usually chemical denaturation is analysed based on the "linear extrapolation model" (LEM, Greene & Pace, 1974). Here it is assumed, that the free energy difference between native and denatured state (ΔG^o) changes linearly with concentration of chemical denaturant (C), defining the equilibrium between native (N) and denatured (D) protein:

$$K = \frac{[D]}{[N]} = \exp\left[-\frac{\Delta G^o}{RT}\right] \qquad \Delta G^o = G_D^o - G_N^o = \Delta G_{H2O}^o - m^*C \quad (2)$$

The parameter m is a measure of the difference in number of bound denaturant-molecules in the native and denatured state, and defines the steepness of the denaturation curve. At 50% denaturation we have K=1 and thus $\Delta G_{H2O}^o = m^* C_{50}$.

Thus, the same value of C_{50} can be obtained by different combinations of stability in the absence of denaturant (ΔG_{H2O}^o) and the slope m. The measured signal S can be analysed based on the following fitting function:

$$S(C) = a^*C + \frac{b^*C}{1 + \frac{1}{K(C)}} \qquad K(C) = exp\left[-\frac{\Delta G_{H2O}^o - m^*C}{RT}\right] \quad (3)$$

In denaturation experiments performed in absence and presence of a ligand, both the value of ΔG^o_{H2O} and m might be altered. If the slope m is the same in presence and absence of ligand, the denaturation constant K can be re-written as

$$K(C) = exp\left[\frac{C_{50} - C}{RT/m}\right]$$

Then the same function as in Eq. (1a) is obtained.

5.2 Temperature induced denaturation

For the analysis of thermal denaturation curves the temperature dependence of the free energy has to be taken into account:

$$\begin{aligned}\Delta G^o &= \Delta H^o - T\Delta S^o \\ &= \Delta H^o(T_m) + \Delta c_p(T - T_m) - T\left[\Delta S^o(T_m) + \Delta c_p\, ln\,(T/T_m)\right]\end{aligned}$$

with ΔH^o, ΔS^o the binding enthalpy and entropy and Δc_p the change in heat capacity upon the transition from the native to the denatured state.

At $T = T_m$ we have K=1 and thus $\Delta H^o(T_m) = T_m \Delta S^o(T_m)$.

Therefore, the free energy can also be expressed in terms of $\Delta H^o(T_m)$, T_m and Δc_p:

$$\Delta G^o = \Delta H^o(T_m)\left(1 - \frac{T}{T_m}\right) + \Delta c_p(T - T_m) - T\Delta c_p\, ln\left(\frac{T}{T_m}\right)$$

For the dependence of the measured signal S on the temperature one gets:

$$S(T) = a^*T + \frac{b^*T}{1 + 1/K(T)} \tag{4}$$

$$K(T) = \exp\left[-\frac{\Delta H^o(T_m)\left(1 - \frac{T}{T_m}\right) + \Delta c_p(T - T_m) - T\Delta c_p\, ln\left(\frac{T}{T_m}\right)}{RT}\right]$$

An analysis based on this equation is usually done for calorimetric data, e.g., Burova et al. (1999). When analysing, e.g., CD-data, often a good fit can be found for a range of combinations of ΔG°(Tm) and Δc_p (e.g., Parisi et al., 2005). Thus, Δc_p is not well defined. Setting $\Delta c_p = 0$ yields.

$$K(T) = exp\left[-\frac{\Delta H^o(T_m - T)}{RTT_m}\right] \tag{5}$$

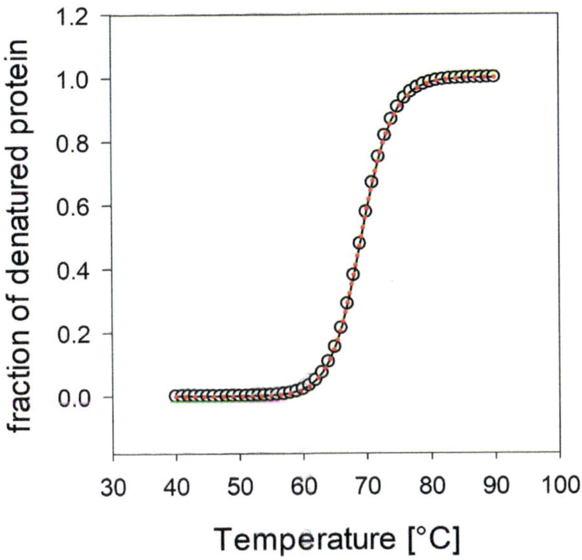

Fig. 2 Role of heat-capacity for analysis of thermal denaturation data. Based on the fit-parameters presented by Burova et al. (1999) for porcine OBP, a denaturation curve is calculated: $T_m = 69.2\,°C$, $\Delta H° = 391.2\,kJ/mol$, $\Delta c_p = 2.98\,kJ/mol$, symbols. The curve was then re-analysed based on the simplified expressions for the temperature dependence of the denaturation constant (Eq. 5, with T*Tm and Tm*Tm in the numerator). In both cases $\Delta H°$ is well reproduced (389 and 390 kJ/mol, black and red line, respectively). Thus, here the heat-capacity change does not have an impact on the analysis.

Using this equation in Eq. (4) and approximating T by T_m, we obtain the simplified function given in Eq. (1b).

As an example, based on the results obtained from CD-data for stability of porcine OBP (Burova et al., 1999), denaturation curves were calculated and analysed by the simplified functions. The results show that indeed for this situation the change in heat capacity is not of importance (Fig. 2).

6. Ligand binding and protein stability

6.1 Shift of stability of OBPs upon binding of odorants

Thermal and chemical stability of OBPs has been reported to increase in the presence of "good ligands." Table 3 shows some examples of strong ligands causing a change in stability.

In thermal denaturation CD-experiments with porcine OBP, for IMBP and DMO a different shift in stability was observed (Table 2), despite the very similar affinities for these ligands measured at room temperature by

Table 3 Effect of ligands on chemical and thermal stability of OPBs.
Thermal stability

OBP	Ligand	[OBP]/μM	[Ligand]/μM	ΔTm/°C	Ref.
Porcine					
WT	DMO	2100	8400	5	Paolini, Tanfani, Fini, Bertoli, and Paolo (1999)
	PYR		8400	5	
	DMO	2	2	5	
WT	IBMP	8	17	4.7	Burova, Choiset, Jankowski, and Haertlé (1999)
	IBMP	14	125	5.9	
	DMO	15	139	9	
Bovine					
WT_{rec}	OCT	500	2000	5	Marabotti, Scire, et al. (2008)
MutGCC	OCT	500	2000	3	
Insect					
Honey bee OBP14 WT	Geraniol	42	200	5	Schwaighofer et al. (2014)
	Eugenol		200	10	
Honey bee OBP14 WT	Many	42	200	−0.5 to 12	Schwaighofer et al. (2014)
Honex bee OBP14 $MutC^{+}$	Many	42	200	−2 to 8.5	
Wasp OBP2 WT	Many	42	200	−0.9 to 6.1	Ahmed, Zhang, Wang, He, and Bai (2017)
Wasp OBP2 1 $MutC^{minus}$	Many	42	200	0 to 4.2	
Wasp OBP3 WT	Many	42	200	1.2 to 4.2	

Table 3 Effect of ligands on chemical and thermal stability of OPBs.—cont'd
Chemical stability

OBP[a]	Ligand	[OBP]/μM	[Ligand]/μM	ΔC_{50}/M	Ref.
Porcine					
WT	1-AMA	5	30	0.2	Parisi et al. (2003)
	DHM	5	20	0.6	
Bovine					
WT	DHM	2	20	0.6	Mazzini et al. (2002)
WT_{rec}	OCT	≈8	10,000	0.6	Stepanenko et al. (2016b)
$MutG^+$	OCT	≈8	10,000	0.6	
$MutGCC^+$	OCT	≈8	10,000	0.4	

Shifts of C_{50} value (ΔC_{50}) or Tm (ΔTm) for different OBPs and different ligands. Values for ΔTm and ΔC_{50} are taken from the literature cited, or extracted from the data shown.
For details see also legend of Table 1.
Ligand: ligand used in the study.
[Ligand] = concentration of ligand employed.
ΔC_{50} for wasp OPB3: a range is given, ranging from minimal to maximal shift, since a large number of substances were tested.

Isothermal Titration calorimetry (ITC) experiments (Burova et al., 1999), 0.3 and 0.2 μM, respectively.

Based on FTIR measurements, a similar shift for DMO and PYR of about 5 °C was observed (Paolini et al., 1999), which is reasonable since the dissociation constant K_D reported for PYR is about 0.5 μM, for DMO it is 0.2 μM. Based on fluorescence spectroscopy for DMO a similar shift of 5 °C was observed, despite much lower ligand/protein ratio and total concentration. FTIR experiments indicate that the presence of OCT stabilises the WT-bovine OPB more than its monomeric mutant (ΔT_m = 5 °C compared to 3 °C, Marabotti, Scire, et al., 2008).

Chemical stability of porcine OBP is also increased in the presence of DHM and 1-AMA (Parisi et al., 2003). The effect of DHM is stronger, despite a lower concentration in the assay. For 1-AMA, which is an artificial fluorescent ligand, K_D values between 0.5 and 2 μM are reported. For DHM a similar IC_{50} value as for IBMP was obtained in 1-AMA competition experiments, its K_D being 0.3 μM, as calculated from ITC experiments. For details on ligand-binding assays, see chapter "Ligand binding assays" by Pelosi.

OCT induced a similar shift in C_{50} in WT-bovine OPB and the monomeric G+ mutant. The mutant GCC, which is further stabilized by a disulphide bridge is less affected (Stepanenko et al., 2016b), similar to the smaller shift in T_m compared to WT.

In case of insect OBPs three studies are available where for various odorants the effect on thermal stability was tested employing CD spectroscopy (Ahmed et al., 2017; Schwaighofer, Kotlowski, et al., 2014; Schwaighofer, Pechlaner, et al., 2014). As mentioned above, in these measurements the transition observed does not reflect complete denaturation, since the secondary structure level after the transition is still considerable. Thus, the shift in the transition rather reflects a shift between two conformations, and might be determined by the difference in affinity of the ligand to the two conformations, rather than the affinity to one, the native, conformation.

6.2 Choosing the right ligand concentration

The concentrations of proteins and ligands chosen in the studies reported above vary widely, also the ligand to protein ratios. The concentration of ligand ($[x_{tot}]$) necessary to ensure significant saturation of the protein depends on the dissociation constant (K_D) and the total protein concentration ($[P_{tot}]$):

The concentration of protein-ligand complex ($[Px]$) is given by

$$[Px] = \frac{[P_{tot}] + [x_{tot}] + K_D}{2} - \sqrt{\left(\frac{[P_{tot}] + [x_{tot}] + K_D}{2}\right)^2 - [P_{tot}][x_{tot}]} \quad (6)$$

Thus, both ligand and protein concentration, and not only the ratio, define the saturation level $[Px]/[P_{tot}]$ of the protein. For example, with a K_D of 1 μM and at a ligand to protein ratio of 1:1, the fraction of ligated protein is about 50% at 2 μM protein, and 96% at 500 μM protein.

6.3 Analysis of denaturation in presence of ligands

Under equilibrium conditions and if only two states exist, the native (N) and the denatured (D) one, the affinity ($K_x = 1/K_D$) of the ligand (x) at the condition where half of the protein is denatured can be estimated (Celej, Dassie, Freire, Bianconi, & Fidelio, 2005; Schön, Brown, Hutchins, & Freire, 2013). The basics are recapitulated shortly in the following.

In presence of ligand, the total native protein $[N_{tot}]$ consists of two populations: without ligand [N] and with ligand bound [NX], which in turn is defined by the free ligand concentration [x] and K_x.

$$[N_{tot}] = [N] + [NX] = [N](1 + K_x[x])$$

Binding of the ligand shifts the equilibrium between native and denatured state, leading to an apparent denaturation constant K_{app}, which depends on free ligand concentration [x] and K_x:

$$\frac{[D]}{[N_{tot}]} = \frac{K}{1 + K_x[x]} = K_{app} \qquad \frac{K}{K_{app}} = 1 + K_x[x] \tag{7}$$

The constant K refers to the protein's denaturation constant in the absence of ligand (K=[D]/[N]). When 50% of the protein is denatured in the presence of ligand ($C = C_{50}^x$ or $T = T_m^x$) we have $K_{app} = 1$ and therefore $K_{50} = 1 + K_x[x]$. Thus, at the condition where [D] = [N] + [NX] ($K_{app} = 1$), namely $C = C_{50}^x$ or $T = T_m^x$, the ratio between the unligated species (K_{50} = [D]/[N]) is >1. The exact value can be calculated based on the denaturation curve obtained in absence of the ligand. To do so, K_{50} is calculated by employing Eq. (3), setting $C = C_{50}^x$, or Eq. (4) or (5), setting $T = T_m^x$.

The free ligand concentration is then given by

$$[x] = [x]_{tot} - \frac{K_{50} - 1}{2K_{50}}[P_{tot}] \tag{8}$$

From this the affinity constant can be calculated:

$$K_x = \frac{K_{50} - 1}{[x]} \tag{9}$$

Note, that in this model it is assumed that the ligand binds only to N, implying $K_{50} > 1$.

The binding affinity K_x obtained by this approach is the one observed at $C = C_{50}^x$ and $T = T_m^x$, respectively.

However, both, the presence of denaturant and an increased temperature, could alter K_x. It has been shown that the affinity of the bovine OBP mutant GCC for 1-AMA remains approximately constant up to 2.5 M GdnHCl, where the denaturation process starts (Polverini et al., 2011).

In case of thermal denaturation, a change of K_x with temperature has to be expected. Calorimetric experiments show that the binding enthalpies for odorants are, to all of my knowledge, always negative (Borysik, Briand, Taylor, & Scott, 2010; Burova et al., 1999; Gómez-Velasco et al., 2020; Nespoulous et al., 2004). This implies that the affinity decreases with increasing temperature. Thus, in general an otherwise measured affinity (e.g., via ITC at 25 °C) is expected to be higher than the one estimated via a change in thermal stability.

It has to be considered, that the analysis reported above is based on equilibrium conditions with respect to denaturation and ligand binding. This might not always be fulfilled, in particular in thermal denaturation experiments. If heating was be performed too fast, slow und therefore incomplete dissociation of ligand would lead to increased levels of native protein, resulting in an overestimation of the actual affinity (Brandts & Lin, 1990). It has been shown that dissociation of IBMP and DMO from dimeric bovine OBP is very slow, if induced by infinite dilution (Pevsner, Hou, Snowman, & Snyder, 1990). The observation that endogenous ligand is sometimes co-purified, also points toward slow ligand dissociation.

7. Protocols for measuring protein stability

Ideally, each denaturation curve is measured at least in triplicates, with each measurement performed with a different protein preparation. From these a mean value and standard error is calculated.

7.1 Measuring denaturation curves based on tryptophan fluorescence

Excitation of protein fluorescence is usually done at 280 nm, since here the extinction coefficient of both tryptophan and tyrosine is near its maximum. If only tryptophan is to be excited, usually a wavelength of 295 nm is chosen. Then, tyrosine will not appreciably contribute to the emission spectrum, but the extinction coefficient and with this the emission intensity of tryptophan drops to about 30%. A lower fluorescence intensity can be compensated to some extent by increasing the excitation slit width. However, if this is too large, photobleaching might occur. Alternatively, the protein concentration can be increased, but the absorption at the excitation wavelength should be kept below 0.05 in order to avoid inner filter effects.

1. Equipment
    - **1.1.** Spectrofluorometer with thermostatted cuvette holder; for thermal denaturation, in addition a temperature control with a sensor for the cuvette is required. Magnetic stirring is also desirable.
    - **1.2.** Quartz fluorescence cuvettes (for thermal denaturation: with tight stopper). For chemical denaturation it is useful to have at least two cuvettes which can be used alternatingly.
    - **1.3.** If at hand: thermostatting device, e.g., thermo shaker (with shaking disabled).
    - **1.4.** For chemical denaturation with GdnHCl: Refractometer.

2. Reagents
   2.1. Buffer: for chemical denaturation any buffer selected for its pK value and compatibility with the protein is suitable, with the exception of imidazole buffer which absorbs in the near-UV range. For thermal denaturation preferably choose a buffer with small ionisation enthalpy like phosphate buffer or MOPS, since then the pH does not strongly depend on the temperature.
   2.2. Purified OBP or CSP in the selected buffer at the concentration of at least 50 μM for chemical denaturation or 5–10 μM for thermal denaturation.
   2.3. Chemical denaturation: 6 M GdnHCl in selected buffer, concentration checked via refractive index (see Section 7.5).
   2.4. Detergent to clean cuvettes, e.g., 2% Hellmanex, pure ethanol (not denatured) and distilled water to rinse cuvettes.
3. Protocol
   3.1. For chemical denaturation, prepare protein solution at several concentrations of GdnHCl, e.g., in 0.25 M steps up to 5 M with a protein concentration of 5–10 μM. Steps of 0.25 M GdnHCl will deliver only very few data points in the transition region. Once the transition range is identified, the recording of additional points with a pitch of 0.1 M is advised.

        Incubate samples long enough to reach equilibrium (see Section 3.1). In Parisi et al. (2005), e.g., the OBP was incubated for 24 h. Mix thoroughly, solutions of GdnHCl at high concentrations are rather viscous.
   3.2. Switch on the instrument 15–30 min prior use to allow the lamp to stabilize. Set excitation wavelength to 295 nm, both bandwidths to 5 nm, sensitivity to high, scan speed to about 100 nm/min, 1 nm data pitch. Set temperature between 20 and 25 °C for isothermal experiments, or to the selected starting value for thermal ramps.
   3.3. Record a spectrum of buffer between 310 and 380 nm. Only the Raman scattering peak at about 325 nm should be visible. Additional bands indicate a dirty cuvette or contaminated buffer. Perform this procedure for both cuvettes.
   3.4. For chemical denaturation, fill the cuvette with the first sample and wait until the solution has reached the desired temperature (this step can be omitted if the solution was equilibrated to the same temperature beforehand, e.g., in a thermo shaker),

then record the spectrum between 310 and 380 nm. Check spectrum for unusual "bumps," which can be caused by air bubbles or dust particles crossing the light beam. If present, measure the spectrum again. Repeat the procedure with all samples, starting from the lowest GdnHCl concentration. Use the two cuvettes alternatingly; while using one, soak the other in 2% Hellmanex, then rinse with water and ethanol, and finally dry.

3.5. For thermal denaturation prepare the protein solution at a concentration of 5–10 μM. Make sure that stopper is tightly fastened. Run a temperature ramp between 20 and 80 °C (water bath) or 20 and 95 °C (Peltier element). Typically, the temperature is changed in 2 °C steps, with equilibration intervals of 3–5 min. Alternatively, a quasi-continuous ramp can be chosen, e.g., measuring the fluorescence at a single wavelength in 0.1 °C steps, and using a heating rate of 1 °C/min. Lower heating rates might be necessary (see Section 3.1).

Check the volume in the cuvette after the measurement to ensure that no significant evaporation has occurred.

3.6. For renaturation curves reverse the temperature ramp or dialyze the sample against a lower concentration of denaturant. If it is possible to start with a sufficiently high protein concentration, also simple dilution is an option, if the protein concentration is not affecting the equilibrium.

4. Analysis and anticipated results

4.1. Subtract buffer spectrum from sample spectra if required. Select a suitable wavelength (e.g., maximum of emission in the native or the denatured state) and calculate the mean of the intensities of five wavelengths around this value. If the spectrum is very smooth (not much noise) this is not necessary.

4.2. Plot the intensities at the selected wavelength against GdnHCl concentration or temperature. A curve with a clear transition should be obtained. Both in the pre- and the post-transition range a slope is expected. In case of GdnHCl this is caused by a direct effect of the denaturant on tryptophan fluorescence. In case of thermal denaturation, a visible negative slope is expected, due to the general decrease of quantum yield with increasing temperature.

4.3. For analysis see Section 5. If indications of intermediate states are seen, a parametric plot as used in Stepanenko et al. (2014) can be helpful.

7.2 Measuring denaturation curves based on far UV circular dichroism

Circular dichroism allows to follow changes in the secondary structure (far UV) and the environment of aromatic residues (near UV). Here, we focus on the secondary structure changes, induced by either chemical denaturation or temperature increase, with the aim to determine stability of the protein. For this purpose, analysis of the secondary structure elements is not required. For this aspect see (Greenfield, 2015).

1. Equipment
   - **1.1.** Spectropolarimeter with thermostatted cuvette holder; for thermal denaturation a temperature control with sensor for the cuvette/cuvette holder is also needed.
   - **1.2.** Quartz absorbance cuvettes (for thermal denaturation: with tight stopper) with a path length of 0.1 cm. For chemical denaturation it is useful to have at least two cuvettes which can be used alternatingly.
   - **1.3.** For Chemical Denaturation with GdnHCl: Refractometer
2. Reagents
   - **2.1.** Buffer: if the buffer (e.g., TRIS or HEPES) absorbs in the far UV range, use low concentrations, e.g., 10 mM. Avoid large amounts of salts or other additives like DTT. See Greenfield (2006) for limiting wavelengths for different buffers. For thermal denaturation preferably use a buffer with small ionisation enthalpy like phosphate or MOPS, since in such cases the pH is not strongly temperature dependent.
   - **2.2.** Purified OBP or CSP in selected buffer, at concentration of at least 2 mg/mL for chemical denaturation, or about 0.2 mg/mL for thermal denaturation.
   - **2.3.** Chemical denaturation: 6 M GdnHCl in chosen buffer, concentration checked via refractive index (see Section 7.5).
   - **2.4.** Detergent to clean cuvettes, e.g., 2% Hellmanex, pure ethanol (not denatured) and distilled water to rinse cuvettes.
3. Protocol
   - **3.1.** For chemical denaturation, prepare protein solution at several concentrations of GdnHCl, e.g., in 0.25 M steps up to 5 M with a protein concentration of 0.2 mg/mL. Steps of 0.25 M GdnHCl will deliver only few data points in the transition region. Once this region is identified, the recording of additional points with a pitch of 0.1 M is advised.

Incubate samples sufficiently long to reach equilibrium (see Section 3.1). In Parisi et al. (2005), e.g., the OBP was incubated for 24 h. Mix thoroughly, solutions of GdnHCl at high concentrations are rather viscous.

3.2. Switch-on the instrument prior to measurement, allowing sufficient time to stabilize lamp and purge sample chamber with N_2 to remove oxygen. Set N_2-flow according to the instrument's manual, depending on the selected wavelength range. Set bandwidth to 1–1.5 nm, sensitivity high, scan speed around 100 nm/min, 0.5 nm pitch, 2 s per point, time constant 100 ms. Set temperature between 20 and 25 °C for isothermal experiments, or to the selected starting value for thermal ramps.

3.3. Record a spectrum of the buffer between 190 and 260 nm (thermal denaturation) or 200 nm and 260 nm (chemical denaturation). The signal should not be larger than about ±1 mDeg. Larger values indicate a dirty cuvette. If applicable, perform this procedure for two cuvettes. The photomultiplier voltage should not exceed about 500 V (see instruments manual for exact values) in the whole wavelength range.

3.4. For chemical denaturation fill the cuvette with the samples and record spectrum between 200 and 260 nm. Proceed with all samples, starting from the lowest GdnHCl concentration. Note, that with increasing GdnHCl concentration the absorbance below 210 nm will increase, leading to a high noise level. Alternate the two cuvettes; while using one, soak the other in 2% Hellmanex, then rinse with water and ethanol, and dry.

3.5. For thermal denaturation prepare protein solution with a concentration of about 0.2 mg/mL. Make sure that the stopper is tightly fastened. Run a temperature ramp between 20 and 80 °C (water bath) or 20 and 95 °C (Peltier element). Typically, the temperature is changed in 2 °C steps, with equilibration intervals of 3–5 min. Alternatively, a quasi-continuous ramp can be chosen, e.g., measuring the CD-signal at a single wavelength in 0.1 °C steps, and a heating rate of 1 °C/min. Lower heating rates might be necessary (see Section 3.1). Check volume in cuvette after measurement to ensure that no significant evaporation has occurred.

3.6. For renaturation curves reverse the temperature ramp or dialyze the sample against a lower concentration of denaturant. If it is possible to start with a sufficiently high protein concentration, also simple dilution is an option, if the protein concentration is not affecting the equilibrium.

4. Analysis and anticipated results
   4.1. If necessary, smooth all spectra. Usually, the instrument's software offers several possibilities. Make sure that you did not over-smooth the spectrum by subtracting the smoothed spectrum from the original one. The resulting data points should scatter symmetrically around zero.
   4.2. Subtract buffer spectrum from sample spectra. In the range between 250 and 260 nm the resulting value should be close to zero.
   4.3. Choose a suitable wavelength, e.g., 220 nm for an α-helical protein and plot the value against GdnHCl concentration or temperature. A curve with a clear transition range should be obtained.
   4.4. While the measurement is based on the difference in the extinction coefficients ($\Delta\varepsilon$), often the signal is presented as *mDeg* or some related normalized variant of this and termed "ellipticity" (Greenfield, 2015). Of importance: normalization with respect to concentration is usually based on the concentration of amino-acids, not of the protein.
   4.5. It is possible to extract the fraction of secondary structure elements from CD spectra (Greenfield, 2015; Parisi et al., 2005), but this aspect will not be discussed here.
   4.6. For analysis of the resulting curve see Section 5.

7.3 Measuring denaturation curves based on infrared spectroscopy

Due to rather complicated preparation of the sample, usually IR-based methods are employed only for thermal denaturation studies in the context of protein stability measurements. Similar to circular dichroism experiments, we do not focus on the determination of secondary structure elements, thus the issue of band fitting etc. is not addressed here.

1. Equipment
   1.1. FTIR spectrometer; for thermal denaturation a temperature control with a sensor for the cuvette is also needed.
   1.2. CaF_2 windows
   1.3. Small volume concentrating device, e.g., centrifugal concentrators.
2. Reagents
   2.1. Buffer: preferably a buffer with small ionisation enthalpy, since in such case the pH is not strongly temperature dependent, e.g., phosphate or MOPS buffer.
   2.2. About 500 μL of purified OBP or CSP, about 3 mg/mL in selected buffer.
   2.3. Same buffer prepared in D_20.

3. Protocol
    3.1. Purge instrument with dry air at a dew point of −40 °C during the experiment, and switch it on 24 h prior to the experiment.
    3.2. Equilibrate protein solution in buffer prepared in H_20, concentrate to 40 μL using, e.g., centrifugal concentrators, fill sample with buffer in D_2O to 350 μL, reduce again to about 50 μL, and repeat this step three more times. In the last step, reduce the volume to about 40 μL.
    3.3. Check concentration after this step, if possible. Some protein might get lost in this procedure.
    3.4. Set parameters to record a spectrum every 5 °C. Allow sufficient time for temperature equilibration of the sample.
    3.5. Place buffer in a cell with CaF_2 windows, using 25-μm spacers, and measure a spectrum of the buffer under the same conditions as for the sample at the starting temperature.
    3.6. Place the sample into the spectrometer and start the temperature ramp.
    3.7. For renaturation curves reverse the temperature ramp.
4. Analysis and anticipated results
    4.1. After completion of the experiment, subtract the buffer spectrum. If performed correctly, the D_2O bending absorption close to 1220 cm^{-1} should disappear.
    4.2. Since the changes in absorbance are not very large, usually the results are displayed as difference spectra. Thus, for each spectrum at a given temperature, the spectrum of the native state (starting temperature) is subtracted, or some other selected reference. For better visualisation, the resulting spectra can be shifted along the y-axis.
    4.3. Alternatively, 2nd derivative spectra are calculated to enhance the spectral regions where changes occur.
    4.4. In order to characterise possible intermediates, more sophisticated methods of analysis are sometimes employed, like 2D-spectra (see, e.g., Marabotti, Scire, et al., 2008; Schwaighofer, Kotlowski, et al., 2014).
    4.5. In order to obtain a denaturation curve, plot the absorbance difference at a suitable wavelength or the position of the maximum of the amide I band against the temperature.
    4.6. For further analysis of the curve obtained refer to Section 5.

7.4 Measurements in the presence of ligands

1. When performing the stability measurements in the presence of ligands, extra incubation time of the protein/ligand mixture might be necessary. The time required depends on the binding kinetics of the ligand.
2. If the ligand added is dissolved in organic solvent, then the pure solvent should also be tested.
3. Reaction vessels should be carefully closed and filled with solutions, also during the incubation period, if very volatile ligands are used.
4. For analysis refer to Sections 5 and 6.3.

7.5 Related techniques

Being hygroscopic, GdnHCl solutions based on weighted amounts tend to contain lower concentrations than expected. Since the refraction index is proportional to the concentration, this can be employed to check the actual concentration.

The following equation describes the dependence of the molar concentration M on the refractive index increment ΔN

$$M = 57.147{*}\Delta N + 38.86{*}\Delta N^2 - 91.60{*}\Delta N^3$$

with ΔN being the difference in refractive index between the GdnHCl solution and the solvent (Nozaki, 1972).

7.6 Pros and cons

	Pros	Cons
FTIR	– Yields information about secondary and tertiary structural changes, as well as on aggregation phenomena – Also side chain specific alterations can be monitored, e.g., protonation of carboxyl-groups – Signals not temperature dependent	– Apparatus expensive – Experience needed – Experimental preparation takes time, therefore mostly used for thermal denaturation – Relatively large protein concentrations required

Continued

	Pros	Cons
CD	– Easy to measure – Costs of apparatus: medium scale – Concentration and amount of protein easily affordable – Signals not temperature dependent	– Constraints with respect to buffer composition – Presence of denaturants distorts spectrum
Fluorescence	– Easy to measure – Cost of apparatus lowest of the three – Concentration and amount of protein easily affordable – Spectrum only slightly distorted by denaturants (some quenching)	– Signal is temperature dependent – Inner-filter effect can be an issue

7.7 Alternative methods/procedures

An alternative to spectroscopic approaches is the use of differential scanning calorimetry for the study of thermal denaturation (see Table 1). This method yields more thermodynamic information, but it might be not easy to get access to such a machine. The amount of required sample is rather high, with 1.5 mL of 0.2 mg/mL protein. For more information see, e.g., Brandts and Lin (1990).

7.8 Troubleshooting and Optimization

Problem	Solution
Data scatter strongly	– Check whether longer incubation time to reach equilibrium is required or adhere strictly to the same time schedule – Check volumes titrated, avoid volumes <5 μL to reduce titration errors – In case of chemical denaturation check whether samples are mixed properly
CD signal very noisy	– Measure buffer only, voltage should be lower than about 500 V – Check lamp; with age intensity is reduced – Check gain/sensitivity of instrument; some offer different settings for different amplitudes of signal

Problem	Solution
CD signal low	– Increase protein concentration, if the protein is well folded, the signal should be >10 mDeg (roughly)
Fluorescence signal noisy	– Increase protein concentration, but stay below OD = 0.05 – Increase excitation slit width, but double check whether photobleaching occurs
Fluorescence signal decreases upon repeated measurement	– Photobleaching occurs, reduce slit width

7.9 Summary

Denaturation curves of proteins, with the aim of determining stability to thermal or chemical denaturation are relatively easy to measure, in particular employing fluorescence or circular dichroism. A first level of analysis, delivering an apparent stability parameter such as T_m or C_{50}, defining the point of 50% denaturation, is also easily done. However, if quantitative thermodynamic analysis is required, some effort is necessary to establish whether the curves represent an equilibrium situation. Furthermore, quite often intermediate states are populated, which affect the parameters of the analysis, even if at the first glance a model based on a simple two state model seems to be applicable. Presence of ligands can increase the stability, and in principle, based on such phenomena, ligand binding constants can be determined. This requires again, that the system is in equilibrium, which might not be the case when the ligand has a slow off-rate. Furthermore, comparison with directly determined affinity constants is hampered by possible effects of denaturant on the binding constant, and the general effect of temperature on the affinity. Nevertheless, if investigated under the same conditions, the relative shift of different ligands might give an estimate of their relative affinity. However, since the relation between ΔT or ΔC_{50} and the affinity of the ligand is not linear, it is difficult to judge how large the differences are, in particular if kinetic properties play also a role.

References

Ahmed, T., Zhang, T., Wang, Z., He, K., & Bai, S. (2017). Molecular cloning, expression profile, odorant affinity, and stability of two odorant-binding proteins in *Macrocentrus cingulum* Brischke (Hymenoptera: Braconidae). *Archives of Insect Biochemistry and Physiology*, *94*(2).

Ausili, A., Scire, A., Damiani, E., Zolese, G., Bertoli, E., & Tanfani, F. (2005). Temperature-induced molten globule-like state in human alpha(1)-acid glycoprotein: An infrared spectroscopic study. *Biochemistry*, *44*(49), 15997–16006.

Barth, A. (2007). Infrared spectroscopy of proteins. *Biochimica et Biophysica Acta*, *1767*(9), 1073–1101.

Bianchet, M. A., Bains, G., Pelosi, P., Pevsner, J., Snyder, S. H., Monaco, H. L., et al. (1996). The three-dimensional structure of bovine odorant binding protein and its mechanism of odor recognition. *Nature Structural Biology*, *3*(11), 934–939.

Borysik, A. J., Briand, L., Taylor, A. J., & Scott, D. J. (2010). Rapid odorant release in mammalian odour binding proteins facilitates their temporal coupling to odorant signals. *Journal of Molecular Biology*, *404*(3), 372–380.

Brandts, J. F., & Lin, L. N. (1990). Study of strong to ultratight protein interactions using differential scanning calorimetry. *Biochemistry*, *29*(29), 6927–6940.

Bucci, B. K., Kruse, S. W., Thode, A. B., Alvarado, S. M., & Jones, D. N. M. (2006). Effect of n-alcohols on the structure and stability of the Drosophila odorant binding protein LUSH. *Biochemistry*, *45*(6), 1693–1701.

Burova, T. V., Choiset, Y., Jankowski, C. K., & Haertlé, T. (1999). Conformational stability and binding properties of porcine odorant binding protein. *Biochemistry*, *38*(45), 15043–15051.

Burova, T. V., Rabesona, H., Choiset, Y., Jankowski, C. K., Sawyer, L., & Haertlé, T. (2000). Why has porcine VEG protein unusually high stability and suppressed binding ability? *Biochimica et Biophysica Acta*, *1478*(2), 267–279.

Celej, M. S., Dassie, S. A., Freire, E., Bianconi, M. L., & Fidelio, G. D. (2005). Ligand-induced thermostability in proteins: Thermodynamic analysis of ANS-albumin interaction. *Biochimica et Biophysica Acta*, *1750*(2), 122–133.

Gómez-Velasco, H., Rojo-Domínguez, A., & García-Hernández, E. (2020). Enthalpically-driven ligand recognition and cavity solvation of bovine odorant binding protein. *Biophysical Chemistry*, *257*, 106315.

Greene, R. F., & Pace, C. N. (1974). Urea and guanidine hydrochloride denaturation of ribonuclease, lysozyme, alpha-chymotrypsin, and beta-lactoglobulin. *The Journal of Biological Chemistry*, *249*(17), 5388–5393.

Greenfield, N. J. (2006). Using circular dichroism spectra to estimate protein secondary structure. *Nature Protocols*, *1*(6), 2876–2890.

Greenfield, N. J. (2015). Circular dichroism (CD) analyses of protein-protein interactions. *Methods in Molecular Biology*, *1278*, 239–265.

Khalil, M., Boubegtiten-Fezoua, Z., Hellmann, N., & Hellwig, P. (2016). Extraordinary stability of hemocyanins from *L. polyphemus* and *E. californicum* studied using infrared spectroscopy from 294 to 20 K. *Physical Chemistry Chemical Physics (PCCP)*, *18*(41), 28732–28739.

Marabotti, A., Lefèvre, T., Staiano, M., Crescenzo, R., Varriale, A., Rossi, M., et al. (2008). Mutant bovine odorant-binding protein: Temperature affects the protein stability and dynamics as revealed by infrared spectroscopy and molecular dynamics simulations. *Proteins*, *72*(2), 769–778.

Marabotti, A., Scire, A., Staiano, M., Crescenzo, R., Aurilla, V., Tanfani, F., et al. (2008). Wild-type and mutant bovine odorant-binding proteins to probe the role of the quaternary structure Organization in the Protein Thermal Stability. *Journal of Proteome Research*, 7(12), 5221–5229.

Mazzini, A., Maia, A., Parisi, M., Sorbi, R. T., Ramoni, R., Grolli, S., et al. (2002). Reversible unfolding of bovine odorant binding protein induced by guanidinium hydrochloride at neutral pH. *Biochimica et Biophysica Acta*, *1599*(1–2), 90–101.

Mazzini, A., Polverini, E., Parisi, M., Sorbi, R. T., & Favilla, R. (2007). Dissociation and unfolding of bovine odorant binding protein at acidic pH. *Journal of Structural Biology*, *159*(1), 82–91.

Nagnan-Le Meillour, P., Joly, A., Le Danvic, C., Marie, A., Zirah, S., & Cornard, J.-P. (2019). Binding specificity of native odorant-binding protein isoforms is driven by phosphorylation and O-N-acetylglucosaminylation in the pig *Sus scrofa*. *Frontiers in Endocrinology, 9*.

Nespoulous, C., Briand, L., Delage, M.-M., Tran, V., & Pernollet, J.-C. (2004). Odorant binding and conformational changes of a rat odorant-binding protein. *Chemical Senses, 29*(3), 189–198.

Nozaki, Y. (1972). The preparation of guanidine hydrochloride. *Methods in Enzymology, 26*, 43–50.

Paolini, S., Tanfani, F., Fini, C., Bertoli, E., & Paolo, P. (1999). Porcine odorant-binding protein: Structural stability and ligand affinities measured by fourier-transform infrared spectroscopy and fluorescence spectroscopy. *Biochimica et Biophysica Acta, 1431*(1), 179–188.

Parisi, M., Mazzini, A., Sorbi, R. T., Ramoni, R., Grolli, S., & Favilla, R. (2003). Unfolding and refolding of porcine odorant binding protein in guanidinium hydrochloride: Equilibrium studies at neutral pH. *Biochimica et Biophysica Acta, 1652*(2), 115–125.

Parisi, M., Mazzini, A., Tibor Sorbi, R., Ramoni, R., Grolli, S., & Favilla, R. (2005). Role of the disulphide bridge in folding, stability and function of porcine odorant binding protein: Spectroscopic equilibrium studies on C63A/C155A double mutant. *Biochimica et Biophysica Acta, 1750*(1), 30–39.

Pelosi, P., Zhu, J., & Knoll, W. (2018). Odorant-binding proteins as sensing elements for odour monitoring. *Sensors*, 18(10).

Pevsner, J., Hou, V., Snowman, A. M., & Snyder, S. H. (1990). Odorant-binding protein. Characterization of ligand binding. *The Journal of Biological Chemistry, 265*(11), 6118–6125.

Polverini, E., Lardi, P., Mazzini, A., Sorbi, R. T., Virna, C., Ramoni, R., et al. (2011). Characterization of a Deswapped triple mutant bovine odorant binding protein. *International Journal of Molecular Sciences, 12*(4), 2294–2314.

Sandler, B. H., Nikonova, L., Leal, W. S., & Clardy, J. (2000). Sexual attraction in the silkworm moth: Structure of the pheromone-binding-protein-bombykol complex. *Chemistry & Biology*, 7(2), 143–151.

Schön, A., Brown, R. K., Hutchins, B. M., & Freire, E. (2013). Ligand binding analysis and screening by chemical denaturation shift. *Analytical Biochemistry, 443*(1), 52–57.

Schwaighofer, A., Kotlowski, C., Araman, C., Chu, N., Mastrogiacomo, R., Becker, C., et al. (2014). Honey bee odorant-binding protein 14: Effects on thermal stability upon odorant binding revealed by FT-IR spectroscopy and CD measurements. *European Biophysics Journal, 43*(2–3), 105–112.

Schwaighofer, A., Pechlaner, M., Oostenbrink, C., Kotlowski, C., Araman, C., Mastrogiacomo, R., et al. (2014). Insights into structural features determining odorant affinities to honey bee odorant binding protein 14. *Biochemical and Biophysical Research Communications, 446*(4), 1042–1046.

Scirè, A., Marabotti, A., Staiano, M., Briand, L., Varriale, A., Bertoli, E., et al. (2009). Structure and stability of a rat odorant-binding protein: Another brick in the wall. *Journal of Proteome Research, 8*(8), 4005–4013.

Staiano, M., D'Auria, S., Varriale, A., Rossi, M., Marabotti, A., Fini, C., et al. (2007). Stability and dynamics of the porcine odorant-binding protein. *Biochemistry, 46*(39), 11120–11127.

Steinbrecht, R. A. (1998). Odorant-binding proteins: Expression and function. *Annals of the New York Academy of Sciences, 855*, 323–332.

Stepanenko, O. V., Marabotti, A., Kuznetsova, I. M., Turoverov, K. K., Fini, C., Varriale, A., et al. (2008). Hydrophobic interactions and ionic networks play an important role in thermal stability and denaturation mechanism of the porcine odorant-binding protein. *Proteins, 71*(1), 35–44.

Stepanenko, O. V., Roginskii, D. O., Stepanenko, O. V., Kuznetsova, I. M., Uversky, V. N., & Turoverov, K. K. (2016a). Structure and stability of recombinant bovine odorant-binding protein: I Design and analysis of monomeric mutants. *PeerJ*, *4*, e1933.

Stepanenko, O. V., Roginskii, D. O., Stepanenko, O. V., Kuznetsova, I. M., Uversky, V. N., & Turoverov, K. K. (2016b). Structure and stability of recombinant bovine odorant-binding protein: II Unfolding of the monomeric forms. *PeerJ*, *4*, e1574.

Stepanenko, O. V., Roginskii, D. O., Stepanenko, O. V., Kuznetsova, I. M., Uversky, V. N., & Turoverov, K. K. (2016c). Structure and stability of recombinant bovine odorant-binding protein: III. Peculiarities of the wild type bOBP unfolding in crowded milieu. *PeerJ*, *4*, e1642.

Stepanenko, O. V., Stepanenko, O. V., Staiano, M., Kuznetsova, I. M., Turoverov, K. K., & D'Auria, S. (2014). The quaternary structure of the recombinant bovine odorant-binding protein is modulated by chemical denaturants. *PLoS One*, *9*(1), e85169.

Tegoni, M., Campanacci, V., & Cambillau, C. (2004). Structural aspects of sexual attraction and chemical communlication in insects. *Trends in Biochemical Sciences*, *29*(5), 257–264.

Tegoni, M., Ramoni, R., Bignetti, E., Spinelli, S., & Cambillau, C. (1996). Domain swapping creates a third putative combining site in bovine odorant binding protein dimer. *Nature Structural Biology*, *3*(10), 863–867.

Tsitsanou, K. E., Drakou, C. E., Thireou, T., Vitlin Gruber, A., Kythreoti, G., Azem, A., et al. (2013). Crystal and solution studies of the "plus-C" odorant-binding protein 48 from *Anopheles gambiae*: Control of binding specificity through three-dimensional domain swapping. *The Journal of Biological Chemistry*, *288*(46), 33427–33438.

Vincent, F., Spinelli, S., Ramoni, R., Grolli, S., Pelosi, P., Cambillau, C., et al. (2000). Complexes of porcine odorant binding protein with odorant molecules belonging to different chemical classes. *Journal of Molecular Biology*, *300*(1), 127–139.

Wafer, L., Kloczewiak, M., Polleck, S. M., & Luo, Y. (2017). Isothermal chemical denaturation of large proteins: Path-dependence and irreversibility. *Analytical Biochemistry*, *539*, 60–69.

Ziemba, B. P., Murphy, E. J., Edlin, H. T., & Jones, D. N. M. (2013). A novel mechanism of ligand binding and release in the odorant binding protein 20 from the malaria mosquito *Anopheles gambiae*. *Protein Science*, *22*(1), 11–21.

CHAPTER TEN

Ligand-binding assays with OBPs and CSPs

Chiara D'Onofrio[a], Valeriia Zaremska[a], Jiao Zhu[a,b], Wolfgang Knoll[a,c], Paolo Pelosi[a,*]

[a]AIT Austrian Institute of Technology GmbH, Biosensor Technologies, Tulln, Austria
[b]Faculty of Biology, Institute of Molecular Physiology, Johannes Gutenberg-Universität, Mainz, Germany
[c]CEST Competence Center for Electrochemical Surface Technology, Tulln, Austria
*Corresponding author: e-mail address: ppelosi.obp@gmail.com

Contents

Methods in Enzymology, Volume 642
ISSN 0076-6879
https://doi.org/10.1016/bs.mie.2020.05.006

Abstract

Assessing the ligand-binding properties of OBPs and CSPs is essential for understanding their physiological function. It also provides basic information when these proteins are used as biosensing elements for instrumental measurement of odors. Although different approaches have been applied in the past to evaluate the affinity of receptors and soluble binding proteins to their ligands, using a fluorescent reporter represents the method of choice for OBPs and CSPs. It offers the advantages of working at the equilibrium, being simple, fast and inexpensive, without requiring the use of radioactive tracers. However, as an indirect method, the fluorescence competitive binding approach presents drawbacks and sometimes requires an elaborate analysis to explain unexpected results. Here, after a brief survey of the different approaches to evaluate affinity constants, we focus on the fluorescence binding assay as applied to OBPs and CSPs, discussing situations that may require closer inspection of the results.

Abbreviations

1-AMA 1-aminoanthracene
1-NPN *N*-phenyl-1-naphthylamine
CSP chemosensory protein
OBP odorant-binding protein

1. Introduction

Odorant-binding proteins (OBPs) and Chemosensory proteins (CSPs) are the two main families of carrier proteins mediating chemical communication in vertebrates and insects (Pelosi, Iovinella, Zhu, Wang, & Dani, 2018). The study of their affinities toward small organic molecules, including pheromones and environmental odorants, is of interest from a biological point of view, as well as for potential technological applications.

In fact, their functional characterization is the basis to understand the physiological role of these soluble proteins in chemical communication, and in particular their ligand-binding characterization represents a basis of information on which to build models and suggest hypotheses. Although the fine detection and discrimination of odorants is the task of membrane-bound olfactory receptors (Buck & Axel, 1991; Clyne et al., 1999; Fleischer, Pregitzer, Breer, & Krieger, 2018; Ihara, Yoshikawa, & Touhara, 2013; Leal, 2013; Robertson, 2019) soluble binding proteins, such as OBPs and CSPs, are also selective toward odorants and pheromones (Pelosi, Zhou, Ban, & Calvello, 2006), and in some way they assist olfactory receptors in their task by making the chemical message stronger and more specific (Chang et al., 2015; Forstner, Breer, & Krieger, 2009; Sun et al., 2013). Mapping the binding specificities of OBPs tuned to specific pheromones may also shed light on the structures of their physiological ligands. This approach, named "reverse chemical ecology", proved useful when a direct analysis of the pheromones is difficult or not feasible, as in the case of extinct or endangered species (Choo et al., 2018; Leal, 2017; Zhu et al., 2017).

On the other hand, these proteins, thanks to their exceptional stability, easy synthesis and purification, represent ideal candidates as biosensing elements to be used in artificial devices for environmental monitoring, food quality assessment and clinical analyses (Kotlowski et al., 2018; Larisika et al., 2015; Mulla et al., 2015; Pelosi, Zhu, & Knoll, 2018a). For all such applications, it is fundamental to define for each protein its spectrum of ligands within the types of analytes of interest. Moreover, mutagenesis, that can be easily applied to these proteins (see chapter "Site-directed mutagenesis of odorant-binding proteins" by Zhu et al.), has been used to narrow or even to change their ligand spectra.

For all such tasks, it is of the utmost importance to have a fast, simple and reliable method for measuring the affinities of a protein to a large number of potential ligands. Unfortunately, an ideal method, which is simple and at the same time free from errors and artifacts is not available. In the following sections, we introduce the protocols that have been used for measuring binding constants with soluble proteins, comparing advantages and drawbacks for each of them.

2. Different approaches to measure ligand-binding

A first distinction can be made between protocols that include a step where the free ligand has to be physically separated from the complex before evaluating the relative concentrations (thus possibly disturbing the equilibrium and introducing errors), and those where the levels of bound and free ligand are measured while the system is at the equilibrium. If a separation step is involved, dissociation might occur during this process, whose extent depends on the dissociation rate constant (off-rate) of the complex and the time required for the separation. Generally speaking, in order to avoid large errors, such methods cannot be applied to systems with dissociation constants higher than nanomolar, since then usually dissociation rate constants are in the range of 1–100 s^{-1} (Hulme & Trevethick, 2010; Pang & Zhou, 2017). A graphic summary of the main approaches is reported in Fig. 1.

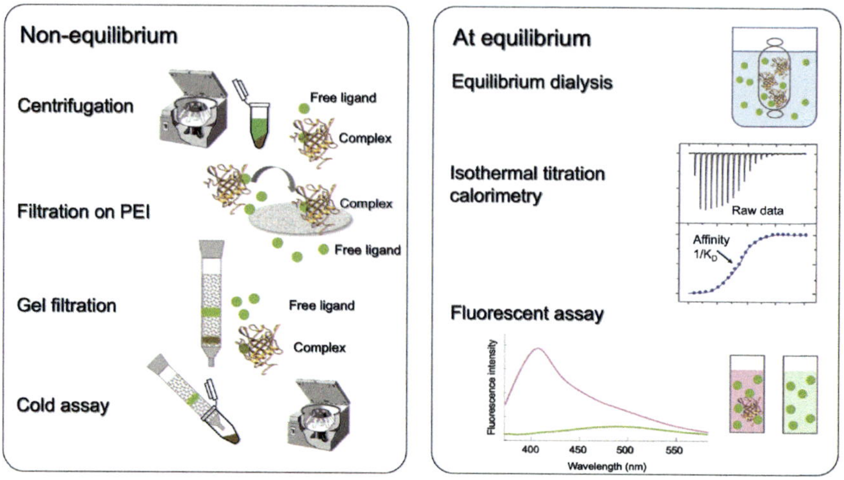

Fig. 1 Different approaches to measure affinity of small ligands to a protein. Methods involving separation of the free ligand from the complex are affected by errors originating from the disturbed equilibrium. Such errors depend on the dissociation rate constant of the complex. As a general rule, such methods can confidently be applied to systems with dissociation constants in the nanomolar range or lower. Approaches that do not require separation can be time consuming and complicated, except for the fluorescent binding assay, that is currently the method of choice for OBPs and CSPs. PEI is polyethylenimine, an anion-exchange resin.

2.1 Methods perturbing the equilibrium

Different techniques have been applied to separate the excess free ligand from the complex after reaching the equilibrium. The most common are listed here below.

(a) In the case of membrane proteins or in general when the binding protein is part of large aggregates, a simple centrifugation step can be applied. This method was adopted in the past to study binding of drugs and neurotransmitters to membrane-bound receptors, such as opioid, acetylcholine and other receptors (Alexander, Davis, & Lefkowitz, 1975; Bennett & Snyder, 1975; Dohlman, Caron, & Lefkowitz, 1987; Jensen & DeSombre, 1973; Pert & Snyder, 1973; Simantov & Snyder, 1977; Snyder, 1979; Snyder, 2004; Snyder & Childers, 1979), but cannot be used with soluble proteins like OBPs and CSPs. Evaluation of the bound ligand was performed by scintillation counting, as in such cases a radioactive version of the ligand was used.

(b) Rapid filtration on glass wool disks soaked in a solution of polyethylenimine was used in past with OBPs and provided reliable results when filtration and washing was very rapidly done with a semi-automatic machine (Dal Monte, Centini, Anselmi, & Pelosi, 1993; Hérent, Collin, & Pelosi, 1995; Pevsner, Hou, Snowman, & Snyder, 1990; Pevsner, Trifiletti, Strittmatter, & Snyder, 1985). The separation is based on the fact that acidic proteins, like most OBPs, are trapped by ionic interaction on the surface of filters, thanks to the basic nature of polyethylenimine, together with their bound ligand. The free fraction of the ligand, on the other hand, is washed away in the buffer (Bruns, Lawson-Wendling, & Pugsley, 1983). Again, the method requires a radioactive ligand, whose activity can be measured directly on the filters. Alternative approaches to avoid the use of a radioactive chemical can be suggested, like the use of a fluorescent reporter, but they have not been applied so far.

(c) Chromatography on a gel-filtration resin separates high molecular weight components (typically the binding protein with its ligand attached) from the free ligand (a small organic compound in the case of OBPs). This method also suffers from the limitation of using a radioactive tracer. Besides, it is very time consuming and therefore not amenable to rapid screening of a number of ligands, and additionally prone to progressive dissociation of the complex during the chromatographic separation, thus leading to large errors. Usually, such methods work well if the dissociation constant of the complex is in the nanomolar range or lower (Hulme & Trevethick, 2010; Pang & Zhou, 2017). However, it was successfully applied to mammalian OBPs (Bignetti et al., 1985; Pelosi, Baldaccini, & Pisanelli, 1982; Pelosi, Pisanelli, Baldaccini, & Gagliardo, 1981; Pelosi & Tirindelli, 1989; Topazzini, Pelosi, Pasqualetto, & Baldaccini, 1985), despite their relatively poor affinity with micromolar dissociation constants. It was shown later that these OBPs, unlike their insect counterparts, exhibit an unusually slow rate of dissociation, with less than 1% of the ligand being released after 1 h (Pevsner et al., 1990). This peculiar phenomenon is attributed to the fact that there is a gated entrance to the binding pocket, involving rotation of the side groups of Phe36 or Tyr89 (Bianchet et al., 1996; Tegoni, Ramoni, Bignetti, Spinelli, & Cambillau, 1996).

(d) A more efficient version of approach (c) was more recently described for measuring binding of pheromones to insect OBPs (Leal, Chen, & Erickson, 2005). It is known as "cold assay", because it does not use radioactive tracers. The main differences with respect to the old method are (1) a fast filtration through mini columns is accomplished in a centrifuge, and (2) the analysis of the bound ligands is performed by gas-chromatography/mass spectrometry, thus avoiding the use of radioactive tracers (Leal, Chen, & Erickson, 2005; Leal, Chen, Ishida, et al., 2005; Xu, Xu, Leal, & Ames, 2011; Zhou et al., 2009; Ishida et al., 2014).

Other methods or variations of the above protocols have been used to evaluate affinities of binding proteins, and only occasionally with OBPs. More detailed and specific information can be found in our recent review (Pelosi, Zhu, & Knoll, 2018b).

2.2 Methods working at the equilibrium

Few approaches have been adopted to measure binding constants without the need to separate bound from free ligands. Working at the equilibrium makes these methods more accurate, as they do not suffer from the progressive dissociation occurring during the procedures described above. However, other difficulties, such as long times of analysis and large amounts of protein required for the experiments, have prevented a wide use of these techniques. The only exception is the method based on fluorescence spectrophotometry, currently the most widely adopted, which however also suffers from potential errors and artifacts.

(a) Equilibrium dialysis. This method still involves separation of bound from free ligand, but the two species are constantly at equilibrium through a dialysis membrane. Therefore, this approach does not suffer from errors due to dissociation of the complex during separation, and perhaps remains the most reliable approach to measure dissociation constants. In principle, equilibrium dialysis could be the method of choice, except for the facts that it is very slow (hours for each measurement) and also requires large amounts of protein. The concentration of protein to be used should be at least in the order of the dissociation constant (Hulme & Trevethick, 2010), therefore this approach is not practical with proteins having affinities in the micromolar range or higher, as it is the case for most OBPs. Besides, the samples taken from the two compartments have to be analyzed for the concentration of ligand. This can be done by GC/MS after extraction in organic solvent, with additional long times of analysis. Alternatively, radioactive tracers can be used, thus introducing safety problems and concerns.

(b) Isothermal titration calorimetry (ITC). In this method, the protein is titrated with the ligand and the heat released or taken up during formation of the complex is measured after each addition. The method is accurate and reliable, if the process under investigation is accompanied by a sufficiently high binding enthalpy. However, it requires expensive instruments and large quantities of the protein. It has been used with OBPs (Briand, Nespoulous, Huet, Takahashi, & Pernollet, 2001; Burova, Choiset, Jankowski, & Haertlé, 1999), but does not qualify for a general practical approach to functionally characterize a binding protein. ITC is described in more detail in chapter "Bacterial expression and purification of vertebrate odorant-binding proteins" by Brulé et al. and chapter "Stability of OBPs" by Hellmann.

(c) Thermophoresis. This interesting technique has been recently developed, and instruments have become commercially available only in the last couple of years. Thermophoresis is based on the observation that proteins, small molecules and their complexes differentially redistribute in a gradient of temperature (Duhr & Braun, 2006). This technique proved to be simple and successful in evaluating binding of aptamers to proteins, as well as detecting interactions between two proteins (Baaske, Foreman, & Vollmer, 2014; Fayad, Morin, & Nehmé, 2017; Mueller et al., 2017; Seidel et al., 2013; Wienken, Baaske, Rothbauer, Braun, & Duhr, 2010). Using fluorescence to monitor the migration of the molecular species under observation, microscale thermophoresis (MST) has been used in evaluating dissociation constants of small ligands to carrier proteins (Bartoschik et al., 2018; Fisher et al., 2017; Gao, Or, Toop, & Wheeldon, 2017; Lang, Gao, & Wheeldon, 2018). Although attractive and promising, the main drawback for a wide adoption of thermophoresis has been so far the high costs of the equipment, as compared to a spectrofluorimeter, the current best alternative for measuring binding events. So far, thermophoresis has not yet been used to investigate binding of ligands to OBPs or CSPs.

(d) Fluorescence binding assay. This has been for the last two decades the method of choice adopted with OBPs and CSPs, and will be described in detail, highlighting its advantages as well as its drawbacks and limits of application.

3. Fluorescence binding assay

3.1 Principles of measurement

The evaluation of the binding of a fluorescent ligand to a protein is based on the dependence of fluorescence on the hydrophobicity of the environment.

Typically, the binding pocket of a carrier protein for hydrophobic molecules, such as lipids, terpenoids and most common odorants, is lined with the side chains of non-polar amino acid, typically valine, leucine, isoleucine, phenylalanine and others, that provide a hydrophobic environment, while the external medium is usually aqueous. Consequently, the fraction of fluorescent ligand bound in the cavity presents an emission spectrum different from that of the same ligand exposed to the external aqueous solution. Generally, upon binding, the fluorescent spectrum undergoes a blue shift accompanied by a significant increase in fluorescence intensity. This makes the evaluation of the bound fraction relatively easy, although subtraction of the blank signal (in the absence of the protein) is always mandatory.

The method therefore involves titration of a protein solution with the fluorescent ligand and recording of the emission spectrum. Equilibrium is often reached within a few seconds and at the most within minutes, so that no long waiting time is needed, and the procedure is fast enough to allow a large number of measurements to be made in a single day.

The sensitivity of this assay depends on the fluorescence yield of the probe used, but usually allows measurements of dissociation constants of the micromolar order. This falls within the values reported for most OBPs and CSPs, but would be inadequate to evaluate binding constants in the nanomolar range, as is the case for many receptors. In such cases, the use of radioactive tracers is difficult to avoid.

3.2 Types of measurement

The method can be adapted to different protocols, depending on which is the specific object of the measurement.

1. Evaluation of the affinity between a protein and a fluorescent ligand. This only involves a direct titration of the protein with the ligand and evaluation of the bound fraction of the ligand, if the emission spectrum of the complex is sufficiently separated from (or is of much higher intensity than) the spectrum generated by the free ligand.
2. Measurement of affinity constants of non-fluorescent ligands. In this case, the best method to adopt is that of competitive binding, where the affinity of the ligand is evaluated from its capacity of displacing the fluorescent reporter from the complex. Accordingly, a solution of the protein and the fluorescent ligand, both at a fixed concentration, is titrated with a solution of a second ligand. Displacement of the bound fluorescent ligand leads to a concomitant, dose-dependent decrease in fluorescence

intensity. From the generated curve the value of the second ligand concentration able to displace half of the initial fluorescent reporter $[IC]_{50}$ is evaluated. This value can be used to calculate the dissociation constant of the second ligand.

3. Quenching of intrinsic fluorescence. Tryptophan residues, when present in a protein, generate a fluorescence peak, upon excitation at or near its absorbance maximum, 280 nm. The wavelength of the emission peak again varies with the environment, ranging from 330 to 355 nm, depending on whether the tryptophan is located inside a hydrophobic cavity or exposed to the aqueous solvent. If the tryptophan is located in the binding cavity, the fluorescence intensity of its emission peak can be quenched by any ligand entering the cavity by the effect of energy transfer. Typically, aromatic molecules absorbing at the wavelength of tryptophan emission can significantly quench the fluorescence signal in a dose-dependent fashion. Thus, a simple titration can indicate whether or not a small organic compound is able to bind to the protein. For determination of a dissociation constant, however, we should consider that the quenching cannot be assumed to reach zero at full occupancy of the protein site; moreover, the intensity of the quenching depends on unknown factors, such as the distance between tryptophan and ligand and the geometry of their interactions. Also, this approach cannot be applied to monitor the entrance of aliphatic molecules or in general any molecule lacking a group capable of energy transfer. On the other hand, it is possible that in principle highly hydrophobic molecules might produce an increase of the fluorescence signal by decreasing the polarity of the tryptophan environment, although strong effects cannot be expected, as generally the binding pockets of OBPs are lined with hydrophobic residues.

These three approaches will be discussed in more detail and with reference to OBPs and CSPs here below, and experimental protocols will be provided.

4. Binding of fluorescent probes to OBPs and CSPs

Several fluorescent reporters have been tested with OBPs and CSPs, but only two probes have been adopted in the large majority of studies.

1-Aminoantracene (1-AMA) was first successfully used with vertebrate OBPs (Paolini, Tanfani, Fini, Bertoli, & Pelosi, 1999). It can be excited at different wavelengths (256, 280, 375), corresponding to different peaks in its absorption spectrum. Emission occurs around 537 nm when the probe is in aqueous solvent, but it is shifted to 481 nm with a strong increase in its intensity when bound to an OBP. The same probe exhibited poorer performance with insect OBPs (Campanacci et al., 2001), and was unsuccessful with CSPs.

N-Phenyl-1-naphthylamine (1-NPN), instead, proved to be a good fluorescent reporter both with insect OBPs and with CSPs, while did not perform as well as 1-AMA with several vertebrate OBPs. However, with a sub-group of vertebrate OBPs, including mouse urinary proteins (MUPs), pig salivary lipocalins (SALs), horse and rabbit lipocalins, all sharing some characteristic sequence motives, 1-NPN gave better results than 1-AMA (Beynon et al., 2002; D'Innocenzo et al., 2006; Marie et al., 2001; Mastrogiacomo, D'Ambrosio, et al., 2014).

Other fluorescent probes have occasionally been adopted with OBPs. Their structures and spectral characteristics are listed in Fig. 2.

A typical protocol for evaluating binding constants between an OBP or CSP and a fluorescent probe (1-AMA or 1-NPN) is described below.

Essentially, a solution of the protein is titrated with the fluorescent ligand and emission spectra are recorded after each addition.

4.1 Equipment

1. Spectrofluorimeter
2. Quartz fluorimeter cuvettes

4.2 Reagents

1. Purified OBP or CSP
2. 50 mM Tris/HCl buffer, pH 7.4
3. 1 mM methanol solution of 1-AMA or 1-NPN

4.3 Protocol

1. Dilute the protein to 2 μM in 1 mL of Tris buffer directly in a spectrofluorimeter cuvette and allow the solution to reach the desired temperature. This could take up to a couple of minutes if the protein stock solution is kept on ice, as generally recommended. Fluorescence intensity is markedly dependent on the temperature and for accurate measurements it would be advisable to use a cuvette holder with temperature control. When using a standard 1×1 cm cuvette, a spacer of about 1 cm should be put under the cuvette, so that no part of the light beam passed outside the solution.

Fluorescent probe	Ex λ (nm)	Em λ (nm)	ref	Fluorescent probe	Ex λ (nm)	Em λ (nm)	ref
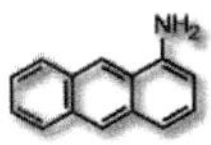 1-Amino anthracene (1-AMA)	256 280 375	537 481	(1)	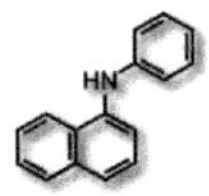 N-Phenyl-1-naphthyl amine (1-NPN)	337	480 410	(2)
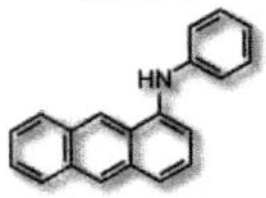 N-Phenyl-1-aminoanthracene (PAA)	410	520 510	(3)	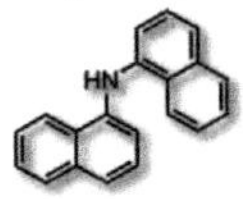 N-naphthyl-1-naphthylamine (1-NNN)	345	420 412	(3)
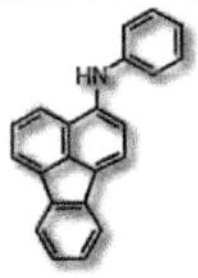 Phenyl-1-aminofluoranthene (PAF)	315	525 515	(3)	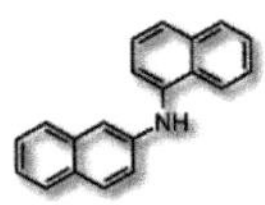 N- naphthyl-2-naphthylamine (2-NNN)	345	420 412	(3)
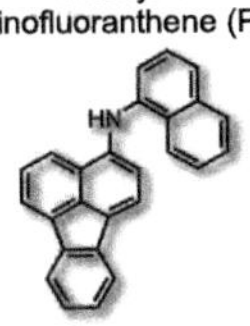 1-Naphthyl-1-amino-fluoranthene (1-NAF)	325	530 509	(3)	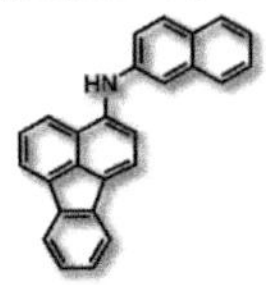 2-Naphthyl-1-amino-fluoranthene (2-NAF)	325	535 510	(3)
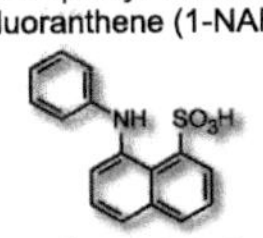 8-Anilinonaphthalene sulfonic acid (ANS)	372	510 468	(4)	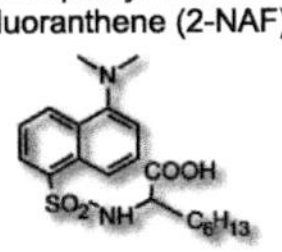 Dansylaminocaprilyc acid (DACA)	345	545 475	(5)
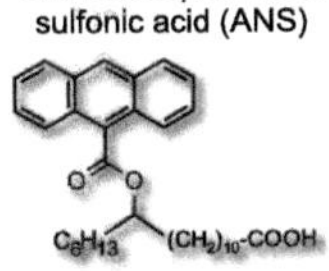 12-(9-anthroyloxy) stearic acid (ASA)	360	445 425	(6)	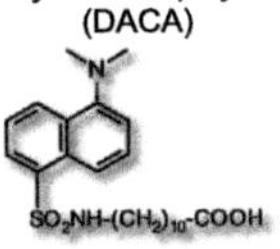 Dansylaminoundecanoic acid (DAUDA)	345	545 490	(7)

Fig. 2 Fluorescent probes used in ligand-binding assays with OBPs and CSPs. In the emission wavelength column, the upper figure indicates the wavelength of the maximal emission intensity in aqueous solution, the lower figure refers to the value when the probe is bound to the protein. (1) (Campanacci et al., 2001; Paolini et al., 1999); (2) (Ban et al., 2003; Ban, Zhang, Yan, & Pelosi, 2002; Beynon et al., 2002; D'Innocenzo et al., 2006; Marie et al., 2001; Mastrogiacomo, D'Ambrosio, et al., 2014) (3) (Mastrogiacomo, Iovinella, & Napolitano, 2014); (4) (Briand, Eloit, et al., 2002; Briand, Swasdipan, et al., 2002; Lartigue et al., 2003; Lobel, Marchese, Krieger, Pelosi, & Breer, 1998) (5) (Briand, Eloit, et al., 2002) (6) (Briand, Swasdipan, et al., 2002); (7) (Briand, Eloit, et al., 2002; Tcatchoff, Nespoulous, Pernollet, & Briand, 2006).

2. Add aliquots of 1 mM methanol solution of 1-AMA or 1-NPN, mix sufficiently, and record the emission spectrum. Generally, 1-AMA is excited at 375 nm and spectra are recorded between 460 and 530 nm. A peak is expected around 490–500 nm. Set both excitation and emission slits at 5 nm and use medium sensitivity (if the signal is too weak, the sensitivity can be increased to high). Read the intensity at the maximum of the emission spectrum generated before addition of the competing ligand, then measure all intensities at this same wavelength. Typically, values recorded at final concentration values of fluorescent probe of 2, 4, 6, 8, 12, 16 μM are enough for drawing a reliable binding curve. For 1-NPN the excitation wavelength is 337 nm, while spectra are recorded between 380 and 450. Spectra with maxima around 400–420 nm are to be expected.
3. Perform blank experiments under the same conditions, but in the absence of protein, and subtract the values from those obtained with the protein.
4. Specific values, recorded in triplicates, are analyzed using a non-linear fitting routine employing appropriate equations (for instance with the software Prism GraphPad). It is worth to remember that if a hyperbolic function is used, the concentration of the complex (as calculated from the fluorescence intensity of the bound ligand) should be plotted against the concentration of the free ligand, not the total ligand. This can be easily appreciated based on the following equations:

Starting from the expression of the equilibrium constant between a protein P and a ligand L, forming the complex PL:

$$K_D = \frac{[P][L]}{[PL]}$$

and the relationship for mass conservation of the protein:

$$[P_0] = [P] + [PL]$$

we obtain by simple substitution the usually employed hyperbolic function describing [PL]:

$$[\mathrm{PL}] = \frac{[L][P_0]}{[L] + K_D}$$

Fig. 3 shows the curves generated when the values of the bound ligand are plotted against the total or the free concentration of ligand.

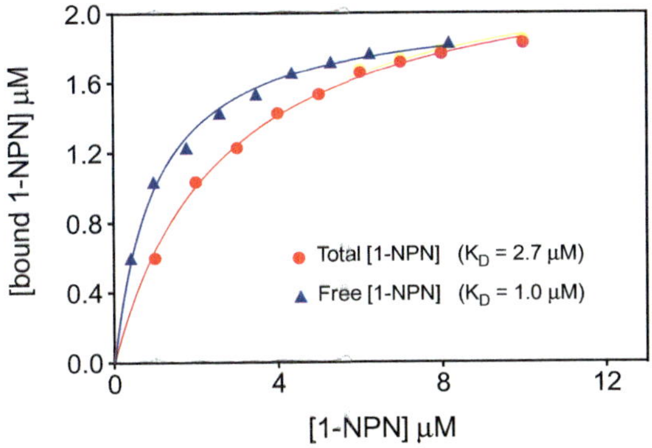

Fig. 3 Binding of 1-NPN to the CSP1 of the honeybee. Values of bound ligand are plotted against free ligand (blue triangles) or total ligand concentrations (red dots). As evident from the curves, the K_D-value obtained from plotting the total ligand concentrations is erroneously higher compared to the correct one (1 μM).

Clear differences can be observed, particularly at low concentrations of ligand, and different values of dissociation constants are obtained. To calculate the concentration of the free ligand after each addition, we need to know the fluorescence intensity at protein saturation and the stoichiometry of the binding. The fluorescence yield of the complex can be very different between proteins, and therefore has to be derived by extrapolating the binding curve to saturation. Thus, it is important to obtain a curve with small slope at high ligand concentrations, in order to keep the error in the estimation of the plateau value small.

The stoichiometry cannot be obtained with such an approach, but for OBPs and CSPs we can just assume a value of 1. However, we should be aware that this might not always be the case (Campanacci et al., 2003), and errors can also be made if the protein is not fully active, through a wrong evaluation of its active concentration.

Alternatively, a different function, allowing to express [PL] as function of total protein and total ligand concentration [Lo] can be used:

$$F_{meas} = F_u * [PL] = F_u * \left[\frac{[P_o] + [L_o] + K_D}{2} - \sqrt{\left(\frac{[P_o] + [L_o] + K_D}{2} \right)^2 - [P_o] * [L_o]} \right]$$

Here, F_{meas} is the measured fluorescence, and F_u the fluorescence intensity characteristic for [PL]. F_u is treated as free parameter, to be determined by the fitting routine. If the stoichiometry is not clear, this can be introduced by replacing [Po] with n*[Po] and treat *n* also as a free parameter. Note, that if $[P_o]\delta\ 0.2*K_D$, the total ligand concentration well represents the free ligand concentration, and the hyperbolic function can be used for analysis.

5. The use of a Scatchard plot, more popular in the past, is not recommended as it can provide less accurate results than a non-linear analysis, because it does not assign the same weight to all data. Differences in the values obtained with the two methods have been observed, but usually do not exceed 20%. However, the Scatchard plot can put into evidence cooperativity effects. If the protein possesses a single binding site without cooperativity effects, the plot should give a straight line. Instead, in the presence of cooperativity effects the plotted values follow a upward or downward curved line, showing positive or negative cooperativity, respectively. These effects might be observed if the entrance of a ligand into the binding site facilitates or inhibits the entrance of a second molecule of the same ligand (positive or negative effect, respectively). Measuring the binding of a probe to a mixture of two OBPs, sometimes a Scatchard plot of unusual shape was observed, that could not be analyzed. This behavior cannot be explained by any of the current models, and it was hypothesized that the two proteins would interact with each other. This hypothesis was supported by the observation that the two OBPs were co-expressed in the same sensillum of the antenna (Qiao et al., 2011; Wang et al., 2013).
6. For most OBPs and CSPs, dissociation constants fall within the lower micromolar range with either 1-AMA or 1-NPN (or both). Values in the upper micromolar range can still be of use for evaluating the affinities of other ligands in competitive binding assays. Even higher dissociation constants would correspond to binding too weak to provide reliable and useful results.

5. Competitive binding assays

Ligands which are not fluorescent can be assayed for their affinities for a protein using an indirect approach, namely competitive binding. Accordingly, a solution of protein and fluorescent probe are titrated with the ligand to be tested. The reduction of fluorescence caused by the added ligand is a measure of the affinity of the ligand for the protein. We expect that in ideal conditions and in the absence of other effects, the fluorescence of the probe is decreasing in a concentration-dependent function asymptotically approaching zero.

The most accepted measure for the ligand affinity is the concentration required to reduce the initial fluorescence to 50%. This value, referred to as $[IC]_{50}$, should correspond, in the absence of other effects, to displacement of 50% of the probe from the initial complex. This is not always true, but, even in ideal conditions, the limitation of this method lies in the fact that what we observe is not the simple binding of the ligand to the protein, but the displacement of the probe from the complex. In fact, it is possible that in some cases the competing ligand might enter the cavity without displacing the probe. Moreover, other effects, not linked to binding events, might change the fluorescence intensity, as for instance in the case of micelle formation observed with several ligands (Pelosi, Zhu, & Knoll, 2018b; Tan, Zaremska, Lim, Knoll, & Pelosi, 2020).

A typical protocol for evaluating binding constants between an OBP or CSP and a general ligand, using the approach of competitive binding assay, is described below.

5.1 Equipment

1. Spectrofluorimeter
2. Quartz fluorimeter cuvettes

5.2 Reagents

1. Purified OBP or CSP
2. 50 mM Tris/HCl buffer, pH 7.4
3. 1 mM methanol solution of 1-AMA or 1-NPN
4. 1 mM methanol solution of the ligand to be assayed

5.3 Protocol

1. Dilute the protein to 2 μM in 1 mL of Tris buffer directly in a spectrofluorimeter cuvette and allow the solution to reach the target temperature.
2. Add 2 μL of 1 mM methanol solution of 1-AMA or 1-NPN and record the emission spectrum. The value corresponding to the maximum of the peak corresponds to the initial fluorescence value, taken as 100%.

3. Titrate the solution with aliquots from a 1 mM methanol solution of the ligand to final concentration of 2, 4, 6, 8, 12, 16 μM and record each time the value corresponding to the same wavelength.
4. Subtract from each value the intensity recorded with the blank (2 μM probe in Tris buffer without protein).
5. Specific values, recorded in triplicates, are usually normalized to 100% at the initial value using a spreadsheet like Excel. This allows to average several titrations, even if the initial value is somewhat variable, and to obtain adequate standard errors. All other values are consequently expressed as percent of the initial fluorescence intensity. From each curve, the value of $[IC]_{50}$ can be then graphically determined, but there are also equations to calculate such value (Huang, 2003) and software is available, such as Sigmaplot or the simple calculator at https://www.aatbio.com/tools/ic50-calculator. The values of $[IC]_{50}$ can be inserted into the Cheng and Prusoff formula (Cheng & Prusoff, 1973):

$$K_D = \frac{[IC]_{50}}{1 + [F]/K_F}$$

to calculate the apparent dissociation constant of the ligand. Here, K_F is the dissociation constant of the complex protein/fluorescent probe, and [F] is the concentration of free fluorescent probe at $[IC]_{50}$. This last value can be calculated by adding the concentration of the free probe (F_0) at the beginning of the experiment (which can evaluated from the probe binding curve, as the y-value corresponding to the 2 μM concentration on the x-axis) to 50% of the bound probe $(2-F_0)/2$, which is the probe displaced by the ligand at its $[IC]_{50}$ value. An example can clarify this concept: in the graph of Fig. 3, reproducing a typical binding isotherm, the free probe concentration at the beginning of the experiment, after 2 μM probe has been added, is 1.0 μM (evaluated from the curve with red dots obtained by plotting the values of bound probe against total probe). Therefore, the bound probe is 1.0 μM at the beginning of the experiment and is reduced to 0.5 μM when the ligand reaches its $[IC]_{50}$ concentration. Therefore, the value [F] to introduce into the equation is 1.0 (initial) +0.5 (displaced) = 1.5 μM.

6. With ligands which are not able to displace 50% of the probe at the maximum concentration used, it is still possible to extrapolate, if the final value is close to 50%. Otherwise, the chemical tested can be considered as a weak ligand or a non-ligand.

6. Tryptophan quenching

When all fluorescent probes fail to bind the protein, or at least fail to show any change in the emission spectrum in the presence of the protein, binding of ligands can be revealed by monitoring the change of the intrinsic tryptophan fluorescence upon addition of a ligand. Several elements, however, must be considered when planning the experiments and interpreting the results.

1. The first requisite is that the protein contains a tryptophan inside the binding pocket. This fact can be clearly assessed if the three-dimensional structure is known, or can be assumed if a reliable model of the protein can be built. Although in principle even tyrosines can emit a fluorescence signal and therefore could act as reporter for the entrance of a ligand in the binding cavity, the intensity of the signal produced by this amino acid is too weak to be of any use.
2. Not all ligands can be tested in this assay, but only those capable of energy transfer from the tryptophan. These are chemicals containing an extended conjugated system, including all aromatic compounds. Also halogen atoms are capable of fluorescence quenching, as reported in the binding of 12-bromododecanol to a CSP of the moth *Mamestra brassicae* (Campanacci et al., 2003).
3. It is important to pay attention, if using phenols as ligands to test, because these compounds can emit a weak fluorescence, capable of neutralizing or even overcoming their own quenching effect on tryptophan.
4. If the ligands show absorption at the wavelength where tryptophan is excited, inner-filter-effects can occur, in case rather large concentrations are employed. Therefore, the absorbance at the excitation should be checked if in doubt.

Here is a typical protocol for monitoring ligand-binding using intrinsic tryptophan quenching.

6.1 Equipment

1. Spectrofluorimeter
2. Quartz fluorimeter cuvettes

6.2 Reagents

1. Purified OBP or CSP
2. 50 mM Tris/HCl buffer, pH 7.4
3. 1 mM methanol solution of the ligand to be assayed

6.3 Protocol

1. Dilute the protein to 2 μM in 1 mL of Tris buffer directly in a spectrofluorimeter cuvette and allow the solution to reach the target temperature.
2. Set the excitation wavelength to 295 nm, both slits at 5 nm, sensitivity to "high", and record the emission spectrum between 310 and 380 nm. If a tryptophan is located inside a hydrophobic binding pocket, the maximum emission should occur around 330–335 nm.
3. Titrate the solution with aliquots of 1 mM methanol solution of the ligand to be tested to final concentration of 2, 4, 6, 8, 12, 16 μM, and record the emission spectrum after each addition.
4. Plotting the intensities at the wavelength corresponding to the maximum of the peak in the absence of ligand against the concentration of ligand should generate a descending curve, whose steepness is a crude measure of the affinity of the ligand for the protein. Usually, the curve does not tend asymptotically to zero, but can level eventually to any value, depending upon several factors, such as the efficiency of energy transfer between tryptophan and ligand, the closeness of the ligand to tryptophan and the geometry of interaction between the two molecules.

7. Analysis of results

The binding assay using fluorescent reporters is currently regarded as by far the best alternative to measure affinity constants between OBPs or CSPs and small ligands. However, being a competitive binding assay, this method is still far from representing the ideal procedure. In fact, some unexpected results are occasionally observed, that need elaborate analysis and interpretation, before reaching immediate conclusions on the affinity of a certain ligand for the protein.

Two main elements should be considered when trying to link the detected change in the fluorescence spectrum with a binding event:

1. what we measure is not the binding of a ligand to the protein, but the displacement of the fluorescent probe from the complex;
2. a fluorescence change cannot be always interpreted in terms of a change in the equilibrium between protein, probe and ligand, as fluorescence is affected by other factors, such as non-equilibrium effects, micelle formation and binding to different binding sites on the protein.

Here we report cases with unexpected behavior and analyze the reasons and the molecular mechanisms behind these phenomena.

7.1 The probe fluorescence increases with the concentration of the competing ligand

Sometimes, the addition of a ligand to the complex between OBP (or CSP) and the fluorescent probe produces an increase in fluorescence, rather than the expected decrease (in the case of a good ligand) or at least no variation (in the case of a non-ligand). In other cases, the addition of a ligand produces a decrease of fluorescence at low concentrations, then an increase at higher values. This phenomenon has been explained with the hypothesis that the ligand can form micelles, which can entrap molecules of the fluorescent probe (Leal & Leal, 2015; Pelosi, Zhu, & Knoll, 2018b; Sun et al., 2012). Micelles, being hydrophobic in their inside, provide an environment for fluorescent compounds similar to that found in most OBPs and CSPs, thus producing a similar shift and increase in the fluorescence spectrum. Most of the typical ligands exhibiting this effect are amphiphilic chemicals, such as long-chain fatty acids and alcohols. The phenomenon has also been observed with some long-chain hydrocarbons. In such cases, it can be hypothesized that part of the hydrocarbon enters the binding pocket, thus making an amphiphilic complex, whose polar head is the protein, thus capable of forming micelles. If a chemical able to produce such effects is also a good ligand for the protein, we first observe a quenching of the fluorescence, corresponding to a binding event. Increasing the concentration of the ligand then produces an increase in fluorescence, as the free fraction of the ligand starts producing micelles. Indeed, experiments in absence of protein support this hypothesis. It is worth recalling, in such context, that micelles form above a certain concentration, called critical micelle concentration (CMC), a value specific for each chemical compound. Fig. 4A shows some examples of this phenomenon.

For practical purposes, to decide whether a tested chemical is a good ligand, and in such case evaluate its affinity for the protein, we can distinguish three situations: (a) the competition curve increases since the first addition: in this case, it is easy to conclude that the chemical has no measurable affinity for the protein; (b) the curve decreases reaching at least 50% quenching, then starts increasing: we can evaluate the $[IC]_{50}$ value and from this calculate the dissociation constant with good confidence; (c) the curve starts decreasing, but then increases before reaching 50% quenching: we can

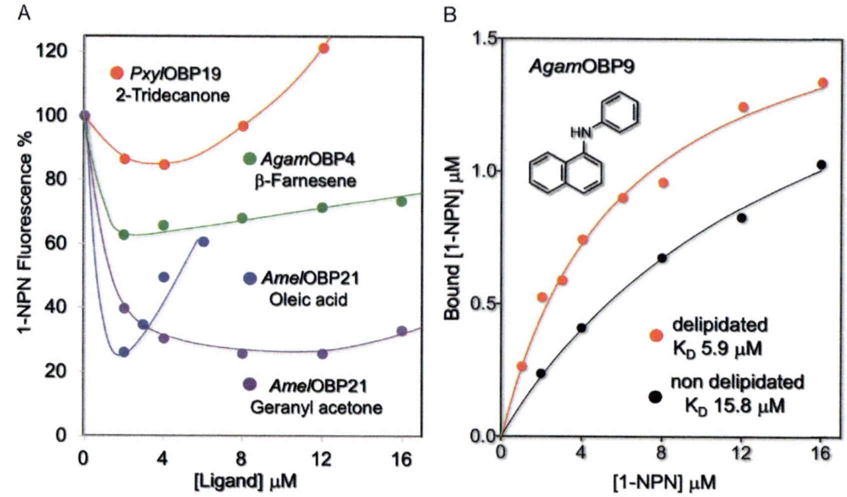

Fig. 4 (A) Examples of ligands causing an increase in fluorescence, most likely due to formation of micelles. The shape of the competition curve varies according to the type of ligand and to its affinity for the protein. (B) Typical case of a protein containing endogenous ligands, that affect binding performance. After delipidation, the dissociation constant of the complex with 1-NPN decreases significantly. Pxyl: *Plutella xylostella*; Agam: *Anopheles gambiae*; Amel: *Apis mellifera*.

attempt to extrapolate the decreasing phase and evaluate a value for $[IC]_{50}$, but the accuracy of the calculated dissociation constant can be questionable, depending on how reliable the extrapolation is.

In any case, when the fluorescence curve decreases at low concentrations of the ligand, we can confidently assume that the chemical is a ligand for our protein. We can try and get better results working at lower concentrations of the fluorescent probe (if the sensitivity of the system still allows for reliable measurements) and limit the concentration range of the ligand to values below the CMC.

7.2 Effect of the probe on the apparent affinity of a ligand

It has been mentioned that while for insect OBPs and CSPs 1-NPN is the probe of choice, for mammalian OBPs both 1-AMA and 1-NPN could be used as suitable fluorescent reporters, depending on the protein. Some OBPs have been reported to bind both probes equally well, but strong differences in the measured affinities of other ligands could still be observed when using one or the other probe. In particular, a recent study showed that the a mutant of the pig OBP1 (Wei, Brandazza, & Pelosi, 2008) binds 1-AMA and 1-NPN with the same good affinities (K_D 0.54 and 0.63 μM, respectively). However, several ligands efficiently displace 1-AMA from the complex with the protein, but have little effect in displacing 1-NPN (Tan et al., 2020). This strange and unexpected behavior is not easy to explain. One hypothesis is that, in the case of 1-NPN, ligands could be able to enter the binding cavity without displacing the fluorescent probe. So far, there is no evidence to support this model and other alternatives could be suggested. These results indicate that some good ligands could be overlooked, as no apparent displacement of the fluorescent reporter is observed. Therefore, it is always advisable to use a second probe and test again chemicals that are suspected to be good ligands, but did not produce displacement of the first probe.

7.3 Elements affecting the binding results

Several aspects should be considered when preparing ligand-binding experiments, and perhaps it might be useful to provide here a sort of checking list with suggestions and criteria to follow.

7.3.1 Purity of the protein

The purity of a protein is generally assessed by gel electrophoresis and UV spectrum. Ideally, a single band should be present in the gel, but minor impurities may not represent a serious problem. On the other hand, there could be modifications in the structure of the protein that are not visible on the gel. For instance, it is not totally infrequent that few amino acids are lost from the C-terminus or the N-terminus, due to limited proteolysis. Although OBPs and CSPs are exceptionally stable owing to their compact structure (Paolini et al., 1999; Schwaighofer et al., 2014), the two termini of the proteins are usually non-structured and hang loose from the core of the polypeptide, thus being more exposed to proteolytic enzymes.

The UV spectrum is important to evaluate the amount of DNA in the protein sample and is essential for determining the protein concentration. While DNA does not seem to interfere with the binding, its presence in the sample prevents an accurate estimate of the protein concentration. It is worth recalling that the absorbance at 280 nm provides the most accurate method to measure protein concentrations in the case of single pure proteins. Other methods, such as the Bradford assay, may be affected by errors up to a factor of 2, depending on the nature of the protein.

7.3.2 Modifications at the N-terminus and C-terminus

While there are no indications that the N-terminus might be important for the binding activity, with insect OBPs the C-terminus may play relevant functions, also depending on its length. The *Bombyx mori* pheromone-binding protein (PBP1) was reported to undergo a major conformational change of its C-terminus at acidic pH, forming a seventh α-helix, which entered the binding site (Damberger et al., 2000). Similar phenomena have been observed with other OBPs exhibiting a "long" C-terminus. Instead, in proteins endowed with a shorter C-terminal segment, this may behave as a "lid" closing the binding pocket. In any case, modifying the C-terminus in insect OBPs may affect the binding performance of the protein (Tegoni, Campanacci, & Cambillau, 2004) (see also chapter "Structure of odorant binding proteins and chemosensory proteins determined by X-ray crystallography" by Gaubert et al.; chapter "Solution structure of insect CSP and OBPs by NMR" by Leone et al.).

Instead, the presence of a His-tag at the N-terminus was reported not to modify the affinity of OBPs for their ligands in a couple of cases (Guo, Huang, Pelosi, & Wang, 2012; Zhu et al., 2016). General directions however cannot be given, and each protein should be considered as a different specific case.

7.3.3 Activity of the protein

Another factor to be considered is the stability of the protein. Partial denaturation, oxidation or limited proteolysis with loss of small segments may affect the activity of the protein. These factors are difficult to assess, particularly in cases when the sequence of the protein is not affected, but only partial denaturation is the cause reducing the efficiency of binding. Instead, when structural modifications are suspected, mass spectrometry techniques, such as MALDI and electrospray, can help checking the molecular weight and, in case of differences with calculated values, identifying the region in the sequence where the modification has occurred.

7.3.4 Purity of the probes

Both 1-AMA and 1-NPN are amines, and therefore are subjected to oxidation. The first, being a primary amine, is less stable in this respect. It is mandatory to keep the methanol solutions in well closed vials at −20 °C. Solutions of both probes at the concentration of 1 mM should be colorless or just pale yellow. When the color becomes yellow or brown, new solutions should be prepared, as the oxidation products can seriously affect the measurements.

7.3.5 Purity of the solvent

Methanol or ethanol used to dissolve the probes and the ligands should be of highest purity, preferably of spectroscopic grade. Solvents kept in plastic bottles should not be used, as they can release phthalates, which are good ligands for several OBPs and CSPs (Zhou et al., 2004).

7.3.6 Endogenous ligands

Proteins expressed in bacteria, more often that those purified from tissues, contain endogenous ligands tightly bound and co-purified with the protein. These ligands are generally fatty acids or other lipids and can affect the binding properties of the protein. When this is the case, it is necessary to delipidate the protein before binding assays, because the ligand present in the protein may compete with the fluorescent probe and with the tested chemicals. Fig. 4B reports a typical example of a protein containing endogenous ligands, whose affinity to the fluorescent probe was significantly improved after delipidation.

Removal of endogenous ligands can be performed by simple extraction with an organic solvent or by elution through a hydrophobic column, such as Sephadex LH-20. If an organic solvent is used, dichloromethane presents the advantage that, being heavier than water, the separation of the aqueous layer is easier. A problem encountered in such phase partition is the formation of a foamy layer at the interface between the two phases. This layer may contain most of the protein and therefore should be reduced. Centrifugation can be necessary, but in any case, some protein loss usually occurs. Consequently, it is necessary to measure the protein concentration again after such treatment.

In some cases, notably when the ligand is a fatty acid, such treatment may not be sufficient, as a carboxylic acid is fully ionized at the neutral pH commonly used for proteins solutions, therefore it would not be soluble in organic solvents. To perform the extraction, in such cases it is necessary to lower the pH to about 4–5 using acetic acid, then extract with solvent, and finally restore the neutral pH in the protein solution to be used for binding experiments.

8. Discussion and conclusion

Ligand-binding assays with soluble proteins, such as OBPs and CSPs are currently performed using a fluorescent reporter and measuring the affinities of other ligands in competitive binding experiments. Although other approaches are feasible and have been adopted in some cases, the fluorescence method is the protocol of choice, owing to several practical advantages. In fact, it is the only method allowing fast and simple screening of many ligands, with a limited use of protein and without the use of unsafe materials, like radioactive tracers. We have also seen that the large majority of OBPs and CSPs bind to different extents 1-AMA or/and 1-NPN, two probes that represent excellent tools because of the large separation between the spectra in buffer and when bound to the protein. Besides, most OBPs and CSPs present a tryptophan residue inside the binding pocket, thus offering an alternative way to monitor the entrance of particular ligands into the binding pocket.

The protocols using fluorescent reporters, although suitable for OBPs and CSPs, which usually bind their ligands with dissociation constants in the micromolar range, is not practical with proteins with stronger affinities, like olfactory receptors, as lower concentrations would have to be used with the consequence that the fluorescent signals could be too weak to be monitored with the typically available equipment.

The fluorescent approach, however, has to be applied with caution if adopted for competition experiments. In these a decrease in fluorescence is generally assumed to be the result of displacement of the probe from the complex. However, in some cases, more complex situations may be present, where results would need careful examination before drawing conclusions, and analysis of the data would require additional information about the interaction between partners, e.g., kinetics of association and dissociation, solution properties of the ligands, etc.

So far, no simple, inexpensive and fast method for measuring direct binding of a ligand to a protein, in particular to an OBP or CSP, is available. Current research on biomimetic gas sensors uses OBPs and CSPs as specific detectors in optical and electronic devices (see chapter "Electrochemical biosensing with odorant binding proteins" by Szunerits et al.; chapter "Electronic biosensors based on EGOFETs" by Mulla and chapter "Stability of OBPs" by Hellmann; chapter "Gravimetric biosensors" by Cali et al.; chapter "Plasmonic biosensors relying on biomolecular conformational changes—Case of odorant binding" by Bozdogan et al.). Such instruments, which are aimed at monitoring markers of diseases, aroma components of foods or noxious chemicals present in the environment, can also be adapted for studying the binding properties of a new protein. We are still far from producing a reliable instrument of this type, but perhaps this could represent a promising approach for a label-free characterization of the binding properties of OBPs and CSPs, as well as of other carrier proteins and receptors.

References

Alexander, R. W., Davis, J. N., & Lefkowitz, R. J. (1975). Direct identification and characterisation of β-adrenergic receptors in rat brain. *Nature*, *258*(5534), 437–440. https://doi.org/10.1038/258437a0.

Baaske, M. D., Foreman, M. R., & Vollmer, F. (2014). Single-molecule nucleic acid interactions monitored on a label-free microcavity biosensor platform. *Nature Nanotechnology*, *9*(11), 933–939. https://doi.org/10.1038/nnano.2014.180.

Ban, L., Scaloni, A., D'Ambrosio, C., Zhang, L., Yan, Y., & Pelosi, P. (2003). Biochemical characterization and bacterial expression of an odorant-binding protein from *Locusta migratoria*. *Cellular and Molecular Life Sciences: CMLS*, *60*(2), 390–400. https://doi.org/10.1007/s000180300032.

Ban, L., Zhang, L., Yan, Y., & Pelosi, P. (2002). Binding properties of a locust's chemosensory protein. *Biochemical and Biophysical Research Communications*, *5*, 50–54.

Bartoschik, T., Galinec, S., Kleusch, C., Walkiewicz, K., Breitsprecher, D., Weigert, S., et al. (2018). Near-native, site-specific and purification-free protein labeling for quantitative protein interaction analysis by MicroScale thermophoresis. *Scientific Reports*, *8*(1), 1–10. https://doi.org/10.1038/s41598-018-23154-3.

Bennett, J. P., & Snyder, S. H. (1975). Stereospecific binding ofd-lysergic acid diethylamide (LSD) to brain membranes: Relationship to serotonin receptors. *Brain Research*, *94*(3), 523–544. https://doi.org/10.1016/0006-8993(75)90234-6.

Beynon, R. J., Veggerby, C., Payne, C. E., Robertson, D. H. L., Gaskell, S. J., Humphries, R. E., et al. (2002). Polymorphism in major urinary proteins: Molecular heterogeneity in a wild mouse population. *Journal of Chemical Ecology*, *28*(7), 1429–1446. https://doi.org/10.1023/A:1016252703836.

Bianchet, M. A., Bains, G., Pelosi, P., Pevsner, J., Snyder, S. H., Monaco, H. L., et al. (1996). The three-dimensional structure of bovine odorant binding protein and its mechanism of odor recognition. *Nature Structural Biology*, *3*(11), 934.

Bignetti, E., Cavaggioni, A., Pelosi, P., Persaud, K. C., Sorbi, R. T., & Tirindelli, R. (1985). Purification and characterisation of an odorant-binding protein from cow nasal tissue. *European Journal of Biochemistry*, *149*(2), 227–231.

Briand, L., Eloit, C., Nespoulous, C., Bézirard, V., Huet, J.-C., Henry, C., et al. (2002). Evidence of an odorant-binding protein in the human olfactory mucus: Location, structural characterization, and odorant-binding properties. *Biochemistry*, *41*(23), 7241–7252. https://doi.org/10.1021/bi015916c.

Briand, L., Nespoulous, C., Huet, J.-C., Takahashi, M., & Pernollet, J.-C. (2001). Ligand binding and physico-chemical properties of ASP2, a recombinant odorant-binding protein from honeybee (*Apis mellifera* L.). *European Journal of Biochemistry*, *268*(3), 752–760. https://doi.org/10.1046/j.1432-1327.2001.01927.x.

Briand, L., Swasdipan, N., Nespoulous, C., Bézirard, V., Blon, F., Huet, J.-C., et al. (2002). Characterization of a chemosensory protein (ASP3c) from honeybee (*Apis mellifera* L.) as a brood pheromone carrier. *European Journal of Biochemistry*, *269*(18), 4586–4596. https://doi.org/10.1046/j.1432-1033.2002.03156.x.

Bruns, R. F., Lawson-Wendling, K., & Pugsley, T. A. (1983). A rapid filtration assay for soluble receptors using polyethylenimine-treated filters. *Analytical Biochemistry*, *132*(1), 74–81. https://doi.org/10.1016/0003-2697(83)90427-X.

Buck, L., & Axel, R. (1991). A novel multigene family may encode odorant receptors: A molecular basis for odor recognition. 13. *Cell*, *65*, 175–187.

Burova, T. V., Choiset, Y., Jankowski, C. K., & Haertlé, T. (1999). Conformational stability and binding properties of porcine odorant binding protein. *Biochemistry*, *38*(45), 15043–15051. https://doi.org/10.1021/bi990769s.

Campanacci, V., Krieger, J., Bette, S., Sturgis, J. N., Lartigue, A., Cambillau, C., et al. (2001). Revisiting the specificity of *Mamestra brassicae* and *Antheraea polyphemus* pheromone-binding proteins with a fluorescence binding assay. *Journal of Biological Chemistry, 276*(23), 20078–20084. https://doi.org/10.1074/jbc.M100713200.

Campanacci, V., Lartigue, A., Hällberg, B. M., Jones, T. A., Giudici-Orticoni, M.-T., Tegoni, M., et al. (2003). Moth chemosensory protein exhibits drastic conformational changes and cooperativity on ligand binding. *Proceedings of the National Academy of Sciences of the United States of America, 100*(9), 5069–5074. https://doi.org/10.1073/pnas.0836654100.

Chang, H., Liu, Y., Yang, T., Pelosi, P., Dong, S., & Wang, G. (2015). Pheromone binding proteins enhance the sensitivity of olfactory receptors to sex pheromones in *Chilo suppressalis*. *Scientific Reports, 5*(1), 13093. https://doi.org/10.1038/srep13093.

Cheng, Y.-C., & Prusoff, W. H. (1973). Relationship between the inhibition constant (KI) and the concentration of inhibitor which causes 50 per cent inhibition (I50) of an enzymatic reaction. *Biochemical Pharmacology, 22*(23), 3099–3108. https://doi.org/10.1016/0006-2952(73)90196-2.

Choo, Y.-M., Xu, P., Hwang, J. K., Zeng, F., Tan, K., Bhagavathy, G., et al. (2018). Reverse chemical ecology approach for the identification of an oviposition attractant for *Culex quinquefasciatus*. *Proceedings of the National Academy of Sciences of the United States of America, 115*(4), 714–719. https://doi.org/10.1073/pnas.1718284115.

Clyne, P. J., Warr, C. G., Freeman, M. R., Lessing, D., Kim, J., & Carlson, J. R. (1999). A novel family of divergent seven-transmembrane proteins: Candidate odorant receptors in *Drosophila*. *Neuron, 22*(2), 327–338. https://doi.org/10.1016/S0896-6273(00)81093-4.

Dal Monte, M., Centini, M., Anselmi, C., & Pelosi, P. (1993). Binding of selected odorants to bovine and porcine odorant-binding proteins. *Chemical Senses, 18*, 713–721.

Damberger, F., Nikonova, L., Horst, R., Peng, G., Leal, W. S., & Wüthrich, K. (2000). NMR characterization of a pH-dependent equilibrium between two folded solution conformations of the pheromone-binding protein from *Bombyx mori*. *Protein Science, 9*(5), 1038–1041. https://doi.org/10.1110/ps.9.5.1038.

D'Innocenzo, B., Salzano, A. M., D'Ambrosio, C., Gazzano, A., Niccolini, A., Sorce, C., et al. (2006). Secretory proteins as potential semiochemical carriers in the horse. *Biochemistry, 45*(45), 13418–13428. https://doi.org/10.1021/bi061409p.

Dohlman, H. G., Caron, M. G., & Lefkowitz, R. J. (1987). A family of receptors coupled to guanine nucleotide regulatory proteins. *Biochemistry, 26*(10), 2657–2664. https://doi.org/10.1021/bi00384a001.

Duhr, S., & Braun, D. (2006). Why molecules move along a temperature gradient. *Proceedings of the National Academy of Sciences of the United States of America, 103*(52), 19678. https://doi.org/10.1073/pnas.0603873103.

Fayad, S., Morin, P., & Nehmé, R. (2017). Use of chromatographic and electrophoretic tools for assaying elastase, collagenase, hyaluronidase, and tyrosinase activity. *Journal of Chromatography A, 1529*, 1–28. https://doi.org/10.1016/j.chroma.2017.11.003.

Fisher, E., Zhao, Y., Richardson, R., Janik, M., Buell, A. K., Aigbirhio, F. I., et al. (2017). Detection and characterization of small molecule interactions with Fibrillar protein aggregates using microscale thermophoresis. *ACS Chemical Neuroscience, 8*(9), 2088–2095. https://doi.org/10.1021/acschemneuro.7b00228.

Fleischer, J., Pregitzer, P., Breer, H., & Krieger, J. (2018). Access to the odor world: Olfactory receptors and their role for signal transduction in insects. *Cellular and Molecular Life Sciences, 75*(3), 485–508. https://doi.org/10.1007/s00018-017-2627-5.

Forstner, M., Breer, H., & Krieger, J. (2009). A receptor and binding protein interplay in the detection of a distinct pheromone component in the silkmoth *Antheraea polyphemus*. *International Journal of Biological Sciences, 5*(7), 745–757.

Gao, Y., Or, S., Toop, A., & Wheeldon, I. (2017). DNA nanostructure sequence-dependent binding of organophosphates. *Langmuir, 33*(8), 2033–2040. https://doi.org/10.1021/acs.langmuir.6b03131.

Guo, H., Huang, L.-Q., Pelosi, P., & Wang, C.-Z. (2012). Three pheromone-binding proteins help segregation between two *Helicoverpa* species utilizing the same pheromone components. *Insect Biochemistry and Molecular Biology, 42*(9), 708–716. https://doi.org/10.1016/j.ibmb.2012.06.004.

Hérent, M.-F., Collin, S., & Pelosi, P. (1995). Affinities of nutty and green-smelling pyrazines and thiazoles to odorant-binding proteins, in relation with their lipophilicity. *Chemical Senses, 20*(6), 601–608. https://doi.org/10.1093/chemse/20.6.601.

Huang, X. (2003). Fluorescence polarization competition assay: The range of resolvable inhibitor potency is limited by the affinity of the fluorescent ligand. *Journal of Biomolecular Screening, 8*(1), 34–38. https://doi.org/10.1177/1087057102239666.

Hulme, E. C., & Trevethick, M. A. (2010). Ligand binding assays at equilibrium: Validation and interpretation. *British Journal of Pharmacology, 161*(6), 1219–1237. https://doi.org/10.1111/j.1476-5381.2009.00604.x.

Ihara, S., Yoshikawa, K., & Touhara, K. (2013). Chemosensory signals and their receptors in the olfactory neural system. *Neuroscience, 254*, 45–60. https://doi.org/10.1016/j.neuroscience.2013.08.063.

Ishida, Y., Tsuchiya, W., Fujii, T., Fujimoto, Z., Miyazawa, M., Ishibashi, J., et al. (2014). Niemann–pick type C2 protein mediating chemical communication in the worker ant. *Proceedings of the National Academy of Sciences of the United States of America, 111*(10), 3847–3852. https://doi.org/10.1073/pnas.1323928111.

Jensen, E. V., & DeSombre, E. R. (1973). Estrogen-receptor interaction: Estrogenic hormones effect transformation of specific receptor proteins to a biochemically functional form. *Science, 182*(4108), 126–134. https://doi.org/10.1126/science.182.4108.126.

Kotlowski, C., Larisika, M., Guerin, P. M., Kleber, C., Kröber, T., Mastrogiacomo, R., et al. (2018). Fine discrimination of volatile compounds by graphene-immobilized odorant-binding proteins. *Sensors and Actuators B: Chemical, 256*, 564–572. https://doi.org/10.1016/j.snb.2017.10.093.

Lang, X., Gao, Y., & Wheeldon, I. (2018). *Quantifying small molecule binding interactions with DNA nanostructures*. In Y. L. Lyubchenko (Ed.), *Nanoscale imaging: Methods and protocols* (pp. 145–155). Springer. https://doi.org/10.1007/978-1-4939-8591-3_9.

Larisika, M., Kotlowski, C., Steininger, C., Mastrogiacomo, R., Pelosi, P., Schütz, S., et al. (2015). Electronic olfactory sensor based on *A. mellifera* odorant-binding protein 14 on a reduced graphene oxide Field-effect transistor. *Angewandte Chemie International Edition, 54*(45), 13245–13248. https://doi.org/10.1002/anie.201505712.

Lartigue, A., Gruez, A., Spinelli, S., Rivière, S., Brossut, R., Tegoni, M., et al. (2003). The crystal structure of a cockroach pheromone-binding protein suggests a new ligand binding and release mechanism. *Journal of Biological Chemistry, 278*(32), 30213–30218. https://doi.org/10.1074/jbc.M304688200.

Leal, W. S. (2013). Odorant reception in insects: Roles of receptors, binding proteins, and degrading enzymes. *Annual Review of Entomology, 58*(1), 373–391. https://doi.org/10.1146/annurev-ento-120811-153635.

Leal, W. S. (2017). Reverse chemical ecology at the service of conservation biology. *Proceedings of the National Academy of Sciences of the United States of America, 114*(46), 12094–12096. https://doi.org/10.1073/pnas.1717375114.

Leal, W. S., Chen, A. M., & Erickson, M. L. (2005). Selective and pH-dependent binding of a moth pheromone to a pheromone-binding protein. *Journal of Chemical Ecology, 31*(10), 2493–2499. https://doi.org/10.1007/s10886-005-7458-4.

Leal, W. S., Chen, A. M., Ishida, Y., Chiang, V. P., Erickson, M. L., Morgan, T. I., et al. (2005). Kinetics and molecular properties of pheromone binding and release. *Proceedings of the National Academy of Sciences of the United States of America*, *102*(15), 5386–5391. https://doi.org/10.1073/pnas.0501447102.

Leal, G. M., & Leal, W. S. (2015). Binding of a fluorescence reporter and a ligand to an odorant-binding protein of the yellow fever mosquito, Aedes aegypti. *F1000Research*, *3*, 305. https://doi.org/10.12688/f1000research.5879.2.

Lobel, D., Marchese, S., Krieger, J., Pelosi, P., & Breer, H. (1998). Subtypes of odorant-binding proteins. Heterologous expression and ligand binding. *European Journal of Biochemistry*, *254*(2), 318–324. https://doi.org/10.1046/j.1432-1327.1998.2540318.x.

Marie, A. D., Veggerby, C., Robertson, D. H. L., Gaskell, S. J., Hubbard, S. J., Martinsen, L., et al. (2001). Effect of polymorphisms on ligand binding by mouse major urinary proteins. *Protein Science*, *10*(2), 411–417. https://doi.org/10.1110/ps.31701.

Mastrogiacomo, R., D'Ambrosio, C., Niccolini, A., Serra, A., Gazzano, A., Scaloni, A., et al. (2014). An odorant-binding protein is abundantly expressed in the nose and in the seminal fluid of the rabbit. *PLoS One*, *9*(11), e111932.

Mastrogiacomo, R., Iovinella, I., & Napolitano, E. (2014). New fluorescent probes for ligand-binding assays of odorant-binding proteins. *Biochemical and Biophysical Research Communications*, *446*(1), 137–142. https://doi.org/10.1016/j.bbrc.2014.02.067.

Mueller, A. M., Breitsprecher, D., Duhr, S., Baaske, P., Schubert, T., & Längst, G. (2017). *MicroScale thermophoresis: A rapid and precise method to quantify protein–nucleic acid interactions in solution.* In M. Kaufmann, C. Klinger, & A. Savelsbergh (Eds.), *Functional genomics: Methods and protocols* (pp. 151–164). Springer. https://doi.org/10.1007/978-1-4939-7231-9_10.

Mulla, M. Y., Tuccori, E., Magliulo, M., Lattanzi, G., Palazzo, G., Persaud, K., et al. (2015). Capacitance-modulated transistor detects odorant binding protein chiral interactions. *Nature Communications*, *6*(1), 1–9. https://doi.org/10.1038/ncomms7010.

Pang, X., & Zhou, H.-X. (2017). Rate constants and mechanisms of protein–ligand binding. *Annual Review of Biophysics*, *46*(1), 105–130. https://doi.org/10.1146/annurev-biophys-070816-033639.

Paolini, S., Tanfani, F., Fini, C., Bertoli, E., & Pelosi, P. (1999). Porcine odorant-binding protein: Structural stability and ligand affinities measured by Fourier-transform infrared spectroscopy and fluorescence spectroscopy. *Biochimica et Biophysica Acta (BBA) - Protein Structure and Molecular Enzymology*, *1431*(1), 179–188. https://doi.org/10.1016/S0167-4838(99)00037-0.

Pelosi, P., Baldaccini, N. E., & Pisanelli, A. M. (1982). Identification of a specific olfactory receptor for 2-isobutyl-3-methoxypyrazine. *Biochemical Journal*, *201*(1), 245–248. https://doi.org/10.1042/bj2010245.

Pelosi, P., Iovinella, I., Zhu, J., Wang, G., & Dani, F. R. (2018). Beyond chemoreception: Diverse tasks of soluble olfactory proteins in insects: Soluble olfactory proteins in insects. *Biological Reviews*, *93*(1), 184–200. https://doi.org/10.1111/brv.12339.

Pelosi, P., Pisanelli, A. M., Baldaccini, N. E., & Gagliardo, A. (1981). Binding of [3H]-2-isobutyl-3-methoxypyrazine to cow olfactory mucosa. *Chemical Senses*, *6*(2), 77–85.

Pelosi, P., & Tirindelli, R. (1989). Structure/activity studies and characterization of an odorant-binding protein. *Chemical Senses*, *1*, 207–226.

Pelosi, P., Zhou, J.-J., Ban, L. P., & Calvello, M. (2006). Soluble proteins in insect chemical communication. *Cellular and Molecular Life Sciences*, *63*(14), 1658–1676. https://doi.org/10.1007/s00018-005-5607-0.

Pelosi, P., Zhu, J., & Knoll, W. (2018a). Odorant-binding proteins as sensing elements for odour monitoring. *Sensors*, *18*(10), 3248. https://doi.org/10.3390/s18103248.

Pelosi, P., Zhu, J., & Knoll, W. (2018b). From radioactive ligands to biosensors: Binding methods with olfactory proteins. *Applied Microbiology and Biotechnology*, *102*(19), 8213–8227. https://doi.org/10.1007/s00253-018-9253-5.

Pert, C. B., & Snyder, S. H. (1973). Opiate receptor: Demonstration in nervous tissue. *Science*, *179*(4077), 1011–1014. https://doi.org/10.1126/science.179.4077.1011.

Pevsner, J., Hou, V., Snowman, A. M., & Snyder, S. H. (1990). Odorant-binding protein. Characterization of ligand binding. *Journal of Biological Chemistry*, *265*(11), 6118–6125.

Pevsner, J., Trifiletti, R. R., Strittmatter, S. M., & Snyder, S. H. (1985). Isolation and characterization of an olfactory receptor protein for odorant pyrazines. *Proceedings of the National Academy of Sciences of the United States of America*, *82*(9), 3050–3054. https://doi.org/10.1073/pnas.82.9.3050.

Qiao, H., He, X., Schymura, D., Ban, L., Field, L., Dani, F. R., et al. (2011). Cooperative interactions between odorant-binding proteins of Anopheles gambiae. *Cellular and Molecular Life Sciences*, *68*(10), 1799–1813. https://doi.org/10.1007/s00018-010-0539-8.

Robertson, H. M. (2019). Molecular evolution of the major arthropod chemoreceptor gene families. *Annual Review of Entomology*, *64*(1), 227–242. https://doi.org/10.1146/annurev-ento-020117-043322.

Schwaighofer, A., Kotlowski, C., Araman, C., Chu, N., Mastrogiacomo, R., Becker, C., et al. (2014). Honey bee odorant-binding protein 14: Effects on thermal stability upon odorant binding revealed by FT-IR spectroscopy and CD measurements. *European Biophysics Journal*, *43*(2–3), 105–112. https://doi.org/10.1007/s00249-013-0939-4.

Seidel, S. A. I., Dijkman, P. M., Lea, W. A., van den Bogaart, G., Jerabek-Willemsen, M., Lazic, A., et al. (2013). Microscale thermophoresis quantifies biomolecular interactions under previously challenging conditions. *Methods*, *59*(3), 301–315. https://doi.org/10.1016/j.ymeth.2012.12.005.

Simantov, R., & Snyder, S. H. (1977). Opiate receptor binding in the pituitary gland. *Brain Research*, *124*(1), 178–184. https://doi.org/10.1016/0006-8993(77)90877-0.

Snyder, S. H. (1979). Receptors, neurotransmitters and drug responses. *New England Journal of Medicine*, *300*(9), 465–472.

Snyder, S. H. (2004). Opiate receptors and beyond: 30 years of neural signaling research. *Neuropharmacology*, *47*, 274–285. https://doi.org/10.1016/j.neuropharm.2004.06.006.

Snyder, S. H., & Childers, S. R. (1979). Opiate receptors and opioid peptides. *Annual Review of Neuroscience*, *2*(1), 35–64. https://doi.org/10.1146/annurev.ne.02.030179.000343.

Sun, Y. F., De Biasio, F., Qiao, H. L., Iovinella, I., Yang, S. X., Ling, Y., et al. (2012). Two odorant-binding proteins mediate the behavioural response of aphids to the alarm pheromone (E)-ß-farnesene and structural analogues. *PLoS One*, 7(3), e32759. https://doi.org/10.1371/journal.pone.0032759.

Sun, M., Liu, Y., Walker, W. B., Liu, C., Lin, K., Gu, S., et al. (2013). Identification and characterization of pheromone receptors and interplay between receptors and pheromone binding proteins in the diamondback moth, *Plutella xyllostella*. *PLoS One*, *8*(4), e62098. https://doi.org/10.1371/journal.pone.0062098.

Tan, J., Zaremska, V., Lim, S., Knoll, W., & Pelosi, P. (2020). Probe-dependence of competitive fluorescent ligand binding assays to odorant-binding proteins. *Analytical and Bioanalytical Chemistry*, *412*, 547–554. https://doi.org/10.1007/s00216-019-02309-9.

Tcatchoff, L., Nespoulous, C., Pernollet, J.-C., & Briand, L. (2006). A single lysyl residue defines the binding specificity of a human odorant-binding protein for aldehydes. *FEBS Letters*, *580*(8), 2102–2108. https://doi.org/10.1016/j.febslet.2006.03.017.

Tegoni, M., Campanacci, V., & Cambillau, C. (2004). Structural aspects of sexual attraction and chemical communication in insects. *Trends in Biochemical Sciences*, *29*(5), 257–264. https://doi.org/10.1016/j.tibs.2004.03.003.

Tegoni, M., Ramoni, R., Bignetti, E., Spinelli, S., & Cambillau, C. (1996). Domain swapping creates a third putative combining site in bovine odorant binding protein dimer. *Nature Structural Biology*, *3*(10), 863–867. https://doi.org/10.1038/nsb1096-863.

Topazzini, A., Pelosi, P., Pasqualetto, P. L., & Baldaccini, N. E. (1985). Specificity of a pyrazine binding protein from cow olfactory mucosa. *Chemical Senses*, *10*(1), 45–49.

Wang, B., Guan, L., Zhong, T., Li, K., Yin, J., & Cao, Y. (2013). Potential Cooperations between odorant-binding proteins of the scarab beetle *Holotrichia oblita* Faldermann (Coleoptera: Scarabaeidae). *PLoS One*, *8*(12), e84795. https://doi.org/10.1371/journal.pone.0084795.

Wei, Y., Brandazza, A., & Pelosi, P. (2008). Binding of polycyclic aromatic hydrocarbons to mutants of odorant-binding protein: A first step towards biosensors for environmental monitoring. *Biochimica et Biophysica Acta (BBA) - Proteins and Proteomics*, *1784*(4), 666–671. https://doi.org/10.1016/j.bbapap.2008.01.012.

Wienken, C. J., Baaske, P., Rothbauer, U., Braun, D., & Duhr, S. (2010). Protein-binding assays in biological liquids using microscale thermophoresis. *Nature Communications*, *1*(1), 1–7. https://doi.org/10.1038/ncomms1093.

Xu, W., Xu, X., Leal, W. S., & Ames, J. B. (2011). Extrusion of the C-terminal helix in navel orangeworm moth pheromone-binding protein (AtraPBP1) controls pheromone binding. *Biochemical and Biophysical Research Communications*, *404*(1), 335–338. https://doi.org/10.1016/j.bbrc.2010.11.119.

Zhou, J.-J., Robertson, G., He, X., Dufour, S., Hooper, A. M., Pickett, J. A., et al. (2009). Characterisation of *Bombyx mori* odorant-binding proteins reveals that a general odorant-binding protein discriminates between sex pheromone components. *Journal of Molecular Biology*, *389*(3), 529–545. https://doi.org/10.1016/j.jmb.2009.04.015.

Zhou, J.-J., Zhang, G.-A., Huang, W., Birkett, M. A., Field, L. M., Pickett, J. A., et al. (2004). Revisiting the odorant-binding protein LUSH of *Drosophila melanogaster*: Evidence for odour recognition and discrimination. *FEBS Letters*, *558*(1–3), 23–26. https://doi.org/10.1016/S0014-5793(03)01521-7.

Zhu, J., Arena, S., Spinelli, S., Liu, D., Zhang, G., Wei, R., et al. (2017). Reverse chemical ecology: Olfactory proteins from the giant panda and their interactions with putative pheromones and bamboo volatiles. *Proceedings of the National Academy of Sciences of the United States of America*, *114*(46), E9802–E9810. https://doi.org/10.1073/pnas.1711437114.

Zhu, J., Ban, L., Song, L.-M., Liu, Y., Pelosi, P., & Wang, G. (2016). General odorant-binding proteins and sex pheromone guide larvae of *Plutella xylostella* to better food. *Insect Biochemistry and Molecular Biology*, *72*, 10–19. https://doi.org/10.1016/j.ibmb.2016.03.005.

CHAPTER ELEVEN

Fine structure immunocytochemistry—An important tool for research on odorant-binding proteins

Rudolf Alexander Steinbrecht*
Max-Planck-Institute for Ornithology, Seewiesen, Germany
*Corresponding author: e-mail address: steinbre@orn.mpg.de

Contents

Abstract

Fine structure immunocytochemistry enables the *in situ* localization of odorant-binding proteins with the high resolution of the electron microscope. Protocols for successfully labeling these proteins in insect chemosensory sensilla are given and some pitfalls pinpointed as well as ways to avoid them. The major achievements accomplished by these methods are briefly reviewed.

Methods in Enzymology, Volume 642
ISSN 0076-6879
https://doi.org/10.1016/bs.mie.2020.04.064

1. Introduction

Except for the conspicuous compound eyes, insect sensory organs are in general housed in separate small organules, the sensilla. The general bauplan of sensilla is similar regardless of the sensory modality: a few receptor cells (mostly 1–4, exceptionally >20) and usually 3 auxiliary cells are connected to a cuticular apparatus, often in the form of a bristle or hair (reviewed by Steinbrecht, 1999). Also the ontogeny of sensilla follows the same basic pattern in olfactory, gustatory, mechanosensory and hygrosensitive sensilla; i.e., in a series of asymmetric cell divisions a mother epidermal cell divides into prospective receptor cells and prospective auxiliary cells (reviewed by Keil, 1997). Although the sensory function of insect sensilla had been inferred from light microscope investigations already more than 100 years ago, the detailed anatomy and a more precise knowledge of modality specific adaptations required the high resolution of the electron microscope, so that only in the second half of the 20th century a more detailed functional morphology of insect sensilla in general, and of chemoreceptive sensilla in particular emerged (reviews by Altner & Prillinger, 1980; Keil & Steinbrecht, 1984; Steinbrecht, 1987, 1999; Schneider & Steinbrecht, 1968).

In parallel to the morphological progress, electrophysiological methods allowed for the physiological investigation of the function of single receptor cells (for review see Schneider, 1992). A particular milestone was the detection of pheromone receptors in the antennae of male moths, which impress by an extreme number of identical sensilla on large antennae (Schneider, Lacher, & Kaissling, 1964). In analogy to the insect compound eye the antenna, therefore, often is addressed as compound nose. The high abundance of identical receptors also encouraged biochemical investigations and in 1981 Richard Vogt identified in homogenates of insect antennae a protein of 15 kDa, which preferentially binds the pheromone of this species and which he called pheromone-binding protein (Vogt & Riddiford, 1981).

For the morphologist this detection produced a series of new questions. Is this protein characteristic for pheromone-perceiving sensilla, or does it occur generally in sensilla or also elsewhere? Where is the protein synthesized? Where degraded? Immunocytochemical labeling should give answers to these questions, provided a specific antibody was available. With this aim, Gunde Ziegelberger and Mamiko Ozaki purified the pheromone-binding protein (PBP) of the saturniid moth *Antheraea polyphemus*, injected it into rabbits and produced an antiserum, which I then could use in order to label this protein on sections of antennae of this species and study the distribution with the resolution of the electron microscope (Steinbrecht, Ozaki, & Ziegelberger, 1992).

Thus, the start of OBP immunocytochemistry dates back almost 30 years. Even though the methods used are standard and have not substantially changed in the following, it might be useful to review the procedures in this chapter.

2. Methods for immunocytochemical localization of OBPs

2.1 Antibody production and purity checks

Once a purified native or recombinant OBP is available in sufficient quantity, antibodies can be readily produced, e.g., in rabbits, mice or goats (Griffiths, 1993). The procedure used for obtaining an antibody against the pheromone-binding protein (PBP) of *Antheraea polyphemus* is described in Steinbrecht et al. (1992). Nowadays, antibody production can be ordered with commercial companies.

Before starting immunocytochemistry, the specificity of the antibody should be tested by immunoblotting of isolated tissue protein. For this, homogenates of tissues (e.g., male moth antennae in the case of PBP) and of tissues that do not contain this protein (e.g., female moth antennae, heads without antennae, legs) are subjected to SDS-polyacrylamide gel electrophoresis (SDS-PAGE). Proteins are transferred to nitrocellulose membranes and these are incubated with diluted antibody (e.g., anti-PBP). Visualization occurs by a second antibody conjugated with a reporter (e.g., an enzyme which produces a visible stain). For details see Section 3.1. Only those electrophoretic lanes that contain the searched-for protein show a stained band (Fig. 1).

2.2 General principles of immunocytochemistry

Two general strategies must be distinguished with immunocytochemical methods, pre- and post-embedding labeling. For pre-embedding methods, either native or gently fixed tissues are incubated with the antibody and thereafter the specimen is embedded, sectioned and studied in the electron microscope. Post-embedding methods, on the other hand, start with fixing, embedding and sectioning the tissues, whereas the reaction with the

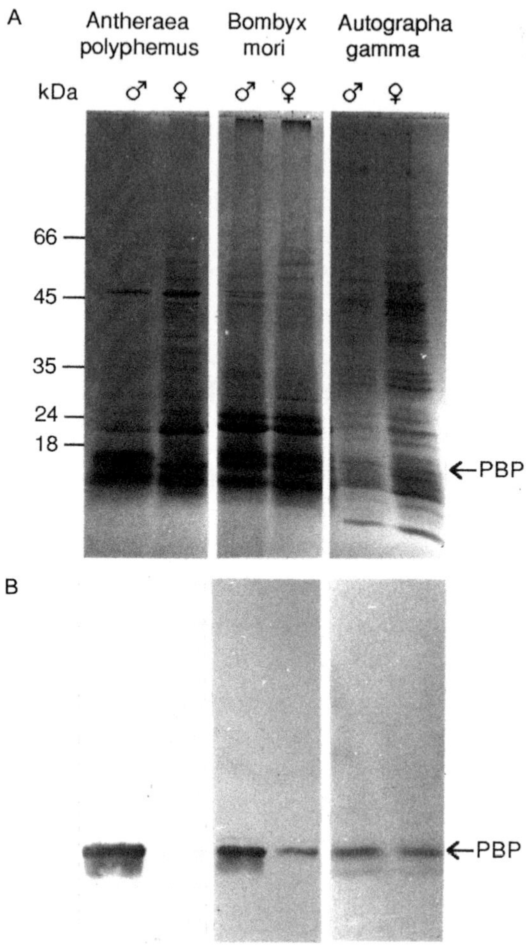

Fig. 1 Analysis of male and female antennal proteins from three moth species. (A) SDS-PAGE stained with Coomassie Blue. (B) Immunoblot stained with anti-PBP (*A. polyphemus*) dilution 1:3000. *From Steinbrecht, R. A., Ozaki, M., & Ziegelberger, G. (1992). Immunocytochemical localization of pheromone-binding protein in moth antennae.* Cell and Tissue Research, 270, *287–303 with permission.*

antibodies is done on the sections where the molecules of interest are exposed. The advantage of pre-embedding methods resides in the possibility of an antibody reaction with native or only slightly modified tissue, whereas post-embedding methods require a sequence of treatments, which might partly or completely abolish the antigenicity of the investigated protein. The disadvantage of pre-embedding methods, on the other hand, is that the antibody needs to access the molecules of interest even in deeper tissue layers. Antibodies are large molecules and may not be able to diffuse to places where the antigen is located. To enable antibody diffusion throughout the entire tissue, permeabilization procedures must be performed which bear the risk of extraction and structural impairment. This problem does not occur with the sectioned tissue in post-embedding methods, especially if optimized preparation protocols are used. For an in depth review of these aspects refer to the book of Griffiths (1993).

In insect sensilla, the cuticle is a severe barrier for diffusion of fixatives and even more so of antibodies. Therefore, pre-embedding labeling is difficult and post-embedding immunolabeling has been exclusively used for the immunocytochemistry of odorant-binding proteins.

2.3 Survey of successful protocols (for details see Section 3)

2.3.1 *Fixation*

Due to the already mentioned barrier function of the insect cuticle, fixation of insect sensilla had been difficult from the onset. Insect cuticle is practically impermeable for glutaraldehyde solutions and osmium tetroxide solutions diffuse only slowly (Keil & Steinbrecht, 2010). Moreover, osmium tetroxide may abolish the antigenicity of proteins almost completely (Griffiths, 1993). As a compromise, mild fixation (i.e., a lower concentration of aldehydes than usually used for pure structural research) with a mixture of glutaraldehyde and formaldehyde has been used with penetration of the fixative from artificially produced openings, i.e., antennae were split or sectioned into small pieces, so that the fixative can diffuse from the cuticle-free surfaces into the tissue (Steinbrecht et al., 1992). Omitting osmium tetroxide as a second fixative resulted not only in reduced contrast but also in a suboptimal preservation of the tissue fine structure. The tolerance of a certain OBP to various fixation agents can be tested by dot blotting, if it is available in purified form (see Section 3.1).

The introduction of cryofixation followed by freeze-substitution (Steinbrecht, 1980) resulted in a much better preservation of the fine structure of insect sensilla than chemical fixation and preserved the antigenicity of the OBPs. Nevertheless, also cryofixation changes the proteins due to the indispensable follow-up procedures of freeze-substitution and embedding (Steinbrecht, 1993).

2.3.2 *Embedding*

The standard embedding methods for electron microscopy in epoxy resins are superior for morphological studies but cannot be used for immunocytochemistry, because they cross-link with proteins by forming covalent bonds, which interferes with antibody binding. Acrylic resins, such as

LR White/Gold or the different variants of Lowicryl resin, do not cross-react with tissue components and, therefore, are now widely used for post-embedding immunocytochemistry (Griffiths, 1993; Möbius, 2016).

2.3.3 Immunolabeling

This is the step, where tissue sections are incubated with the specific antibodies. The resin embedded tissues are sectioned and sections picked up on grids covered with a formvar film, following standard techniques in electron microscopy. These specimens are then brought into contact with various solutions, including the antibodies, usually presented as drops on a parafilm surface (for details see Section 3.5). If an incubation has to last for more than a few minutes, it is done in a moist chamber. The binding between antibody and antigen (i.e., target protein) needs salt and a neutral or slightly alcaline pH and, therefore, is performed in buffered saline. To prevent unspecific binding, pre-incubation steps are conducted before the incubation with antibodies: (1) Free aldehyde groups are blocked and (2) a mixture of different proteins is added to "saturate" sample regions prone to unspecific binding of protein. After incubation of sections with the antibody, a secondary antibody coupled to a gold colloid is used as a reporter to visualize the binding of the primary, OBP-specific, antibody.

2.4 General remarks

OBPs are small soluble proteins which are present at high concentration. Moreover, these proteins can be isolated or recombinantly produced preserving their native conformation and used for the generation of antibodies, which fit to the conformation of the OBPs in the tissue sections. Therefore, immunocytochemistry of OBPs is comparatively straightforward and does not pose severe problems. Nevertheless, a few points have to be considered: First, any soluble protein that is only weakly fixed or cross-linked may be translocated or even completely washed out during sample preparation (Möbius, 2016) and/or contact with the immunolabeling solutions (Stierhof, Schwarz, Dürrenberger, Villiger, & Kellenberger, 1991). The comparison of different preparation procedures is, therefore, advisable. Second, unspecific background may simulate false positive labeling. Therefore, control experiments are required, such as replacing the primary antibody by pre-immune serum. In critical cases, when a weak positive signal has to be discerned from unspecific background labeling, more sophisticated quantitative methods are advisable as described by Laue, Han, Dittmann, and Setlow (2018).

3. Detailed protocols from the Steinbrecht lab

3.1 Immunoblotting and dot blotting

Polyacrylamide gel electrophoresis (PAGE) is carried out according to standard methods under reducing conditions on 15% slab gels (Laemmli, 1970). Electrophoresis apparatus and reagents may be purchased, e.g., from BioRad Laboratories, Hercules, Calif., USA. Proteins are stained with Coomassie brilliant blue R250; alternatively proteins are transferred to nitrocellulose membranes for 2 h by the semi-dry blotting procedure of Kyhse-Andersen (1984). After blocking with 0.05% Tween 20 in phosphate-buffered saline (PBS) for 1 h at room temperature (Batteiger, Newhall, & Jones, 1982), the nitrocellulose membranes are incubated with the specific antibody in appropriate dilution for 2–3 h. The binding of this antibody is detected by a second antibody, which is conjugated to horseradish peroxidase; using n-chloronaphthol as substrate a black precipitate results. In case the primary antibody has been raised in rabbits, a goat IgG-horseradish peroxidase conjugate is used as secondary antibody. All washing procedures and dilutions are performed with PBS-Tween.

For dot blotting, purified proteins are dissolved in PBS and aliquots applied to a nitrocellulose membrane using a suction device (BioDot, BioRad Laboratories, Calif., USA). Some of the dots are fixed in raising concentrations of chemical fixative before performing the immunoreactions as described above.

3.2 Chemical fixation

As chemical fixative a mixture of paraformaldehyde (2.8%) and glutaraldehyde (3%) in PBS (pH 7.2) is used. Specimens are cut under fixative into small pieces and fixation is carried out at 4 °C for 2–8 h. Thereafter dehydration is carried out in rising alcohol concentrations.

3.3 Cryofixation and freeze-substitution

For cryofixation specimens are plunged into super-cooled propane (−180 °C). For better handling, specimens are fixed on thin (0.1 mm) tungsten wire. A simple, self-made apparatus for controlled immersion is described in Steinbrecht (1993). For freeze-substitution, the specimens are transferred into pure acetone at −79 °C. Cooling for freeze-substitution can be accomplished most simply in a Dewar filled with dry ice-acetone slush. A device for programmed time-temperature protocols is described in Marshall and Kent (1991); a similar apparatus is also commercially available from, e.g., Leica Microsystems, Germany. After 2 days the temperature is slowly raised over night to −30 °C and the medium replaced by 3% glutaraldehyde in acetone (GA-acetone) for another 24 h. GA-acetone is prepared by adding 6 mL of 50% aqueous glutaraldehyde to 100 mL pure analytical grade acetone and storing this medium for at least 2 days over molecular sieve. After warming to room temperature the substitution medium is replaced by two changes of pure acetone and further by three changes of 100% ethanol.

3.4 Embedding and sectioning

Embedding in LR White (Agar Scientific, UK) is done for both, chemically and cryofixed specimens in the same way through mixtures of 2:1, 1:1, 1:2 of 100% ethanol and LR White. Specimens remain in pure LR White overnight at 4 °C and are embedded in gelatine capsules, filled completely with the resin monomer. After sealing the capsules with the cap, polymerization is accomplished at 60 °C. To avoid extensive heating of the sample (polymerization is an exothermal reaction) the gelatine capsules are fitted into an aluminum block as a heat sink.

Ultrathin sections are cut in the routine way on an ultramicrotome using a diamond knife. Sections are picked up on Formvar-coated single-hole grids and can be stored (most probably for years) until immunolabeling.

3.5 Immunocytochemical labeling

3.5.1 Normal procedure

For immunocytochemistry, grids are floated on 30 μL droplets deposited on a parafilm surface. The sequence of steps is as follows:

2×2 min on PBS containing 50 mM glycine for masking free aldehyde groups,

2×10 min on PBG (PBS containing 0.05% gelatine and 0.5% bovine serum albumine),

overnight on the primary antibody diluted with PBG,

6×2 min on PBG,

1 h on the secondary antibody, coupled to 10 nm colloidal gold, diluted 1:20 with PBG,

2×2 min on PBS-glycine,

2×2 min on H_2O.

Between each step, the fluid which remains on the grid surface is removed with filter paper without drying the surface completely. Avoid trapping of fluid by the forceps (e.g., use anti-capillary forceps or wipe the forceps dry after each step), because it will flow back onto the grid.

3.5.2 *Silver intensification*

As 10 nm gold granules are not well visible at low magnifications, some grids may be silver intensified according to the method of Danscher (1981). For this, grids are floated 15 min on a photographic developer containing silver lactate in dim light. This increases the size of the gold particles to some 40 nm.

3.5.3 *Double labeling*

Double labeling with two antisera on the same grid can be performed by two procedures. According to the method of Bienz, Egger, and Pasamontes (1986), silver intensification for 17–20 min is performed after the first labeling with primary and secondary antibody. This renders the first labeling complex inactive, so that the second labeling—again with primary and secondary antibodies—can follow without interference. The two labels are easily discriminated by the different grain size (40 nm versus 10 nm).

In case the antisera are produced in different animals, e.g., rabbit and mouse, double labeling can also be accomplished simultaneously. First the primary antibodies are applied one after the other and then after washing the secondary antibodies are applied in a mixture; e.g., goat-anti-rabbit IgG conjugated with 30 nm gold and goat-anti-mouse IgG conjugated with 10 nm gold. These sections are not silver intensified.

3.5.4 *Contrast enhancement*

Staining the grids after immunocytochemistry with 2% uranyl acetate or with 2% uranyl acetate followed by 0.1% lead citrate (Reynolds, 1963) as in routine electron microscopy increases the visibility of cellular components.

4. Major achievements of OBP immunocytochemistry

Already the first immunocytochemical study using an antibody against the PBP of *Antheraea polyphemus* reports a full series of new discoveries (Steinbrecht et al., 1992). (i) It could be shown that PBP was present in the sensillum lymph in all pheromone-sensitive sensilla of male *Antheraea polyphemus* (Fig. 2A and C). (ii) There was no labeling of PBP in basiconic

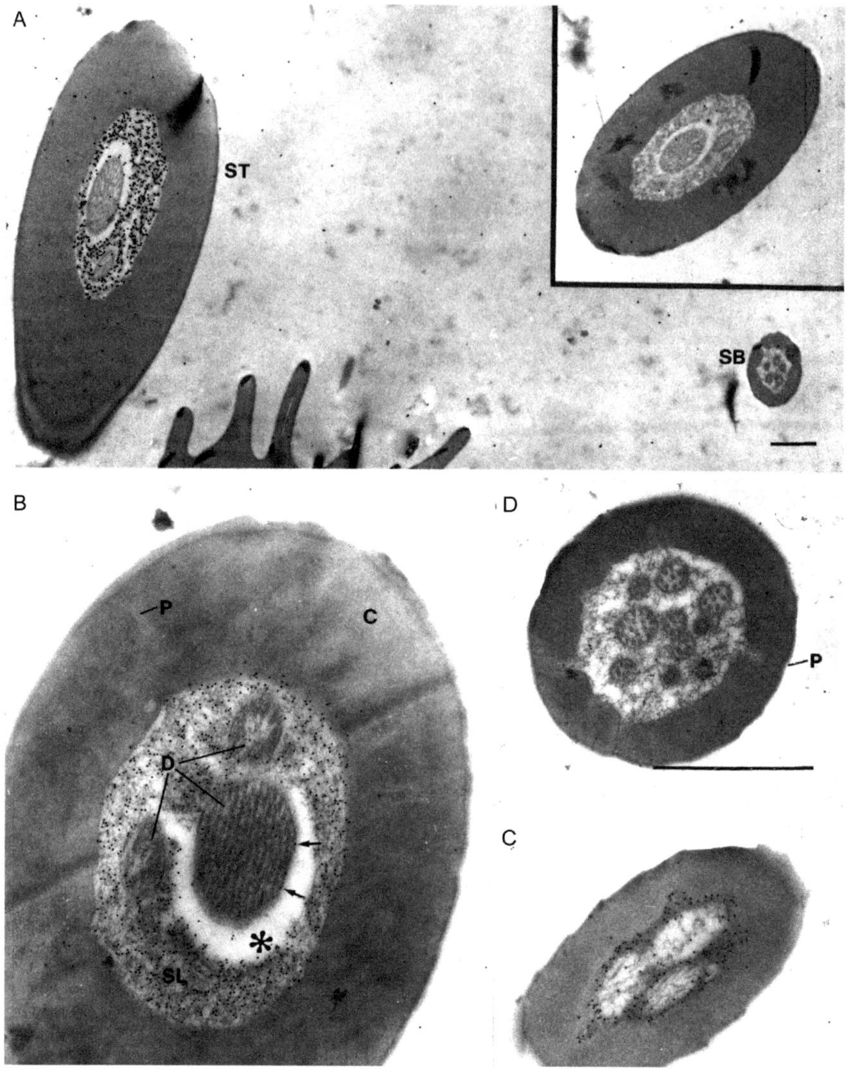

Fig. 2 *Antheraea polyphemus*, antennal sensilla after labeling with anti-PBP. (A) Cross-sections of long sensillum trichodeum (ST) and s. basiconicum (SB) on the same micrograph; only the pheromone-sensitive sensillum trichodeum is labeled. Inset: when pre-immune serum is used instead of primary antibody, no specific labeling of s. trichodea is observed. (B and C) Oblique sections of s. trichodea through proximal and distal parts of sensory hair, respectively. The sensillum lymph (SL) is heavily labeled; the dendrites (D) and the cuticle (C) show only a few unspecific gold particles. In (B), the sensillum lymph has been artificially separated from the large dendrite by shrinkage (asterisk), but still some gold particles are found on the dendritic membrane (arrows). In (D) only very few gold particles are found on sensillum lymph of sensillum basiconicum. P, stimulus conducting pore-tubule systems. All specimens freeze-substituted in glutaraldehyde acetone, except C (chemically fixed in formaldehyde/glutaraldehyde). Scale bar in (A)–(D) 1 μm. *From Steinbrecht, R. A., Ozaki, M., & Ziegelberger, G. (1992). Immunocytochemical localization of pheromone-binding protein in moth antennae.* Cell and Tissue Research, 270, *287–303 with permission.*

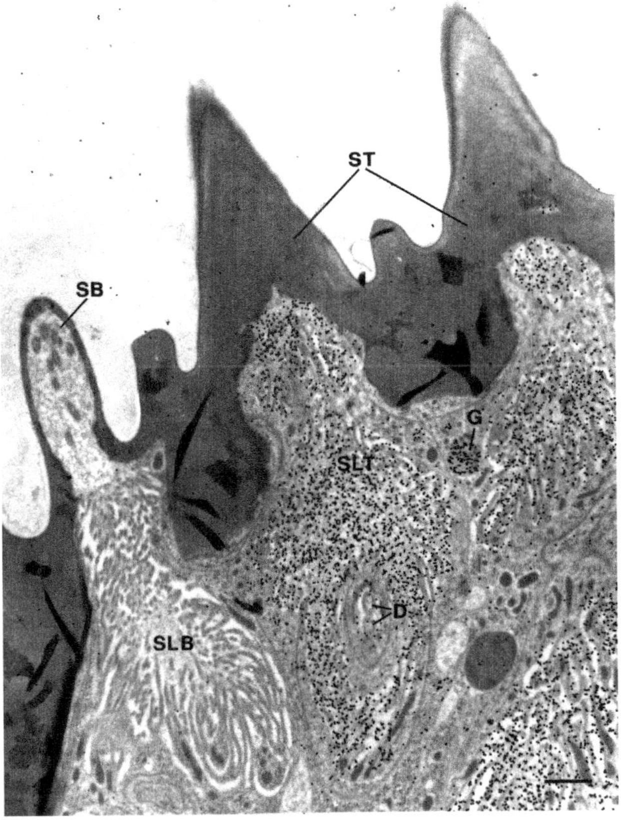

Fig. 3 *Bombyx mori*; oblique section through base of two s. trichodea (ST) and one s. basiconicum (SB). The sensillum lymph of the pheromone-sensitive sensilla trichodea (SLT) is labeled by anti-PBP (Apo), that of the sensillum basiconicum (SLB) is not. D, dendritic outer segments surrounded by inner sensillum-lymph cavity; G, heavily labeled granule in auxiliary cell. Specimen freeze-substituted in GA-acetone; anti-PBP 1:30,000; silver intensified. Bar 1 μm. *From Steinbrecht, R. A., Ozaki, M., & Ziegelberger, G. (1992). Immunocytochemical localization of pheromone-binding protein in moth antennae.* Cell and Tissue Research, 270, *287–303 with permission.*

sensilla, which do not respond to the pheromone (Fig. 2D). (iii) The antiserum against PBP of *Antheraea polyphemus* also labels PBPs of *Bombyx mori* and other moths (Fig. 3). (iv) A new role of the auxiliary cells was discovered: the tormogen and trichogen cell, but not the thecogen cell, synthesize PBP (Figs. 3 and 4). (v) On the other hand, reabsorption of PBP from the sensillum lymph occurs in all sensillar cells, also in the receptor cells, via coated vesicles (Figs. 3 and 4).

Apparently, the specificity of anti-PBP of *Antheraea polyphemus* was not extremely narrow. This can be explained by structural similarity of

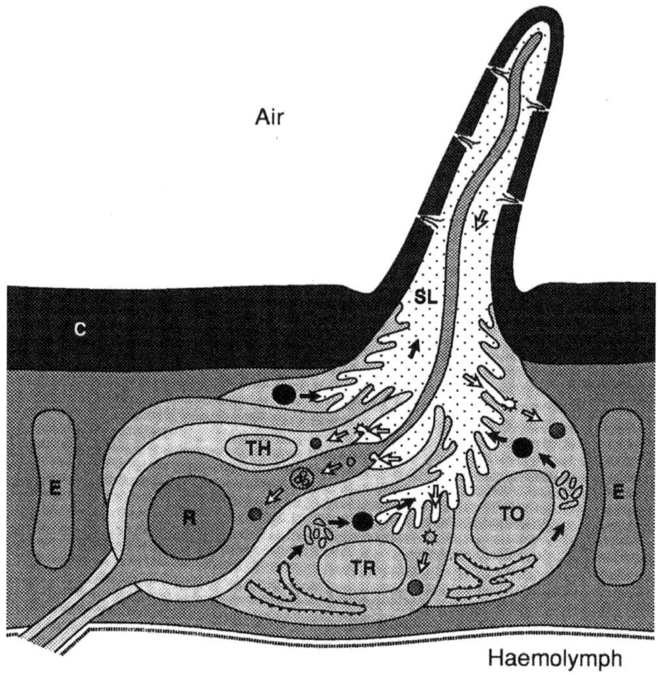

Fig. 4 Schematic drawing of possible pathways of biosynthesis and degradation of PBP in pheromone-sensitive sensilla trichodea. Biosynthesis and exocytosis (filled arrows) occurs in the trichogen (TR) and tormogen (TO) cell following the conventional pathway of protein secretion via granular endoplasmic reticulum, Golgi apparatus and secretory granules (black). Endocytosis and turnover (open arrows) probably occur in all sensillar cells via coated or non-coated pits and vesicles, multi-vesicular bodies and lysosomes (stippled). C, cuticle and sensory hairs with pores; E, epidermal cell; R, receptor cell with axon and sensory process, the outer segment of which extends into the hair and is surrounded by sensillum lymph (SL); Th, thecogen cell, completely ensheathing receptor cell soma and inner dendritic segment. *From Steinbrecht, R. A., Ozaki, M., & Ziegelberger, G. (1992). Immunocytochemical localization of pheromone-binding protein in moth antennae.* Cell and Tissue Research, 270, *287–303 with permission.*

different PBPs (Raming, Krieger, & Breer, 1990; Vogt, Callahan, Rogers, & Dickens, 1999). The broad range of cross reactivity of the *Antheraea polyphemus* serum was later used for "fishing" a great number of other pheromone-binding proteins and even of chemosensory proteins (CSPs) in other species (e.g., Nagnan-Le Meillour et al., 2000).

When more antibodies became available, e.g., antibodies against the general odorant-binding proteins 1 and 2 in moths or antibodies against various OBPs in the fruitfly or the locust, this opened the door for a large scale comparative analysis of the distribution of the various OBPs and CSPs in a great number of insect species starting with moth antennae and continuing on various sensilla of *Drosophila melanogaster* or *Locusta migratoria* (e.g., Jin et al., 2005; Park et al., 2000; Shanbhag, Hekmat-Scafe, et al., 2001; Shanbhag, Park, Pikielny, & Steinbrecht, 2001; Steinbrecht, Laue, & Ziegelberger, 1995; Zhang, Maida, & Steinbrecht, 2001).

Section series through insect olfactory sensilla offer the advantage that different antibodies can be applied to consecutive sections. Individual sensilla can be unequivocally mapped on such consecutive sections due to their distribution pattern, so that the labeling characteristics of individual sensilla can be readily studied for as many antibodies as available (Maida, Mameli, Müller, Krieger, & Steinbrecht, 2005). For example, labeling antennal sections of the silk moth *Bombyx mori* with anti-PBP or anti-GOBP2 nicely demonstrated that some OBPs may exclude each other; PBP is found in the pheromone-sensitive sensilla trichodea, but not in sensilla basiconica that are sensitive for plant odors, whereas the distribution of GOBP2 follows the opposite scheme (Steinbrecht et al., 1995). GOBP1, on the other hand, is found in s. trichodea as well as in s. basiconica, albeit in different concentrations (Fig. 5) (Maida et al., 2005).

Labeling section series with several antibodies showed that several OBPs can be co-localized in the same sensillum (Hekmat-Scafe, Steinbrecht, & Carlson, 1997; Maida et al., 2005). Of particular interest was the discovery of several OBPs co-localized in a sensillum that houses only a single receptor cell in *Drosophila* (Fig. 6). Using double labeling on the same section it could also be demonstrated that these different OBPs are present in the same secretory granules (Fig. 7), i.e., that their synthesis does not follow completely separate pathways (Shanbhag et al., 2005).

The general localization of OBPs in the sensillum lymph of insect olfactory sensilla has been taken as an indicator of their participating in stimulus conduction and/or stimulus inactivation (see Kaissling, 2001, for a detailed discussion), which has been also suggested by electrophysiological experiments (Van den Berg & Ziegelberger, 1991) and ligand binding assays (Maida, Krieger, Gebauer, Lange, & Ziegelberger, 2000). OBPs may even contribute to the excitation of receptor cells (Pophof, 2002, 2004). Immunocytochemistry has, however, also demonstrated similar proteins at locations, where such a function can be excluded. PBPRP2 of *Drosophila*

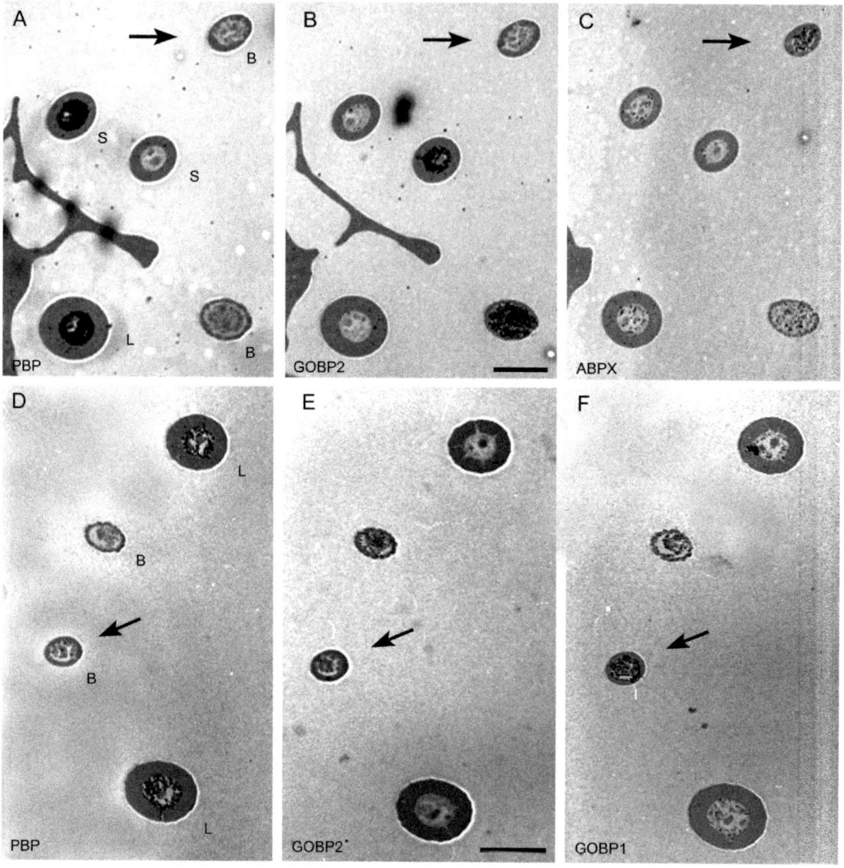

Fig. 5 Co-expression of odorant-binding proteins in antennal sensilla of male *Bombyx mori* is demonstrated by multiple labeling in successive sections. Individual sensilla can be easily identified and compared for their labeling properties. Sections in (A) and (D) were labeled with anti-PBP, those in (B) and (E) with anti-GOBP2, (C) with anti-ABPX and (F) with anti-GOBP1. Long sensilla trichodea (L) heavily express PBP, and no GOBP2; labeling for the other two OBPs is only slightly above background. Short s. trichodea (S) and s. basiconica (B) have variable labeling properties. Note that those sensilla that neither express PBP nor GOBP2 (arrows) exhibit the strongest labeling for ABPX or GOBP1. Scale bars 3 µm. *From Maida, R., Mameli, M., Müller, B., Krieger, J., & Steinbrecht, R. A. (2005). The expression pattern of four odorant-binding proteins in male and female silk moths, Bombyx mori.* Journal of Neurocytology, 34, *149–163 with permission.*

melanogaster, for example, has been localized in the non-innervated lumen of gustatory sensilla (Shanbhag, Park, et al., 2001) (Fig. 8). Moreover, even the subcuticular space of normal epidermal cells on *Drosophila* antennae is labeled by anti-PBPRP2 (Park et al., 2000) (Fig. 9).

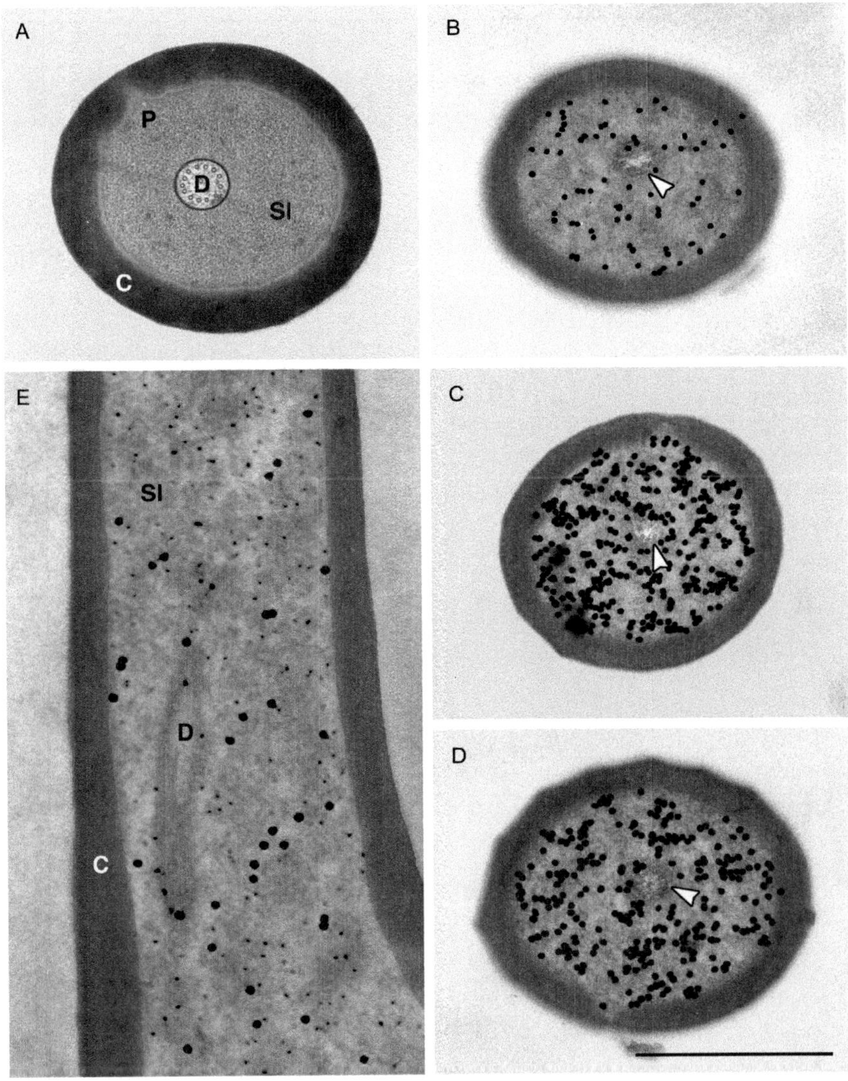

Fig. 6 *Drosophila melanogaster*, sensillum trichodeum sub-type T1 in cross (A–D) and longitudinal (E) section close to the hair base. Even though containing only one olfactory receptor neuron, sub-type T1 expresses the three odorant-binding proteins LUSH, OS-E and OS-F, as do s. trichodea housing two or three receptor neurons. (A) In an Epon-section after cryosubstitution in acetone/OsO4, the single unbranched outer dendritic segment (D) is clearly visible surrounded by sensillum lymph (Sl). Pores (P) in the cuticular wall (C) are infrequent in this sensillum type. (B–E) In LR White-sections prepared for immunocytochemistry, membranes have much less contrast and dendrites (arrow heads) are discernible mainly by their microtubular cytoskeleton. (B–D) Sections from a section series through an identical sensillum, labeled for LUSH (B), OS-E (C), and OS-F (D) demonstrate the co-localization of these three OBPs in the same sensillum. (E) Double-labeling for LUSH (30 nm gold particles) and OS-F (10 nm gold particles) shows the co-localization of these two proteins on the same section. Scale bar 1 μm. *From Shanbhag, S. R., Smith, D. P., & Steinbrecht, R. A. (2005). Three odorant-binding proteins are co-expressed in sensilla trichodea of Drosophila melanogaster.* Arthropod Structure & Development, 34, *153–165 with permission.*

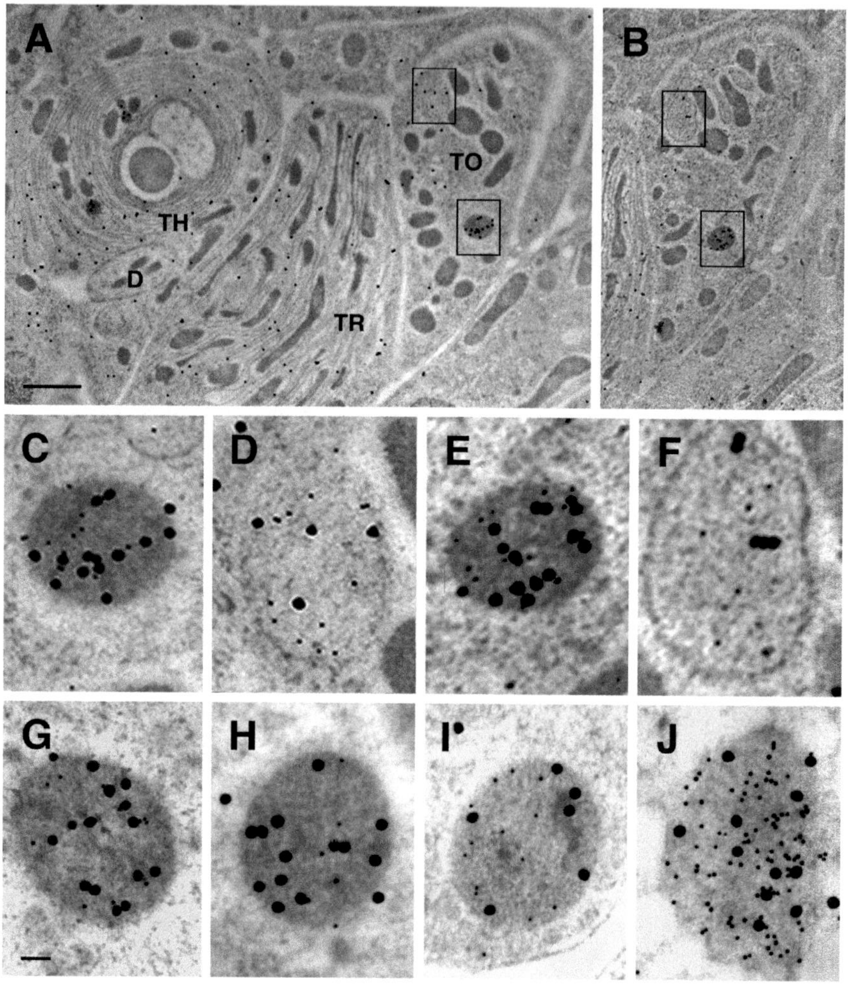

Fig. 7 Co-localization of LUSH, OS-E, OS-F as demonstrated by double-labeling. (A) Low magnification shows the three auxiliary cells (TO, TR, TH) of a T-1 sensillum double labeled for LUSH (30 nm gold particles) and OS-F (10 nm gold particles). As the small gold particles are not visible at this magnification, the two boxed granules are enlarged in (C) and (D) demonstrating a different morphology of the granules as well as different ratios of the two proteins present. (B) A different section of the same sensillum double labeled for OSE (small grains) and LUSH (large grains). The same two granules (boxes) are enlarged in (E) and (F) showing a similar ratio for LUSH and OSE as shown for LUSH and OSF in (C) and (D). (G–J) other intracellular granules from s. trichodea after double labeling for LUSH (large particles) and OSE (small particles) in (G–I) or OSF (small particles) in (J). Note the different ratios of the proteins in different granules. Scale bars 1 μm in (A) for (A and B), 100 nm in (G) for (C–J). *From Shanbhag, S. R., Smith, D. P., & Steinbrecht, R. A. (2005). Three odorant-binding proteins are co-expressed in sensilla trichodea of Drosophila melanogaster.* Arthropod Structure & Development, 34, *153–165 with permission.*

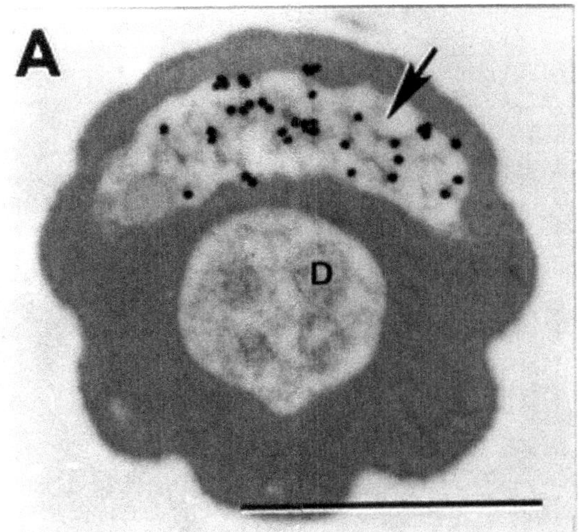

Fig. 8 *Drosophila melanogaster*. Immunolabeling for PBPRP2 in labellar taste sensilla. A section through the shaft of a taste bristle. Gold particles are present only in the crescent-shaped lumen (arrow) and not in the lumen containing the receptor-cell dendrites (D). Scale bar 0.5 μm. *From Shanbhag, S. R., Park, S.-K., Pikielny, C. W., & Steinbrecht, R. A. (2001). Gustatory organs of* Drosophila melanogaster*: Fine structure and expression of the putative odorant-binding protein PBPRP2.* Cell Tissue Res., 304, *423–437 with permission.*

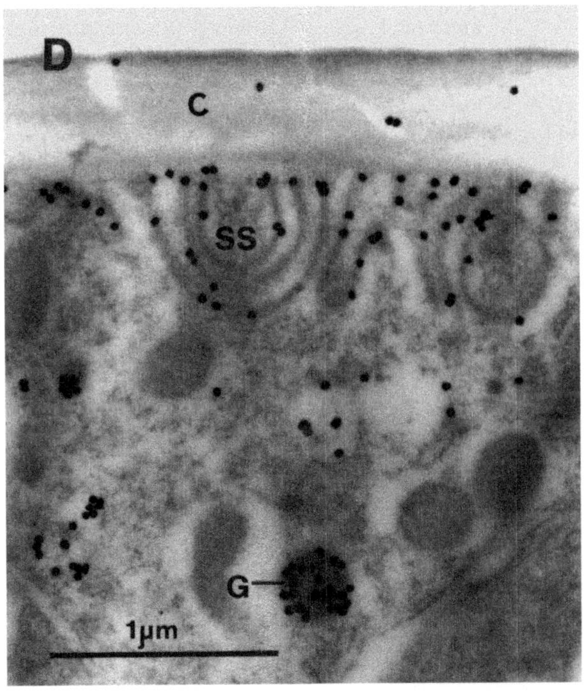

Fig. 9 *Drosophila melanogaster*. Cuticle (C) of a maxillary palp with underlying epidermis. Except for some intracellular granules (G), labeling for PBPRP2 is particular dense in the subcuticular space (SS) in between the apical epidermal microvilli and lamellae. *From Park, S. K., Shanbhag, S. R., Wang, Q., Hasan, G., Steinbrecht, R. A., & Pikielny, C. W. (2000). Expression patterns of two putative odorant-binding proteins in the olfactory organs of Drosophila melanogaster have different implications for their functions.* Cell and Tissue Research, 300, *181–192 with permission.*

Acknowledgments

I want to express my deep gratitude to all my former collaborators and guests who in the past had contributed to our success, in particular to Barbara Müller, Michael Laue, Gunde Ziegelberger, Rosario Maida, Shubha Shanbhag, Shan-gan Zhang, Mamiko Ozaki. Renate Alton helped to prepare this manuscript like many papers before—thank you, Renate. I gratefully acknowledge the help of Michael Laue, who critically read and improved the manuscript.

References

Altner, H., & Prillinger, L. (1980). Ultrastructure of invertebrate chemo-, thermo-, and hygroreceptors and its functional significance. *International Review of Cytology*, *67*, 69–139.

Batteiger, B., Newhall, W. J., & Jones, R. B. (1982). The use of tween 20 as a blocking agent in the immunological detection of proteins transferred to nitrocellulose membranes. *Journal of Immunological Methods*, *55*, 297–307.

Bienz, K., Egger, D., & Pasamontes, L. (1986). Electron microscopic immunocytochemistry. Silver enhancement of colloidal gold marker allows double labeling with the same primary antibody. *Journal of Histochemistry and Cytochemistry*, *34*, 1337–1342.

Danscher, G. (1981). Localization of gold in biological tissue. *Histochemistry*, *71*, 81–88.

Griffiths, G. (1993). *Fine structure immunocytochemistry*. Berlin: Springer.

Hekmat-Scafe, D. S., Steinbrecht, R. A., & Carlson, J. R. (1997). Coexpression of two odorant-binding protein homologs in *Drosophila*: Implications for olfactory coding. *Journal of Neuroscience*, *17*, 1616–1624.

Jin, X., Brandazza, A., Navarrini, A., Ban, L., Zhang, S., Steinbrecht, R. A., et al. (2005). Expression and immunolocalisation of odorant-binding and chemosensory proteins in locusts. *Cellular and Molecular Life Sciences*, *62*, 1156–1166.

Kaissling, K. E. (2001). Olfactory perireceptor and receptor events in moths: A kinetic model. *Chemical Senses*, *26*, 125–150.

Keil, T. A. (1997). Comparative morphogenesis of sensilla: A review. *International Journal of Insect Morphology and Embryology*, *26*, 151–160.

Keil, T. A., & Steinbrecht, R. A. (1984). Mechanosensitive and olfactory sensilla of insects. In R. C. King & H. Akai (Eds.), *Insect ultrastructure* (pp. 477–516). Boston, MA: Springer.

Keil, T. A., & Steinbrecht, R. A. (2010). Insects as model systems in cell biology. In T. Müller-Reichert (Ed.), *Vol. 96, Methods in cell biology* (pp. 363–394), Elsevier.

Kyhse-Andersen, J. (1984). Electroblotting of multiple gels: A simple apparatus without buffer tank for rapid transfer of proteins from polycrylamide to nitrocellulose. *Journal of Biochemical and Biophysical Methods*, *10*, 203–209.

Laemmli, U. K. (1970). Cleavage of structural proteins during the assembly of the head of bacteriophage T4. *Nature*, *227*, 680–685.

Laue, M., Han, H. M., Dittmann, C., & Setlow, P. (2018). Intracellular membranes of bacterial endospores are reservoirs for spore core membrane expansion during spore germination. *Scientific Reports*, *8*, 1–12.

Maida, R., Krieger, J., Gebauer, T., Lange, U., & Ziegelberger, G. (2000). Three pheromone-binding proteins in olfactory sensilla of the two silkmoth species *Antheraea polyphemus* and *Antheraea pernyi*. *European Journal of Biochemistry*, *267*, 2899–2908.

Maida, R., Mameli, M., Müller, B., Krieger, J., & Steinbrecht, R. A. (2005). The expression pattern of four odorant-binding proteins in male and female silk moths, *Bombyx mori*. *Journal of Neurocytology*, *34*, 149–163.

Marshall, A. T., & Kent, M. (1991). A device for freeze-substituting a large number of samples under controlled conditions. *Journal of Microscopy (Oxford)*, *162*, 335–340.

Möbius, W. (2016). Immunoelectron microscopy: High-resolution immunocytochemistry. In R. A. Bradshaw & P. D. Stahl (Eds.), *Encyclopedia of cell biology* (pp. 32–43). Waltham, MA: Academic Press.

Nagnan-Le Meillour, P., Cain, A. H., Jacquin-Joly, E., Francois, M. C., Ramachandran, S., Maida, R., et al. (2000). Chemosensory proteins from the proboscis of *Mamestra brassicae*. *Chemical Senses*, *25*, 541–553.

Park, S. K., Shanbhag, S. R., Wang, Q., Hasan, G., Steinbrecht, R. A., & Pikielny, C. W. (2000). Expression patterns of two putative odorant-binding proteins in the olfactory organs of *Drosophila melanogaster* have different implications for their functions. *Cell and Tissue Research*, *300*, 181–192.

Pophof, B. (2002). Moth pheromone binding proteins contribute to the excitation of olfactory receptor cells. *Naturwissenschaften*, *89*, 515–518.

Pophof, B. (2004). Pheromone-binding proteins contribute to the activation of olfactory receptor neurons in the silkmoths *Antheraea polyphemus* and *Bombyx mori*. *Chemical Senses*, *29*, 117–126.

Raming, K., Krieger, J., & Breer, H. (1990). Primary structure of a pheromone-binding protein from *Antheraea pernyi*: Homologies with other ligand-carrying proteins. *Journal of Comparative Physiology B*, *160*, 503–509.

Reynolds, E. S. (1963). The use of lead citrate at high pH as an electron-opaque stain in electron microscopy. *The Journal of Cell Biology*, *17*, 208–212.

Schneider, D. (1992). 100 years of pheromone research. *Naturwissenschaften*, *79*, 241–250.

Schneider, D., Lacher, V., & Kaissling, K. E. (1964). Die Reaktionsweise und das Reaktionsspektrum von Riechzellen bei *Antheraea pernyi* (Lepidoptera, Saturniidae). *Zeitschrift für Vergleichende Physiologie*, *48*, 632–662.

Schneider, D., & Steinbrecht, R. A. (1968). Checklist of insect olfactory sensilla. *Symposia of the Zoological Society London*, *23*, 279–297.

Shanbhag, S. R., Hekmat-Scafe, D., Kim, M. S., Park, S. K., Carlson, J. R., Pikielny, C., et al. (2001). Expression mosaic of odorant-binding proteins in *Drosophila* olfactory organs. *Microscopy Research and Technique*, *55*, 297–306.

Shanbhag, S. R., Park, S.-K., Pikielny, C. W., & Steinbrecht, R. A. (2001). Gustatory organs of *Drosophila melanogaster*: Fine structure and expression of the putative odorant-binding protein PBPRP2. *Cell and Tissue Research*, *304*, 423–437.

Shanbhag, S. R., Smith, D. P., & Steinbrecht, R. A. (2005). Three odorant-binding proteins are co-expressed in sensilla trichodea of *Drosophila melanogaster*. *Arthropod Structure & Development*, *34*, 153–165.

Steinbrecht, R. A. (1980). Cryofixation without cryoprotectants. Freeze substitution and freeze etching of an insect olfactory receptor. *Tissue & Cell*, *12*, 73–100.

Steinbrecht, R. A. (1987). Functional morphology of pheromone-sensitive sensilla. In G. D. Prestwich & G. J. Blomquist (Eds.), *Pheromone biochemistry* (pp. 353–384). Academic Press.

Steinbrecht, R. A. (1993). Freeze-substitution for morphological and immunocytochemical studies in insects. *Microscopy Research and Technique*, *24*, 488–504.

Steinbrecht, R. A. (1999). Olfactory receptors. In E. Eguchi & Y. Tominaga (Eds.), *Atlas of arthropod sensory receptors* (pp. 156–176). Tokyo: Springer.

Steinbrecht, R. A., Laue, M., & Ziegelberger, G. (1995). Immunolocalization of pheromone-binding protein and general odorant-binding protein in olfactory sensilla of the silk moths *Antheraea* and *Bombyx*. *Cell and Tissue Research*, *282*, 203–217.

Steinbrecht, R. A., Ozaki, M., & Ziegelberger, G. (1992). Immunocytochemical localization of pheromone-binding protein in moth antennae. *Cell and Tissue Research*, *270*, 287–303.

Stierhof, Y.-D., Schwarz, H., Dürrenberger, M., Villiger, W., & Kellenberger, E. (1991). Yield of immunolabel compared to resin sections and thawed cryosections. In M. A. Hayat (Ed.), *Vol. 3, Colloidal gold: Principles methods and applications* (pp. S87–115). San Diego: Academic Press. Hrsg.

Van den Berg, M. J., & Ziegelberger, G. (1991). On the function of the pheromone binding protein in the olfactory hairs of *Antheraea polyphemus*. *Journal of Insect Physiology*, *37*, 79–85.

Vogt, R. G., Callahan, F. E., Rogers, M. E., & Dickens, J. C. (1999). Odorant binding protein diversity and distribution among the insect orders, as indicated by LAP, an OBP-related protein of the true bug *Lygus lineolaris* (Hemiptera, Heteroptera). *Chemical Senses*, *24*, 481–495.

Vogt, R. G., & Riddiford, L. M. (1981). Pheromone binding and inactivation by moth antennae. *Nature*, *293*, 161–163.

Zhang, S.-G., Maida, R., & Steinbrecht, R. A. (2001). Immunolocalization of odorant-binding proteins in noctuid moths (Insecta, Lepidoptera). *Chemical Senses*, *26*, 885–896.

CHAPTER TWELVE

Effect of OBPs on the response of olfactory receptors

Bing Wang[a,†], Song Cao[a,†], Wei Liu[a], Guirong Wang[a,b,c,*]

[a]State Key Laboratory for Biology of Plant Diseases and Insect Pests, Institute of Plant Protection, Chinese Academy of Agricultural Sciences, Beijing, China
[b]Guangdong Laboratory of Lingnan Modern Agriculture, Shenzhen, China
[c]Genome Analysis Laboratory of the Ministry of Agriculture, Agricultural Genomics Institute at Shenzhen, Chinese Academy of Agricultural Sciences, Shenzhen, China
[*]Corresponding author: e-mail address: wangguirong@caas.cn

Contents

Abstract

The technique of two-electrode voltage-clamp (TEVC) recording from the heterologous expression system of olfactory receptors (ORs) in *Xenopus laevis* oocytes has been widely used to deorphanize insect ORs, that is to identify specific ligands for each of them.

† These authors contributed equally to this work.

Methods in Enzymology, Volume 642
ISSN 0076-6879
https://doi.org/10.1016/bs.mie.2020.04.065

However, there is a controversial issue on whether ORs are activated by the odorant/OBP complex or the odorant alone. The mechanism of interaction among odorants, odorant-binding proteins (OBPs) and ORs remains largely unknown, due to the limitations in the use of scientific and innovative methods. In this chapter, the modified *Xenopus* oocytes expression system combined with TEVC technique is used to approach this issue. We describe the experimental strategies and provide detailed protocols for recording the signals generated by ORs in response to odorant/OBP complex at different concentrations. Results obtained by this approach have revealed that the presence of OBPs in the system affects the selectivity and sensitivity responses of ORs. Such studies help understanding the molecular mechanism of odorant detection in peripheral nervous system.

1. Introduction

In the last four decades, a great advance in electrophysiological technologies has contributed to functional characterization of voltage-dependent ion channels (Guan, Chen, & Zhang, 2013; Sucher, Deitcher, Baro, Warrick, & Guenther, 2000). The heterologous expression system in *Xenopus laevis* oocytes was previously confirmed to be sensitive enough for identification of mRNA and its translation in living cells (Gurdon, Lane, Woodland, & Marbaix, 1971), and was subsequently used to express genes encoding voltage- and ligand-gated ionic channels (Hollmann, O'Shea-Greenfield, Rogers, & Heinemann, 1989; Miledi, 1982; Moriyoshi et al., 1991; Sucher et al., 2000). Two-electrode voltage-clamp (TEVC) recording technology, invented by Cole and his colleagues in the 1930s and refined by Hodgkin et al. in the 1940s (Guan et al., 2013; Hodgkin, Huxley, & Katz, 1952; Sucher et al., 2000), is a powerful electrophysiological tool combined with the *X. laevis* oocytes system to uncover the biophysical and pharmacological properties, as well as structure-function relationships of ion channels and transporters (Guan et al., 2013; Sucher et al., 2000; Wang, Zhang, & Schroeder, 2017). Meanwhile, this method is also widely used in insect neurobiological research, particularly in functional study of receptors involved in chemoreception (Carey, Wang, Su, Zwiebel, & Carlson, 2010; Hallem, Ho, & Carlson, 2004; Wetzel et al., 2001).

Chemoreception in insect is an essential physiological process mediating insect behaviors by detecting a variety of environmental chemical stimuli (Pelosi, Iovinella, Zhu, Wang, & Dani, 2017). The hydrophobic odorant molecules get into the sensillar lymph and are solubilized and ferried by odorant-binding proteins (OBPs) to olfactory receptors (ORs) or other chemosensory receptors expressed on the dendritic membrane of sensory neurons (Breer, Fleischer, Pregitzer, & Krieger, 2019; Leal, 2013, 2019; Pelosi et al., 2017; Wang et al., 2013). In this process, transmembrane ORs are of vital importance for peripheral signal transduction. The insect OR genes are not an exception and can be expressed in *Xenopus* oocytes and deorphanized using the common technique of TEVC approach. The first and best documented example is an OR of *Drosophila melanogaster* (Wetzel et al., 2001).

In general, cRNA, encoding the candidate OR and the odorant receptor co-receptor (Orco), are injected together into *Xenopus* oocyte. These transformed cells respond with good sensitivity to ligands dissolved in DMSO, even in the absence of OBPs (Terrado, Pinnelli, Sanes, & Plettner, 2019). In fact, it has been suggested that OBPs (or PBPs particularly studied in Lepidoptera) are not strictly required in the natural olfactory system, but might increase the sensitivity to odorants (Biessmann et al., 2010; Leal, 2019; Matsuo, Sugaya, Yasukawa, Aigaki, & Fuyama, 2007; Pelletier, Guidolin, Syed, Cornel, & Leal, 2010; Xu, Atkinson, Jones, & Smith, 2005; Zhu et al., 2019). In moths, pheromone-binding proteins (PBPs), a sub-group of OBPs, usually act as sex pheromone-transporters in the sensillum lymph of antenna, and play an important role in initial reception of pheromones (Leal, 2019). So far, there is a controversial issue on whether ORs are activated by a pheromone-PBP complex or a pheromone molecule per se (Kaissling, 2013; Leal, 2019). Therefore, several methods have been developed both *in vivo* and *in vitro* system to study interactions between pheromones, PBPs and pheromone receptors (PRs) of Lepidoptera moths (Chang et al., 2015; Grosse-Wilde, Gohl, Bouché, Breer, & Krieger, 2007; Grosse-Wilde, Svatos, & Krieger, 2006; Pophof, 2002, 2004; Sun et al., 2013; Syed, Ishida, Taylor, Kimbrell, & Leal, 2006). Earlier studies conducted on the silkmoths *Bombyx mori* and *Antheraea polyphemus* showed that the PBP/pheromone complex may enhance the responses of PRs or of olfactory receptor neurons (ORNs) (Forstner, Breer, & Krieger, 2009; Grosse-Wilde et al., 2006; Pophof, 2002, 2004; Syed et al., 2006). *In vivo* tip recordings technology was applied for monitoring the effect of PBPs on the response of ORNs housed in male trichoid sensilla of isolated antennae of *B. mori* and *A. polyphemus*. These experiments have shown that the specificity of ORNs is enhanced when the bombykol is used in complex with its PBP. Subsequently, the *in vivo Drosophila* "empty neuron" expression system was used to co-express BmorPBP/BmorOR1 in order to detect the effect of PBPs on the response of ORs. It was observed that the presence of BmorPBP enhanced the sensitivity of BmorOR1 in response to bombykol. Similar effects were also recorded in modified *in vitro* methods using HEK293 cell expression system combined with calcium imaging technique (Forstner et al., 2009; Grosse-Wilde et al., 2006; Miazzi et al., 2019). All studies mentioned, performed with different approaches, have suggested a complex interplay between PBPs and PRs/PR-expressing neurons. However, there is still no evidence OBP/odorant complex might activate ORs.

The goal of TEVC recording in olfactory system is to deorphanize chemosensory receptors, especially for *in vitro* characterizations of ORs. It is obvious that this method is more practical in functional screenings of large numbers of ORs compared to other heterologous OR-deorphanizing systems (Wang, Liu, He, & Wang, 2016). The basic principle of TEVC recording has been discussed in detail in previous studies (Dascal, 2001; Guan et al., 2013; Hodgkin et al., 1952; Wang et al., 2017). There are several advantages with the *Xenopus* oocytes system. Firstly, it is easy to synthesize cRNA of OR and Orco, and to inject into oocytes thanks to their large size. Secondly, the expression levels of receptors in the membrane is quite high owing to the robustness of oocytes. Thirdly, hydrophobic odorants can be conveniently dissolved in dimethyl sulfoxide (DMSO) to stimulate the OR-expressing oocytes (Montagné, de Fouchier, Newcomb, & Jacquin-Joly, 2015). However, in the past the *Xenopus* oocytes system combined with TEVC technique has been generally used to record the response of the OR/Orco complex alone, without the additional elements, such as OBPs and degrading enzymes, which are present in the antennal sensilla. More recent studies, instead, included OBPs in such protocols with the aim of studying the effect of these soluble carriers on the sensitivity and selectivity of ORs (Chang et al., 2015; Sun et al., 2013; Xu, Hooper, Pickett, & Leal, 2012). The first of these works found that the presence of BmorPBP1 negatively affected the responses of BmorOR1/Orco to the pheromone (Xu et al., 2012). On the other hand, using a similar approach, we observed that PBPs could increase the selectivity and sensitivity of PRs to their sex pheromone components in two Lepidopteran species, *Plutella xyllostella* and *Chilo suppressalis* (Chang et al., 2015; Sun et al., 2013). Such contrasting results could depend on environmental factors, such as the pH used, but could also be explained when accounting for different physiological roles of OBPs (Leal, 2019).

This chapter introduces a step-by-step protocol to investigate the effects of OBPs on the response of ORs expressed in *Xenopus* oocytes. Data analysis, trouble-shooting advice, instructions and notes are provided in detail. This report summarizes the recent work from our laboratory and suggests potential developments in understanding the complex interplay between odorants, ORs and OBPs.

2. Materials, equipment and reagents

2.1 Expression and purification of OBPs

For materials, equipment, reagents and protocols for expression and purification of OBPs refer to Chapter "Bacterial expression and purification of vertebrate odorant-binding proteins," by Brulé et al.

2.2 Fluorescence binding assays of OBPs

For materials, equipment and reagents used in this part refer to Chapter "Ligand binding assays with OBPs and CSPs," by Pelosi et al.

2.3 cRNA synthesis of *OR/Orco* genes

- Block heater at 37 °C and 70 °C (e.g., Ruicheng Instrument Co., Ltd., Hangzhou, China)
- Centrifuge (e.g., Eppendorf, Hamburg, Germany)
- Plasmid DNA extraction kit (e.g., TIANprep Midi Plasmid Kit, TIANGEN, Beijing, China)
- Phenol:Chloroform:Isoamyl Alcohol 25:24:1 (e.g., Cat. No. P3803, Sigma-Aldrich, St Louis, MO, USA)
- mMESSAGE mMACHINE Kit: T7 (AM1340) or SP6 (AM1344) kits (Thermo Fisher Scientific, Waltham, MA), store at −20 °C

2.4 Frog dissection and oocyte isolation

- Adult female *Xenopus laevis*: Sexually mature *X. laevis* frogs (with unfertilized eggs) were purchased from Xenopus Resource Center (Qingdao, China). All animal procedures are performed under anesthesia, and wounds are to prevent infection. All efforts are made to minimize suffering.

Note: The protocols and all animal procedures here described and used in our work were approved by the Institutional Animal Care and Use Committee of Chinese Academy of Agricultural Sciences, Beijing, China.

- Sterile surgical blade (e.g., Jinhuan, Shanghai, China)
- Sterile suture line (e.g., Jinhuan, Shanghai, China)

- Sterile forceps (e.g., Zhenye, Dongguan, China)
- Shaker (e.g., Voshin Instruments, Wuxi, China), 70 rounds/min
- Stereomicroscope (e.g., Leica, Germany)
- Incubator set to 18 °C (e.g., Levntec, Hangzhou, China)
- 10 × Ringer's buffer (960 mM NaCl, 20 mM KCl, 50 mM $MgCl_2 \cdot 6H_2O$ and 50 mM HEPES): Dissolve 56.1 g NaCl, 1.5 g KCl, 10.2 g $MgCl_2 \cdot 6H_2O$ and 11.9 g HEPES directly into 800 mL distilled water, mix thoroughly with the aid of a magnetic stirrer, adjust the pH to 7.6 by adding 1 M NaOH solution and set volume to 1 L. Filter sterilize the liquid mixture using air pump equipment, including Buchner flask, filterable membrane and vacuum pump. Check the osmolality (between 2020 and 2050 mOsm/kg), and store at 4 °C.
- Gentamycin stock solution (10 mg/mL): Dissolve 0.1 g gentamycin sulfate salt (e.g., Cat. No. E003632, Sigma-Aldrich, St Louis, MO, USA) into 10 mL distilled water to the concentration of 10 mg/mL and filter it with 0.22 μm PES (Polyethersulfone) syringe filters. Aliquot into 1 mL for each tube and store at −20 °C.
- Tetracycline stock solution (50 mg/mL): Dissolve 0.5 g tetracycline hydrochloride (e.g., Cat. No. 0422, Amresco, Ohio, USA) into 10 mL distilled water at a concentration of 50 mg/mL and filter it with 0.22 μm PES syringe filters. Aliquot into 1 mL for each tube and store at −20 °C.
- Streptomycin stock solution (100 mg/mL): Dissolve 1 g streptomycin sulfate (e.g., Cat. No. 0382, Amresco, Ohio, USA) into 10 mL distilled water at a concentration of 100 mg/mL and filter it with 0.22 μm PES syringe filters. Aliquot into 1 mL for each tube and store at −20 °C.
- Sodium pyruvate stock solution (275 mg/mL): Dissolve 5.5 g sodium pyruvate (e.g., Cat. No. P2256, Sigma-Aldrich, St Louis, MO, USA) into 20 mL distilled water at a concentration of 275 mg/mL and filter it with 0.22 μm PES syringe filters. Aliquot into 1 mL for each tube and store at −20 °C.
- 100 × $CaCl_2$ stock solution (60 mM): Dissolve 4.41 g $CaCl_2 \cdot 2H_2O$ into 500 mL distilled water at a concentration of 60 mM and filter sterilize. Store at 4 °C.
- Dialyzed horse serum (e.g., Cat. No. 26050-070, Thermo Fisher Scientific, Waltham, MA). Store at −20 °C. Before use, thaw at room temperature and aliquot into 50 mL for each tube.
- 1 × Ringer's buffer (without Ca^{2+}): Working solution is prepared by adding 100 mL of 10 × Ringer's solution into distilled water to final volume of 1 L and filter it (prepare just before use).

- $1\times$ Ringer's buffer (with Ca^{2+}): Working solution is prepared by adding 100 mL of $10\times$ Ringer's solution and 10 mL of $100\times$ $CaCl_2$ stock solution into distilled water to final volume of 1 L and filter it.
- Washing buffer: Working solution is prepared by adding 1 mL of gentamycin stock solution (10 mg/mL) into 1 L of $1\times$ Ringer's buffer (without Ca^{2+}), mix thoroughly and store at 4°C (prepare just before use).
- Incubation medium: Working solution is prepared by adding 1 mL tetracycline stock solution (50 mg/mL), 1 mL streptomycin stock solution (100 mg/mL), 2 mL sodium pyruvate stock solution (275 mg/mL), 50 mL dialyzed horse serum into $1\times$ Ringer's buffer (with Ca^{2+}) to final volume of 1 L (prepare just before use).
- Anesthetizing solution: Add 1.5 g Ethyl-3-aminobenzoate methanesulfonate salt (e.g., Cat. No. A-5040, Sigma-Aldrich, St Louis, MO, USA) into 1 L water, mix briefly (prepare just before use).
- Enzyme solution: Add 20 mg Collagenase Type I (e.g., Cat. No. 17100-017, Thermo Fisher Scientific, Waltham, MA) and 2 mL washing buffer into a 15 mL tube and mix thoroughly (prepare just before use).

2.5 Oocyte microinjection and culture

- Glass capillaries (e.g., #1B120F-3, WPI, Sarasota, FL, USA)
- Capillary puller (e.g., P-1000 Micropipette Puller, Sutter Instrument, Novato, USA)
- 60-mm Petri dish (e.g., Thermo Fisher Scientific, Waltham, MA)
- 24 wells plates (e.g., Thermo Fisher Scientific, Waltham, MA)
- Microinjector equipped with a control panel and a control pedal (e.g., Nanoliter injector 2010, World Precision Instruments, Sarasota, FL, USA)
- Stereomicroscopes (e.g., Leica, Germany)
- Right hand micromanipulators with magnetic base (e.g., Scientifica, UK)
- Mineral oil (e.g., Cat. No. M8410, Sigma-Aldrich, St Louis, MO, USA)

2.6 Electrophysiological recording

- Glass capillaries (e.g., 1B120F-3, WPI, Sarasota, FL, USA)
- Capillary puller (e.g., P-1000 Micropipette Puller, Sutter Instrument, Novato, USA)
- Electrode holders (a straight style and a 45° with handle) (e.g., Warner Instruments, Hamden, CT)
- Silver wires (e.g., AG25-10, Warner Instruments, Hamden, CT)

- Stereomicroscopes (e.g., Leica, Germany)
- Left and right micromanipulators (e.g., MM-33L/R, Warner Instruments, Hamden, CT)
- Magnetic bases (e.g., MB/B, Warner Instruments, Hamden, CT)
- 50 mL perfusion tubes (e.g., Zhengkang Medical Apparatus, Jiangsu, China)
- A shelf to fix perfusion tubes, custom-made
- Slotted bath oocyte recording chamber (e.g., RC-3Z, Warner Instruments, Hamden, CT)
- Water pump (e.g., Lange, Baoding, China)
- Faraday cage (e.g., Felles Photonic Instruments, Shanghai, China)
- Vibro-Isolating table, custom-made
- A TEVC including the following parts: e.g., OC-725C amplifier (Warner Instruments, Hamden, CT), Digidata 1440A low-noise data acquisition system (Molecular Devices, model: Digidata 1440A), and Clampex 10.3 software (Molecular Devices) for data analysis.
- 3 M KCl solution: Dissolve 11.175 g KCl into 50 mL distilled water at a concentration of 3 M and filter it with 0.22 μm PES syringe filters. Store at room temperature.
- Sodium hypochlorite solution (NaClO) (e.g., Macklin, Shanghai, China)
- Odorants (with purity >96%) (e.g., Sigma-Aldrich, St Louis, MO, USA)
- 1 × Ringer's buffer (with Ca^{2+})
- OBP solution (1 μM): Add the purified OBP solution into 1 × Ringer's buffer (with Ca^{2+}) to a final concentration of 1 μM (prepare just before use)
- Odorant stock solution (1 M): 1 M stock solutions were prepared in dimethyl sulfoxide (DMSO) and stored at −20 °C until use
- Odorant working solution (without OBP): A serial dilution of odorant stock solution is prepared for electrophysiological recordings. Odorant stock solutions are brought to room temperature and diluted in 1 × Ringer's solution (with Ca^{2+} and containing 0.1% DMSO) to a concentration of 100 μM in a total volume of 50 mL. Vibrate the solution to completely mix. Prepare 10 μM of the odorant working solution by adding 5 mL of 100 μM solution into 45 mL of 1 × Ringer's solution (with Ca^{2+} containing 0.1% DMSO). Similarly, prepare the 1 and 0.1 μM of the odorant solution. Working solutions with other concentrations are prepared if needed.
- Odorant and OBP solution: To analyze the ability of OBP on the response of ORs, odorant solution with OBP are prepared. Firstly, dilute

5 μL of 1 M odorant stock solution in 50 mL of OBP solution (1 μM) to a final concentration of 100 μM. Then prepare 10, 1 and 0.1 μM of the solutions with the method mentioned above. Working solutions with other concentrations are determined and prepared during the recording.

3. Protocol

3.1 Expression and purification of OBPs

The sequences of OBPs encoding mature proteins without signal peptides are amplified and cloned into expression vectors (e.g., pET30a, Novagen, Madison, WI) which are used to transform competent cells (e.g., BL21 *Escherichia coli*, Transgen, Beijing, China). Individual colonies are identified by PCR. Constructed DNA sequences are confirmed by DNA sequencing. The detailed protocols for OBPs expression and purification are described in Chapter "Bacterial expression and purification of vertebrate odorant-binding proteins," by Brulé et al.

3.2 Fluorescence binding assays of OBPs

To establish the effect of OBPs on the OR responses to odorants, the affinities of OBPs alone to candidate ligands are measured in fluorescence competitive binding assays. These procedures are described in Chapter "Ligand binding assays with OBPs and CSPs," by Pelosi et al.

3.3 cRNA synthesis of *OR/Orco* genes

1. The open reading frame (ORF) encoding each OR is amplified and subcloned into an expression vector, such as pT7Ts vector, with the Kozak sequence (GCCACC) added between the restriction enzyme site and the initiation codon in the forward primer. The constructed vector is used to transform competent cells (e.g., BL21 *E. coli*, Transgen, Beijing, China). After overnight culture on LB plate containing 50 μg/mL Amp^+, a single clone is identified by PCR and sequencing.
2. The positive single clone used to inoculate LB medium containing 50 μg/mL Amp^+ at 37 °C overnight (12–16 h) with shaking for 220 rpm. Then, plasmid DNA is extracted using a Plasmid DNA extraction kit (e.g., TIANprep Midi Plasmid Kit, TIANGEN, Beijing, China). The DNA template is then used to synthesize cRNA.

Note: Make sure the plasmid DNA template is free from contaminants, that might inhibit the RNA polymerase.

3. The cRNA is synthesized from the linearized vector using a mMESSAGE mMACHINE T7 kit (Cat. no. AM1345, Ambion, Austin, TX). Plasmid DNA is linearized by appropriate restriction enzyme, and extracted with Phenol:Chloroform:Isoamyl Alcohol (25:24:1), precipitated with 100% ethanol, washed with 80% ethanol and resuspended in RNase-free water.

Note: Plasmid DNA must be linearized with a restriction enzyme downstream of the insert to be transcribed. The plasmid template must be completely digested before the transcription reaction, because even a small amount of circular template can adversely affect the final result.

(a) The *in vitro* transcription reaction is set up at room temperature containing the following reagents: 10 μL of 2× CAP/NTP, 2 μL of 10× Reaction Buffer, 0.1–1 μg of linearized DNA template, 2 μL of Enzyme mix, to a final reaction volume of 20 μL with RNase-free water. Mix, spin down and incubate at 37 °C for 2 h.

(b) Stop the reaction and precipitate the RNA by adding 30 μL Nuclease-free water and 30 μL Lithium Chloride (LiCl) precipitation solution. Chill the tube for ≥30 min at −20 °C to precipitate the RNA.

(c) Wash the RNA pellet twice with 70% ethanol and resuspend in RNase-free water. The quantity and integrity of cRNA are determined using a Nanodrop ND-1000 spectrophotometer (e.g., Nano-Drop products, Wilmington, DE, USA) and gel electrophoresis analysis. The cRNA band should be single and bright. The cRNA is diluted using RNasc-free water at a final concentration of 2 μg/μL. Aliquots of 1 μL are stored at −80 °C.

3.4 Frog dissection and oocyte culture

The protocol for frog dissection is taken from Goldin (1992) and Goldin and Sumikawa (1992) and illustrated in Fig. 1.

1. Put an adult female African clawed frog into 1 L anesthetizing solution and cover with a layer of ice in a 5 L beaker.
2. Keep the frog in the ice/water for 30 min, and make sure that the frog is anesthetized completely.
3. Transfer and keep the frog lying on its back in a bucket full of ice.
4. Operation and oocyte collection.
   - Wipe the ventral skin of a frog with medical iodine at least twice.
   - Make a small longitudinal incision (~1 cm) at one side of the lower abdomen with a sterile surgical blade. Cut the skin and ovarian epidermis, and let unfertilized eggs exposed.

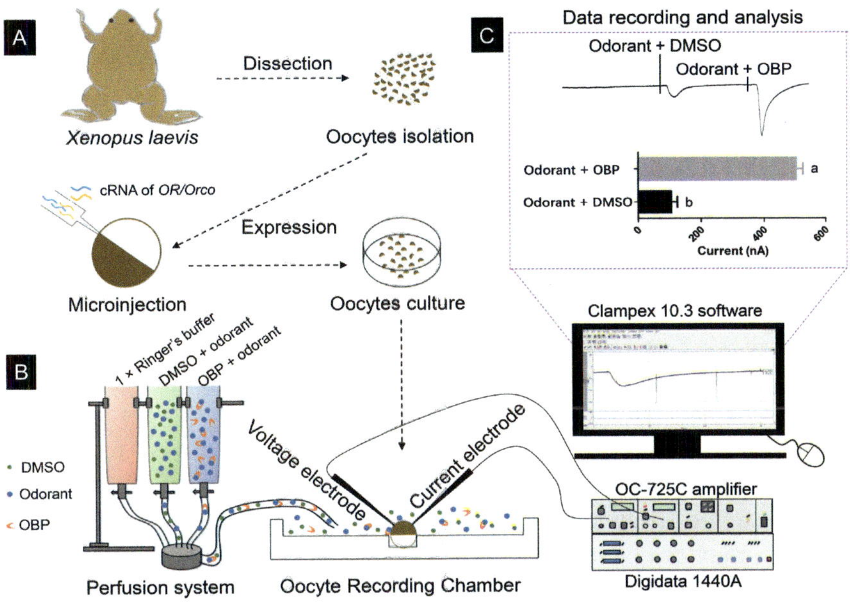

Fig. 1 Graphical workflow of modified two-electrode voltage-clamp (TEVC) technology combined with the *Xenopus* oocytes expression system. This set-up is currently used to measure the affinity of odorants and olfactory receptors (ORs), and to study the effect of odorant binding proteins (OBPs) on such interactions. (A) Frog dissection performed to collect oocyte messes. Single mature and healthy oocytes are isolated and microinjected with a mixture of *OR* and *Orco* cRNA. The injected cells are then cultured for heterologous expression at 18 °C for 2–6 days. (B) The cultured oocytes are transferred to the oocyte recording chamber. Whole-cell currents are recorded using TEVC technology. During this process, the oocyte is washed using a perfusion system and treated with either odorants dissolved in DMSO or OBP/odorant complexes. Between each stimulus, the oocytes are washed with 1 × Ringer's buffer until the current returns to baseline. (C) The signal amplification is accomplished by an OC-725C oocyte clamp at a holding potential of −80 mV. The signal is lowpass filtered at 50 Hz and digitized at 1 kHz. Data acquisition is carried out with Digidata 1440A and Clampex 10.3 software. The data are analyzed using GraphPad Prism 5.0 software.

- Use a sterile forcep to take out egg masses from both ovary sides and keep them in a 90-mm Petri dish containing washing buffer (make sure the washing buffer covers the ovary completely).
- Use sterile line to suture the wound of the ovarian epidermis and skin separately (3–5 sutures/cut, 6 node/suture).

5. Put the frog in a container with shallow sterile water and let it breathe in air. Wait for its complete recovery from anesthesia. Send it back to the frog feeding facility after 48 h of observation.
6. Cut the egg masses into small pieces of 3–4 mm in size, and wash them with washing buffer until the solution is clear.

7. Prepare enzyme lysate buffer (according the amount of ovary).
8. Transfer oocytes into 15 mL tubes containing enzyme lysate buffer in 1: 3 volume proportion to the oocytes.
9. Rotate the tubes with a shaker at room temperature for ~60 min at the speed of 70 round/min until follicular membrane and intercellular fascial tissue are fully defolliculated.
10. Check oocytes under stereomicroscope. Make sure most of oocytes have been defolliculated.

Note: Overtreatment of oocytes with collagenase solution will reduce the viability of oocytes.

11. Wash the oocytes repeatedly with washing buffer for 5–6 times until collagenase solution is totally removed.

Note: During the process, oocytes should not be exposed to the air. Generally, follicular membrane and intercellular fascial tissue outside the oocytes can be easily removed after digestion with enzyme solution. If necessary, they can be cleaned using forceps with sharp tips under a stereomicroscope paying attention not to damage the membrane of eggs.

12. Wash oocytes 3–5 times using incubation medium and transfer the healthy oocytes to 90-mm dishes containing the incubation medium. Mark the date on each dish and keep them at 18 °C overnight waiting for selection and injection. During this step unhealthy oocytes may die and be discarded.
13. The next day, choose mature and healthy oocytes (stage V–VI) under stereomicroscope and transfer them to a 60-mm dish.

3.5 Oocytes microinjection

The protocol for oocytes microinjection is taken from Goldin (1992) and Goldin and Sumikawa (1992) and illustrated in Fig. 1.

1. Injection needles are made by borosilicate glass capillaries using a micropipette puller. Each borosilicate capillary generates two injection tips. Usually, the following conditions are used for pulling glass capillaries suitable for *X. laevis* oocytes injection: Heat = 575, Pull = 35, Velocity = 50, Time = 255.

Note: When preparing the glass capillaries, test the RAMP value first. Install a glass capillary onto a puller, and run RAMP TEST to determine the RAMP value for melting the glass. After discarding the first value, make three RAMP tests and calculate the mean to use for HEAT value setting.

For each batch of glass capillaries, one test is enough and the RAMP value can be used for all other capillaries.

2. 1 μL cRNA of *OR* gene and 1 μL cRNA of *Orco* gene are thawed on ice and mixed.
3. The tip of the pulled capillary is cut with a blade under a stereomicroscope. The aperture should be about 20 μm in diameter.
4. The injection device includes microinjectors with dispenser and a three-dimensional micromanipulator. Fill the glass capillary with mineral oil and install it onto the dispenser.
5. Drop the 2 μL cRNA mixture onto a piece of clean parafilm and draw it into the glass needle with the injector under stereomicroscope. Make sure that the mRNA is free from RNase and no air has been drawn into the needle.
6. Put a polypropylene mesh at the bottom of the 60-mm Petri dish to keep the oocytes from moving. Insert the microprobe into an oocyte and apply pressure to inject 27.6 nL cRNA mixture of *OR* and *Orco* under a stereomicroscope. After injection of an oocyte, wait for ~1 s and then withdraw the needle gently.
7. After injection of each target gene, transfer the oocytes into 24-well plates (1 to 3 oocytes in each well) and incubate for 2–6 days for heterologous expression at 18 °C in the incubation medium.

3.6 Electrophysiological recording

The protocol for TEVC recording is taken from Guan et al. (2013) and Wang et al. (2017) (see also Fig. 1).

1. Preparation of silver wires for recording. Immerse silver wires (~4 cm long and 0.25 mm thick) in NaClO solution to half length, until the immersed part becomes uniformly black. Silver wires used in the slotted bath recording chamber (~6 cm long and 0.25 mm thick) also need to be oxidized by NaClO solution.
2. Preparation of odorant solutions. The odorant solutions with/without OBPs are prepared as described above. All ligand solutions are freshly prepared.
3. Load 1 × Ringer's buffer (with Ca^{2+}) in each perfusion tubes and adjust the height of tubes so that the flow rate is about 5 mL/min. Use a water pump in a drainage system and make sure the solution covers the oocyte during washing.

4. Preparation of glass electrodes. Adjust each parameter according to the instruction manual and pull capillaries to a fine taper with proper parameters. The following parameters are used to pull micropipettes for electrophysiological recording: Heat = 565, Pull = 0, Velocity = 35, Time = 255.
5. Install the electrode holders. Fill 3 M KCl solution in each glass electrode (add ~2/3 volume of the glass capillaries). Insert the oxidized silver wire. Then, put a glass electrode in each holder and install them on a micromanipulator. Adjust the position of the glass electrodes using the micromanipulators.
6. Measure the resistance of the two electrodes. Lower the electrodes with the micromanipulators into the bath solution and adjust "Vm OFFSET" and "Ve OFFSET" to zero. Press buttons "Vm Electrode Test" and "Ve Electrode Test" respectively, and the resistance of the electrodes will be displayed. The electrode resistance should be in the range of 0.1–1 MΩ. If the electrode resistance is not within this range, adjust the parameters of the puller referring to the manual instructions.
7. Turn on the Digidata 1440A data acquisition system, click button "View Only" on the computer screen and make sure there are no external interferences.

Note: External interferences may come from inefficient shielding and grounding, from the two holders, the water pump, the water bath or the perfusion system. To reduce the coupling capacitance between the two electrodes, a grounded metal shield is needed and it should not contact the bath solution.

8. Transfer an oocyte expressing the OR of interest gently to the bath chamber using a sucking pipette.
9. Lower the tips of the two electrodes into the bath using the micromanipulators, and position the electrode tips near the surface of the oocyte (do not contact).
10. Impale the oocyte: under a stereomicroscope, operate the tips of the electrodes against the surface of the oocyte gently using the micromanipulator, and impale the oocyte by further advancing the electrode tip slowly until it pops in (first impalement with the voltage electrode and then with the current electrode). Open the Clampex 10.3 software. The resting membrane potential of the oocyte is displayed normally if the voltage is in the range of −10 to −50 mV for a good oocyte.

11. Switch to voltage clamp recording mode: turn the "CLAMP" to "SLOW" and "GAIN" to "MAX" on Digidata 1440A acquisition system. Then, start to record.
12. Record the currents. When the current becomes stable on the Clampex 10.3 software, click button "Record" to record the currents. In the bath chamber, oocytes are washed by odorant solution alone or OBP/odorants complex according to the following experimental design.

Note: Chemical-induced currents are recorded with the TEVC at holding potential of −80 mV. Signals are amplified with an OC-725C amplifier (Warner Instruments, Hamden, CT), lowpass filtered at 50 Hz and digitized at 1 kHz. Data acquisition is carried out with Digidata 1440A and Clampex 10.3 software (Axon Instruments Inc., Union City, CA).

13. The first aim is to check if OBPs modify the response of ORs to a certain concentration of odorant (usually 10^{-4} M). In this stage, differences in currents are measured between treatments of oocytes with odorants supplied in 0.1% DMSO or in 1 μM OBP. The second aim is to confirm that OBP can increase the sensitivity of ORs to an odorant. In this stage, oocytes are challenged with increasing concentrations of odorant solutions (with or without OBP). During the recording, odorant solutions (with or without OBP) are supplied for about 15 s at a flow rate of 5 mL/min. After each treatment, the oocytes are washed with 1 × Ringer's buffer (with Ca^{2+}) until the current returns to the baseline. Record the drop-out current value, and the starting time for each stimulus. The concentrations of odorant solutions used in dose-response experiments are determined and prepared according to the response sensitivity (e.g., a serial dilution from 10^{-7} to 10^{-3}). The dose-response experiment of each odorant solution (with DMSO or with OBP) are performed at least six replicates.
14. Stop the recording. After the button "Stop" has been pressed to end recording, turn the "CLAMP" to "OFF" and "GAIN" to "0" and withdraw the electrodes carefully to avoid crumbing the tip of the glass electrodes. Remove the tested oocyte and start with another oocyte following the above procedure.
15. Finish the recording and clean the setup. Close the software first, then turn off the digitizer and the amplifier. Empty the remaining liquid in the perfusion system, and wash the system thoroughly with clean water to avoid obstructing the perfusion tubes and chamber. Empty the waste liquid collected from the outlet of the water pump.

3.7 Data analysis

1. Current value calculation. Open a recording file using the software pCLAMP 10.2 and calculate the dropout currents as response values activated by each stimulus.
2. Statistical analysis. For comparison of paired responses from OR, the Student's *t*-test is adopted, while multiple comparisons are assessed by one-way ANOVA analysis using SPSS software (SPSS Inc., Chicago, IL, USA). The dose-response data are analyzed using GraphPad Prism 5.0 software (GraphPad Software Inc., San Diego, CA), and the EC_{50} value is calculated. These values are used to compare the sensitivity of ORs to ligands when dissolved in DMSO or in the presence of OBPs.

4. Results and discussions

The *Xenopus* oocytes expression system combined with TEVC technique is economical, simple, and reliable, and has been widely used for deorphanizing insect ORs in the last two decades (Bastin-Héline et al., 2019; Choo et al., 2018; Cui et al., 2018; Jiang et al., 2014; Mitchell et al., 2012; Mitsuno et al., 2008; Nakagawa, Sakurai, Nishioka, & Touhara, 2005; Sakurai et al., 2004; Wang, Carey, Carlson, & Zwiebel, 2010; Wanner et al., 2007; Wetzel et al., 2001; Zhang et al., 2016; Zhang et al., 2017). The *in vitro* function of ORs/Orco complex alone showed in many cases have broad selectivity to several compounds. However, the endogenous molecular and cellular environment of antennal sensilla is more complicated than the *in vitro* set-up. In particular, whether OBPs are necessary for odorant detection in the peripheral nervous system of antenna has been a controversial issue (Larter, Sun, & Carlson, 2016; Scheuermann & Smith, 2019; Xiao, Sun, & Carlson, 2019; Xu et al., 2005; Zhu et al., 2016; Zhu et al., 2019). Hence, the method for *in vitro* heterologous expression system has been modified to include the presence of OBPs (Fig. 1). Two recent studies in our lab have addressed the interactions between PRs and PBPs using the modified *Xenopus* oocytes expression system combined with TEVC technique (Chang et al., 2015; Sun et al., 2013). It should be emphasized that detailed information on the affinities of PRs and of PBPs to the same odorants is essential to investigate the effect of PBPs on the responses of PRs using the above described approach (Fig. 2A, B).

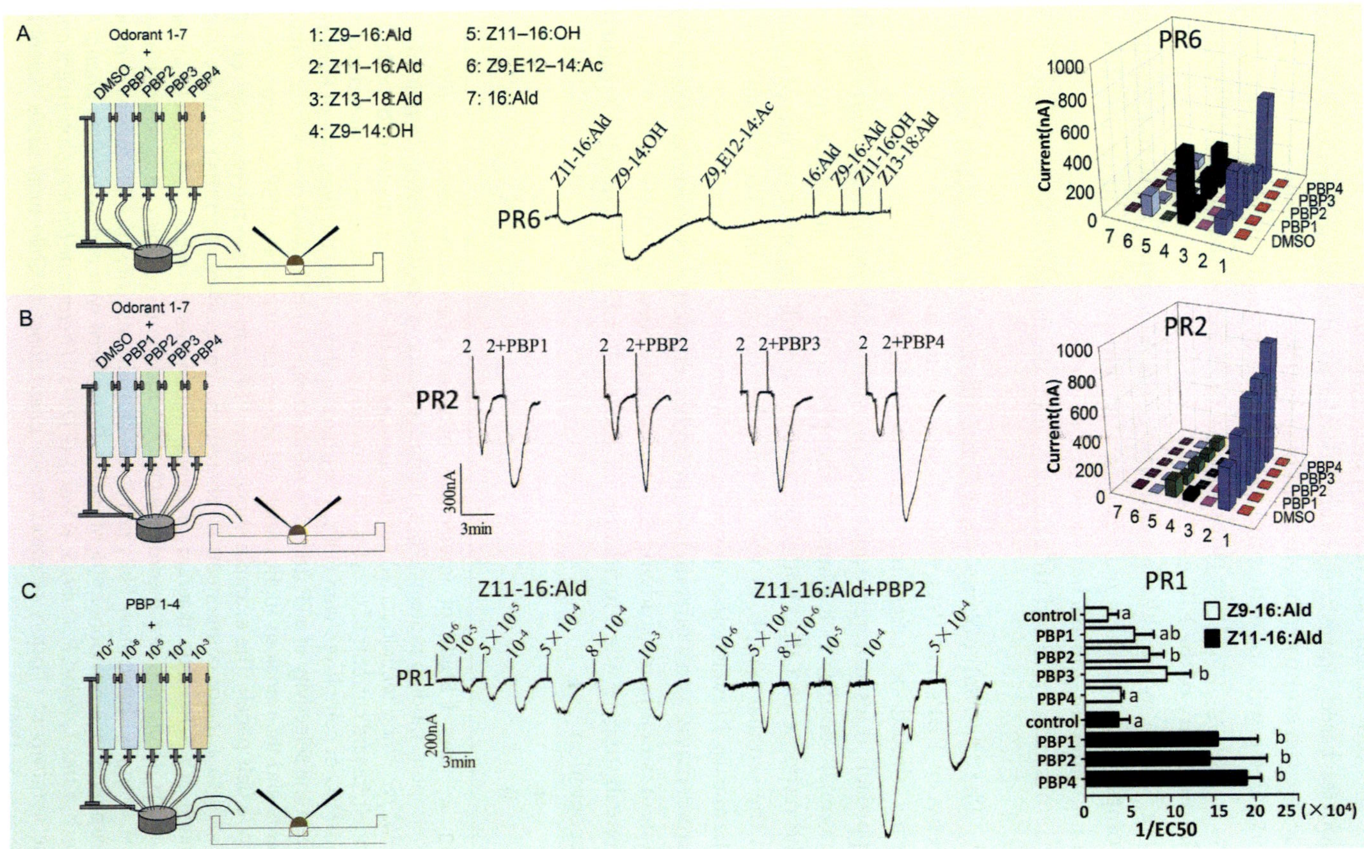

Fig. 2 See figure legend on next page.

We can design different experimental strategies in modified *Xenopus* oocytes-based expression system in order to address some issues. For example, we can compare the effect of different PBPs to the responses of the same PR. The results show that the contribution of each PBP may be specific for each PR (Fig. 2A, B). In particular, we can observe that some PBPs increase the sensitivity of the PR-expressing oocyte cells to the major sex pheromone (Fig. 2B, C), while others could change the selectivity of PR to pheromone components and their analogs (Fig. 2A). However, recent results obtained using different approaches show that neurons (receptors) in most of sensilla can normally respond to odorants after deletion of OBPs *in vivo* (Xiao et al., 2019).

The *Xenopus* oocytes system has allowed to investigate interactions between PBPs and PRs, while being robust and reliable. However, some issues should be taken into account, when considered that the oocyte approach does not reproduce the physiological conditions present in the antennal sensilla. For example, the pH could play an important role in the response of ORs, but at present we are still ignorant on the pH value in the sensillar lymph, as well as on other components, like salts and other chemicals. Nevertheless, given such limitations, the *Xenopus* oocytes system remains as a convenient and highly efficient heterologous expression system, suitable for studying the interactions between OBPs and ORs, thus contributing to the elucidation of the molecular mechanisms of odorant recognition.

Fig. 2—Cont'd Study of interactions between pheromone receptors (PRs) and pheromone binding proteins (PBPs) using the *Xenopus* oocytes system combined with TEVC recording technique. (A) Schematic diagram of PBP/odorants-treated mode (left panel). Inward currents of oocytes are recorded upon stimulations of odorants dissolved in DMSO or complexed with OBPs. The example shows how the presence of PBPs can affect the selectivity of a PR (tuned to minor components) toward pheromone components and their analogs (right panel). (B) Schematic diagram of PBP/odorants-treated mode (left panel). The example shows how PBPs can increase the responses of a PR (tuned to a major component) to pheromone components and their analogs (right panel). (C) Schematic diagram of concentration-treated mode (left panel). Inward currents of oocytes are recorded with OBPs/odorants complex in a serial dilution. The example shows how PBPs can increase the sensitivity of a PR toward a major sex pheromone component (right panel). *Reproduced with permission from Chang, H., Liu. Y., Yang. T., Pelosi. P., Dong. S., & Wang. G. (2015). Pheromone binding proteins enhance the sensitivity of olfactory receptors to sex pheromones in Chilo suppressalis.* Scientific Reports, 5, *13093.*

5. Conclusions

The *Xenopus* oocytes system combined with TEVC recording technique is a powerful electrophysiological tool which has been widely used to express *in vitro* and characterize the functions of many known neurotransmitter receptors and voltage-activated ion channels (Goldin & Sumikawa, 1992). Owing to advantages of the oocyte (such as good expression, high acquisition efficiency, no expression of endogenous receptor protein, etc.), this model is suitable and popular for heterologous expression of membrane receptors, and insect ORs are no exception. In recent years, this method has been successfully modified and developed to study the interactions between ORs and OBPs (Chang et al., 2015; Sun et al., 2013; Xu et al., 2012). In this chapter, we describe the detailed protocol, including information on reagents and equipment, and provide trouble-shooting advice, based on recent work performed in our lab.

Acknowledgments

We are grateful for the advice and assistance provided by Prof. Paolo Pelosi. We thank Prof. Loic Briand for his critical review comments. We thank Dr. Hetan Chang for his advice. This project was supported by National Natural Science Foundation of China (31725023, 31872039) and Shenzhen Science and Technology Program (KQTD20180411143628272).

References

Bastin-Héline, L., de Fouchier, A., Cao, S., Koutroumpa, F., Caballerovidal, G., Robakiewicz, S., et al. (2019). A novel lineage of candidate pheromone receptors for sex communication in moths. *eLife*, *8*, e49826.

Biessmann, H., Andronopoulou, E., Biessmann, M. R., Douris, V., Dimitratos, S. D., Eliopoulos, E., et al. (2010). The *Anopheles gambiae* odorant binding protein 1 (AgamOBP1) mediates indole recognition in the antennae of female mosquitoes. *PLoS One*, *5*(3), e9471.

Breer, H., Fleischer, J., Pregitzer, P., & Krieger, J. (2019). Molecular mechanism of insect olfaction: Olfactory receptors. In J. F. Picimbon (Ed.), *Olfactory concepts of insect control—Alternative to insecticides*. Cham: Springer.

Carey, A. F., Wang, G., Su, C. Y., Zwiebel, L. J., & Carlson, J. R. (2010). Odorant reception in the malaria mosquito *Anopheles gambiae*. *Nature*, *464*(7285), 66–71.

Chang, H. T., Liu, Y., Yang, T., Pelosi, P., Dong, S. L., & Wang, G. R. (2015). Pheromone binding proteins enhance the sensitivity of olfactory receptors to sex pheromones in *Chilo suppressalis*. *Scientific Reports*, *5*(1), 13093.

Choo, Y. M., Xu, P., Hwang, J. K., Zeng, F., Tan, K., Bhagavathy, G., et al. (2018). Reverse chemical ecology approach for the identification of an oviposition attractant for *Culex quinquefasciatus*. *Proceedings of the National Academy of Sciences of the United States of America*, *115*(4), 714–719.

Cui, W. C., Wang, B., Guo, M. B., Liu, Y., Jacquin-Joly, E., Yan, S. C., et al. (2018). A receptor-neuron correlate for the detection of attractive plant volatiles in *Helicoverpa assulta* (Lepidoptera: Noctuidae). *Insect Biochemistry and Molecular Biology, 97*, 31–39.

Dascal, N. (2001). Voltage clamp recordings from *Xenopus* oocytes. In *Current protocols in neuroscience*. editorial board, Jacqueline N. Crawley. Chapter 6, Unit 6.12.

Forstner, M., Breer, H., & Krieger, J. (2009). A receptor and binding protein interplay in the detection of a distinct pheromone component in the silkmoth *Antheraea polyphemus*. *International Journal of Biological Sciences, 5*(7), 745–757.

Goldin, A. L. (1992). Chapter 15. Maintenance of *Xenopus laevis* and oocyte injection. In *Vol. 207. Methods in enzymology* (pp. 266–279). Academic Press.

Goldin, A. L., & Sumikawa, K. (1992). Chapter 16. Preparation of RNA for injection into *Xenopus* oocytes. In *Vol 207. Methods in enzymology* (pp. 279–297). Academic Press.

Grosse-Wilde, E., Gohl, T., Bouché, E., Breer, H., & Krieger, J. (2007). Candidate pheromone receptors provide the basis for the response of distinct antennal neurons to pheromonal compounds. *European Journal of Neuroscience, 25*(8), 2364–2373.

Grosse-Wilde, E., Svatos, A., & Krieger, J. (2006). A pheromone-binding protein mediates the bombykol-induced activation of a pheromone receptor *in vitro*. *Chemical Senses, 31*(6), 547–555.

Guan, B., Chen, X., & Zhang, H. (2013). Two-electrode voltage clamp. In N. Gamper (Ed.), *Vol. 998. Methods in Molecular Biology (Methods and Protocols). Ion channels* (pp. 79–89). Totowa, NJ: Humana Press.

Gurdon, J. B., Lane, C. D., Woodland, H. R., & Marbaix, G. (1971). Use of frog eggs and oocytes for the study of messenger RNA and its translation in living cells. *Nature, 233*(5316), 177–182.

Hallem, E. A., Ho, M. G., & Carlson, J. R. (2004). The molecular basis of odor coding in the *Drosophila* antenna. *Cell, 117*(7), 965–979.

Hodgkin, A. L., Huxley, A. F., & Katz, B. (1952). Measurement of current-voltage relations in the membrane of the giant axon of loligo. *The Journal of Physiology, 116*(4), 424–488.

Hollmann, M., O'Shea-Greenfield, A., Rogers, S. W., & Heinemann, S. (1989). Cloning by functional expression of a member of the glutamate receptor family. *Nature, 342*(6250), 643–648.

Jiang, X. J., Guo, H., Di, C., Yu, S., Zhu, L., Huang, L. Q., et al. (2014). Sequence similarity and functional comparisons of pheromone receptor orthologs in two closely related *Helicoverpa* species. *Insect Biochemistry and Molecular Biology, 48*, 63–74.

Kaissling, K. E. (2013). Kinetics of olfactory responses might largely depend on the odorant-receptor interaction and the odorant deactivation postulated for flux detectors. *Journal of Comparative Physiology A, 199*(11), 879–896.

Larter, N. K., Sun, J. S., & Carlson, J. R. (2016). Organization and function of *Drosophila* odorant binding proteins. *eLife, 5*, e20242.

Leal, W. S. (2013). Odorant reception in insects: Roles of receptors, binding proteins, and degrading enzymes. *Annual Review of Entomology, 58*(1), 373–391.

Leal, W. S. (2019). Chapter 9. Molecular biology of reception. In J. D. Allison & R. T. Carde (Eds.), *Pheromone communication in moths: Evolution, behavior, and application* (pp. 127–138). Berkeley: University of California Press.

Matsuo, T., Sugaya, S., Yasukawa, J., Aigaki, T., & Fuyama, Y. (2007). Odorant-binding proteins OBP57d and OBP57e affect taste perception and host-plant preference in *Drosophila sechellia*. *PLoS Biology, 5*(5), e118.

Miazzi, F., Hoyer, C., Sachse, S., Knaden, M., Wicher, D., Hansson, B. S., et al. (2019). Optimization of insect odorant receptor trafficking and functional expression via transient transfection in HEK293 cells. *Chemical Senses, 44*(9), 673–682.

Miledi, R. (1982). A calcium-dependent transient outward current in *Xenopus laevis* oocytes. *Proceedings of the Royal Society B: Biological Sciences, 215*(1201), 491–497.

Mitchell, R. F., Hughes, D. T., Luetje, C. W., Millar, J. G., Soriano-Agatón, F., Hanks, L. M., et al. (2012). Sequencing and characterizing odorant receptors of the cerambycid beetle *Megacyllene caryae. Insect Biochemistry and Molecular Biology, 42*(7), 499–505.

Mitsuno, H., Sakurai, T., Murai, M., Yasuda, T., Kugimiya, S., Ozawa, R., et al. (2008). Identification of receptors of main sex-pheromone components of three Lepidopteran species. *European Journal of Neuroscience, 28*(5), 893–902.

Montagné, N., de Fouchier, A., Newcomb, R. D., & Jacquin-Joly, E. (2015). Advances in the identification and characterization of olfactory receptors in insects. *Progress in Molecular Biology and Translational Science, 130*, 55–80.

Moriyoshi, K., Masu, M., Ishii, T., Shigemoto, R., Mizuno, N., & Nakanishi, S. (1991). Molecular cloning and characterization of the rat NMDA receptor. *Nature, 354*(6348), 31–37.

Nakagawa, T., Sakurai, T., Nishioka, T., & Touhara, K. (2005). Insect sex-pheromone signals mediated by specific combinations of olfactory receptors. *Science, 307*(5715), 1638–1642.

Pelletier, J., Guidolin, A., Syed, Z., Cornel, A. J., & Leal, W. S. (2010). Knockdown of a mosquito odorant-binding protein involved in the sensitive detection of oviposition attractants. *Journal of Chemical Ecology, 36*(3), 245–248.

Pelosi, P., Iovinella, I., Zhu, J., Wang, G., & Dani, F. R. (2017). Beyond chemoreception: Diverse tasks of soluble olfactory proteins in insects. *Biological Reviews, 93*(1), 184–200.

Pophof, B. (2002). Moth pheromone binding proteins contribute to the excitation of olfactory receptor cells. *Naturwissenschaften, 89*(11), 515–518.

Pophof, B. (2004). Pheromone-binding proteins contribute to the activation of olfactory receptor neurons in the silkmoths *Antheraea polyphemus* and *Bombyx mori. Chemical Senses, 29*(2), 117–125.

Sakurai, T., Nakagawa, T., Mitsuno, H., Mori, H., Endo, Y., Tanoue, S., et al. (2004). Identification and functional characterization of a sex pheromone receptor in the silkmoth *Bombyx mori. Proceedings of the National Academy of Sciences of the United States of America, 101*(47), 16653–16658.

Scheuermann, E. A., & Smith, D. P. (2019). Odor-specific deactivation defects in a drosophila odorant-binding protein mutant. *Genetics, 213*(3), 897–909.

Sucher, N. J., Deitcher, D. L., Baro, D. J., Warrick, R. M. H., & Guenther, E. (2000). Genes and channels: Patch/voltage-clamp analysis and single-cell RT-PCR. *Cell & Tissue Research, 302*(3), 295–307.

Sun, M. J., Liu, Y., Walker, W. B., Liu, C. C., Lin, K. J., Gu, S. H., et al. (2013). Identification and characterization of pheromone receptors and interplay between receptors and pheromone binding proteins in the diamondback moth, *Plutella xyllostella. PLoS One, 8*(4), e62098.

Syed, Z., Ishida, Y., Taylor, K., Kimbrell, D., & Leal, W. (2006). Pheromone reception in fruit flies expressing a moth's odorant receptor. *Proceedings of the National Academy of Sciences of the United States of America, 103*(44), 16538–16543.

Terrado, M., Pinnelli, G. R., Sanes, J., & Plettner, E. (2019). Binding interactions, structure-activity relationships and blend effects in pheromone and host olfactory detection of herbivorous Lepidoptera. In J. F. Picimbon (Ed.), *Olfactory concepts of insect control-alternative to insecticides*. Cham: Springer.

Wang, G., Carey, A. F., Carlson, J. R., & Zwiebel, L. J. (2010). Molecular basis of odor coding in the malaria vector mosquito *Anopheles gambiae. Proceedings of the National Academy of Sciences of the United States of America, 107*(9), 4418–4423.

Wang, B., Guan, L., Zhong, T., Li, K. B., Yin, J., & Cao, Y. Z. (2013). Potential cooperations between odorant-binding proteins of the scarab beetle *Holotrichia oblita* faldermann (Coleoptera: Scarabaeidae). *PLoS One*, *8*(12), e84795.

Wang, B., Liu, Y., He, K., & Wang, G. R. (2016). Comparison of research methods for functional characterization of insect olfactory receptors. *Scientific Reports*, *6*(1), 32806.

Wang, C., Zhang, J., & Schroeder, J. I. (2017). Two-electrode voltage-clamp recordings in *Xenopus laevis* oocytes: Reconstitution of abscisic acid activation of SLAC1 anion channel via PYL9 ABA receptor. *Bio-Protocol*, 7(2), 2114.

Wanner, K. W., Nichols, A. S., Walden, K. K., Brockmann, A., Luetje, C. W., & Robertson, H. M. (2007). A honey bee odorant receptor for the queen substance 9-oxo-2-decenoic acid. *Proceedings of the National Academy of Sciences of the United States of America*, *104*(36), 14383–14388.

Wetzel, C., Behrendt, H., Gisselmann, G., Störtkuhl, K., Hovemann, B., & Hatt, H. (2001). Functional expression and characterization of a *Drosophila* odorant receptor in a heterologous cell system. *Proceedings of the National Academy of Sciences of the United States of America*, *98*(16), 9377–9380.

Xiao, S., Sun, J. S., & Carlson, J. R. (2019). Robust olfactory responses in the absence of odorant binding proteins. *eLife*, *8*, e51040.

Xu, P. X., Atkinson, R., Jones, D. N., & Smith, D. P. (2005). *Drosophila* OBP lush is required for activity of pheromone-sensitive neurons. *Neuron*, *45*(2), 193–200.

Xu, P. X., Hooper, A. M., Pickett, J. A., & Leal, W. S. (2012). Specificity determinants of the silkworm moth sex pheromone. *PLoS One*, 7(9), e44190.

Zhang, R. B., Wang, B., Grossi, G., Falabella, P., Liu, Y., Yan, S. C., et al. (2017). Molecular basis of alarm pheromone detection in aphids. *Current Biology*, *27*(1), 55–61.

Zhang, D. D., Wang, H. L., Schultze, A., Fross, H., Francke, W., Krieger, J., et al. (2016). Receptor for detection of a Type II sex pheromone in the winter moth *Operophtera brumata*. *Scientific Reports*, *6*, 18576.

Zhu, G. H., Xu, J., Cui, Z., Dong, X. T., Ye, Z. F., Niu, D. J., et al. (2016). Functional characterization of SlitPBP3 in *Spodoptera litura* by CRISPR/Cas9 mediated genome editing. *Insect Biochemistry and Molecular Biology*, *75*, 1–9.

Zhu, G. H., Zheng, M. Y., Sun, J. B., Khuhro, S. A., Yan, Q., Huang, Y., et al. (2019). CRISPR/Cas9 mediated gene knockout reveals a more important role of PBP1 than PBP2 in the perception of female sex pheromone components in *Spodoptera litura*. *Insect Biochemistry and Molecular Biology*, *115*, 103244.

CHAPTER THIRTEEN

Site-directed mutagenesis of odorant-binding proteins

Jiao Zhu[a,b], Valeriia Zaremska[a,c], Chiara D'Onofrio[a,c], Wolfgang Knoll[a,d], Paolo Pelosi[a,*]

[a]AIT Austrian Institute of Technology GmbH, Biosensor Technologies, Tulln, Austria
[b]Faculty of Biology, Institute of Molecular Physiology, Johannes Gutenberg-Universität, Mainz, Germany
[c]Department of Nanobiotechnology, University of Natural Resources and Life Sciences, Vienna, Austria
[d]CEST Competence Center for Electrochemical Surface Technology, Tulln, Austria
*Corresponding author: e-mail address: ppelosi.obp@gmail.com

Contents

Abstract

Modifying the affinity of odorant-binding proteins (OBPs) to small ligands by replacement of specific residues in the binding pocket may lead to several technological applications. Thanks to their compact and stable structures, OBPs are currently regarded as the best candidates to be used in biosensing elements for odorants and volatiles detection. The wide and rich information on the structure of these proteins both in their apo-forms and in complexes with specific ligands provides guidelines to design reliable mutants to

Methods in Enzymology, Volume 642
ISSN 0076-6879
https://doi.org/10.1016/bs.mie.2020.05.014

monitor specific targets. The same engineered proteins may also find applications in the slow release of pheromones and other chemicals in the environment, as well as in the fine purification of drugs, including the resolution of racemates. Apart from such useful applications, site-directed mutagenesis represents an interesting approach to dissect the specific interactions between small chemicals and amino acid residues in the binding pocket. These studies can lead to design of better ligands, such as pheromone analogues with desired physico-chemical characteristics. In this chapter we examine the different uses of mutagenesis applied to OBPs and report a couple of protocols that have been successful in our hands.

Abbreviations

OBP odorant-binding protein
GOBP general odorant-binding protein
CSP Chemosensory protein
GC/MS gas-chromatography/mass spectrometry
LB Luria-Bertani medium

1. Introduction

Odorant-binding proteins (OBPs) of both vertebrates and insects, as well as chemosensory proteins (CSPs), are the object of increasing attention both in basic science and applied research (Mulla et al., 2015; Pelosi, Iovinella, Felicioli, & Dani, 2014; Pelosi, Iovinella, Zhu, Wang, & Dani, 2018; Pelosi, Mastrogiacomo, Iovinella, Tuccori, & Persaud, 2014; Pelosi, Zhu, & Knoll, 2018a,b; Sun, Xiao, & Carlson, 2018).

On one side, there is an active interest to improve our knowledge on the perception of odors and on the chemical communication between living organism. In this respect, OBPs and other olfactory binding proteins are easier to study both structurally and functionally than olfactory receptors, and can provide information on the chemical signals exchanged between individuals of the same species or used by living organisms to monitor the environment.

On the other hand, the small, compact and stable structure of OBPs (see Chapters "Structure of odorant binding proteins and chemosensory proteins determined by X-ray crystallography" by Cambillau; "Solution structure of insect CSP and OBPs by NMR" by Picone), together with their easy and inexpensive synthesis (see Chapter "Bacterial expression and purification of vertebrate odorant-binding proteins" by Briand), makes them very attractive candidates as biorecognition elements in biosensors for odor monitoring, food quality evaluation or clinical analysis (see Chapters "Electrochemical biosensing with odorant binding proteins" by Szunerits; "Electronic biosensors, based on graphene FETs" by Knoll; "Electronic biosensors based on EGOFETs" by Mulla; "Gravimetric biosensors" by Persaud; "Plasmonic biosensors relying on biomolecular conformational changes: Case of odorant binding proteins" by Dostalek).

Basic olfactory research, as well as technological applications dealing with monitoring and delivering of odors, require a solid base of detailed information on the structure and function of OBPs. In this respect, site-directed mutagenesis can provide the necessary tools to better understand the physiological roles of OBPs, as well as for designing proteins tailored to specific requirements in the biosensor field or for other biotechnological uses (Pelosi, Iovinella, et al., 2014; Pelosi, Mastrogiacomo, et al., 2014).

Thanks to their small and compact structure, OBPs can be easily engineered with the purpose of modifying their ligand-binding characteristics. In fact, their compact folding, often stabilized by disulfide bridges (particularly in the case of insect OBPs) makes the design of mutants relatively easy and reliable, as the overall scaffolding of the protein is not likely to be affected by single amino acid substitutions (see Chapters "Structure of odorant binding proteins and chemosensory proteins determined by X-ray crystallography" by Cambillau; "Solution structure of insect CSP and OBPs by NMR" by Picone). Moreover, the presence of a single binding site, that excludes any allosteric effect, is a great advantage when trying to predict the effect of a mutation on the affinity to specific ligands. Finally, the large number of known structures (several dozens, most of them of insect OBPs, see Chapters "Structure of odorant binding proteins and chemosensory proteins determined by X-ray crystallography" by Cambillau; "Solution structure of insect CSP and OBPs by NMR" by Picone), both as apo-proteins and in complexes with their ligands, provides an excellent data base to model the folding of new OBPs, as well as to predict their interactions with ligands.

2. Applications of site-directed mutagenesis

There are several motivations for engineering natural OBPs at specific sites through mutagenesis. The following are a few examples taken from the literature where targeted mutations have been introduced into OBPs with the aim of solving specific problems or understanding some observed behaviors of these proteins.

2.1 Suggesting structures of pheromones in the absence of biological samples

Reverse chemical ecology refers to a recently developed strategy to study chemical communication between living organisms applying a reverse strategy compared to the traditional approach. Classical chemical ecology starts with the identification of pheromones and other semiochemicals, that are subsequently used in biochemical, physiological and behavioral studies. On the basis of the results obtained, then, techniques of biochemistry and molecular biology are applied to eventually identify the proteins mediating the observed phenomena. In reverse chemical ecology, instead, we start with the sequences and structures of binding proteins and receptors to formulate hypotheses on the nature of the pheromones and other physiological ligands (Choo et al., 2018; Leal, 2017; Zhu et al., 2017). This approach is particularly useful when dealing with extinct or endangered species, difficulties in collecting samples, due for instance to limited time windows, fast degradation of the semiochemicals or complications in approaching wild species.

In such cases, reverse chemical ecology can be applied, starting with the sequence of the binding protein (for example, an OBP), which is found in the genomes, available even for several extinct species (see Chapter "Reconstructing protein coding sequences from ancient DNA" by Hofreiter). The protein is then expressed in a heterologous system (OBPs can be produced in large amounts in bacteria, see also Chapter "Bacterial expression and purification of vertebrate odorant-binding proteins" by Briand) and used in binding assays to formulate hypotheses on the best ligands. In such problems, the possibility of preparing specific mutants where single amino acid residues are replaced in the binding pocket allows to finely probe the mode of binding of ligands and eventually suggest the structure of the unknown pheromones. A successful example is provided by the study on the OBPs of the giant panda, which suggested some long-chain linear aldehydes and acids as putative components of the sex pheromone blend for this species (Zhu et al., 2017). In particular, the OBP3 of the giant panda binds some unsaturated linear aldehydes of 14 and 16 carbon atoms, together with some plant terpenoids. Replacement of Asn90 with Leu completely abolished affinity to linear aldehydes, but did not affect binding of terpenoids, thus indicating a strict specificity for the fitting of linear aldehydes into the binding cavity of OBP3 (Zhu et al., 2017).

2.2 Supporting a ligand-binding mechanism

Replacement of single residues in the binding cavity may confirm a model of interactions between a ligand and an OBP suggested by docking simulation. In fact, it has been observed that different ligands can fit inside the same cavity assuming different orientations and therefore interacting with different residues. Molecules of odorants and pheromones, being hydrophobic and volatile, usually contain a single polar group, which likely establishes the strongest interaction (typically a H-bond) with a suitable group of a specific amino acid. Replacing such amino acid with another of different nature would disrupt binding and reduce the affinity of the ligand for the protein, if the model was correct. There are a few examples in the literature where single residue substitutions drastically reduced or even abolished affinity of specific ligands for an OBP, the above mentioned case of the panda OBP3 being one of them. In a clear and elegant study, site-specific mutagenesis showed that a single lysine was essential for the binding of nonanal to the human OBP. In fact, replacing Lys112 with Ala drastically reduced affinity of the protein to nonanal and, to a minor extent, to homologous aldehydes (Tcatchoff, Nespoulous, Pernollet, & Briand, 2006). In a similar study, replacing a lysine with methionine in the C-terminal segment of OBP7 of the moth *Helicoverpa armigera* reduced the affinity of this protein for linear long-chain aldehydes, components of the specie's sex pheromone (Sun, Huang, Pelosi, & Wang, 2013). In both cases, it was suggested that the ligand aldehydes could form a Schiff base with the lysine inside the binding pocket. The GOBP2 of the moth *Plutella xylostella* binds the moth's sex pheromone main component (Z)11-hexadecenal with good affinity. Several mutants, where two or three aromatic residues were replaced with leucine, failed to bind this compound as well as some structurally related aldehydes (Zhu et al., 2016).

2.3 Modifying the binding spectrum

Ligand-binding studies sometimes reveal that an OBP or a CSP can bind with similar affinities ligands belonging to different chemical classes. The panda OBP3 again provides a suitable example, binding long-chain linear aldehydes (sex pheromones for some insects and putative pheromone components for the panda), as well as terpenoids and derivatives, characterized by branched chains (Zhu et al., 2017). Such apparently strange behavior is justified by the assumption that ligands of different chemical classes may bind to the same cavity, but with different orientations. In fact, often the binding pockets of these proteins are relatively large compared with the size of their ligands, so that small chemicals can be accommodated using alternative positions. Moreover, the structures of vertebrate OBPs and those of CSPs are rather flexible, compared to the more compact folding of insect OBPs, constrained by three interlocked disulfide bridges (Leal, Nikonova, & Peng, 1999; Scaloni, Monti, Angeli, & Pelosi, 1999). Consequently, they can easily adapt to accept ligands of different shapes. In the case of the panda OBP3 a single substitution in the binding pocket, replacing Asn90 with leucine, drastically reduced the affinity of linear unsaturated aldehydes for the protein, but retained the binding of terpenoids (Zhu et al., 2017). These results, besides supporting the idea that ligands of the two classes bind in different positions of the same pocket, also provide a rationale for narrowing the specificity of an OBP or CSP. Such strategy represents a useful approach to produce binding proteins with improved specificity for uses in biosensing elements or other biotechnological applications.

2.4 Improving the specificity of OBPs

Within the same strategy described above, we can consider situations where, rather than changing the target of a binding protein, we want to narrowly focus on its specificity. This could be the case when designing biosensors for specific volatiles, such as indicators of spoiled food, markers for certain diseases or specific chemicals to be detected, such as drugs, explosives or pollutants. Given the diversity of binding proteins and their ligands, general guidelines cannot be provided, and solutions are usually found in a trial and error approach. Certainly, detailed knowledge of the three-dimensional structure of the protein and the interactions of ligands with residues inside the binding pocket represent a solid base of information for the design of targeted mutants. An example, taken from our recent experience, can better illustrate these concepts. The aim of this study was to elucidate the mechanism underlying chiral discrimination in binding proteins. We first observed that the mutant of pig OBP1, bearing the amino acid substitution F88W, and described below (Wei, Brandazza, & Pelosi, 2008), binds with different affinities the two enantiomers of carvone. We also noticed that the isopropenyl group of carvone closely interacts with two isoleucine residues (Ile42 and Ile100) in the binding pocket, and speculated that the asymmetric side chain of this amino acid could be involved in chiral discrimination.

This hypothesis was checked by preparing mutants where either of the two isoleucines or both of them were replaced by leucines. In fact, substitution of Ile100 abolished discrimination between the two enantiomers, while replacement of Ile42 only reduced the effect (Zaremska, Tan, Lim, Knoll, & Pelosi, 2020). In this case, a detailed information on the X-ray structure of the protein complexed with some ligands (Vincent et al., 2000) greatly helped in formulating the correct hypothesis. The results of this study suggest that we could endow an OBP with the capability of discriminating between enantiomers of ligands by introducing isoleucine residues inside the binding pocket. Such engineered proteins might find applications in the fabrication of highly selective biosensors as well as in the derivatization of resins for chiral analysis or resolution of racemates in the pharmaceutical industry.

2.5 Introducing a tryptophan as a fluorescent reporter

Binding of small ligands to OBPs and CSPs is generally measured using a fluorescent reporter added to the protein and recording the decrease of fluorescence caused by a competitor ligand (see Chapter "Ligand-binding assays with OBPs and CSPs" by Pelosi). In some instances, however, it might be desirable to record the intrinsic fluorescence of the protein and its variations consequent to ligand binding. This could be the case when using OBPs as biosensing elements in optical devices: avoiding an external fluorescent probe would afford a direct monitoring of the binding and produce better manageable systems. Measuring the intrinsic fluorescence of a binding protein may also be the alternative with proteins that do not apparently bind the commonly adopted fluorescent probes (Pelosi, Zhu, & Knoll, 2018a).

The intrinsic fluorescence of a protein is produced by its tryptophan residues. Tyrosine also emit fluorescence, but their contribution is almost negligible and of no use for practical applications. Both insect OBPs and CSPs often present a tryptophan residue inside their binding pockets, and in a number of cases such fortunate coincidence has been used to monitor the entrance of ligands in the binding cavity (Pelosi, Zhu, & Knoll, 2018a). Vertebrate OBPs also contain a conserved tryptophan, part of the lipocalin signature (Flower, North, & Sansom, 2000); however, this residue is located too far away from the binding pocket to be of practical use. There are also members of the insect OBP and CSP families which do not contain a tryptophan in the binding cavity or even do not contain such residue at all.

Site-directed mutagenesis, therefore, can be applied to all such cases to introduce a tryptophan, which could act as a fluorescent reporter from the binding cavity. A successful example is provided by the pigOBP1, which was modified by replacing a residue of phenylalanine in position 88, located right inside the binding pocket, with a tryptophan (Wei et al., 2008). The intrinsic fluorescence of the mutated protein was efficiently quenched by polyaromatic hydrocarbons, thus providing a tool to be used in optical biosensing of these pollutant compounds. Moreover, the mutant showed improved affinity to large aromatic compounds and better specificity, being able to discriminate between the enantiomers of carvone (Mulla et al., 2015; Wei et al., 2008; Zaremska et al., 2020).

2.6 Adding a group for easy attachment to biosensors

The easiest way of adding an anchor for attachment of a recombinant protein to the surface of a biosensor is to express the protein with a His-tag at its N-terminus or C-terminus. This can be done by simply inserting the gene into the appropriate plasmid containing the sequence encoding the His-tag. This approach has been adopted for anchoring the honeybee OBP14 to the graphene surface of a field-effect transistor and record responses to volatile compounds (Kotlowski et al., 2018).

Another alternative is to add a cysteine residue, again at the N-terminus or at the C-terminus. Usually, in both OBPs and CSPs all cysteines are paired into disulfide bridges, therefore this additional cysteine will be in its free reduced form and ready to establish a covalent bond with a gold surface. Alternatively, the free sulfhydryl group can be reacted with phenylene diisothiocyanate, whose second isothiocyanate group is used for attachment to a derivatized surface. In exceptional cases, OBPs can present an odd number of cysteines, and therefore attention should be paid that the additional terminal cysteine should not pair with the odd cysteine already present in the sequence. When introducing an additional cysteine, this residue could be the last amino acid of the sequence, but cannot be the first, as the initial methionine is required for starting the expression of the protein. Therefore it should be inserted as the second residue.

Whether a His-tag or a cysteine are used for attachment, these modifications cannot be put at the N- or C-terminus indiscriminately. It has been shown that in insect OBPs the C-terminus segment can interfere with binding in different ways, depending on its length. With some OBPs, the terminal segment might in appropriate conditions enter the binding site or cover the entrance to the pocket, thus preventing or affecting the entrance of a ligand (Damberger et al., 2000; Tegoni, Campanacci, & Cambillau, 2004; see also Chapters "Structure of odorant binding proteins and chemosensory proteins determined by X-ray crystallography" by Cambillau; "Solution structure of insect CSP and OBPs by NMR" by Picone).

3. Feasibility of site-specific mutagenesis on OBPs

Odorant-binding proteins are particularly suitable to be engineered by site-specific mutagenesis thanks to some specific properties:

1. OBPs are stable, compact proteins with a conserved three-dimensional folding across widely divergent sequences (Pelosi, Iovinella, et al., 2018; Pelosi, Zhou, Ban, & Calvello, 2006). This applies both to vertebrate and insect OBPs, which, despite their common name, constitute two families of proteins completely different in structure. In fact, vertebrate OBPs are part of the lipocalin superfamily, that comprises proteins sharing down to 10% of their amino acids and endowed with a variety of different functions, but sharing a common three-dimensional folding (Flower et al., 2000). Many different roles have also been described for insect OBPs (Pelosi, Iovinella, et al., 2018), that also display high sequence divergence across species, again with percent of identical amino acids as low as 10%. In particular, the structure of insect OBPs is further stabilized by three interlocked disulfide bridges, that prevent the scaffolding of the protein from collapsing, even after threatening amino acid substitutions. Such exceptional stability, that can be verified by the fact that both vertebrate and insect OBPs can withstand prolonged boiling (Paolini, Tanfani, Fini, Bertoli, & Pelosi, 1999; Schwaighofer et al., 2014; see also Chapter "Stability of OBPs" by Hellmann), suggest that introducing mutations, particularly in their binding pocket, is not likely to affect the overall shape and assembly of these proteins.
2. A large number of three-dimensional structures have been solved for both classes of OBPs (Pelosi, Iovinella, et al., 2018; Tegoni et al., 2004; see also Chapters "Structure of odorant binding proteins and chemosensory proteins determined by X-ray crystallography" by Cambillau; "Solution structure of insect CSP and OBPs by NMR" by Delia Picone). This wide and solid basis of information allows confident modeling of other members of these families. A reliable model, in fact, when the structure is not available, is the first requirement for designing specific mutants.

3. Their simple structure with a single well-defined binding pocket allows to reasonably predict the effects of mutagenesis on their binding properties. Moreover, being OBPs easy to purify and thanks to simple and fast binding assay protocols (Pelosi, Zhu, & Knoll, 2018a; see also Chapter "Ligand-binding assays with OBPs and CSPs" by Pelosi), a relatively large number of mutants can be prepared and tested in order to find the ideal one in a sort of trial and error approach.

4. Strategies for site-specific mutagenesis with OBPs

There are a number of protocols for introducing mutations into a DNA sequence, but all of them are variations of a common strategy (Ke & Madison, 1997; Kirsch & Joly, 1998; Kretz et al., 2004; Liu & Naismith, 2008; Smith & Murrell, 2011). We always start with a construct, where the wild-type sequence is already inserted into an expression plasmid, for instance a pET vector, and perform PCR amplifications with primers bearing the desired mutation. Fig. 1 summarizes three typical cases, that can be commonly encountered when introducing specific mutations in a gene.

In any case, a common protocol can be adopted, along with the following steps:

1. A first PCR is performed with a primer bearing the desired substitution and another primer complementary either to another region of the gene sequence or to a region of the vector. For instance, we can use T7 as forward primer and a specific oligonucleotide containing the mutation as reverse primer. It is advisable that the resulting amplification product be neither too short, nor too long, as this double-strand DNA will be

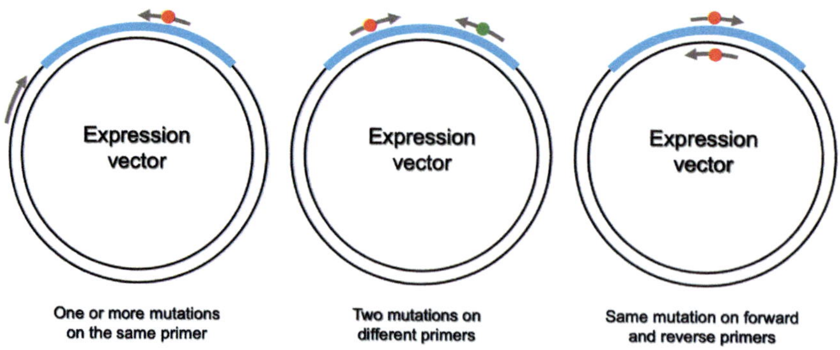

Fig. 1 Examples of strategies for introducing one or more mutations in a gene.

used in a second PCR as primer to amplify the all plasmid. A length of 100–200 bp is usually suitable and always gave good results in our hands.

2. The PCR product is separated on a 1% agarose gel and the amplification band is cut and purified using a standard kit. This step could be avoided, and the second PCR can be performed directly in the same tube without any separation. However, checking and purifying the first product (a) provides information on the success of the first PCR and (b) gives better control on the composition of the second PCR mix.
3. A second PCR is performed on the same plasmid containing the wild-type sequence, using the double-strand product of the first PCR as both forward and reverse primers. Due to the unusual length of this primer, the annealing temperature in this step can coincide with the elongation temperature. During this phase, the entire plasmid is amplified. It is necessary, therefore, to use for both PCR steps a polymerase that does not add overhanging bases, in order to close the circular plasmid. Phusion, pfu and similar products can be adopted, but not normal Taq.
4. Instead of using two primers in different regions of the sequence, forward and reverse primers can be designed on the same segment, both containing the mutation, and can partially or totally overlap. Along with this approach, in a single step the entire plasmid is amplified. This strategy proves particularly suitable when introducing or deleting a segment in the middle of a sequence (Liu & Naismith, 2008).
5. The product of the second PCR is digested with the enzyme *Dpn*I, which cuts methylated DNA (therefore the original plasmid, obtained by bacterial growth) but not the modified plasmid synthesized *in-vitro*. This step is essential to prevent the wild-type plasmid interfering with the transformation. In fact, the yield of amplification is quite low and there is so little of the amplified plasmid that usually cannot be seen on a gel. Even small amounts of the wild-type plasmid, therefore, can seriously compete with the amplified product in transforming the cells.
6. The crude digestion product is used to transform bacterial cells, such as XL-1, DH-5α or similar, that are then plated and incubated overnight. Individual colonies are then grown, and plasmids are extracted and sequenced. There is no alternative to sequencing for checking whether the plasmids contain the desired mutation. Generally, out of three colonies, one or two contain the mutation. The results strongly depend on the efficiency of the digestion with *Dpn*I.
7. The positive plasmids are ready for transformation into BL-21 cells or similar and protein expression.

5. Specific approaches

Depending on how many residues we want to replace and where they are located in the sequence, we can adopt different variations of the general strategy above described.

1. Single mutation. Both protocols described above can be applied, either the one using a primer bearing the mutation and the other in a different region of the sequence, or that with forward and reverse primers overlapping, both containing the mutation.
2. Few mutations very close to one another. Again, the same protocol can be applied. In this case, the primer should contain all the mutations we want to introduce, keeping in mind that there is a limit on the number of mismatches before the annealing temperature becomes too low. It is a good rule not to use annealing temperatures lower than 50 degrees.
3. Two or more mutations located in two different regions of the sequence. In such cases, both the forward primer and the reverse primer can include one mutation each or even additional replacements, along with the recommendations suggested above.
4. Deleting the C-terminal segment. This can be easily accomplished by inserting a stop codon in the desired position. Accordingly, the reverse primer can be designed for introducing the stop codon, while a T7 or any other oligonucleotide designed in the middle of the gene can be used as forward primer.
5. Deleting the N-terminal segment. In this case, a methionine can be inserted at the beginning of the desired sequence, but at the same time the starting methionine of the wild type should be replaced with another amino acid. If the segment to be deleted is very short, then both mutations can be inserted in the same primer, otherwise the forward primer will contain a mutation for replacing the initial methionine, while the reverse primer will introduce the new one. Attention should be paid to the fact that the first PCR product is not too short, because this could introduce problems in the gel purification (in case such approach is preferred) or in a melting temperature which is too low and is likely to generate unwanted annealing. An alternative way would be to amplify the desired fragment with appropriate primers and then clone the product into a vector.
6. Finally, when multiple substitutions have to be introduced and none of the above approaches seems suitable or practical, a new gene could be

synthesized and cloned into an expression vector. This alternative is now rather inexpensive and practical, as gene synthesis is reliable and free from errors, particularly with relatively short sequences, as those coding OBPs (400–500 bp) and CSPs (300–400 bp).

A large number of papers report general protocols for introducing single or multiple mutations into a gene sequence in simple and reliable ways.

Specifically, site-directed mutagenesis has been applied to several OBPs with success, following different variations of the same common strategy (Ramoni et al., 2002; Rusconi et al., 2012; Spinelli et al., 2012; Sun et al., 2013; Tcatchoff et al., 2006; Wang et al., 2013; Wei et al., 2008; Yin et al., 2015; Zhu et al., 2017, 2016). General techniques can be adopted for the mutagenesis of OBPs. However, when designing the mutations to be introduced, OBPs present particular advantages, due to their small size, compact structure and conserved folding, as discussed above. Besides, the detailed structural information available on the interactions between ligands and amino acid residues in the binding pocket provides a great help in identifying the key residues to be replaced for specific requirements.

Here, just as typical examples, we report two specific protocols that have been successful in our hands, the first for a mammalian OBP, the second for an insect OBP.

6. Site-directed mutagenesis on a mammalian OBP

The pigOBP1 (Dal Monte, Centini, Anselmi, & Pelosi, 1993; Paolini et al., 1998; Vincent et al., 2000; Zaremska et al., 2020) binds the two enantiomers of menthol with different affinities, but cannot discriminate between the two carvones. Introducing a tryptophan in the binding site (Wei et al., 2008), chiral discrimination is extended to the enantiomeric carvones as well (Mulla et al., 2015). Looking at the three-dimensional structure of the protein, we noticed the presence of two isoleucine residues in the binding pocket, opposite to the introduced tryptophan and likely to interact with the hydrophobic regions of menthol and carvone. To investigate the effect of these two residues on chiral discrimination, we replaced both isoleucines with leucines with the effect of losing chiral discrimination for both pairs of ligands (Zaremska et al., 2020) (Fig. 2).

The two isoleucines that have been replaced are located in positions 42 and 100 of the amino acid sequence. Being relatively far from each other, we prepared two primers, a forward one with the mutation at

QEPQPEQDPFELSGKWITSYIGSSDLEKIGENAPFQVFMRSIEFDDKESKVYL
NFFSKENGICEEFSLIGTKQEGNTYDVNYAGNNKWVVSYASETALIISNINVD
EEGDKTIMTGLLGKGTDIEDQDLEKFKEVTRENGIPEENIVNIIERDDCPAK

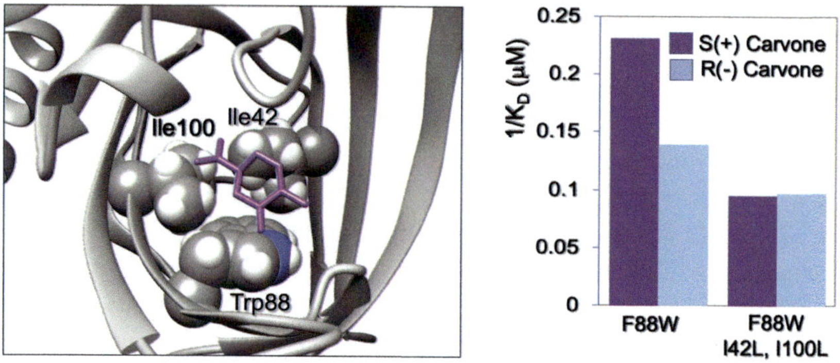

Fig. 2 Two isoleucine residues in the binding pocket of the pigOBP1-F88W are responsible for the property of this protein to differentiate between the enantiomers of carvone and menthol. Their replacement with leucines abolishes chiral discrimination.

position 42, the reverse with a mutation at position 100 of the mature protein sequence:

Forward primer: ATGCGTAGCCTTGAATTTGATG.

The underlined triplet codes for leucine and replaces the original ATT codon of isoleucine.

Reverse primer: GATCTTAGAGAGTATCAGGGCAG.

The reverse-complement of the underlined triplet codes for leucine and replaces the original ATC codon of isoleucine.

6.1 Equipment

1. Thermocycler
2. Bench centrifuge
3. System for agarose gel electrophoresis
4. Water bath

6.2 Reagents

1. Plasmid containing the gene to be mutated
2. Phusion polymerase (NEB)
3. 10 mM dNPTs mix
4. *Dpn*I restriction enzyme (NEB)
5. LB medium
6. Antibiotic
7. Agarose
8. Kit for plasmid extraction

6.3 Protocol

6.3.1 *First PCR*

In a first PCR a segment of pigOBP1 is amplified using the above primers, each containing one of the two mutations to be introduced.

Mixture		Temperature cycles
Water	26 µL	95 °C (5′)
5 × Buffer	10	
10 mM dNPTs mix	2	95 °C (30″)
Template (pET5b/pigOBP1-m2)	1	50 °C (30″) 35 cycles
Primer forward 10 µM	5	72 °C (1′)
Primer reverse 10 µM	5	
Phusion DNA polymerase (NBE)	1	72 °C (10′)
Total	50	

6.3.2 *Second PCR*

The same tube was left in the thermocycler and subjected directly to a second PCR, according to the following temperature cycle. During this step, the whole plasmid is amplified.

98 °C (5′)	
98 °C (1′) 72 °C (7′) 72 °C (10′)	20 cycles

6.3.3 *Digestion with* DpnI

The crude PCR product is digested with the enzyme *Dpn*I without further purification.

PCR product 17 µL
10 × Buffer 2 µL
*Dpn*I (NEB) 1 µL
Incubated at 37 °C for 2 h.

6.3.4 *Transformation of DH5α* E. coli *cells*

An aliquot of the digestion product (2 μL) is used to transform DH5α *E. coli* cells. The transformation product is plated on LB/agar/antibiotic and the plate left overnight at 37 °C. We expect to find a reasonable number of colonies (10–50), but not too many. A plate full of colonies suggests that the *Dpn*I enzyme did not succeed in cutting all the original (biologically amplified, therefore methylated) plasmid, which then transformed the cells with better efficiency than the mutated plasmid.

6.3.5 *Analysis of colonies*

Three or more colonies are grown in LB/antibiotic, plasmids are extracted using a commercial mini-prep kit and sequenced. In our experiment, four of the six plasmids sequenced, contained both mutations (Zaremska et al., 2020). There is no better way to check for mutations than sequencing.

7. Site-directed mutagenesis on an insect OBP

In this work we managed to switch the affinity of an insect OBP from sex pheromones (linear unsaturated aldehydes) to terpenoids and plant volatiles, by replacing two residues in the binding pocket.

Insect OBPs, in particular those of Lepidoptera, can be clearly distinguished, on the basis of their sequences, into three main sub-classes: PBPs (pheromone-binding proteins), GOBPs (general-odorants binding proteins) and OBPs. The identification of PBPs as a group of OBPs tuned to sex pheromones is supported by functional and physiological studies (Pelosi et al., 2006; Vogt, Prestwich, & Lerner, 1991; Vogt & Riddiford, 1981; Vogt, Rybczynski, & Lerner, 1991). GOBPs, instead, first suggested to be binding proteins for general odorant (Vogt, Prestwich, & Lerner, 1991; Vogt, Rybczynski, & Lerner, 1991), were later recognized as a second group of pheromone-binding proteins, although some of them can also bind plant natural odorants (Vogt, Prestwich, & Lerner, 1991; Vogt, Rybczynski, & Lerner, 1991; Zhou et al., 2009).

In our work (Zhu et al., 2016), we found that the GOBP2 of the backdiamond moth *Plutella xylostella* is tuned to the species sex pheromones (*Z*)11-hexadecenal and structurally similar aldehydes, while it shows poor affinities to plant volatiles and other terpenoids. Comparing the sequences of the three PBPs, the two GOBPs and a few representative OBPs of *P. xylostella* (Fig. 3) we noticed the presence of aromatic residues in both PBPs and GOBPs, that were replaced by aliphatic nonpolar residues in the

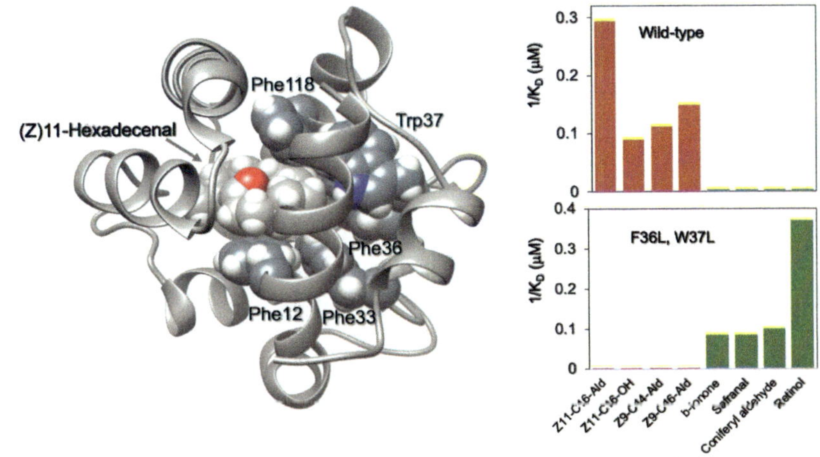


```
Pxy1GOBP2_ABY71035          TAEVMSHVTAHFGKTLEECREESGLSGEIMEEFHHFWREDFEVVHRELGCAIICMSNKFQLMQDDARMHH
Pxy1GOBP1_ABY71034          TVEVMKDVTLGFGEALEQCREQSQLTEEMMEEFYHFWREDFKFEARAVGCAIHCMSRYFNLLGEQQRMHH
Pxy1PBP2_AGH13203           SADIMQKLTIGFTEALEHCKKDLQISNEVMQDFYNFWREDYALVNREMGCVLMCMAARLDLVTEDLKMHH
Pxy1PBP1_BAG71422           SADVMKGLSENFGKALGDCKKELDLPDSIMTEFYNFWKDDYVLSDRSTGCAIICLSSKLDLLDPDGNLHH
Pxy1PBP3_ACI28451           SKETMKDITSGFFKVLNECKHELNLPDHLVGDFYHYWRQEYALLDRDLGCAILCMSRKLELIDASGKLHH
Pxy1_ANG08530                   MTRAQLKKTLSMPKKQCMTKHGVTEDMVGGIEGGKFVEDRKVMCYIACIYQTIQVMKNDRLNKD
Pxy1_ANG08535                     LTDEQKEKLKKHKTECLAETKPEVEHVDKLKNGDYTTENEALKK---YAHCMMIKSELMTKDG----
Pxy1_AMR99731           LSPPLLFESLKSEPTISTLYTTDKKADKITNECLVEMYPRNLYNYPLNIDKNDVPCILHCVLKKFGIMSNDGKIII
Pxy1_ANG08534            SSVSFHDGNDEHAEDDMMGIMTHCNETFRFDPAYWQSLNESGTFPDEN-DKNPKCFIRCVLEGTRVASLDSVFDA

Pxy1GOBP2_ABY71035      ENMHDYIKSFPKGDLLSETMVR--LIHNCEKKYDDIDDECSRVVKTAACFKKDAQAEGIAPELTMIEAVLEKY
Pxy1GOBP1_ABY71034      DNTHKFIQSFPNGEVLSHQMVG--IIHTCEQQHDAETDHCWRILRVAECFKRESQAQGLAPSMEMLMAEFIMEADV
Pxy1PBP2_AGH13203       GNAHEFAKKHGADDTMAKQLVT--IIHECEQGAASVPDECARTLEMAKCFKTRIHELKWAPSMEVVLEEVITELKN
Pxy1PBP1_BAG71422       GNAKDFALKHGADEGMAGQLVG--MIHECEKAAPDNPDACLKVLDIANCFKKKIHELKWAPSMDVVVAEVLADV
Pxy1PBP3_ACI28451       GNTQEFAEKHGADNSMASKLVE--VLHSCEKQHEAVSDECQRALEVAKCFRSSVHELGWAPTIDVIIEEVLTDM
Pxy1_ANG08530           LVIKQVDLLYP--PELKAPTKA--AVKQCMPVQDKYPDMCEGVFWSVKCLYEQDPASFVFP
Pxy1_ANG08535           KFRKDVALAKVPNPADKPMVEK--LIDTCLANKGDTP--QQTAWNYVKCYHEKDPKHAIFLKLG
Pxy1_AMR99731           KNYYRRVQQIHRYDPRILISDVGDTCAQKIKSMDLSRDVCKKAKVFNECTQLFAVSLKDSGR
Pxy1_ANG08534           ARAAEVFAGERGGRPMDDLESL---ARNCADDR-RETCKCERAYGFMKCLMEAEIEKYEVYKADDS
```

Fig. 3 Site-directed mutagenesis, applied to the GOBP2 of the backdiamond moth *Plutella xylostella*, switches the affinity of this protein from linear aldehydes to branched isoprenoids. The cartoon of *P. xylostella* GOBP2 shows in space-filling mode the six aromatic residues conserved in the three PBPs and the two GOBPs of this species. The same amino acids are highlighted in blue in the alignment reported in the lower part of the figure. It also shows a molecule of the main pheromone component (*Z*)11-hexadecenal, sandwiched between two aromatic residues. Replacement of Phe36 and Trp37 with leucines switches the affinity of the protein from linear aldehydes to branched terpenoids (upper right graphs).

other OBPs. Reasoning that linear chain unsaturated aldehydes can better interact with planar aromatic residues, while terpenoids could fit better in a region dominated by branched residues (such as valine, isoleucine and leucine), we decided to replace one, two or more aromatic residues with leucine.

As a result, we observed a much-reduced affinity of our mutants to sex pheromones, while they could strongly bind terpenes and plant-derived volatiles. An example is shown in Fig. 3, where a double mutant showed a completely reversed spectrum of binding as compared to the wild type.

The protocol adopted is very similar to the one reported above, except for the introduction of a purification step between the two PCR amplifications. Here we report the details on the preparation of one of the five mutants described in the original paper (Zhu et al., 2016), in particular the replacement of both Phe36 and Trp37 with leucine residues. To accomplish this task, we designed a single oligonucleotide bearing both mutations. This was used as a reverse primer together with T7 as forward primer:

Forward primer T7: TAATACGACTCACTATAGGG.

Reverse primer: GAAGTCTTCACGCAAGAGGTGGTGGAA CTCCTC.

The reverse-complements of the two underlined triplets code both for leucine and replace the original codons for Phe36 and Trp37.

The required equipment (7.1) and reagents (7.2) are the same as those listed for the previous protocol (6.1 and 6.2).

7.1 Protocol

7.1.1 First PCR

Amplification of a segment of PxylGOBP2 by PCR.

Mixture		Temperature cycles
Water	34 μL	95 °C (3′)
5 × Buffer	10	
10 mM dNPTs mix	2	95 °C (30″)
Template (pET30/PxylGOBP2)	1	50 °C (30″) 35 cycles
Primer forward 10 μM	1	72 °C (1′)
Primer reverse 10 μM	1	
Pfu DNA polymerase (Promega)	1	72 °C (6′)
Total	50	

7.1.2 Purification of the PCR product

The entire PCR product is loaded onto a 1% agarose gel and separated by electrophoresis. The amplified band is excised from the gel and purified using a commercial kit.

7.1.3 *Second PCR*

A second PCR is performed using fresh reagents and the purified double-strand DNA from the first PCR as a pair of primers. During this step, the whole plasmid is amplified.

Mixture		Temperature cycles
Water	2 μL	95 °C (30″) 9 cycles
5 × Buffer	4	68 °C (6′)
10 mM dNPTs mix	2	
Template (pET30/PxylGOBP2)	1	68 °C (16′)
1st PCR product	10	
Pfu DNA polymerase (Promega)	1	
Total	20	

7.1.4 *Digestion with* Dpn*I*

As in Section 6.3.3.

7.1.5 *Transformation of DH5α* E. coli *cells*

As in Section 6.3.4.

7.1.6 *Analysis of colonies*

As in Section 6.3.5.

8. Analysis of results

Introducing a mutation into a gene of an OBP or CSP represents only the first (and perhaps the easiest) step toward answering the question that prompted the use of site-directed mutagenesis. The next step is the bacterial expression of the protein. Although, in most cases the mutants are successfully expressed and in yields as good as with the wild type, in rare instances the mutation can affect the expression of the protein for different reasons difficult to analyze. In such cases, the first suggestion is to use a codon optimized for bacteria, but this may not be enough to solve the problem. Other strains of bacterial cells can be tried to improve the yield, but such approach is usually inefficient if no expression is detected in the first place.

Different plasmids can also be tried, in particular those expressing the OBP as a fusion protein. The first and easiest choice is to add a His-tag at the N-terminus. In some cases, this solved the problem, otherwise our OBP or CSP can be expressed as a fusion protein with maltose-binding protein, thioredoxin or other soluble polypeptides (Saez & Vincentelli, 2014; Vincentelli et al., 2011).

The most important step, after a mutation has been successfully introduced in the sequence and the protein has been expressed and purified, is the measurement of its affinity toward the ligands of interest. The results will confirm or disprove the initial hypothesis, suggesting further mutations to be tested. In some cases, it is possible that the mutation may have the effect of completely disrupting the folding of the protein with the effect of a total loss of binding. Such event is not very likely with OBPs and CSPs, due to their compact folding, but may be expected when modifying a cysteine involved in a disulfide bridge or replacing a charged amino acid with a nonpolar residue. It is not possible to provide general rules or suggestions that can be applied with confidence. Although OBPs and CSPs represent well-defined families with conserved three-dimensional folding, nevertheless members within each family may be different from one another, sometimes with identities between sequences of the same species as low as 10%. Besides, even small differences in the sequence of a protein may have a strong impact on the structure and consequently on its ligand-binding properties.

9. Discussion and conclusions

In conclusion, there are several reasons for modifying the performance of an OBP or a CSP by replacing one or few amino acids in the binding pocket or by introducing modifications to endow the protein with other features, such as the addition of a His-tag for directional immobilization onto a biosensor surface.

We have shown that site-directed mutagenesis can be done using simple and standard protocols, widely described in the literature for other proteins. More important, designing mutations and identifying the target residues to be changed is made easier with OBPs and CSPs, thanks to their compact and conserved structure and based on a large amount of information concerning the modes of fitting of ligands inside the binding pockets.

The large number of structures available, most for insect OBPs (see Chapters "Structure of odorant binding proteins and chemosensory proteins determined by X-ray crystallography" by Cambillau; "Solution structure of insect CSP and OBPs by NMR" by Picone), allows the modeling of other OBPs with a certain confidence. On the other hand, analyzing the effects of mutations can provide support for the model proposed.

An important field which can greatly benefit from a wide application of mutagenesis to OBPs is that of biosensors (see Chapters "Electrochemical biosensing with odorant binding proteins" by Szunerits; "Electronic biosensors, based on graphene FETs" by Knoll; "Electronic biosensors based on EGOFETs" by Mulla; "Gravimetric biosensors" by Persaud; "Plasmonic biosensors relying on biomolecular conformational changes: Case of odorant binding proteins" by Dostalek). In recent years, OBPs have been the proteins of choice as biosensing elements in devices for detecting and discriminating odors. So far, most of the studies have been utilizing native OBPs to detect their physiological ligands. This approach certainly represents a limitation on the number of target volatiles to be monitored, while engineering OBPs for improving their specificity to the natural ligand or modifying their spectrum of binding can open a wider range of possible applications. In the future, when enough information on the structural elements of OBPs crucial for binding specific ligands will be available, we might even dream of designing entirely new OBPs. The novel proteins may reproduce the conserved scaffolding of a typical OBP with its disulfide bridges and whatever else is essential to preserve the compact folding of the polypeptide. This basic and essential skeleton can then be furnished with appropriate residues in the binding pocket to create the specific environment to bind a desired ligand. At the same time, the external amino acid residues can be selected according to the environment where the binding protein is supposed to perform its function.

References

Choo, Y.-M., Xu, P., Hwang, J. K., Zeng, F., Tan, K., Bhagavathy, G., et al. (2018). Reverse chemical ecology approach for the identification of an oviposition attractant for *Culex quinquefasciatus*. *Proceedings of the National Academy of Sciences of the United States of America*, *115*(4), 714–719. https://doi.org/10.1073/pnas.1718284115.

Dal Monte, M. D., Centini, M., Anselmi, C., & Pelosi, P. (1993). Binding of selected odorants to bovine and porcine odorant-binding proteins. *Chemical Senses*, *18*(6), 713–721. https://doi.org/10.1093/chemse/18.6.713.

Damberger, F., Nikonova, L., Horst, R., Peng, G., Leal, W. S., & Wüthrich, K. (2000). NMR characterization of a pH-dependent equilibrium between two folded solution conformations of the pheromone-binding protein from *Bombyx mori*. *Protein Science*, *9*(5), 1038–1041. https://doi.org/10.1110/ps.9.5.1038.

Flower, D. R., North, A. C. T., & Sansom, C. E. (2000). The lipocalin protein family: Structural and sequence overview. *Biochimica et Biophysica Acta (BBA)—Protein Structure and Molecular Enzymology*, *1482*(1), 9–24. https://doi.org/10.1016/S0167-4838(00)00148-5.

Ke, S.-H., & Madison, E. L. (1997). Rapid and efficient site-directed mutagenesis by single-tube 'megaprimer' PCR method. *Nucleic Acids Research*, *25*(16), 3371–3372. https://doi.org/10.1093/nar/25.16.3371.

Kirsch, R. D., & Joly, E. (1998). An improved PCR-mutagenesis strategy for two-site mutagenesis or sequence swapping between related genes. *Nucleic Acids Research*, *26*(7), 1848–1850. https://doi.org/10.1093/nar/26.7.1848.

Kotlowski, C., Aspermair, P., Khan, H. U., Reiner-Rozman, C., Breu, J., Szunerits, S., et al. (2018). Electronic biosensing with flexible organic transistor devices. *Flexible and Printed Electronics*, *3*(3), 034003. https://doi.org/10.1088/2058-8585/aad433.

Kretz, K. A., Richardson, T. H., Gray, K. A., Robertson, D. E., Tan, X., & Short, J. M. (2004). Gene site saturation mutagenesis: A comprehensive mutagenesis approach. In D. E. Robertson & J. P. Noel (Eds.) *Vol. 388, Methods in enzymology, Academic Press*. https://doi.org/10.1016/S0076-6879(04)88001-7.

Leal, W. S. (2017). Reverse chemical ecology at the service of conservation biology. *Proceedings of the National Academy of Sciences of the United States of America*, *114*(46), 12094–12096. https://doi.org/10.1073/pnas.1717375114.

Leal, W. S., Nikonova, L., & Peng, G. (1999). Disulfide structure of the pheromone binding protein from the silkworm moth, *Bombyx mori*. *FEBS Letters*, *464*(1–2), 85–90. https://doi.org/10.1016/S0014-5793(99)01683-X.

Liu, H., & Naismith, J. H. (2008). An efficient one-step site-directed deletion, insertion, single and multiple-site plasmid mutagenesis protocol. *BMC Biotechnology*, *8*(1), 91. https://doi.org/10.1186/1472-6750-8-91.

Mulla, M. Y., Tuccori, E., Magliulo, M., Lattanzi, G., Palazzo, G., Persaud, K., et al. (2015). Capacitance-modulated transistor detects odorant binding protein chiral interactions. *Nature Communications*, *6*(1), 1–9. https://doi.org/10.1038/ncomms7010.

Paolini, S., Scaloni, A., Amoresano, A., Marchese, S., Napolitano, E., & Pelosi, P. (1998). Amino acid sequence, post-translational modifications, binding and labelling of porcine odorant-binding protein. *Chemical Senses*, *23*(6), 689–698. https://doi.org/10.1093/chemse/23.6.689.

Paolini, S., Tanfani, F., Fini, C., Bertoli, E., & Pelosi, P. (1999). Porcine odorant-binding protein: Structural stability and ligand affinities measured by Fourier-transform infrared spectroscopy and fluorescence spectroscopy. *Biochimica et Biophysica Acta*, *1431*, 179–188.

Pelosi, P., Iovinella, I., Felicioli, A., & Dani, F. R. (2014). Soluble proteins of chemical communication: An overview across arthropods. *Frontiers in Physiology*, *5*, 320. https://doi.org/10.3389/fphys.2014.00320.

Pelosi, P., Iovinella, I., Zhu, J., Wang, G., & Dani, F. R. (2018). Beyond chemoreception: Diverse tasks of soluble olfactory proteins in insects: Soluble olfactory proteins in insects. *Biological Reviews*, *93*(1), 184–200. https://doi.org/10.1111/brv.12339.

Pelosi, P., Mastrogiacomo, R., Iovinella, I., Tuccori, E., & Persaud, K. C. (2014). Structure and biotechnological applications of odorant-binding proteins. *Applied Microbiology and Biotechnology*, *98*(1), 61–70. https://doi.org/10.1007/s00253-013-5383-y.

Pelosi, P., Zhou, J.-J., Ban, L. P., & Calvello, M. (2006). Soluble proteins in insect chemical communication. *Cellular and Molecular Life Sciences*, *63*(14), 1658–1676. https://doi.org/10.1007/s00018-005-5607-0.

Pelosi, P., Zhu, J., & Knoll, W. (2018a). Odorant-binding proteins as sensing elements for odour monitoring. *Sensors*, *18*(10), 3248. https://doi.org/10.3390/s18103248.

Pelosi, P., Zhu, J., & Knoll, W. (2018b). From radioactive ligands to biosensors: Binding methods with olfactory proteins. *Applied Microbiology and Biotechnology*, *102*(19), 8213–8227. https://doi.org/10.1007/s00253-018-9253-5.

Ramoni, R., Vincent, F., Ashcroft, A. E., Accornero, P., Grolli, S., Valencia, C., et al. (2002). Control of domain swapping in bovine odorant-binding protein. *Biochemical Journal*, *365*(3), 739–748. https://doi.org/10.1042/bj20011631.

Rusconi, B., Maranhao, A. C., Fuhrer, J. P., Krotee, P., Choi, S. H., Grun, F., et al. (2012). Mapping the *Anopheles gambiae* odorant binding protein 1 (AgamOBP1) using modeling techniques, site directed mutagenesis, circular dichroism and ligand binding assays. *Biochimica et Biophysica Acta (BBA)—Proteins and Proteomics, 1824*(8), 947–953. https://doi.org/10.1016/j.bbapap.2012.04.011.

Saez, N. J., & Vincentelli, R. (2014). High-throughput expression screening and purification of recombinant proteins in *E. coli*. *Methods in Molecular Biology, 1091*, 33–53. https://doi.org/10.1007/978-1-62703-691-7_3.

Scaloni, A., Monti, M., Angeli, S., & Pelosi, P. (1999). Structural analysis and disulfide-bridge pairing of two odorant-binding proteins from *Bombyx mori*. *Biochemical and Biophysical Research Communications, 266*(2), 386–391. https://doi.org/10.1006/bbrc.1999.1791.

Schwaighofer, A., Kotlowski, C., Araman, C., Chu, N., Mastrogiacomo, R., Becker, C., et al. (2014). Honey bee odorant-binding protein 14: Effects on thermal stability upon odorant binding revealed by FT-IR spectroscopy and CD measurements. *European Biophysics Journal, 43*(2–3), 105–112. https://doi.org/10.1007/s00249-013-0939-4.

Smith, T. J., & Murrell, J. C. (2011). Mutagenesis of soluble methane monooxygenase. In A. C. Rosenzweig & S. W. Ragsdale (Eds.) *Vol. 495, Methods in enzymology*, Academic Press, (Chapter nine). https://doi.org/10.1016/B978-0-12-386905-0.00009-7.

Spinelli, S., Lagarde, A., Iovinella, I., Legrand, P., Tegoni, M., Pelosi, P., et al. (2012). Crystal structure of *Apis mellifera* OBP14, a C-minus odorant-binding protein, and its complexes with odorant molecules. *Insect Biochemistry and Molecular Biology, 42*(1), 41–50. https://doi.org/10.1016/j.ibmb.2011.10.005.

Sun, Y.-L., Huang, L.-Q., Pelosi, P., & Wang, C.-Z. (2013). A lysine at the C-terminus of an odorant-binding protein is involved in binding aldehyde pheromone components in two *Helicoverpa* species. *PLoS One, 8*(1), e55132. https://doi.org/10.1371/journal.pone.0055132.

Sun, J. S., Xiao, S., & Carlson, J. R. (2018). The diverse small proteins called odorant-binding proteins. *Open Biology, 8*(12), 180208. https://doi.org/10.1098/rsob.180208.

Tcatchoff, L., Nespoulous, C., Pernollet, J.-C., & Briand, L. (2006). A single lysyl residue defines the binding specificity of a human odorant-binding protein for aldehydes. *FEBS Letters, 580*(8), 2102–2108. https://doi.org/10.1016/j.febslet.2006.03.017.

Tegoni, M., Campanacci, V., & Cambillau, C. (2004). Structural aspects of sexual attraction and chemical communication in insects. *Trends in Biochemical Sciences, 29*(5), 257–264. https://doi.org/10.1016/j.tibs.2004.03.003.

Vincent, F., Spinelli, S., Ramoni, R., Grolli, S., Pelosi, P., Cambillau, C., et al. (2000). Complexes of porcine odorant binding protein with odorant molecules belonging to different chemical classes. *Journal of Molecular Biology, 300*(1), 127–139. https://doi.org/10.1006/jmbi.2000.3820.

Vincentelli, R., Cimino, A., Geerlof, A., Kubo, A., Satou, Y., & Cambillau, C. (2011). High-throughput protein expression screening and purification in *Escherichia coli*. *Methods, 55*(1), 65–72. https://doi.org/10.1016/j.ymeth.2011.08.010.

Vogt, R. G., Prestwich, G. D., & Lerner, M. R. (1991). Odorant-binding-protein subfamilies associate with distinct classes of olfactory receptor neurons in insects. *Journal of Neurobiology, 22*(1), 74–84. https://doi.org/10.1002/neu.480220108.

Vogt, R. G., & Riddiford, L. M. (1981). Pheromone binding and inactivation by moth antennae. *Nature, 293*(5828), 161–163. https://doi.org/10.1038/293161a0.

Vogt, R. G., Rybczynski, R., & Lerner, M. R. (1991). Molecular cloning and sequencing of general odorant-binding proteins GOBP1 and GOBP2 from the tobacco hawk moth *Manduca sexta*: Comparisons with other insect OBPs and their signal peptides. *Journal of Neuroscience, 11*(10), 2972–2984. https://doi.org/10.1523/JNEUROSCI.11-10-02972.1991.

Wang, S.-Y., Gu, S.-H., Han, L., Guo, Y.-Y., Zhou, J.-J., & Zhang, Y.-J. (2013). Specific involvement of two amino acid residues in cis-nerolidol binding to odorant-binding protein 5 AlinOBP5 in the alfalfa plant bug, *Adelphocoris lineolatus* (Goeze). *Insect Molecular Biology*, *22*(2), 172–182. https://doi.org/10.1111/imb.12012.

Wei, Y., Brandazza, A., & Pelosi, P. (2008). Binding of polycyclic aromatic hydrocarbons to mutants of odorant-binding protein: A first step towards biosensors for environmental monitoring. *Biochimica et Biophysica Acta (BBA)—Proteins and Proteomics*, *1784*(4), 666–671. https://doi.org/10.1016/j.bbapap.2008.01.012.

Yin, J., Zhuang, X., Wang, Q., Cao, Y., Zhang, S., Xiao, C., et al. (2015). Three amino acid residues of an odorant-binding protein are involved in binding odours in *Loxostege sticticalis* L. *Insect Molecular Biology*, *24*(5), 528–538. https://doi.org/10.1111/imb.12179.

Zaremska, V., Tan, J., Lim, S., Knoll, W., & Pelosi, P. (2020). Isoleucine residues determine chiral discrimination of odorant-binding protein. *Chemistry*. https://doi.org/10.1002/chem.202000872. Online ahead of print.

Zhou, J.-J., Robertson, G., He, X., Dufour, S., Hooper, A. M., Pickett, J. A., et al. (2009). Characterisation of *Bombyx mori* odorant-binding proteins reveals that a general odorant-binding protein discriminates between sex pheromone components. *Journal of Molecular Biology*, *389*(3), 529–545. https://doi.org/10.1016/j.jmb.2009.04.015.

Zhu, J., Arena, S., Spinelli, S., Liu, D., Zhang, G., Wei, R., et al. (2017). Reverse chemical ecology: Olfactory proteins from the giant panda and their interactions with putative pheromones and bamboo volatiles. *Proceedings of the National Academy of Sciences of the United States of America*, *114*(46), E9802–E9810. https://doi.org/10.1073/pnas.1711437114.

Zhu, J., Ban, L., Song, L.-M., Liu, Y., Pelosi, P., & Wang, G. (2016). General odorant-binding proteins and sex pheromone guide larvae of *Plutella xylostella* to better food. *Insect Biochemistry and Molecular Biology*, *72*, 10–19. https://doi.org/10.1016/j.ibmb.2016.03.005.

CHAPTER FOURTEEN

Silencing of OBP genes: Generation of loss-of-function mutants of PBP by genome editing

Yusuke Shiota[a,b], **Takeshi Sakurai**[c,*]

[a]Department of Evolutionary Neuroethology, Max Planck Institute for Chemical Ecology, Jena, Germany
[b]Research Center for Advanced Science and Technology, The University of Tokyo, Tokyo, Japan
[c]Faculty of Agriculture, Tokyo University of Agriculture, Kanagawa, Japan
*Corresponding author: e-mail address: ts206448@nodai.ac.jp

Contents

Abstract

Pheromone binding proteins (PBPs) are small soluble proteins (about 15 kDa) that play striking roles in the detection of sex pheromones in insects. Many studies including structural analysis, binding simulation, and *in vitro* assays have been performed to clarify the modes of action of PBPs. Although these studies have provided valuable

Methods in Enzymology, Volume 642
ISSN 0076-6879
https://doi.org/10.1016/bs.mie.2020.05.009

contributions toward the understanding of which key amino acid components contribute to the correct folding of PBPs and their binding affinities to sex pheromones, the functional characteristics of PBPs in the natural environment is still obscure. Recent developments in genome editing have begun to enable the functional examination of PBPs in *in vivo*. Among insect PBPs, BmPBP1 is one of the most well-characterized, there being rich understanding of its structure, biochemical analysis, binding affinity, localization, and the relationship between the type of olfactory receptors and its expression. A recent study has shown that BmPBP1 contributes sensitivity, but not selectivity of sex pheromone detection in the silkmoth *Bombyx mori*. In this chapter, based on a current report of the functional characterization of BmPBP1 using genome editing, we provide one example of a useful analytical method to clarify the functional role of PBP *in vivo*.

1. Introduction

Sex pheromone detection by male moths is an example of a sophisticated olfactory system in animals. The striking feature of this system is its high sensitivity and specificity, allowing the detection of minute quantities of pheromone molecules emitted from conspecific female moths, which in turn triggers electrophysiological responses in the antennae and elicits pheromone source searching behavior in males. Pheromone binding proteins (PBPs) belong to a subfamily of the odorant binding proteins (OBP), unique to moth species, and play an important role in the highly sensitive and specific pheromone detection (Leal, 2013; Pelosi, Zhou, Ban, & Calvello, 2006; Vogt, Große-Wilde, & Zhou, 2015; Vogt & Riddiford, 1981). Similar to OBPs, PBPs are small soluble proteins (about 15 kDa), expressed in accessory cells surrounding olfactory receptor neurons (ORNs) and secreted into the sensillum lymph space of pheromone-sensitive sensilla (Laue & Steinbrecht, 1997). Because of its binding ability to pheromone components and a high-level expression in the sensillum lymph, various functional roles of PBP have been suggested, as follows: (1) enhancement of sensitivity by solubilizing hydrophobic pheromone molecules in aqueous sensillum lymph and transporting them to dendritic membrane where pheromone receptor (PR) is located (Du & Prestwich, 1995; Kaissling, 1985; Vogt & Riddiford, 1981), (2) contribution to discrimination of pheromones by binding to different pheromone components with different affinities (Maida, Ziegelberger, & Kaissling, 2003; Plettner, Lazar, Prestwich, & Prestwich, 2000; Zhang et al., 2017), and (3) involvement in the temporal resolution of antennal responses by rapidly inactivating pheromone molecules (Kaissling, 2009, 2013; Ziegelberger, 1995).

Studies on the biochemical, structural, and histological properties of PBPs have provided useful information regarding the mechanisms underlying binding and release of pheromones, binding selectivity of each PBP, and expression pattern in moth antennae. Although *in vitro* binding assays and structural analyses have revealed essential amino acids involved in proper folding and the composition of binding pockets for target ligands (Klusak, Havlas, Rulisek, Vondrasek, & Svatos, 2003; Sandler, Nikonova, Leal, & Clardy, 2000), there have been conflicting conclusions regarding binding selectivity to pheromone components. For example, different binding assay studies of BmPBP1, a PBP in the silkmoth *Bombyx mori* have shown different affinities to pheromone components. Docking simulation and *in vitro* binding assays showed that BmPBP1 could bind to both pheromones, bombykol and bombykal, with similar affinities (Gräter, Xu, Leal, & Grubmüller, 2006). However, another study showed that BmPBP1 could bind only to bombykol (Hooper et al., 2009). In addition, it was reported that in the HEK293T cell line, which expresses BmOR1, the *B. mori* sex pheromone receptor, BmOR1 was able to respond to both bombykol and bombykal when these compounds were solubilized in dimethyl sulfoxide (DMSO). However, BmOR1 responded only to bombykol when BmPBP1 was used as a solubilizer instead of DMSO (Große-Wilde, Svatos, & Krieger, 2006). The aforementioned study suggests that PBP acts as filter to selectively bind to the pheromone component. On the other hand, *Xenopus oocytes* or Sf21 cells, which express both BmOR1 and the co-receptor BmOrco, were able to respond specifically to bombykol dissolved in DMSO, suggesting that the selectivity of the response to pheromones is dominantly determined by the specificity of the receptor, not those of BmPBP1 (Mitsuno, Sakurai, Namiki, Mitsuhashi, & Kanzaki, 2015; Nakagawa, Sakurai, Nishioka, & Touhara, 2005). Therefore, the contribution of PBP to the discrimination of pheromone components is still unclear.

Thus, to clarify the functional roles of PBPs, tests should be conducted *in vivo*, the natural environment where PBPs function. However, current *in vivo* data regarding the role of PBPs is limited, mainly due to the lack of an efficient method that can directly examine the function of PBPs. Although RNA interference (RNAi)-mediated gene knockdown can be used to investigate the *in vivo* function of PBPs, the efficiency of RNAi is not sufficiently high to examine gene function, as was observed in lepidopteran insects (Daimon, Kiuchi, & Takasu, 2014).

Recent development of genome editing tools has provided an opportunity to examine the function of PBPs *in vivo* in moth species, including *B. mori.* To date, three generations of genome editing tools have been developed, the first being zinc-finger nuclease (ZFN) (Kim, Cha, & Chandrasegaran, 1996); the second, Tal effector nuclease (TALEN) (Christian et al., 2010); and the third, the clustered regularly interspaced palindromic repeats (CRISPR)/Cas system (Jinek et al., 2012). These tools consist of a DNA binding component that can be designed to bind to a specific sequence of genomic DNA, and a DNA cleavage component that causes a double-strand break (DSB) of genomic DNA in the target sequence. DSBs often cause insertion and/or deletion mutations by non-homologous end joining during the repair process. Therefore, it is possible to introduce mutations in a gene of interest by properly designing a DNA binding domain to bind to target sequences. Recently, these tools have been used effectively to generate loss-of-function mutants of PBP genes in several lepidopteran insects, including *B. mori*, as summarized in the following section.

2. Analysis of PBPs using genome editing tools

Although binding assays provide important insights into the selective affinity of PBPs to various chemicals, the function of PBPs should be also tested *in vivo*. This is important because *in vitro* binding assays cannot simulate the exact environment of the surrounding proteins. In this context, genome editing is a useful technique for *in vivo* functional analysis since it mimics the natural protein environment. Recent development of genome editing tools has enabled the investigation of the *in vivo* function of PBPs in several moth species.

The silkmoth *B. mori* is a model lepidopteran insect used in sex pheromone research. Various molecular genetic approaches have been applied extensively to study this insect. The *B. mori* female moths emit two pheromone components: bombykol and bombykal. These components are detected by a pair of ORNs in a pheromone-sensitive sensillum, the sensillum lymph of which contains BmPBP1 (Forstner, Gohl, Breer, & Krieger, 2006; Krieger, Nickisch-rosenegk, Mameli, Pelosi, & Breer, 1996). Several *in vitro* studies have examined BmPBP1 with respect to its 3D structure (Klusak et al., 2003; Sandler et al., 2000), binding assay to pheromone components (Gräter et al., 2006), and functional analysis in a heterologous system, such as HEK293T cell expressing BmOR1 (Mitsuno et al., 2015; Nakagawa et al., 2005).

It is reported that among the three genome editing tools, TALEN is highly efficient at generating insertion/deletion mutant genetic lines in

B. mori (Daimon et al., 2014). A recent study investigated the *in vivo* function of BmPBP1 using *BmPBP1*-knockout moth lines (Shiota et al., 2018). Interestingly, electroantennogram (EAG) responses to both bombykol and bombykal were significantly reduced to a similar extent in *BmPBP1*-knockout male moths, suggesting that this PBP essentially contributes to sensitivity, but cannot distinguish between bombykal and bombykol. Similar EAG results were reported by using genome editing technique in *Spodoptera litura* (Zhu et al., 2016, 2019), *Helicoverpa armigera* (Ye et al., 2017), and *Chilo suppressalis* (Dong et al., 2019). EAG responses of PBP homozygous mutants were significantly lower than those of wild-type moths, indicating the role of PBP *in vivo* in the sensitive detection of pheromones. In the following sections, we discuss TALEN-mediated *BmPBP1* gene knockout and the functional analyses of knockout silkmoths.

2.1 TALEN-mediated genome editing in Lepidoptera species

TALEN is an artificial protein that consists of a DNA recognition/binding domain of transcription activator-like effector (TALE) from *Xanthomonas* and a restriction nuclease, *Fok*I, which functions as a dimer (Christian et al., 2010; Porteus, 2010). TALE contains repeat structures that consist of 34–35 amino acid residues as one unit, known as a "TALE repeat." TALE repeats recognize one of four nucleotides in the target genome. Crystal structure analysis revealed that a single TALE repeat forms two α-helix structures and varies at the 12th and 13th amino acid positions, known as the repeat variable diresidue (RVD), which contributes to the binding and stability of the target nucleotide. Generally, if the RVD consists of asparagine and isoleucine (NI), the TALE repeat recognizes A (adenine) in the desired gene region. Likewise, histidine and aspartic acid (HD) recognize C (cytosine), asparagine and glycine (NG) recognize T (thymine), and asparagine and asparagine (NN) recognize G (guanine) (Boch et al., 2009). Although the number of TALE repeats in *Xanthomonas* ranges from 1 to 33, TALENs with 15–20 TALE repeats are generally used for genome editing tools. This is because TALENs work as a pair, and typically 15 RVD repeats of one TALEN can secure sufficiently high specificity.

3. TALEN construction

Here, we describe a protocol of TALEN-mediated *BmPBP1* gene knockout. We first describe a protocol for the construction of TALENs and then generation of gene knockout moths.

3.1 Equipment

1. Thermal cycler
2. Centrifuge
3. System for agarose gel electrophoresis
4. Incubator
5. Micromanipulator and microinjection system
6. Stereo microscope

3.2 Chemicals

1. DNeasy Blood & Tissue kit (QIAGEN) or other genomic DNA extraction reagents
2. Takara Ex *Taq* HS (Takara Bio) or other PCR enzymes
3. Sequencing reagents
4. Plasmids from Golden Gate TALEN and TAL Effector Kit 2.0 (Addgene)
5. Plasmids from Yamamoto Lab TALEN Accessory Pack (Addgene)
6. pBlue-TAL vector (Addgene)
7. Takara Ligation Kit version 2.1 (Takara Bio)
8. Competent *E. coli* cells
9. HiSpeed Plasmid Midi Kit (Qiagen)
10. Primers for colony PCR and sequencing analysis

    For first assembly (pCR8_F1: 5′-TTGATGCCTGGCAGTTCCCT-3′ and pCR8_R1: 5′-CGAACCGAACAGGCTTATGT-3′)

    For second assembly (sTALseqF: 5′-GTGCATGCTTGGCGTAAC-3′ and *Fok*I-R: 5′-TCGTGAGGCACGTACTTGAG-3′)
11. *Bsa*I restriction enzyme
12. Ampicillin sodium
13. LB medium
14. Agar powder
15. Agarose
16. Primers for detection of TALE repeat insertion
17. PrimeSTAR HS (Takara Bio) or other high-fidelity DNA polymerase
18. Wizard SV gel and PCR clean-up system (Promega) or other DNA fragment purification kits
19. *Xba*I restriction enzyme
20. mMESSAGE mMACHINE T7 Ultra Transcription Kit (Ambion)
21. Potassium phosphate
22. 100% Ethanol

23. 70% Ethanol
24. Phenol/chloroform

4. Protocol

4.1 Construction of vectors for TALENs

Vectors for TALENs for *BmPBP1* gene knockout were constructed according to the protocol reported previously (Cermak et al., 2011; Sakuma et al., 2013). The schematic diagram of the workflow of this section is also described in Fig. 1. For a detailed TALEN construction protocol, refer to the original publications describing TALEN construction (Cermak et al., 2011; Sakuma et al., 2013).

1. Select the target site using the "TALEN Targeter" (https://tale-nt.cac.cornell.edu/).
2. For the generation of knockout silkmoths, use the non-diapausing strain *w1 pnd* for ease of RNA microinjection experiments. To avoid possible polymorphism between sequence of the target site of your strain and *w1 pnd*, we recommend extracting genomic DNA from *w1 pnd* and amplifying the target genome regions by PCR, then determining and comparing DNA sequences of the target site.

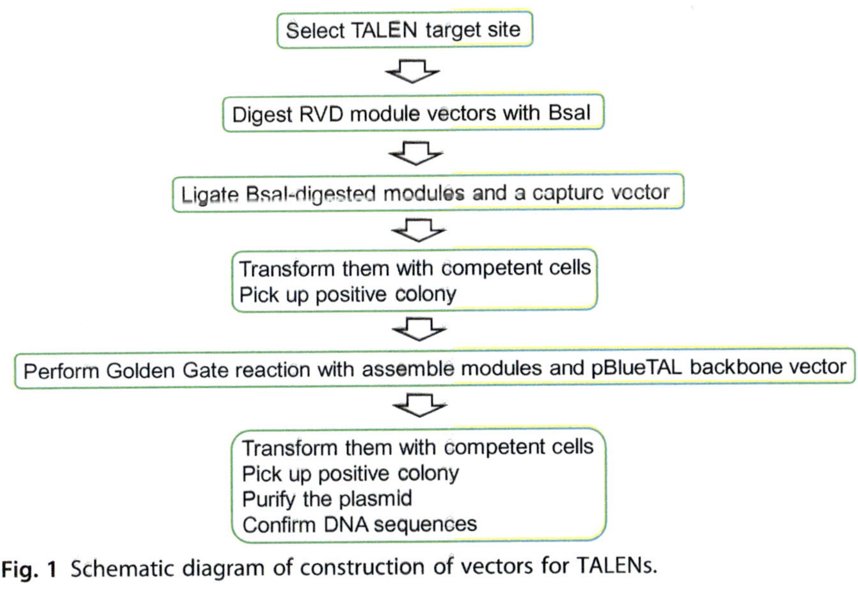

Fig. 1 Schematic diagram of construction of vectors for TALENs.

3. Prepare the plasmid DNA required for TALEN constructs, including RVD module vectors, capture vectors, and the TALEN backbone vector pBlue-TAL, which is optimized for *B. mori*.
4. Digest plasmids for the first step ligation with *Bsa*I overnight at 37 °C. Run reaction products on an agarose gel, cut out DNA bands corresponding to RVD module vectors, capture vectors, or backbone vectors.
5. Prepare mixture of *Bsa*I-digested modules and a capture vector with DNA Ligation Kit ver. 2.1 (Takara Bio) and incubate at 16 °C for 30 min for ligation.
6. Use an aliquot of the digestion product (0.5–1 μL) to transform competent *E. coli* cells. Spread transformed cells on LB/spectinomycin plate and incubate overnight at 37 °C.
7. Screen clones containing designed insert by colony PCR, using *Ex Taq* HS DNA Polymerase with pCR8_F1 and pCR8_R1 primers. Apply the following conditions.

Cycle number	Denature	Anneal	Extend
1	94 °C, 1 min		
2–40	94 °C, 20 s	55 °C, 20 s	72 °C, 40 s
41			72 °C, 2 min

Analyze PCR products on an agarose gel electrophoresis.

8. Pick positive colonies and culture in LB/spectinomycin broth overnight at 37 °C, and then purify plasmid using a commercial Miniprep plasmid purification kit.
9. Confirm proper construction by DNA sequencing of plasmid inserts.
10. Assemble modules into the pBlue-TAL backbone vector by Golden Gate reaction, as described earlier (Cermak et al., 2011; Sakuma et al., 2013).
11. Transform competent *E. coli* cells with an aliquot of reaction products, spread cells on LB/ampicillin plate and incubate overnight at 37 °C.
12. Screen positive clones by colony PCR using *Ex Taq* HS DNA polymerase with sTALseqF and *Fok*I-R primers with the following conditions.

Cycle number	Denature	Anneal	Extend
1	94 °C, 1 min		
2–31	94 °C, 20 s	60 °C, 20 s	72 °C, 1.5 min
32			72 °C, 2 min

Analyze PCR products on an agarose gel electrophoresis.

13. Culture positive colonies at 37 °C overnight in LB/ampicillin broth, purify plasmids, and confirm the insert DNA by sequencing analysis. Store the plasmid at −20 °C until use.

4.2 Injection of TALEN cRNA into silkmoth eggs

1. Digest overnight the plasmid constructed in the previous section with *Xba*I at 37 °C.
2. Synthesize cRNA of TALENs with mMESSAGE mMACHINE T7 Ultra Transcription Kit (Ambion). Dissolve 0.4 mg/mL of each RNA in injection buffer: 0.5 mM potassium phosphate buffer (pH 7.0), 5 mM KCl.
3. Cross newly eclosed *w1 pnd* female moths with male moths of the same strain for at least 2 h. Collect eggs by placing mated female moths on a piece of paper. Immerse the paper with the laid eggs in distilled water for 2–3 min. The floated eggs are aligned onto a slide glass whose surface is covered with glue.
4. Sterilize eggs by putting them in a plastic container with paper immersed with formalin for 10 min.
5. Inject the TALEN cRNA into eggs using a microinjection system within about 4–6 h after they have been laid. Typically, we inject about 2–300 eggs. Since the efficiency of TALEN-mediated mutation is very high, smaller numbers of eggs may be enough to obtain mutation in silkmoth lines.
6. After injection, seal the holes made in the eggs with a small amount of glue.
7. Incubate injected eggs at 25 °C in a humidified box with 16:8/light:dark conditions, until hatched.

4.3 Screening broods for mutated individuals

The schematic diagram of the workflow of this section is also described in Fig. 2.

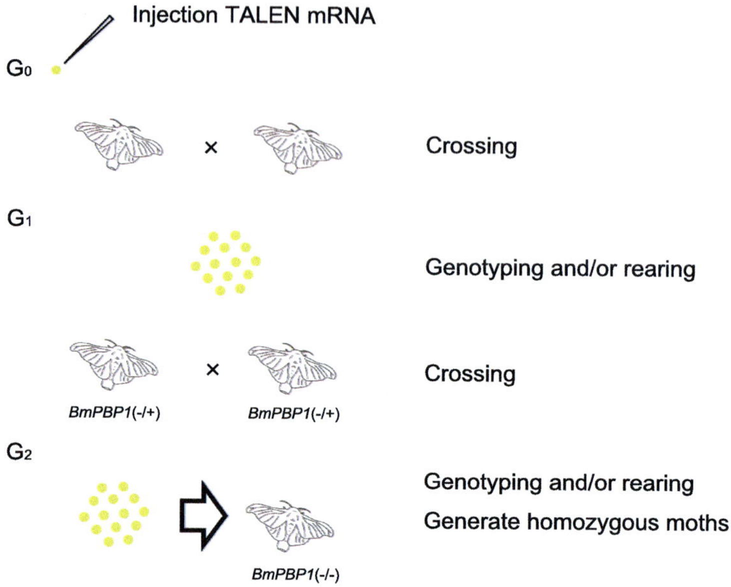

Fig. 2 Schematic diagram of the generation of *BmPBP1*-knockout moths using TALEN-mediated gene targeting.

1. To obtain G_1 eggs, cross G_0 adult moths with wild-type moths
2. Develop eggs to head pigmentation stage, then pick up 10–15 eggs from each brood and purify the genomic DNA from them using commercially available DNA purification kits such as DNeasy Blood & Tissue Kit (Qiagen).
3. Design PCR primers to amplify DNA fragments containing the target region of TALENs. The following primer pair is used for *BmPBP1* knockout. Sense: 5′-CGACCTTCGCAAGGTATGAT-3′, antisense: 5′-AGGCACATTATAGCGCATCC-3′.
4. Prepare PCR reaction mix according to the manufacturer's instructions. In our case, we used Takara *Ex Taq* HS DNA polymerase with buffer and dNTPs as follows:

Component	Amount (μL)
10× *Ex Taq* buffer	2.0
dNTP mixture	1.6
Sense primer	0.2

—cont'd

Component	Amount (μL)
Antisense primer	0.2
Template DNA	2.0
Ex Taq HS	0.1
dH_2O	13.9
Total	20.0

5. Amplify DNA fragments using the following thermal cycling conditions.

Cycle number	Denature	Anneal	Extend
1	95 °C, 5 min		
2–36	95 °C, 30 s	53 °C, 15 s	72 °C, 20 s
37			72 °C, 5 min

6. Confirm that the PCR products of the desired size have been amplified by agarose gel electrophoresis.
7. Purify PCR products using commercially available purification kits, such as Wizard PCR and Gel Clean-Up kit (Promega), according to the manufacturer's instructions.
8. Sequence around the target region of purified PCR products, using sense and antisense primers used to amplify genomic DNA. Carefully check chromatogram of DNA sequences. If insertion or deletion around the target sequence occurs, the PCR products shows multiple chromatogram peaks after a mutation site (Fig. 3B).
9. Grow residual G_1 eggs of positive broods into adults, and then remove a hindwing for genomic DNA extraction using commercially available DNA purification kits such as DNeasy Blood & Tissue Kit (Qiagen).
10. Amplify DNA fragments containing the target site, purify them, and sequence them as described in steps 4, 6, and 7. Subclone PCR products showing multiple chromatogram peaks into cloning vector such as pGEM-easy T vector (Promega). Determine the detailed genotype of each insert by sequencing.

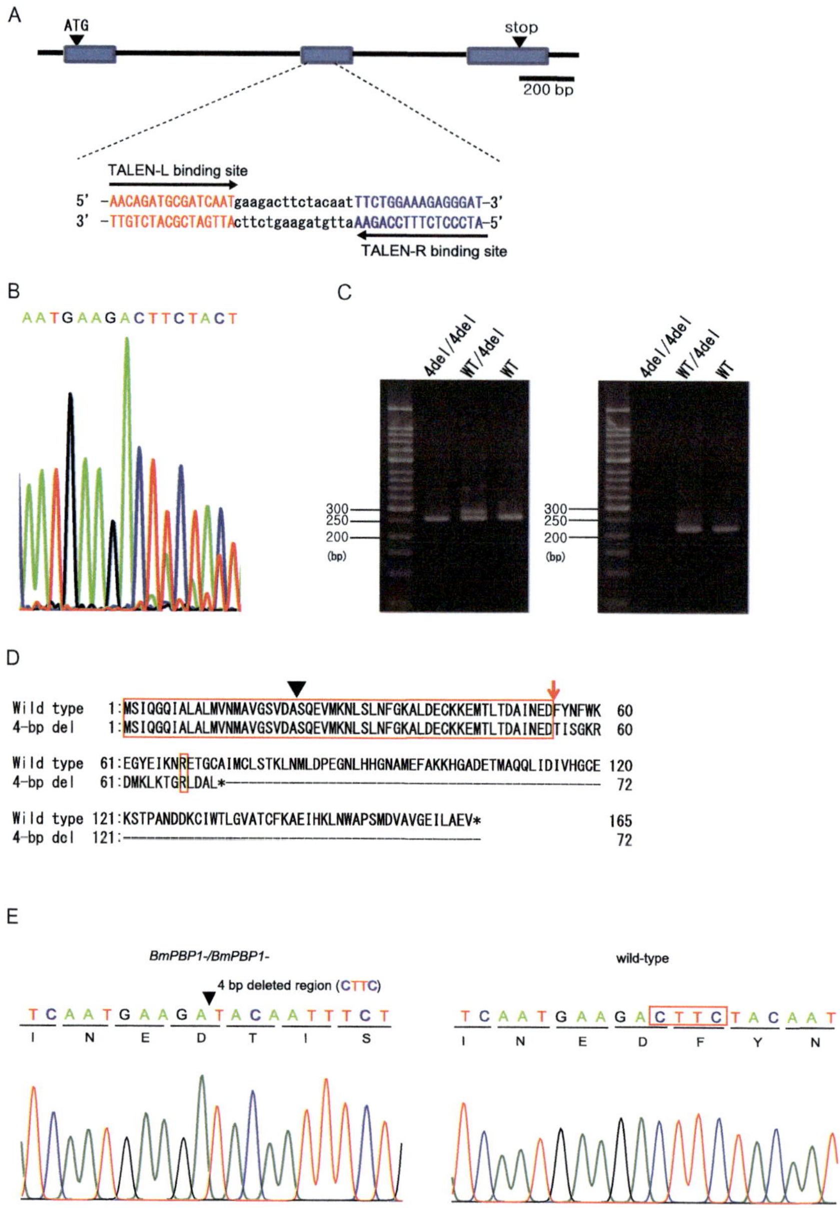

Fig. 3 TALEN construction and analysis of *BmPBP1* gene in the generated moths. (A) Schematic representation of the genomic structure of the *BmPBP1* gene. The sequences of TALEN recognition sites are shown at the bottom of the genomic structure. (B) Sequencing result around target genome in the *BmPBP1* gene. An example of the sequence chromatogram with multiple peaks after a mutation site is shown. (C) A representative genomic PCR result of the 4-bp deletion allele. The PCR products

(Continued)

11. Cross with moths of the same genotype to produce homozygous mutant moths.
12. After establishment of homozygous mutant lines, carry out PCR-based genotyping as described in Section 6.

5. Physiological analysis of the effects of gene knockout *in vivo*

To examine the effect of the knockout of a PBP gene (e.g., *BmPBP1*), one possibility to test whether the protein affects sensitivity and/or selectivity of antennal response is to perform an EAG experiment. The first electrophysiological recording from the insect antennae to odorants was performed in 1957 by the German scientist, Schneider (Schneider, 1957). He successfully recorded the physiological responses of antennae to extracted reagent of female's gland and other tested odorants, using antennae of male *B. mori*. To date, this technique is widely applied to examine the response profile in moth species. ORNs located in the olfactory sensilla produce electrical signals by opening ion channels in response to odorant molecules that go through minute olfactory pores and bind to olfactory receptors that are expressed on the surface of the dendritic membrane. EAG responses are considered as the result of changes of the receptor potential elicited by odorant molecules on the antennae.

Here, we describe the method to confirm the genotype of gene-edited insects. Next, we introduce how to investigate the phenotype of insects *in vivo*.

5.1 Equipment

1. Mounting stands for EAG amplifier
2. Anti-vibration table (recommended, but not necessary in EAG recording in most cases)

Fig. 3—Cont'd were performed in a mutation-homozygous individual (4del/4del), a mutation-heterozygous individual (WT/4del), and a wild-type individual. (D) Deduced amino acid sequences of wild-type (top) and 4-bp deletion *BmPBP1* knockout moths (bottom). The black arrowhead on the sequences indicates a signal peptide cleavage site, and the red arrow indicates the position of a frame shift caused by TALEN-mediated deletions. Red boxes indicate identical amino acids sequences between the two sequences. (E) Analysis of DNA sequencing results in *BmPBP1*-knockout and wild-type male around the 4-bp deletion region. *Figures modified from Shiota, Y., Sakurai, T., Daimon, T., Mitsuno, H., Fujii, T., Matsuyama, S., et al. (2018). In vivo functional characterisation of pheromone binding protein-1 in the silkmoth, Bombyx mori.* Scientific Reports, 8*(1), 1–8.*

3. Labview USB-6525 (required appropriate A/D converter)
4. Laptop PC (or desktop computer and peripherals)
5. Faraday cage (recommended, but not necessary in EAG recording in most cases)
6. Signal amplifier (recommended with filtering function)
7. Filter paper for odor stimulation
8. Humidifier (e.g., bubbling bottle filled with distilled water)
9. Three-way electrical valve
10. Taigon tubes (silicon tubes)
11. Flow meter
12. Ag/AgCl coated electrode
13. Glass Pasteur pipette (for odor stimulation)
14. Heat block
15. Thermal cycler
16. Electrophoresis apparatus

5.2 Chemicals

1. SPECTRA 360 electrode gel (or appropriate ringer solution)
2. *n*-Hexane (for control experiment)
3. Synthesized pheromones (for odor stimulation)
4. 50 mM NaOH
5. 0.2 M Tris-HCl
6. TAE solution.
7. *Ex Taq* Hs (Takara, or appropriate *Taq* polymerase)
8. Ethidium bromide
9. Agarose
10. 200 bp DNA Ladder (NEB)

6. Protocol

6.1 Electroantennogram (EAG)

1. Adjust air flow to 1.0 L/min.
2. Put appropriate amount of electrode gel on each Ag/AgCl electrode.
3. Excise at the base of antennae and a few segments at the tip.
4. Put isolated antennae on the electrodes and immediately cover with electrode gel.
5. Prepare glass cartridge inserted with a piece of filter paper, containing test odor (*n*-hexane for control odor, and pheromones for test odor).

6. Make sure that stable basement voltage is maintained. Correct DC offset of monitored EAG signal.
7. Apply control and test odor to antennae and record the responses. The EAG responses are filtered at 300 Hz and digitized at 1 kHz.
8. Analyze the acquired data with an analytical software/program such as MATLAB (MathWorks).

6.2 Genotyping of tested moths

After EAG experiments, the genotype of each moth is confirmed by PCR-based genotyping as follows:

1. Prepare template genomic DNA from adult moths. Cut one hindwing of moths and put into a 1.5 mL tube.
2. Perform extraction of genomic DNA from cut hindwing. Add 200 μL of 50 mM NaOH solution into the tube.
3. Perform heat denaturation. Incubate sample in a heat block for 15 min at 95 °C.
4. Neutralize solution by adding 200 μL of 0.2 M Tris-HCl (pH 8.0) solution into reaction solution, and vigorously vortex the sample.
5. Set up PCR reaction using the composition described below. The PCR reaction needs to be run using two separate primer sets for each sample.

Component	Amount (μL)
10 × *Ex Taq* buffer	2.0
dNTP mixture	1.6
Sense primer	0.2
Antisense primer	0.2
Template DNA(reaction solution)	2.0
Ex Taq HS	0.1
dH_2O	13.9

6. Place the reaction mixture in a thermal cycler, and run PCR reaction with following program:

Cycle number	Denature	Anneal	Extend
1	95 °C, 5 min		
2–36	95 °C, 30 s	53 °C, 15 s	72 °C, 20 s
37			72 °C, 5 min

7. Run the PCR products on a 4% (wt/vol) agarose gel for 25 min. Stain the gel with ethidium bromide and confirm the presence of intended amplicon and patterns of the band.

7. Analysis of results

We describe a method to generate *BmPBP1*-knockout silkmoths using TALEN-mediated genome editing for clarification of the *in vivo* function of PBPs. Selection of a target site is an important step to enable the generation of a loss-of-function mutant. In this regard, three-dimensional structures of PBPs provide useful information to select candidates target sites. In our case, we designed TALENs to target the sequence in the second exon of the *BmPBP1* gene (Fig. 3A). If a frame shift occurs at the target sequence, an approximately 50 amino acid truncated protein should be produced, whose binding pocket and tertiary structure is expected not to be formed properly due to the lack of necessary amino acid residues. Indeed, the generated mutant BmPBP1 protein was composed of 51 amino acids and contained only three out of the nine residues required for the formation of the bombykol binding pocket, and only one out of the six cysteine residues required for proper folding (Fig. 3B–E) (Klusak et al., 2003; Sandler et al., 2000; Shiota et al., 2018). Thus, thanks to structural and binding analyses of PBPs, we can select a target site based on the possible functionalities of mutant PBPs.

After establishment of PBP gene knockout moths, the function of the PBP can be assessed by examining the effects of loss of PBP on physiological responses. There are two major methods used to analyze antennal responses to odorants: EAG and single sensillum recording (SSR). EAG measures the sum of ORN responses in the antennae and it is relatively easy to measure responses. SSR can examine spike responses of ORNs in a single sensillum at a single ORN resolution. Because bombykol and bombykal are detected by heromone-sensitive sensilla that express BmPBP1 throughout whole antenna, we employed EAG to examine the effects of the loss of

BmPBP1 on pheromone responses (see details in Section 6.1). It is known that EAG peak amplitudes increase in a dose-dependent manner. Therefore, peak EAG amplitudes can be a good index to investigate the sensitivity of antennae to a given stimulus. In our analysis, the peak EAG amplitudes of *BmPBP1*-knockout male antennae to both bombykol and bombykal were significantly reduced compared to those of wild-type male antennae (Fig. 4A and B). Although *BmPBP1*-knockout antennae still produce responses to pheromones, this may be explained by the non-specific binding to water soluble proteins expressed in sensillum lymph. Alternatively, there may be another direct pathway from olfactory pores to the dendritic membrane

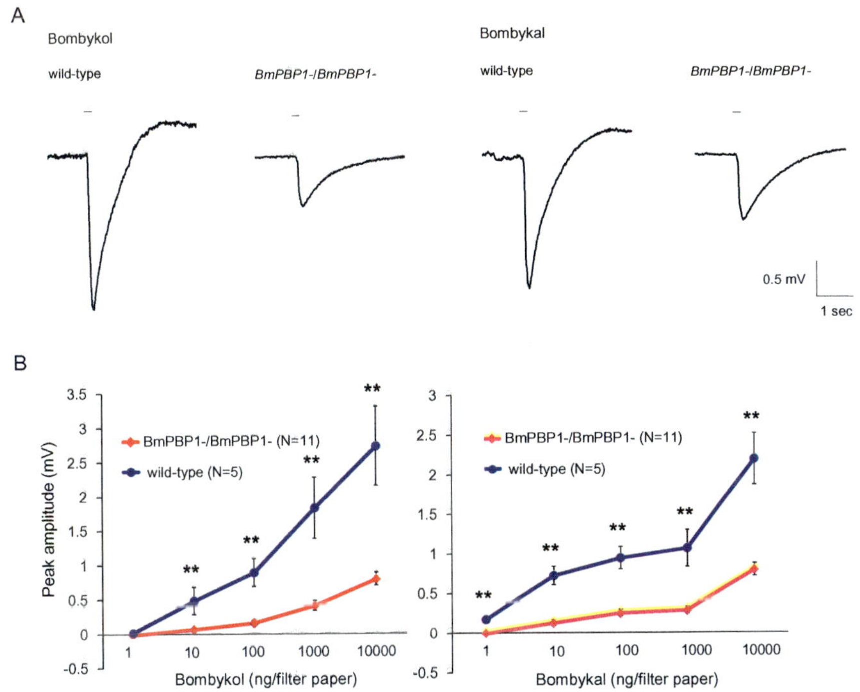

Fig. 4 EAG analysis in generated gene knockout animals. (A) Representative EAG responses in *BmPBP1-/BmPBP1-* and wild-type male moths to 1000 ng bombykol (left) and bombykal (right). (B) Dose-dependent increase of induced EAG responses to bombykol- (left) and bombykal- (right) in *BmPBP1-/BmPBP1-* (red; $n=11$) and wild-type (blue; $n=5$) male moth antennae. Error bars represent mean ± SEM. The asterisks indicate significant differences between the groups (**$P<0.01$), as determined by Student's *t*-test for comparing data pairs. *Figures modified from Shiota, Y., Sakurai, T., Daimon, T., Mitsuno, H., Fujii, T., Matsuyama, S., et al. (2018). In vivo functional characterisation of pheromone binding protein-1 in the silkmoth, Bombyx mori.* Scientific Reports, *8(1), 1–8.*

to transport some subsets of pheromone molecules (Keil, 1982, 1984). The reduction of EAG amplitude was similar in extent between the responses to both bombykol and bombykal. These results indicated that BmPBP1 is required for the sensitive detection of both bombykol and bombykal but does not play a significant role in distinguishing bombykol from bombykal.

8. Summary

In this chapter, we have described the common challenges of the *in vivo* analysis of PBP function. Until recently, the functional characterization of PBPs/OBPs in insects was challenging. However, with the development of various genome editing techniques, researchers are now able to analyze the *in vivo* functions of genes of interest. We believe that these methods will increase our knowledge of how odorant/pheromone binding proteins contribute to the incredible ability of insect olfaction.

Acknowledgment

This work was supported by a Grant-in-Aid for Scientific Research (B) (18H02211), Japan Society for the Promotion of Science (JSPS), Japan, awarded to T.S.

References

Boch, J., Scholze, H., Schornack, S., Landgraf, A., Hahn, S., Kay, S., et al. (2009). Breaking the code of DNA binding specificity of TAL-type III effectors. *Science*, *326*, 1509–1512.

Cermak, T., Doyle, E. L., Christian, M., Wang, L., Zhang, Y., Schmidt, C., et al. (2011). Efficient design and assembly of custom TALEN and other TAL effector-based constructs for DNA targeting. *Nucleic Acids Research*, *39*, e82.

Christian, M., Cermak, T., Doyle, E. L., Schmidt, C., Zhang, F., Hummel, A., et al. (2010). Targeting DNA double-strand breaks with TAL effector nucleases. *Genetics*, *186*, 757–761.

Daimon, T., Kiuchi, T., & Takasu, Y. (2014). Recent progress in genome engineering techniques in the silkworm, *Bombyx mori*. *Development, Growth & Differentiation*, *6*(1), 14–25.

Dong, X. T., Liao, H., Zhu, G. H., Khuhro, S. A., Ye, Z. F., & Dong, S. L. (2019). CRISPR/Cas9-mediated PBP1 and PBP3 mutagenesis induced significant reduction in electrophysiological response to sex pheromones in male Chilo suppressalis. *Insect Science*, *26*(3), 388–399.

Du, G., & Prestwich, G. D. (1995). Protein structure encodes the ligand binding specificity in pheromone binding proteins. *Biochemistry*, *34*(27), 8726–8732.

Forstner, M., Gohl, T., Breer, H., & Krieger, J. (2006). Candidate pheromone binding proteins of the silkmoth *Bombyx mori*. *Invertebrate Neuroscience*, *6*(4), 177–187.

Gräter, F., Xu, W., Leal, W., & Grubmüller, H. (2006). Pheromone discrimination by the pheromone-binding protein of *Bombyx mori*. *Structure*, *14*(10), 1577–1586.

Große-Wilde, E., Svatos, A., & Krieger, J. (2006). A pheromone-binding protein mediates the bombykol-induced activation of a pheromone receptor *in vitro*. *Chemical Senses*, *31*, 547–555.

Hooper, A. M., et al. (2009). High-throughput ESI-MS analysis of binding between the Bombyx mori pheromone-binding protein BmorPBP1, its pheromone components and some analogues. *Chemical Communications, 14*(38), 5725–5727.

Jinek, M., Chylinski, K., Fonfara, I., Hauer, M., Doudna, J. A., & Charpentier, E. (2012). A programmable dual-RNA-guided DNA endonuclease in adaptive bacterial immunity. *Science, 337*, 816–821.

Kaissling, K. E. (1985). Insect olfactory cells: Electrophysiological and biochemical studies. In JP Changeux & F Hucho (Eds.), *Molecular basis of nerve activity* (pp. 173–183). *Proceedings of the international symposium in memory of David Nachmansohn*, Berlin: W. de Gruyter.

Kaissling, K. E. (2009). Olfactory perireceptor and receptor events in moths: A kinetic model revised. *Journal of Comparative Physiology A: Neuroethology, Sensory, Neural, and Behavioral Physiology, 195*(10), 895–922.

Kaissling, K. E. (2013). Kinetics of olfactory responses might largely depend on the odorant-receptor interaction and the odorant deactivation postulated for flux detectors. *Journal of Comparative Physiology A: Neuroethology, Sensory, Neural, and Behavioral Physiology, 199*(11), 879–896.

Keil, T. A. (1982). Contacts of pore tubules and sensory dendrites in antennal chemosensilla of a silkmoth: Demonstration of a possible pathway for olfactory molecules. *Tissue & Cell, 14*, 451–462.

Keil, T. A. (1984). Surface coats of pore tubules and olfactory sensory dendrites of a silkmoth revealed by cationic markers. *Tissue & Cell, 16*, 705–717.

Kim, Y. G., Cha, J., & Chandrasegaran, S. (1996). Hybrid restriction enzymes: Zinc finger fusions to Fok I cleavage domain. *Proceedings of the National Academy of Sciences of the United States of America, 93*, 1156–1160.

Klusak, V., Havlas, Z., Rulisek, L., Vondrasek, J., & Svatos, A. (2003). Sexual attraction in the silkworm moth: Nature of binding of bombykol in pheromone binding protein—An ab initio study. *Chemistry & Biology, 10*, 331–340.

Krieger, J., Nickisch-rosenegk, E., Mameli, M., Pelosi, P., & Breer, H. (1996). Binding proteins from the antennae of *Bombyx mori*. *Insect Biochemistry and Molecular Biology, 26*(3), 297–307.

Laue, M., & Steinbrecht, R. A. (1997). Topochemistry of moth olfactory sensilla. *International Journal of Insect Morphology and Embryology, 26*, 217–228.

Leal, W. S. (2013). Odorant reception in insects: Roles of receptors, binding proteins, and degrading enzymes. *Annual Review of Entomology, 58*, 373–391.

Maida, R., Ziegelberger, G., & Kaissling, K.-E. (2003). Ligand binding to six recombinant pheromone-binding proteins of *Antheraea polyphemus* and *Antheraea pernyi*. *Journal of Comparative Physiology B, 173*, 565–573.

Mitsuno, H., Sakurai, T., Namiki, S., Mitsuhashi, H., & Kanzaki, R. (2015). Novel cell-based odorant sensor elements based on insect odorant receptors. *Biosensors and Bioelectronics, 65*, 287–294.

Nakagawa, T., Sakurai, T., Nishioka, T., & Touhara, K. (2005). Insect sex-pheromone signals mediated by specific combinations of olfactory receptors. *Science, 307*(5715), 1638–1642.

Pelosi, P., Zhou, J. J., Ban, L. P., & Calvello, M. (2006). Soluble proteins in insect chemical communication. *Cellular and Molecular Life Sciences: CMLS, 63*(14), 1658–1676.

Plettner, E., Lazar, J., Prestwich, E. G., & Prestwich, G. D. (2000). Discrimination of pheromone enantiomers by two pheromone binding proteins from the gypsy moth *Lymantria dispar*. *Biochemistry, 39*, 8953–8962.

Porteus, M. (2010). Creating zinc finger nucleases to manipulate the genome in a site-specific manner using a modular-assembly approach. *Cold Spring Harbor Protocols 2010*. https://doi.org/10.1101/pdb.top93.

Sakuma, T., et al. (2013). Efficient TALEN construction and evaluation methods for human cell and animal applications. *Genes to Cells, 18*(4), 315–326.

Sandler, B. H., Nikonova, L., Leal, W. S., & Clardy, J. (2000). Sexual attraction in the silkworm moth: Structure of the pheromone- binding-protein-bombykol complex. *Chemistry and Biology*, 7(2), 143–151.

Schneider, D. (1957). Electrophysiological investigation on the antennal receptors of the silk moth during chemical and mechanical stimulation. *Experientia*, *13*(2), 89–91.

Shiota, Y., Sakurai, T., Daimon, T., Mitsuno, H., Fujii, T., Matsuyama, S., et al. (2018). *In vivo* functional characterisation of pheromone binding protein-1 in the silkmoth, *Bombyx mori*. *Scientific Reports*, *8*(1), 1–8.

Vogt, R. G., Große-Wilde, E., & Zhou, J. J. (2015). The Lepidoptera odorant binding protein gene family: Gene gain and loss within the GOBP/PBP complex of moths and butterflies. *Insect Biochemistry and Molecular Biology*, *62*, 142–153.

Vogt, R. G., & Riddiford, L. M. (1981). Pheromone binding and inactivation by moth antennae. *Nature*, *293*, 161.

Ye, Z. F., Liu, X. L., Han, Q., Liao, H., Dong, X. T., Zhu, G. H., et al. (2017). Functional characterization of PBP1 gene in *Helicoverpa armigera* (Lepidoptera: Noctuidae) by using the CRISPR/Cas9 system. *Scientific Reports*, 7(1), 8470.

Zhang, T., et al. (2017). Binding affinity of five PBPs to *Ostrinia* sex pheromones. *BMC Molecular Biology*, *18*, 4.

Zhu, G. H., Xu, J., Cui, Z., Dong, X. T., Ye, Z. F., Niu, D. J., et al. (2016). Functional characterization of SlitPBP3 in *Spodoptera litura* by CRISPR/Cas9 mediated genome editing. *Insect Biochemistry and Molecular Biology*, *75*, 1–9.

Zhu, G. H., Zheng, M. Y., Sun, J. B., Khuhro, S. A., Yan, Q., Huang, Y., et al. (2019). CRISPR/Cas9 mediated gene knockout reveals a more important role of PBP1 than PBP2 in the perception of female sex pheromone components in *Spodoptera litura*. *Insect Biochemistry and Molecular Biology*, *115*, 103244.

Ziegelberger, G. (1995). Redox-shift of the pheromone-binding protein in the silkmoth *Antheraea polyphemus*. *European Journal of Biochemistry*, *232*(3), 706–711.

CHAPTER FIFTEEN

Electrochemical biosensing with odorant binding proteins

Sabine Szunerits[a,*], Rabah Boukherroub[a], Alina Vasilescu[b]

[a]University of Lille, CNRS, Centrale Lille, Université Polytechnique Hauts-de-France, UMR 8520—IEMN, Lille, France

[b]International Centre of Biodynamics, Bucharest, Romania

*Corresponding author: e-mail address: sabine.szunerits@univ-lille.fr

Contents

Abstract

The development of sensors that mimic the natural smell sensing mechanism and selectively recognizes the odorants remains highly challenging. Electrochemical based sensing approaches aiming at monitoring molecular recognition events between surface receptors and analytes in solution or in the gas phase, are one possible transduction platforms among others for the construction of an artificial nose. The principle of electrochemical detection lies on the shift of the potential/current during the recognition event, which is proportional to the concentration of the analyte, in our case the odorant. A tremendous amount of efforts has been put into making electrochemical sensors

Methods in Enzymology, Volume 642
ISSN 0076-6879
https://doi.org/10.1016/bs.mie.2020.04.071

sensitive and selective to the analyte of interest through the use of nanomaterials, development of different detection schemes and application of innovative receptor ligands for selective detection of the analyte. There have been significant advances in electrochemical based odorant sensing by using odorant binding proteins (OBP) as surface receptors, small soluble proteins present in nasal mucus at millimolar concentrations where the hydrophobic binding pocket gives the ability to reversibly bind odorant molecules. As OBPs are robust and easy to produce receptors, they are good candidates for the design of biosensors. In this chapter, we focus on the progress made on the detection of odorant molecules using OBPs as a bioreceptor and electrochemistry as a transduction method.

1. Introduction

The human olfactory system is still regarded as the most important analytical method for the assurance of odor quality in many industries. In the case of human being sense of smell, the threshold level of odor sensing is substrate dependent: it can be as low as 0.4 ng/L as is the case of methyl mercaptan, and reach 5.8 mg/L for diethyl ether. The level varies between different individuals and greatly varies between various animals. Odorants that are typically found in nature tend to be a complex mixture of many different chemical species at various concentrations rather than one type of molecule. Varying the concentration of compounds in the mixture, changing the functional group chemistry or position, and altering the stereochemistry or conformation of a molecule, among other changes that can be made, all affect the smell of an odorant or odorous mixture (Beets, 1978). Although there are examples of simple single molecules with distinct aromas, the discovery of the olfactory code is an extremely complicated task (Graben & Sachse, 2018). Relating the structure of an odorant molecule to the sensation that is perceived is not an easy task. While odorants are generally molecules with molecular weight around 30–300 $g\,mol^{-1}$, they may vary in structure from simple aliphatic short-chain organic molecules, to complex aromatic molecules with multiple functional groups (Table 1). It illustrates that small changes in the nature of the functional group, the shape, and the size of a molecule can induce large changes in our perception of odor quality.

1.1 What is the underlying principle that a molecule is an odorant?

Relating the structure of an odorant molecule to the sensation that is perceived is difficult, and there are several unexplained findings. The specific

Table 1 Chemical structures of some odorants and their smell perception.

Molecule	Chemical structure	Smell
Ethanol	CH_3–CH_2–OH	Alcoholic
Geosmin	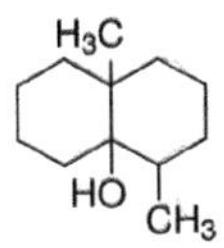	Earthy
2,3,6-Trichloroanisol	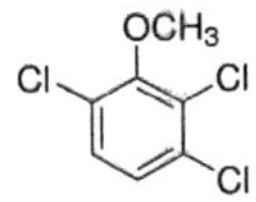	Moldy
Acetic acid	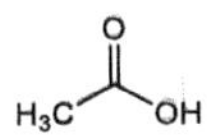	Vinegar
Ethyl acetate	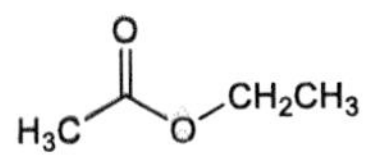	Fruity, ethereal
Amyl acetate	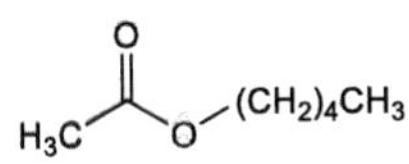	Banana
Butyl mercaptan	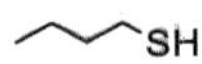	Putrid

features of the molecule such as size, shape and functional groups (i), its interaction with olfactory receptors (ii) and the integration and processing in the brain of the information from the olfactory receptors (iii) are the three complex components of the mechanism that determines the odor perceived for a specific compound (Pelosi, Zhu, & Knoll, 2018). One typical characteristic of odorants relates to their affinity with olfactory receptors (ORs), which translates into a mixture of repulsive and attractive forces that a molecule encounters when bound to the receptor. It is clear that a simplistic molecular conformation and shape model of odorant-receptor interaction does not significantly account for the observations. There are certainly some correlations between the shape of a molecule and odorant receptor response, but likewise there are many cases where very different shapes produce the same pattern of activation of the odorant receptor repertoire because many ORs are broadly tuned. In fact, the human nose hosts a range of more than 300 olfactory receptors (Olender, Nativ, & Lancet, 2013) which respond with different affinity to a specific stimulus. Different odorants are sensed by different combinations of ORs from this network. Thus, a multi-dimensional "image" of a molecule or of a complex odorant mixture is constructed that is further processed by the brain (Mori & Sakano, 2011). The olfactory map is integrated with other sensorial information and compared to stored information from the past, finally delivering a specific "odor" sensation. While significant advances with regards of ORs were made by identification of high affinity ligands for part of them, much more remains to be learned about the processing of the olfactory map in the brain (Pelosi et al., 2018). To date, despite the high interest and various theories advanced so far, the structure-odor relationship remains elusive (Voshall, 2016).

1.2 Biological odorant sensors

For living organisms, the initial step in odor sensing is the capture of the analyte molecules by few extracellular proteins and membrane-bound OR (Abaffy, 2015; de March, Kim, Antonczak, Goddard 3rd, & Golebiowski, 2015). Difference in the structures of these proteins determine their interaction with different odorants and make them odor specific. Due to their specificity for odors, ORs are well adapted to detect a specific odorant in a mixture of analytes. This knowledge contributed to the development of bioelectronic odor sensing systems where biological components (OR, olfactory tissues, olfactory cells, odorant binding proteins (OBP), synthetic polypeptides, etc.) derived from natural systems, are integrated into the sensing device. The vision is to mimic these natural biological systems using artificial bioelectronic devices such that, odorous substances can be detected at a nanomolar range (Becherelli, Zampetti, Pantalei, Bernabei, & Persaud, 2010).

In this chapter, the concept of electrochemical bioelectronic odor sensing systems, noted in the following text as electrochemical-nose, together with surface immobilization of biological matrixes will be discussed (Fig. 1). The use of electrochemical sensors as biosensors, started in 1962 with the construction of a glucometer using glucose oxidase-based sensors. Since then electrochemical sensors have become one of the most widely accepted biosensing technique. The possibility of miniaturization and multiplexing together with the ability of constructing flexible, disposable and cheap electrochemical sensing devices have made electrochemical sensors very attractive. These sensors are also appealing due to their high sensitivity, simplicity of operation, fast response time, and low cost. The improvements

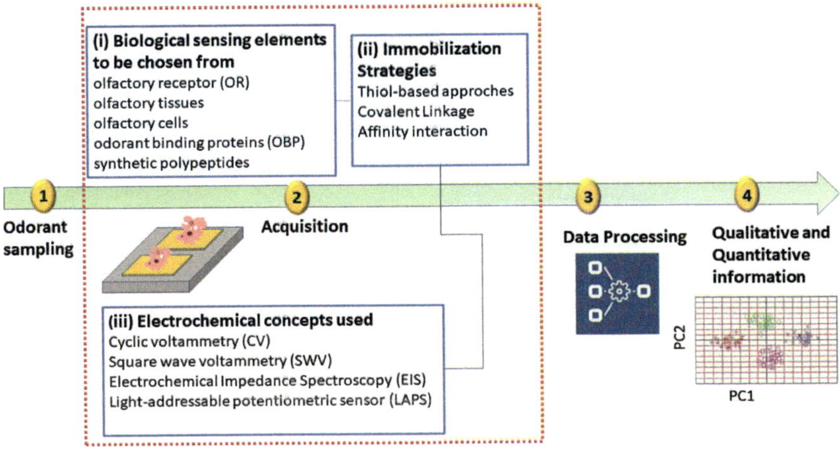

Fig. 1 Design of electrochemical artificial bioelectronic noses (Electrochemical-nose): (1) odorant sampling, (2) acquisition: Before acquisition three steps have to be considered: (i) selection of biological sensing elements to be attached to electrode, (ii) surface immobilization of the bioreceptors by different methods, (iii) choice of the most appropriate electrochemical transducer concept, (3) data processing, (4) qualitative or quantitative information about target odorants.

reported in the detection limits as well in specificity of electrochemical biosensors in the last decade is mainly due to the achievements realized in materials science and subsequent availability of a wide range of nanomaterials and composite materials with good electrical conductivity and/or catalytic activity. Together with the advancement OR receptor proteins and odorant binding proteins, interest of using electrochemical transducer concepts for label-free odor-sensing is slowly emerging (Barou, Sigoillot, Bouvet, Briand, & Meunier-Prest, 2015; Benilova et al., 2008; Guo et al., 2015; Hu et al., 2007; Lee, Jeong, Jun, Kim, & Park, 2009; Lee, Jun, Jo, Kim, & Park, 2009; Liu et al., 2006, 2011; Lu et al., 2014, 2015, 2016). The state of the art of these sensors will be discussed in the following sections.

2. Brief historical outline: From electronic to bioelectronic to electrochemical noses

Identification of odors at low concentration levels is a natural feature of biological smell systems. In humans, the sense of smell is influenced by many factors, such as age, gender, state of health and mood. Quantitative measures of senses are not available and limited to linguistic ability of the individual in communication between one person to another and is highly

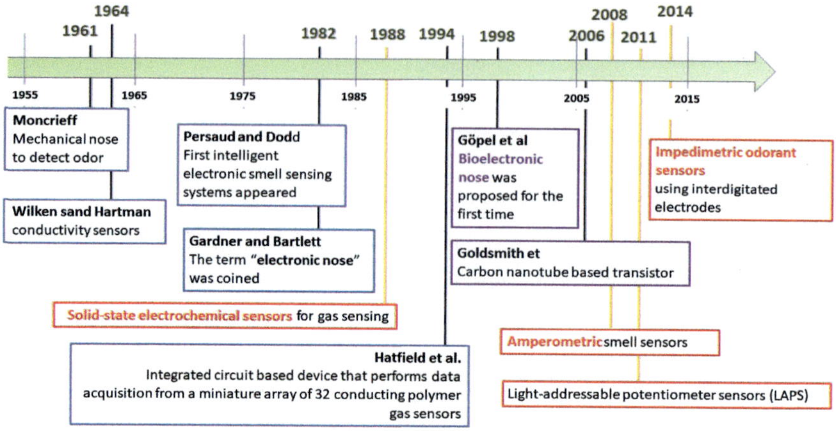

Fig. 2 Milestones in the development of artificial smell systems: from electronic nose (e-nose) to bioelectronic nose: Timeline of Electrochemical based smell sensors.

subjective. Odor sensing systems, widely termed *electronic-noses*, are artificial systems allowing simple odor discrimination and providing measurement of odor intensity without the influence of age, gender, etc. as mentioned above (Fig. 2).

The field of electronic nose is relatively recent dating back to 1961 when Moncrieff developed a mechanical nose to detect odor (Moncrieff, 1961). The proposed concept is based on the use of different coating materials including polyvinyl chloride, gelatin, and vegetable fats, capable of providing different and complementary resistivity data allowing for the discrimination of simple and complex aromas. These studies were limited to the use of a single temperature-sensitive resistor, but postulated that an array with six thermistors, provided with six different coatings, could discriminate large numbers of different aromas. A variety of different sensor types were developed in the following period, such as, the early work by Wilkens and Hartman (1964) based on conductivity sensors measuring a relative resistance change between two electrodes. Different kinds of sensing materials were deposited on these electrodes whose properties change upon interaction with a gas. Three types of materials were commonly used: metal oxides, conducting polymer composites and intrinsically conducting polymers.

The first intelligent electronic smell sensing systems appeared in late 1980's. Persaud and Dodd (1982) reported a successful discrimination of a wide variety of odors using plural semiconductor transducers. The term "electronic nose" was coined in 1988 by Gardner and Bartlett, who defined it as an instrument that is made up of an array of electronic chemical sensors having partial specificity and an appropriate pattern-recognition system, which can recognize simple or complex odors (Craven, Gardener, & Bartlett, 1996; Hatfield, Neaves, Hicks, Persaud, & Travers, 1994) described an integrated circuit based device that performs data acquisition from a miniature array of 32 conducting polymer based gas sensors. David and Gardner designed a circuit capable of measuring signals from arrays of resistive and piezoelectric sensor types in the same board (David & Gardner, 1997).

The *bioelectronic nose* differs from electronic nose (Wilson & Baietto, 2009) as the active sensing material is not made of chemically sensitive elements (e.g., metal oxides, conducting polymers), but of a biological element. Göpel et al. were the first in 1998 to propose the concept of a bioelectronic nose (Göpel et al., 1998). The utilization of biological elements such as proteins of OR as the sensitive element was suggested to improve the sensitivity and selectivity of these smell sensors. One of the major challenges for achieving the bioelectronic nose is the sample preparation and immobilization onto a sensor surface. The first bioelectronic nose was based on the direct immobilization of the olfactory cilia, or of various whole cells expressing ORs. Indeed, there are inherent difficulties associated with the expression of ORs in heterologous cell lines (Lu, Echeverri, & Moyer, 2003) which have enabled the use of recombinant receptors for pharmacological and biophysical studies. Consequently, until now, most bioelectronic noses have been developed for only a few ORs (e.g., rat receptor 17, human OR17-40, *C. elegans* ODR-10, human 2AG1). These receptors served as models to prove the feasibility of the bioelectronic nose concept since the preferential odorant ligand and working concentration ranges are known. The other major challenges for achieving a bioelectronic nose are sample preparation and immobilization onto the sensor surface. Since OR are G protein coupled receptors, they need to stay in their membrane environment to be functional.

Until now, chemo-resistors, field-effect transistors (FET), and quartz crystal microbalance (QCM) have been demonstrated and well-developed as promising candidates for the sensing units of e-nose and bioelectronic nose (Barbosa, Oliveira, & Roque, 2018; Dung et al., 2018; Wasilewski, Gebicki, & Kamysz, 2017, 2018; Zhang et al., 2018). In particular, electrical read out schemes using FET found interest (Goldsmith et al., 2011; Mulla et al., 2015). It was in 2011 when the first electronic approach based on a carbon nanotube based transistor modified with OR proteins was proposed as read out element of OR/odorant binding by Goldsmith et al. (Goldsmith et al., 2011). This work represented a significant progress on a path toward a real bioelectronic nose directly comparable to biological olfactory systems and opened the interest in this field. Since then, a large variety of electrical OR-based biosensors have been proposed (Jin et al., 2012; Lee et al., 2012; Lim et al., 2013; Park, Kwon, et al., 2012; Park, Lim, et al., 2012; Son et al., 2016) such as the one proposed by Lee et al. (2012) mimicking the human olfaction mechanism by using carboxylated polypyrrole nanotubes functionalized with OR proteins for sensing 0.02 ppt of odorant such as helional, piperonal, safrole and phenyl propanol. A carbon nanotube-based FET for the assessment of *Salmonella* contamination with 3-methyl-1-butanol being the specific odorant with a 1 fM detection limit in food using OBP was more lately reported by Son et al. (2016). These types of sensors will be discussed in Chapter 17 "Electronic biosensors based on EGOFETs" by Luisa Torsi et al.

The first ten-years of the bioelectronic nose were based on the use of piezo-electric transducers (Ko & Park, 2005; Sung, Ko, & Park, 2006; Wu, 1999). For this, piezoelectric interfaces such as quartz crystal microbalance (QCM) substrates were modified with isolated OR proteins (ORPs) (Wu, 1999) or were coated with crude membrane extracts, containing the expressed receptor protein (Sung et al., 2006), or with cells containing the ORs (Ko & Park, 2005).

Besides QCM, other mass-sensitive detection methods such as Surface Plasmon Resonance (SPR) have been employed. Vidic et al. immobilized nanosomes on SPR chips and demonstrated the detection of nanomolar concentrations of specific odorants (Vidic et al., 2006). Zhang et al. described a device based on nanocup arrays modified with gold nanoparticles and peptides that specifically bind trinitrotoluene (TNT) (Zhang et al., 2016). Based on localized surface plasmon resonance (LSPR), the sensor detected changes in the refractive index resulting from the binding of TNT to the peptides, and achieved a detection limit of $3.12 \times 10^{-7}\,\text{mg}\,\text{mL}^{-1}$.

Another optical biosensor proposed by Bonnot et al. uses biophotonic ring resonators on a silicon nitride chip which were modified with bovine OBPs (Bonnot et al., 2014). This approach allowed the detection of dimethyl methylphosphonate in 15 min with a detection limit of 6.8 ppb ($35\,\mu\text{g/m}^3$).

Acoustic biosensors have been reported as well, such as the one by Di Pietrantonio et al. (2014), that successfully detected *R*-(−)-1-octen-3-ol (octenol) and *R*-(−)-carvone (carvone).

More lately, Manai et al. described bio-microelectromechanical systems (bioMEMS) for odorant detection made by immobilization of OBPs onto polycrystalline diamond cantilevers (Manai et al., 2014, 2017). OBP's immobilization by affinity between Ni^{2+} from Ni-NTA functionalized diamond surface and a histidine tail of the protein enabled controlled immobilization compared to the random protein orientation resulting from covalent attachments of OBP through standard EDC/NHS coupling to hexanoic acid coated diamond interfaces. Consequently, bioMEMS with OBP immobilized via Ni^{2+}-histidine affinity led to an increased detection sensitivity by 20% for an odorant, 2-isobutyl-3-methoxypyrazine (IBMP), and 2,4-dinitrotoluene (compound similar to the explosive TNT).

The use of electrochemical transducers for gas sensing is still in its infancy. Indeed the first gas electrochemical gas sensors are based on solid state electrochemical sensing of oxygen, carbon dioxide, nitric oxide (NOx) (Inoue, Seki, Eguchi, & Arai, 1990), (Hu et al., 2019) with only some rare examples of odorant sensing (Mori et al., 2012). While solid-state electrochemical sensors have proven high stability and ensured their successful commercialization, these sensors are operated at high temperatures, limiting the integration of bioreceptors. The first electrochemical sensor used electrochemical impedance as detection principle. Measured impedance data can be interpreted using the equivalent circuit, and based on this approach the electrochemical process involved can be separated and the sensing mechanism investigated. A first step toward electrochemical based bioelectronic noses are olfactory cells-based platforms using olfactory cell modified silicon chips as sensing element. Lee, Jun, et al. (2009) proposed the use of microfabricated planar electrodes to measure the extracellular potential generated by odorant stimulation in a cell-based system. One of the first impedimetric odorant biosensors was demonstrated by Lu et al. (2014). It is based on OBPs-modified interdigitated gold electrodes allowing the detection of floral odor and pheromones by electrochemical impedance sensing. The relative decrease of charge resistance was found to be proportional to the logarithm of ligand concentration. This underlines the importance of the sensing matrix in these systems.

Before discussing the electrochemical sensors in detail, different aspects of surface immobilization will be highlighted. Indeed, the first step in any biological odorant sensor is to immobilize the biological element on surfaces without loss of activity. In the following section, we will discuss biomolecule immobilization strategies.

3. Design of a bioelectrical nose: Choice of the biological materials and surface immobilization aspects

The basic elements of any electronic nose are (i) odor delivery system that transfers the molecule of the source material to the sensor array system, (ii) sensor chamber, (iii) transducer that converts the chemical input into an electrical signal and then into digital signal, and (iv) pattern recognition system that takes the digital signal and identifies the odor. The core of the artificial bioelectrical nose is the sensing matrix and the way the matrix is anchored onto the transducer (Fig. 3). Production of the biosensing material is still one of the crucial stages in the construction of bioelectronic noses. However, effective surface immobilization is equally crucial to preserve the three-dimensional structure of the proteins obtained in different expression models. The fundamental aim of each immobilization procedure is the effective linking of the biosensing materials on the transducer without loss of biological activity. In case of ORs, a subfamily of G-proteins coupled receptors, which upon contact with odor molecules result in activation of ionic channels in the olfactory neurons, depolarization of the membrane and change of the action potential, this means being embedded in a cellular

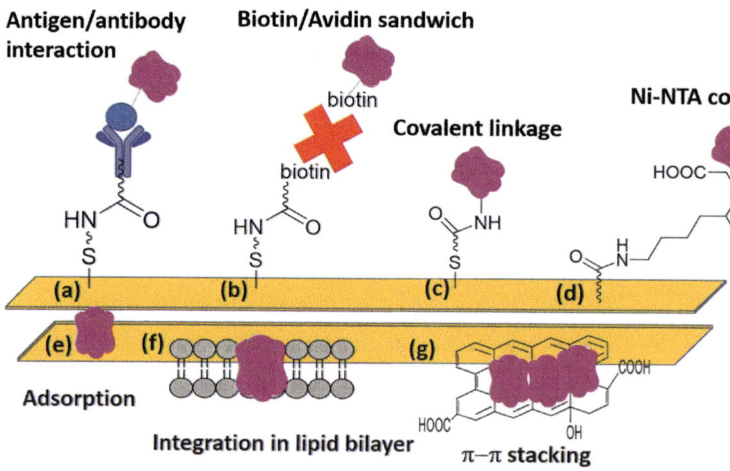

Fig. 3 Surface immobilization strategies in bioelectronic smell sensors: (A) direct linkage using antigen-antibody interactions, (B) direct linking via biotin/avidin interaction, (C) covalent linking via EDC/NHS chemistry, (D) linkage using affinity of histidine-tagged membrane proteins to Ni-NTA modified carbon nanotubes; (E) adsorption (F) embedding of proteins into lipid bilayers, (G) immobilization through π-π stacking.

membrane environment (e.g., lipid layer) to maintain their functionality. Olfactory receptors have some benefits compared to olfactory tissue and cells, such as long-term stability, higher level of activity and easier maintenance. These lipid bilayers, synthesized from phospholipids, are physically connected to the electrochemical interface and suitable proteins are placed into it. This immobilization ensures consistent distribution, control over protein concentration and lack of interferences originating from other membrane proteins present in the fractions of the cell membranes of the expressing systems. The lipid bilayer itself is immobilized on the electrochemical transducer using a variety of immobilization methods including vesicle fusion (Hardy, Nayak, & Zauscher, 2013), Langmuir-Blodgett approaches (Ariga, Yamauchi, Mori, & Hill, 2013) or via direct adsorption or deposition on a polymer cushion. In the case of tethered membranes on a solid support, structural integrity of the protein is the result of the elimination of direct contact with the sensor surface.

Another approach consists of direct linkage of the proteins to the sensor's surface using linkers such as streptavidin-biotin or antibody binding, promoting proper orientation. Goldsmith et al. (2011) proposed immobilization of the recombinant receptors mOR174-9, mOR2031 and mOR256-17, in which histidine reveals affinity with the surface of carbon nanotubes modified with nickel nitrilotriacetic acid (Ni-NTA) (Fig. 3).

Another approach is the immobilization of nanovesicles of 100–200 nm in size, which are formed after the acquisition of ORs-expressing cells that were subjected to the action of cytochalasin B (Jo et al., 2014). Exposure of the nanovesicles to odorants activates a cascade of signals, which result in an increase in the intercellular concentration of corresponding ions inside the nanovesicles, detected by changes in the i/V curves by field effect transistors (FET) (Jin et al., 2012; Park, Lim, et al., 2012).

Next to ORs and nanovesicles, the immobilization of odorant binding proteins (OBPs) has caught increasing attention (Kotlowski et al., 2018; Larisika et al., 2015; Mulla et al., 2015). OBPs are promising sensing elements as they are robust enough to stand up to a wide range of pH and temperature (even to 80–100 °C) without denaturation and losing their binding properties. The easy isolation and purification compared to membrane proteins makes them appealing to discriminate diverse odors. There are about 20 OBP-like genes that have been characterized from the whole genome sequence of the honeybee (*Apis mellifera*) with Acer-ASP2 being an important honeybee OBP with good affinity to many ligands such as floral odorants.

In order for odorants to reach the receptors, the molecule must partition from air to an aqueous phase. In vertebrates, the mucus layer overlying the olfactory epithelium contains an ionic medium bathing the ORs. This medium contains about 1% of small water-soluble proteins called OBPs, because they are capable of binding several types of odorant molecules with low affinity. It is a first filtering mechanism for chemical signals and mediating the activation of the OR. First isolated by Pelosi and co-workers in 1982, their specific role was unclear (Pelosi, Baldaccini, & Pisanelli, 1982). However, their high stability and their diversity make them increasingly popular for biological odor sensors and the development of OBP-based biosensors for gas sensing started to flourish in the last decade. Generally, dissociation constants are in the higher nanomolar or lower micromolar range for strong odorants (Larisika et al., 2015), well adapted for electrochemical read out schemes.

More lately, the immobilization of synthetic peptides was demonstrated. Peptides can be stored for a few months making the material most suitable for commercial applications. Molecular modeling helps in the rational design of the sequence as well as production of ligand/receptor affinities. Peptides self-assembled on the surface of single walled carbon nanotubes (SWCNTs) were modified by Lim et al. by π-π stacking interaction with peptide receptors, bringing the ligand-analyte interaction very close to the transducer (Lim et al., 2013).

4. Bioelectronic noses based on electrochemical transducers

Electrochemical read outs can be divided into three different sensor formats, each will be discussed in more details thereafter.

1. Light-addressable potentiometric sensors (LAPS) measuring electrophysiological processes
2. Sensors based on electrochemical impedance measurements (EIS)
3. Sensors using voltammetry and amperometry detection

4.1 Electrical odor sensors such as light-addressable potentiometric sensors (LAPS)

The mechanism of signal detection and transduction in olfaction is an electrophysiological process, mainly taking place among olfactory epithelium receptor cells and their corresponding mitral cells in olfactory bulb. It is believed that a satisfactory bioelectric nose should be a hybrid system of olfactory neurons and an extracellular potential detection transducer. This was achieved with potentiometric sensors in connection with cell-semiconductor hybrid systems to monitor the extracellular potentials of the olfactory cell. Under light irradiation, the semiconductor absorbs light and produces electron-hole pairs. When the cell produces potential changes by the generation of ionic currents of the Na^+ and K^+, which is equal to a change of bias voltage, fluctuation in the photocurrent is observed. Liu et al. (2006) described a LAPS device made by modifying a silicon semiconductor with poly-l-ornithine and laminin, prior to culturing rat OR neurons on this surface. When exposed to μM concentrations of glutamic acid or acetic acid, the extracellular potential of the OR cells registered changes at a specific frequency. The amplitude of the frequency was correlated with the odorant concentrations.

Lee, Jun, et al. (2009) proposed the use of microfabricated planar electrodes to measure the extracellular potential generated by odorant (octanal) stimulation in a cell-based system (Fig. 4B). Compared to classic electrophysiology setups, planar electrodes allow simplification of the experiments and require less skilled personnel. The planar electrodes were coated with polylysine, then human embryonic kidney (HEK)-293 cells, expressing the OR I7 transfected with gustatory cyclic nucleotide gated (CNG) channel, were cultured on the electrodes. Exposure of cell-coated electrodes to the specific odorant, octanal, led to an intracellular flux of Ca^{2+} ions, which was measured with the planar electrodes. Further on, the researchers found that by stimulating the cells with a biphasic electrical current, the cellular activity increased leading to a five times enhancement of the signal recorded as a response to the odorants (Lee, Jeong, et al., 2009).

4.2 Impedimetric odorant sensors

The first report of an OBP based sensor for the impedimetric detection of volatile compounds in the gas phase describes the immobilization of cysteine-modified recombinant pig OBP by self-assembly on Au IDEs printed on a silicon substrate (Capone et al., 2009). Following exposure to ethanol or methanol vapors in air, the impedance of the system increases. The use of interdigitated electrodes is advanced as a means to amplify the impedimetric signal in comparison with classic three electrode experimental setups. This popular experimental setup for impedimetric sensing, relying on gold IDEs was also used by Liu et al. (2013) for the detection of pheromones and floral odors. The electrodes were coated with a

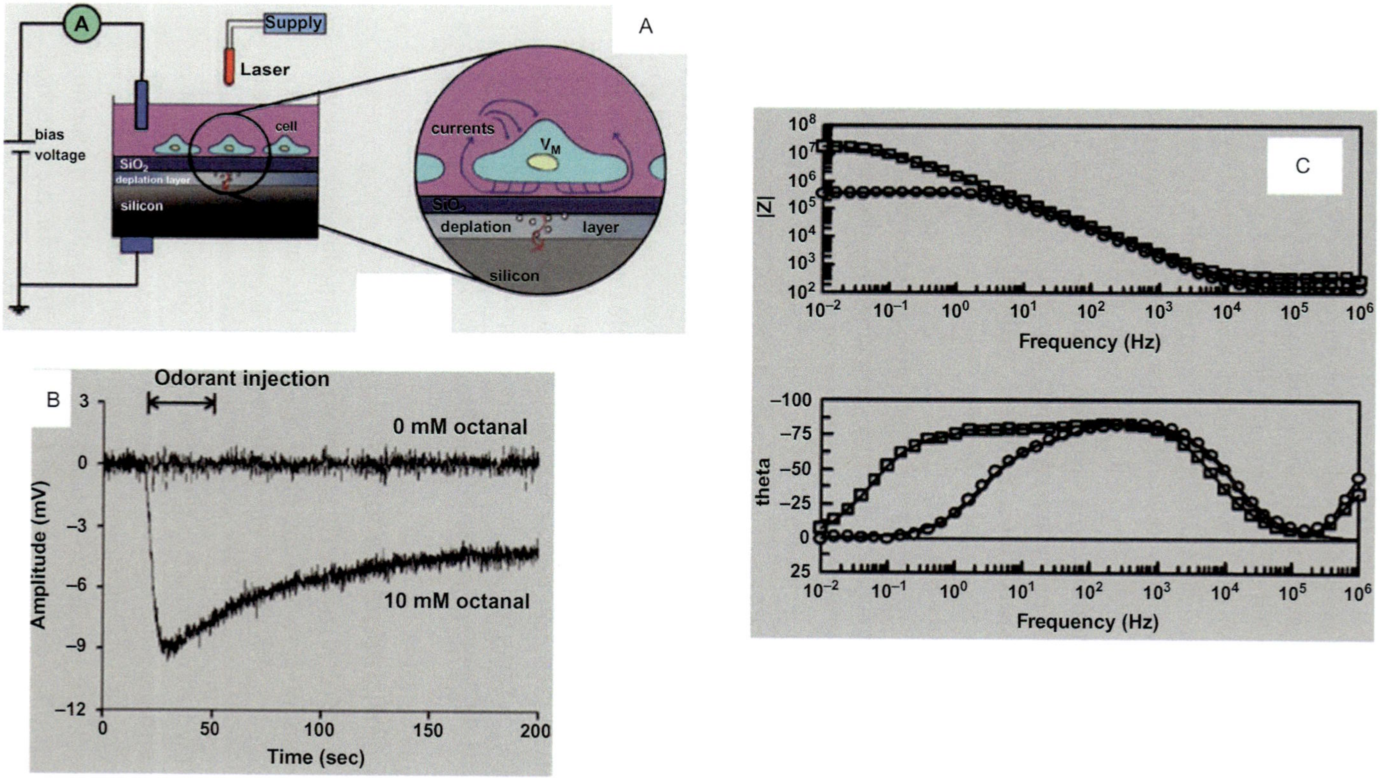

Fig. 4 (A) Scheme of cell-based biosensor using LAPS (according to Ref. Liu et al., 2006). (B) Field potential profile of HEK-293 cells. (C) Electrochemical Impedance Spectroscopy (EIS): Bode plots in the impedance modulus vs. potential modulation frequency and the phase angle θ vs. potential modulation frequency: before (squares), and after (circles) exposure to odorant molecules. *Panel B: reprint from Lee, S. H., Jun, S. B., Jo, H. J., Kim, S. J., & Park, T. H. (2009). Cell-based olfactory biosensor using microfabricated planar electrode.* Biosensors & Bioelectronics, *24, 2659–2664. Panel C: reprint with permission form Hou, Y., Jaffrezic-Renault, N., Martelet, C., Tlili, C., Zhang, A., Pernollet, J. C. et al. (2005). Study of Langmuir and Langmuir-Blodgett films of odorant-binding protein/amphiphile for odorant biosensors.* Langmuir, *21(9), 4058–4065. doi:10.1021/la0471801.*

nitrocellulose membrane to which a recombinant chemosensory protein Ac-ASP3 from honeybee was attached by hydrophobic interactions (Liu et al., 2013). The resistance to charge transfer decreased significantly following the interactions of the chemosensory protein with various odorant ligands (e.g., the bee alarm pheromone isoamyl acetate, the worker bee pheromone geraniol), as a direct result of the conformational changes of the protein subsequent to ligand binding.

Lu et al. (2014) immobilized OBPs from honeybee, i.e. Acer-ASP2, onto an array of interdigitated gold electrodes and followed ligand binding by EIS. The array consisted of 60 pairs of electrodes with a single electrode being 20 μm in width. Nitrocellulose membranes were used for protein immobilization. The same group used this approach for the integration of OBPs from oriental fruit on interdigitated electrodes modified with thiol-PEG-COOH ligands where BdorOBP2 was covalently grafted using EDC/NHS coupling (Lu et al., 2015). In a further development, the group immobilized human odorant-binding proteins (hOBPs) by EDC/NHS coupling on alumina nanopores (20–40 nm) functionalized with 2-carboxyethyl phosphonic acid (Lu et al., 2016). In the presence of odorant molecules such as benzaldehyde, docosahexaenoic acid, and lauric acid, the analysis of EIS spectra of the hOBP–functionalized nanopore arrays emphasized an increase in the series electrical resistance of nanopores that was correlated with odorant concentration.

Hou et al. (2005) have studied the functionalization of gold electrodes with a recombinant rat OBP. Stable OBP layers were obtained when the electrodes were first coated with a self-assembled monolayer of octadecanethiol, and then modified by depositing a mixture of OBP and octadecylamine and stabilizing it by cross-linking with glutaraldehyde. Non-Faradaic EIS tests revealed that the resistance of the mixed OBP-1F/ODA LB film decreases as a result of odorant binding (from 1.18 MΩ to 25 kΩ) (Fig. 4C). This study provides the sensing principle for quantitative detection of odorant.

Yao et al. (2015) used an array of IDEs and Acer-ASP2 OBP from honeybees added in the electrolyte solution for real time monitoring of odorant-protein binding by EIS. The resistance to charge transfer became smaller with the increase in the concentration of odorant ligand. The IDE array consisted of 63 electrodes at the bottom of each well (Yao et al., 2015).

Khadka et al. (2019) integrated odorant receptors Or71a, Or10a and Or22a from *Drosophila melanogaster* into nano-sized liposomes that were subsequently covalently attached to a thiol coated gold electrode to sensitively detect specific odorant ligands, i.e., 4-ethylguaiacol, methyl salicylate and methyl hexanoate, respectively. The electrode was polarized at −0.7 V and the detection was performed in PBS buffer by non-faradaic EIS.

4.3 Voltammetric odorant sensors

A different electroanalytical binding assay for the detection of odorant molecules was proposed by Barou et al. (2015). Recombinant rOBP3 expressed as a His-tagged N-terminal protein was immobilized on an Ni^{2+}-agarose support and the redox probe 2-methyl-1,4-natphtoquinine (MNQ), able to bind rOBP3, was used for the sensing of a model odorant molecule, 2-isobutyl-3-methoxypyrazine (IBMP). The competitive test relied on the displacement of MNQ bound to rOBP3 by the odorant molecule IBMP followed by the quantitative detection of the released MNQ via its electrochemical reduction at the surface of a gold electrode. The measurements were performed by square wave voltammetry and ferricyanide was used as a recycling molecule to amplify the electrochemical signal.

More lately, Lu et al., 2017 proposed reduced graphene oxide (rGO) functionalized electrodes for the sensing of "fat taste" indicator compounds docosahexaenoic acid, linoleic acid, and lauric acid via cyclic voltammetry. Recombinant human OBPII proteins were covalently attached to rGO-coated screen-printed carbon electrodes and the taste indicator compounds were added in the electrochemical cell together with ferricyanide as redox probe. The decrease in the oxidation and reduction peak intensity of $[Fe(CN)_6]^{3-/4-}$ recorded by cyclic voltammetry was correlated with the concentration of the taste indicator present in the electrochemical cell. The sensor exhibited higher affinity to longer alkyl chain molecules. The specific detection of octenol was achieved by Misawa et al. (Misawa et al., 2019) with an amperometric sensor made by reconstructing mosquito ORs in a bilayer lipid membrane in a miniaturized cell equipped with two Ag/AgCl electrodes. A difference of 60 mV was applied between the two electrodes and the ion current changes following the binding of octenol to the ORs were correlated with the concentration of octenol down to 0.01 ppm. The sensor was installed in a mobile robot and octenol sensing was coupled with robot's movements (Misawa et al., 2019) (Table 2).

Table 2 Bioelectronic noses based on electrochemical detection schemes and OBPs.

Receptor	Transducer	Support	Analyte	LOD	Ref
Rat OR cells	LAPS	SiO_2	Glutamic acid Acetic acid	25 μM	Liu et al. (2006)
Human embryonic kidney (HEK)-293 cells expressing the OR I7 and gustatory cyclic nucleotide gated channel (CNG)	Potentiometric	Au planar microelectrode array (MEA)	Octanal	1 mM	Lee, Jun, et al. (2009)
pOBP	EIS	Gold IDE	Methanol ethanol	312–434 μM	Capone et al. (2009)
Ac-ASP3 from honeybee	EIS	Gold IDE	Isoamylacetate geraniol Phenyl-acetaldehyde	1–0.1 μM	Liu et al. (2013)
Rat OBP-1F	Non-faradaic EIS	Gold	Isoamylacetate	Not available	Hou et al., 2005
Honeybee OBPs Acer-ASP2	EIS	Gold IDE	Isoamylacetate	1 nM	Yao et al. (2015)
hOBP	EIS	Al_2O_3 nanopores array	Benzaldehyde docosa-hexaenoic acid lauric acid	0.003–50 μM	Lu et al. (2016)
BdorOBP2 from oriental fruit fly	EIS	Gold IDE	Isoamylacetate β-ionone, benzaldehyde	3–84 μM	Lu et al. (2015)

Continued

Table 2 Bioelectronic noses based on electrochemical detection schemes and OBPs.—cont'd

Receptor	Transducer	Support	Analyte	LOD	Ref
Honeybee OBPs (Acer-ASP2)	EIS	Gold IDE	Linalool geraniol β-ionone 4-allylveratrolephenyl-acetaldehyde Dibutyl-phthalate	1–1000 μM	Lu et al. (2014)
Or7 and Or8 of yellow fever mosquito (*Aedes aegypti*)	Amperometry	Ag/AgCl	1-octen-3-ol	0.01 ppm	Misawa et al. (2019)
Or71a Or10a Or22a from Drosophila	Non faradaic EIS	Gold	4-Ethylguaiacol Methyl -salicylate methyl -hexanoate	0.1 pM -0.1 aM	Khadka et al. (2019)
rOBP3	SWV	Gold	2-Isobutyl-3-methoxypyrazine (IBMP)	Not determined	Barou et al. (2015)
hOBPII	CV	rGO/SPCE	Docosahexaenoic acid Linoleic acid Lauric acid	0.18–5 μM	Lu et al. (2017)

4.4 Equipment

1. Potentiostat/galvanostat (e.g., Metrohm Autolab) with impedance module
2. Counter (Pt plate or wire) and Reference electrodes (Saturated calomel electrode SCE, KCl-saturated Ag/AgCl wires)
3. Interdigitated gold electrodes
4. Au planar microelectrode array (MEA)
5. Al_2O_3 nanopores array
6. Light-addressable potentiometric sensors (LAPS)

4.5 Reagents

1. polyvinyl chloride (Sigma-Aldrich)
2. gelatin (Sigma-Aldrich)
3. 3-methyl-1-butanol (Sigma-Aldrich)
4. carboxylated pyrrole (Sigma-Aldrich)
5. streptavidin (Sigma-Aldrich)
6. biotin (Sigma-Aldrich)
7. 1-ethyl-3-(3-dimethylaminopropyl)carbodiimide (EDC) (Sigma-Aldrich)
8. N-hydroxysuccinimide (NHS) NHS (Sigma-Aldrich)
9. nitrilotriacetic acid (NTA) (Sigma-Aldrich)
10. octanal, octenol, methanol, ethanol (Sigma-Aldrich)
11. phenyl-acetaldehyde (Sigma-Aldrich)
12. polylysine (Sigma-Aldrich)
13. human embryonic kidney (HEK)-293 cells
14. single walled carbon nanotubes (SWCNTs) (Sigma-Aldrich)
15. 2-carboxyethyl phosphonic acid (Sigma-Aldrich)
16. His-tagged N-terminal protein
17. 2-isobutyl-3-methoxypyrazine (IBMP) (Sigma-Aldrich)
18. 2-methyl-1,4-natphtoquinine (MNQ) (Sigma-Aldrich)
19. reduced graphene oxide (rGO) (Sigma-Aldrich)
20. docosahexaenoic acid, linoleic acid, lauric acid (Sigma-Aldrich)
21. Ferricyanide $[Fe(CN)_6]^{3-/4-}$ (Sigma-Aldrich)

5. Protocols

5.1 Detection of octanal with LAPS sensors based on planar microelectrodes

1.1 Culture the HEK 293 cells expressing the OR I7 odorant binding receptor and transfected with gustatory cyclic nucleotide gated

(CNG) channel on poly-D-lysine coated microfabricated planar electrodes and grow them in DMEM medium supplemented with 10% FBS in 5% CO_2.

1.2 Insert the cell-coated electrodes in a measurement cell and connect it to a detection system (e.g., oscilloscope) and to a delivery system for providing odorant solutions to the measurement cell (e.g., syringe pump).

1.3 Prepare odorant solution of different concentrations by serial dilution from a 1M stock solution with the running buffer.

1.4 Stimulate the cells with a biphasic electrical current and expose them to different concentrations of odorant (octanal), record the variation of the field potential of the cells in time due to the intracellular flux of Ca^{2+} ions upon the introduction of odorant.

1.5 Correlate the field amplitude with odorant dose (Lee, Jeong, et al., 2009).

5.2 Detection of floral odors and pheromones with impedimetric biosensor based on odorant-binding proteins from honeybee

2.1 Functionalize interdigitated electrode arrays, appropriately coated (e.g. with nitrocellulose (Lu et al., 2014) or thiols (Hou et al., 2005)) with odorant binding proteins.

2.2 Insert the electrode array in the measurement cell and connect it to the electrochemical station, set to acquire electrochemical impedance data over a frequency range of 100kHz to 0.01Hz.

2.3 For Faradaic EIS, add 5mM $K_4[Fe(CN)_6]/K_3[Fe(CN)_6]$(1:1) in the electrolyte in the sample measurement cell.

2.4 Add odorant ligands in the cell, record the impedance until obtaining a stable signal and using an appropriate equivalent electrical circuit, calculate the resistance to charge transfer (R_{CT}) at electrode surface for different concentrations of odorants.

2.5 Plot the R_{CT} versus the odorant concentration for different odorant molecules.

5.3 Sensing of "fat taste" indicator compounds with voltametric screen-printed biosensor based on OBP

3.1 Modify screen-printed electrodes with electrochemically reduced graphene oxide and covalently attach odorant binding proteins

3.2 Insert the electrodes in the measurement cell, equilibrate with PBS buffer pH 7.4 and connect to the potentiostat. Set the potential range from 1.0 V to −1.0 V and the scan rate at 50 mV/s.

3.3 Add $Fe(CN)_6^{3-/4-}$ (1:1) and the "fat taste" indicator compounds docosahexaenoic acid, linoleic acid, and lauric acid in the cell and record the cyclic voltammograms.

3.4 Plot the cathodic peak current versus the concentration of odorant.

3.5 Wash the cell with ultrapure water after each measurement and immerse in PBS buffer for 5 min to flush residual odorants and recover the OPB activity for the next measurement.

6. Summary

Despite significant progress in the improvements of electrical and bioelectronic nose devices, there are still limitations on their practical applications stemming from the production of the biological ligand to maintaining desired stability and repeatability of measurements. Odorant binding proteins (OBP) and artificial polypeptides seem to be the most convenient biomatrices for commercial applications. As discussed in this chapter, electrochemical techniques compared to mass-based (SPR, QCM) and field effect modulation detection schemes (FET) are still rather rarely employed in this highly exciting field at the interface between materials science, biology and computing.

The contribution of the involvement of the combination of receptors, cell signaling and electrochemistry in many aspects of biology and medicine is documented. It is thus reasonable to apply electrochemical sensing approach to smell identification. Moreover, electrochemical sensors are scalable for mass fabrication and can be made flexible adapting themselves well for potential electronic noses. It is thus expected, that in the near future, dynamic progress in this area will contribute to sensitive and selective electrochemical smell sensors development with portable electrochemical noses being of high interest for medical applications as well as for the integration into food package chain to ensure food quality.

References

Abaffy, T. (2015). Human olfactory receptors expression and their role in non-olfactory tissues– A mini-review. *Journal of Pharmacogenomics & Pharmacoproteomics*, *6*, 4.

Ariga, K., Yamauchi, Y., Mori, T., & Hill, J. P. (2013). 25th anniversary article: What can be done with the Langmuir-Blodgett method? Recent developments and its critical role in materials science. *Advanced Materials*, *25*, 6477.

Barbosa, A. J. M., Oliveira, A. R., & Roque, A. C. A. (2018). Protein- and peptide-based biosensors in artifical olfaction. *Trends in Biotechnology*, *36*, 1244.

Barou, E., Sigoillot, M., Bouvet, M., Briand, L., & Meunier-Prest, R. (2015). Electrochemical detection of the 2-isobutyl-3-methoxypyrazine model odorant based on odorant-binding proteins: The proof of concept. *Bioelectrochemistry*, *101*, 28–34.

Becherelli, R., Zampetti, E., Pantalei, S., Bernabei, M., & Persaud, K. C. (2010). Design of a very large chemical sensor system for mimicking biological olfaction. *Sensors and Actuators B: Chemical*, *146*, 446.

Beets, M. G. J. (1978). Structure-activity relationships. In *Human chemoreception*. London: Applied Science Publishers Ltd. 221.

Benilova, I. V., Minic Vidic, J., Pajot-Augy, E., Soldatkin, A. P., Martelet, C., & Jaffrzeic-Renault, N. (2008). Electrochemical study of human olfactory receptor OR 17–40stimulation by odorants in solution. *Materials Science and Engineering: C*, *28*, 633–939.

Bonnot, K., Cuesta-Soto, F., Rodrigo, M., Varriale, A., Sanchez, N., D'Auria, S., et al. (2014). Biophotonic ring resonator for ultrasensitive detection of DMMP as a simulant for organophosphorus agents. *Analytical Chemistry*, *86*(10), 5125–5130. https://doi.org/10.1021/ac500903s.

Capone, S., De Pascali, C., Francioso, L., Siciliano, P., Persaud, K. C., & Pisanelli, A. M. (2009). Electrical characterization of a pig odorant binding protein by impedance spectroscopy. *Proceedings of IEEE Sensors*, 1758–1762.

Craven, M. A., Gardener, J. W., & Bartlett, P. N. (1996). Electonic boses-development and future prospects. *Trends in Analytical Chemistry*, *15*, 486.

David, D. C., & Gardner, J. W. (1997). High-precisionintelligent interface for a hybrid electronic nose. *Sensors and Actuators, B: Chemical*, *62*, 724–728.

de March, C. A., Kim, S. K., Antonczak, S., Goddard, W. A., 3rd, & Golebiowski, J. (2015). G protein-coupled odorant receptors: From sequence to structure. *Protein Science*, *24*(9), 1543–1548. https://doi.org/10.1002/pro.2717.

Di Pietrantonio, F., Benetti, M., Dinca, V., Cannatà, D., Verona, E., D'Auria, S., et al. (2014). Tailoring odorant-binding protein coatings characteristics for surface acoustic wave biosensor development. *Applied Surface Science*, *302*, 250–255. https://doi.org/10.1016/j.apsusc.2013.10.112.

Dung, T. T., Oh, Y., Choi, S.-J., Kim, U.-D., Oh, M.-K., & Kim, M. (2018). Applications and advances in bioelectonic noses for odour sensing. *Sensors*, *18*, 103.

Goldsmith, B. R., Mitala, J. J., Josue, J., Castro, A., Lerne, R. M. B., Bayburt, T. H., et al. (2011). Biomimetic chemical sensors using nanoelectronic read out of olfactory receptor proteins. *ACS Nano*, *5*, 5408–5416.

Göpel, W., Ziglera, C., Breer, H., Schild, D., Apfelbach, R., Joerges, J., et al. (1998). Bioelectronic noses: A status report part I. *Biosensors & Bioelectronics*, *13*, 479–493.

Graben, V., & Sachse, S. (2018). Fundamental principles of the olfactory codeAuthor links open overlay panelVeitGrabeSilkeSachse. *Biosystems*, *164*, 94–101.

Guo, Z. Z., Zine, N., Lagarde, F., Daligault, J., Persuy, M. A., Pajot-Augy, E., et al. (2015). A novel platform based on immobilized histidine tagged olfactoryreceptors, for the amperometric detection of an odorant moleculecharacteristic of boar taint. *Food Chemistry*, *184*, 1–6.

Hardy, G. J., Nayak, R., & Zauscher, S. (2013). Model cell membranes: Techniques to form complex biomimetic supported lipid bilayers via vesicle fusion. *Current Opinion in Colloid & Interface Science*, *18*, 448–458.

Hatfield, V., Neaves, P., Hicks, P. J., Persaud, K., & Travers, P. (1994). Towards and integratedelectronic nose using conducting polymer sensors. *Sensors and Actuators, B: Chemical*, *26-27*, 221–228.

Hou, Y., Jaffrezic-Renault, N., Martelet, C., Tlili, C., Zhang, A., Pernollet, J. C., et al. (2005). Study of Langmuir and Langmuir-Blodgett films of odorant-binding protein/amphiphile for odorant biosensors. *Langmuir*, *21*(9), 4058–4065. https://doi.org/10.1021/la0471801.

Hu, T., Jaffrezic-Renault, N., Martelet, C., Zhang, A., Minic, J., Gorojankina, T., et al. (2007). A novel detection strategy for odorant molecules based on controlledbioengineering of rat olfactory receptor I7. *Biosensors & Bioelectronics*, *22*, 1550–1555.

Hu, W., Wang, L., Jian, Y., Ren, C., Jin, K., Su, X., et al. (2019). Electronic noses: From advanced materials to sensors aided with data processing. *Advanced Materials Technologies*, *4*, 1800488.

Inoue, T., Seki, N., Eguchi, K., & Arai, H. (1990). Low-temperature operation of solid electrolyte oxygen sensors using perovskite-type oxide electrodes and cathodic reaction kinetics. *Journal of the Electrochemical Society*, *137*, 2523–2527.

Jin, H. J., Lee, S. H., Kim, T., Park, J., Song, H., Park, T., et al. (2012). Nanovesicle-based bioelectronic nose platform mimicking human olfactory signal transduction. *Biosensors & Bioelectronics*, *35*, 335–341.

Jo, W., Kim, J., Yoon, J., Jeong, D., Cho, S., Jeong, H., et al. (2014). Large-scale generation of cell-derived nanovesicles. *Nanoscale*, *6*, 12056.

Khadka, R., Aydemir, N., Carraher, C., Hamiaux, C., Colbert, D., Cheema, J., et al. (2019). An ultrasensitive electrochemical impedance-based biosensor using insect odorant receptors to detect odorants. *Biosensors & Bioelectronics*, *126*, 207–213. https://doi.org/10.1016/j.bios.2018.10.043.

Ko, H. J., & Park, T. H. (2005). Piezoelectric olfactory biosensor: Ligand specificity and dose-dependence of an olfactory receptor expressed in a heterologous cell system. *Biosensors & Bioelectronics*, *20*, 1327.

Kotlowski, C., Aspermair, P., Reiner-Rozman, C., Breu, J., Szunerits, S., Kim, J.-J., et al. (2018). Electronic biosensing with flexible organic transistor devices. *Flexible and Printed Electronics*, *3*, 034003.

Larisika, M., Kotlowski, C., Steininger, C., Mastrogiacomo, R., Pelosi, P., Schütz, S., et al. (2015). Electronic olfactory sensor based on A. mellifera odorant-binding protein 14 on a reduced graphene oxide field-effect transistor. *Angewandte Chemie International Edition*, *54*, 13245–13248.

Lee, S. H., Jeong, S. H., Jun, S. B., Kim, S. J., & Park, T. H. (2009). Enhancement of cellular olfactory signal by electrical stimulation. *Electrophoresis*, 3283–3288.

Lee, S. H., Jun, S. B., Jo, H. J., Kim, S. J., & Park, T. H. (2009). Cell-based olfactory biosensor using microfabricated planar electrode. *Biosensors & Bioelectronics*, *24*, 2659–2664.

Lee, S. H., Kwon, O. S., Song, H. S., Park, S. J., Sung, J. H., Jang, J., et al. (2012). Mimicking the human smell sensing mechanism with an artifical nose platfomr. *Biomaterials*, *33*, 1722–1729.

Lim, J. H., Park, J., Ahn, J. H., Jin, H. J., Hong, S., & Park, T. H. (2013). A peptide receptor-based bioelectronic nose for the real-time determination of seafood quality. *Biosensors & Bioelectronics*, *39*, 244–249.

Liu, Q., Cai, H., Xu, Y., Li, Y., Li, R., & Wang, P. (2006). Olfactory cell-based biosensor: a first step towards a neurochip of bioelectronic nose. *Biosensors & Bioelectronics*, *22*, 318–322.

Liu, Q., Hu, B., Ye, W., Cai, H., Zhang, F., & Wang, P. (2011). Extracellular recording of spatiotemporal patterning in response to odors in the olfactory epithelium by microelectrode arrays. *Biosensors & Bioelectronics*, *27*, 12–17.

Liu, Q., Wang, H., Li, H., Zhang, J., Zhuang, S., Zhang, F., et al. (2013). Impedance sensing and molecular modeling of an olfactory biosensor based on chemosensory proteins of honeybee. *Biosensors and Bioelectronics*, *40*(1), 174–179. https://doi.org/10.1016/j.bios.2012.07.011.

Lu, M., Echeverri, F., & Moyer, B. D. (2003). Endoplasmic reticulum retention, degradation, and aggregation of olfactory G-protein coupled receptors. *Traffic*, *4*, 416–433.

Lu, Y., Huang, Y., Li, S., Zhang, Q., Wu, J., Xiong, Z., et al. (2017). Fat taste detection with odorant-binding proteins (OBPs) on screen-printed electrodes modified by reduced graphene oxide. *Sensors and Actuators, B: Chemical*, *252*, 973–982.

Lu, Y., Li, H., Zhuang, S., Zhang, D., Zhang, Q., Zhou, J., et al. (2014). Olfactory biosensor using odorant binding proteins from honeybee: Ligands of floral odors and pheromones detection by electrochemical impedance. *Sensors and Actuators, B: Chemical, 193*, 420–427.

Lu, Y., Yao, Y., Zhang, Q., Zhang, D., Zhuang, S., Li, H., et al. (2015). Olfactory biosensor for insect semiochemicals analysis by impedance sensing of odorant-binding proteins on interdigitated electrodes. *Biosensors & Bioelectronics, 67*, 662–669.

Lu, Y., Zhang, D., Zhang, Q., Huang, Y., Luo, S., Yao, Y., et al. (2016). Impedance spectroscopy analysis of human odorant binding proteins immobilized on nanopore arrays for biochemical detection. *Biosensors & Bioelectronics, 79*, 251–257.

Manai, R., Habchi, M., Kamouni-Belghiti, D., Persuy, M. A., Rousseau, L., Possas Abreu, M., et al. (2017). Diamond micro-cantilevers as transducers for olfactory receptors - based biosensors: Application to the receptors M71 and OR7D4. *Sensors and Actuators B: Chemical, 238*, 1199–1206.

Manai, R., Scorsone, E., Rousseau, L., Ghassemi, F., Possas Abreu, M., Lissorgues, G., et al. (2014). Grafting odorant binding proteins on diamond bio-MEMS. *Biosensors & Bioelectronics, 60*, 311–317. https://doi.org/10.1016/j.bios.2014.04.020.

Misawa, N., Fujii, S., Kamiya, K., Osaki, T., Takaku, T., Takahashi, Y., et al. (2019). Construction of a biohybrid odorant sensor using biological olfactory receptors embedded into bilayer lipid membrane on a chip. *ACS Sensors, 4*(3), 711–716. https://doi.org/10.1021/acssensors.8b01615.

Moncrieff, R. W. (1961). An instrument for measuring and classifing odor. *Journal of Applied Physiology, 16*, 742–749.

Mori, M., Itagaki, Y., Sadaoka, Y., Nakagawa, S.-I., Kida, M., & Kojima, T. (2012). Detection of offensive odorant in air with a planar-type potentiometric gas sensor based on YSZ with Au and Pt electrodes. *Sensors and Actuators, B: Chemical, 161*, 471.

Mori, K., & Sakano, H. (2011). How is the olfactory map formed and interpreted in the mammalian brain? *Annual Review of Neuroscience, 34*, 467–499.

Mulla, M. Y., Tuccori, E., Magliulo, M., Lattanzi, G., Palazzo, G., Peraud, K., et al. (2015). Capacitance-modulated transistor detects odorant binding protein chiral interactions. *Nature Communications, 6*, 6010.

Olender, T., Nativ, N., & Lancet, D. (2013). HORDE: Comprehensive resource for olfactory receptor genomics. *Methods in Molecular Biology, 1003*, 23–38.

Park, S. J., Kwon, O. S., Lee, S. H., Song, H. S., Park, T. H., & Jang, J. (2012). Ultrasensitive flexible graphene based field-effect transistor (FET)-type bioelectronic nose. *Nano Letters, 12*, 5082–5090.

Park, J., Lim, J. H., Jin, H. J., Namgung, S., Lee, S. H., Park, T. H., et al. (2012). A bioelectronic sensor based on canine olfactory nanovesicle–carbon nanotube hybrid structures for the fast assessment of food quality. *Analyst, 137*, 3249.

Pelosi, P., Baldaccini, N. E., & Pisanelli, A. M. (1982). Identification of a specific olfactory receptor for 2-isobutyl-3-methoxypyrazine. *The Biochemical Journal, 201*, 245–248.

Pelosi, P., Zhu, J., & Knoll, W. (2018). From gas sensors to biomimetic artificial noses. *Chemosensors, 6*(32), 1–16.

Persaud, K., & Dodd, G. (1982). Analysis of discrimination mechanisms in the mammalian olfactory system using a model nose. *Nature, 299*, 352–355.

Son, M., Kim, D., Kang, J., Lim, J. H., Lee, S. H., Ko, H. J., et al. (2016). Bioelectronic nose using odorant binding protein-derived peptide and carbon nanotube field-effect transistor for the assessment of Salmonella contamination in food. *Analytical Chemistry, 88*, 11283.

Sung, J. H., Ko, H. J., & Park, T. H. (2006). Piezoelectric biosensor using olfactory receptor protein expressed in Escherichia coli. *Biosensors & Bioelectronics, 21*, 1981.

Vidic, J. M., Grosclaude, J., Persuy, M.-A., Aioun, J., Salesse, R., & Pajot-Augy, E. (2006). Quantitative assessment of olfactory receptors activity in immobilized nanosomes: A novel concept for bioelectronic nose. *Lab on a Chip*, *6*(8), 1026–1032. https://doi.org/10.1039/B603189G.

Voshall, L. B. (2016). Laying a controversial smell theory to rest. *PNAS*, *112*(21), 6525–6526.

Wasilewski, T., Gebicki, J., & Kamysz, W. (2017). Bioelectronic noce: Current status and perspectives. *Biosensors & Bioelectronics*, *87*, 480–494.

Wasilewski, T., Gebicki, J., & Kamysz, W. (2018). Advances in olfaction-inspired biomaterials appled to bioelectronic noses. *Sensors and Actuators, B: Chemical*, *257*, 511–537.

Wilkens, W. F., & Hartman, J. D. (1964). An electronic analogue for the olfactory process. *Annals of the New York Academy of Sciences*, *116*, 608–612.

Wilson, A. D., & Baietto, M. (2009). Applications and advances in electronic-nose technologies. *Sensors*, *9*, 5099–5148.

Wu, T. Z. (1999). A piezoelectric biosensor as an olfactory receptor for odour detection: Electronic nose. *Biosensors & Bioelectronics*, *14*, 9–18.

Yao, Y., Lu, Y., Zhang, Q., Zhang, D., Zhuang, S., Li, H., et al. (2015). Biomimetic olfactory sensor for dynamically and conveniently monitoring ligands binding to odorant-binding proteins with impedance sensing. *International Journal of Electrochemical Science*, *10*, 5548–5560.

Zhang, X., Cheng, J., Wu, L., Mei, Y., Jaffrezic-Renault, N., & Guo, Z. Z. (2018). An overview of an artificial nose system. *Talanta*, *184*, 93–102.

Zhang, D., Zhang, Q., Lu, Y., Yao, Y., Li, S., Jiang, J., et al. (2016). Peptide functionalized nanoplasmonic sensor for explosive detection. *Nano-Micro Letters*, *8*(1), 36–43. https://doi.org/10.1007/s40820-015-0059-z.

CHAPTER SIXTEEN

Electronic biosensors based on graphene FETs

Patrik Aspermair[a,b,c], Vladyslav Mishyn[b], Sabine Szunerits[b], Wolfgang Knoll[a,c,*]

[a]CEST Competence Center for Electrochemical Surface Technology, Tulln, Austria
[b]University of Lille, CNRS, Centrale Lille, Université Polytechnique Hauts-de-France, UMR 8520—IEMN, Lille, France
[c]AIT Austrian Institute of Technology GmbH, Biosensor Technologies, Tulln, Austria
*Corresponding author: e-mail address: wolfgang.knoll@ait.ac.at

Contents

Abstract

Olfaction is capable of accomplishing incredible tasks: it starts with capturing an odor molecule, delivering it to the odorant receptors, converting it into an electrical stimulus and transmitting the data to the brain. And all of this in milliseconds. The sense of smell is not yet fully decoded and is far from being replicated by modern sensor technologies.

Methods in Enzymology, Volume 642
ISSN 0076-6879
https://doi.org/10.1016/bs.mie.2020.05.016

One approach to convert biological recognition- and binding events in real-time and in a label-free manner to electrical signals is emulated in a "biomimetic electronic smell sensor". It is based on a transistor, in many cases realized as a field-effect transistor (FET) with a biorecognition element, e.g., an odorant binding protein (OBP) converting the binding event of one of its typically many ligands directly into a measurable electrical signal. OBPs are immobilized on these FETs and modulate the current in the presence of smell molecules due to the charge redistribution in the gated channel. Graphene is an elegant candidate to realize such a sensor device because an atomic monolayer of a semiconducting material leads to increased sensitivity. Beside the direct molecule interaction with the substrate upon binding and its excellent biocompatible character, graphene has the advantage of a biological-friendly working point in the sub-Volt regime. Different approaches of preparation and functionalization of graphene field-effect transistors (gFETs) are utilized to tune the performance for odorant sensing. The evaluation of kinetic binding parameters like association and dissociation rate constants and the equilibrium affinity constants of protein-ligand interactions can be derived from the direct electrical read-out of such miniaturized sensor systems. In this article, the state of the art of gFET preparation, functionalization, and operation for odorant sensing will be discussed.

1. Introduction

The development of a field-effect transistor (FET), enhancing with an applied electric field the electrical signals monitored upon biological binding events, is a lab-on-a-chip approach which needs just a few device components and offers short assay times. Graphene based FETs, so-called gFETs, combined with surface immobilized biorecognition molecules, such as antibodies, enzymes or receptor proteins, enable the detection of analytes within minutes or even seconds in the best cases. The one atom thick transducer later allows for the electric semi-conductor to be in direct contact with the biomolecular binding event. Even the smallest binding events influence the charge carriers in the graphene layer and modulate the sensing signal, enabling a direct electrical readout. Biological smell detection systems, such as the dog's nose or the ability of the silk moth, *B. mori*, to detect single odorant molecules are so fascinating by their efficiency, that scientists have been trying to imitate these sensor systems.

Smell, however, is much more than just detecting a single molecule; it is like creating a picture composed of the stimuli of the individual smell sensor pixels, present as receptors in the nose. The human nose has a limited amount of receptors and is able to distinguish more than one trillion different odors (Bushdid, Magnasco, Vosshall, & Keller, 2014). This extremely high number is the result of receptor combinations with different signal magnitudes and stimulation varieties. The nose, like the eyes, contains special receptors, which are all needed to send signals for a specific odor. However, the way the eyes work is easily understood, as they have only four different receptors, three for colors and one for brightness. The emulation of the eyes receptors is well established in CCD cameras or CMOS chips, where each pixel represents a point of a projected image on a chip, able to discriminate between brightness and wavelength of the light.

For olfaction, the question remains on how can one create an electric signal corresponding to the intensity of a specific odor? Once this question is answered, a camera-like matrix system could be established with more than just one of those odorant pixels (*an oxel*), leading to combinations of odorants and creating a digital image of the present odor. One approach of such an odorant detection system in real-time without labeling starts with a field effect transistor, based on graphene. When a particular species of odorant binding proteins (OBP), e.g., from *Apis mellifera* (western honey bee) is immobilized on a FET surface, the binding event induces local field changes, transducing the molecule interaction directly to an electrical response.

2. Different approaches of graphene FET preparation

For the production of a graphene-based field effect transistor (gFET), it is essential to have a fast and processable preparation protocol for the sensors to ensure reproducible electrical signals with a high signal-to-noise ratio. Two different strategies, common for the fabrication of gFETs, are distinguishable through the source of graphene which is either

- chemically derived graphene oxide (GO) using graphite as starting material, further reduced to reduced graphene oxide (rGO), and
- physically derived pristine graphene using notably chemical vapor deposition (CVD) of mono- and multilayers of graphene grown on nickel or copper substrates at high temperatures from a carbon source gas such as ethane or methane, followed by wet transfer to the gFET sensing channel.

Owing to its excellent electrical conductivity properties, CVD graphene is preferred. However, the wet transfer involves the use of protecting polymer layers, which have to be completely removed from the transferred graphene layer to ensure full access to its superior physical properties, a preparation step which remains challenging. The source for CVD graphene

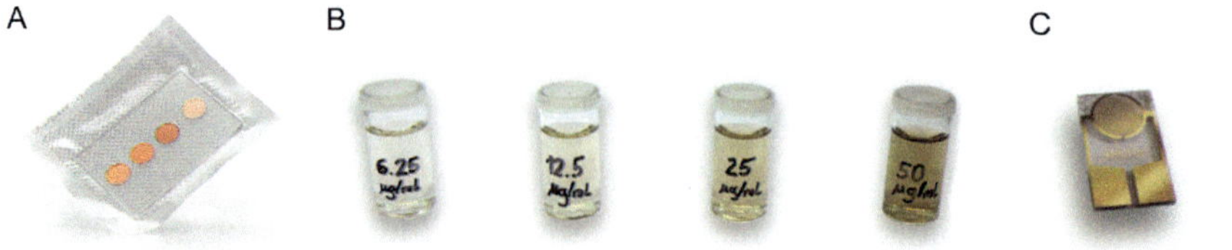

Fig. 1 (A) CVD graphene samples from Graphenea Inc. on copper discs of 12 mm diameter with PMMA coating. (B) Graphene oxide (GO) nano-flake suspensions in Milli-Q water, synthesized at the University of Bayreuth, Germany. The GO solutions were produced from graphite flakes according to a modified Hummers' method (Feicht et al., 2017). (C) Interdigitated electrodes with a sensing area of 3.5 mm^2, used as drain and source electrodes, provided by Micrux Technologies.

used in this work is a monolayer of graphene on a 12 mm Cu-disk from Graphenea Inc. (Fig. 1A).

Graphene oxide flakes were obtained from the University of Bayreuth, depicted at four different dilutions in Fig. 1B (Feicht et al., 2017). The use of chemically derived GO is a cheap and simple alternative for the construction of a gFET.

One viable strategy for graphene FET fabrication is the use of water-soluble graphene oxide (GO) nano-flakes, deposited on the FET substrate by Micrux Technologies, followed by chemical reduction to rGO using hydrazine monohydrate vapor, at 80 °C in an enclosed reduction chamber. Both graphene samples were deposited onto a glass chip commercially available from Micrux Technologies, Spain, with an array of interdigitated electrodes (IDE), which are used as drain and source electrodes of the gFET sensor. The layout of the sensor chip consists of 90 pairs of electrodes, each with a width of 10 µm, separated by 10 µm, and arranged in an array of 3.5 mm in diameter, shown in Fig. 1C.

2.1 Preparation protocol for CVD graphene deposition

The protocols for the deposition of CVD graphene and GO were elaborated and iteratively adjusted to increase the sensor performance of the FET toward the detection of biological binding events of odorants to binding proteins.

2.1.1 Surface cleaning

Adapted from the RCA (Radio Corporation of America) cleaning protocol, SC-1 (standard cleaning):(Kern, 1970) Microelectrodes (ED-IDE1-Au w/o SU8) from Micrux Technologies are cleaned by submerging them in a glass beaker with five parts deionized water, one part ammonia solution (30% NH_4OH) and one part hydrogen peroxide (30% H_2O_2) (e.g., 50 mL dH_2O + 10 mL NH_4OH + 10 mL NH_4OH). The glass beaker is heated to 80 °C in a water bath on a heating plate. During 15 min, all organic residues are removed from the gold and glass surface. Each chip will be then taken with a tweezer, rinsed thoroughly with deionized H_2O, followed by absolute EtOH, then blow dried with compressed air, and stored for later use in a plastic Petri dish.

2.1.2 Self-assembled monolayer deposition

For the deposition of graphene oxide (GO) and CVD graphene, the glass surface between the IDE is functionalized with 3-aminopropyltriethoxysilane (APTES) for GO or triethoxyphenylsilane (TEPS) for CVD graphene. To this end, a 2% APTES or TEPS solution was prepared in absolute EtOH (15 mL EtOH, 0.3 mL APTES/TEPS), while handling the bottles under argon to avoid exposure to oxygen. It is necessary to work in plastic dishes, because silane reagents attach to glass. The solution is poured over the chips in the plastic Petri dish and incubated in ambient conditions for 1 h to allow for the formation of a self-assembled monolayer (SAM). Subsequently, the IDE chips were rinsed with absolute EtOH to remove excess silane. It is important to not point the jet of the washing EtOH directly onto the sensing area and carefully blow dry with compressed air. All chips are placed in a glass Petri dish to withstand 120 °C.

2.1.3 Annealing of the silane SAM

The glass Petri dish with all chips is placed in an oven at ambient pressure and 120 °C for 2 h, in order to anneal the APTES or TEPS on the surface. The IDE are then coated with a linker for either GO or CVD graphene and the deposition is followed immediately. There is no rinsing or cleaning before the deposition of GO or CVD graphene.

2.1.4 Fabrication of CVD graphene FETs

CVD graphene can be deposited to the modified IDE by a chemical wet transfer process, as shown in Fig. 2.

CVD graphene deposition steps:

1. O_2 plasma etching (back of the sample)
   50 W/100 mT/25 sccm/1 min
2. Etching of Cu foil

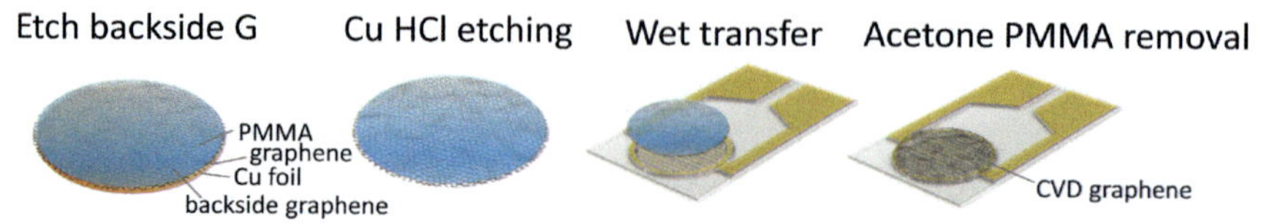

Fig. 2 The process of deposition of CVD graphene on gold-based interdigitated electrodes (IDE). First, the backside of the copper disc is plasma etched to remove residual graphene, followed by copper foil etching in diluted hydrochloride (HCl) solution. The remaining floating graphene sheet, supported by a PMMA layer (60 nm, 495 K, A2), is transferred to the IDE chip and annealed for 30 min at 90 °C. The PMMA is removed in the final step using hot acetone washings.

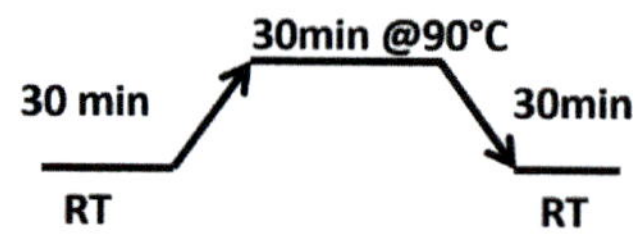

Fig. 3 Annealing step after the deposition of the CVD graphene.

Cu foil floating in a solution of HCl 2 mL/H_2O_2 2 mL and /H_2O 250 mL for 2 days.

3. Transfer onto the host glass substrate
   - transfer of the floating PMMA/graphene to clean H_2O water,
   - place the sample under the floating PMMA/graphene layer,
   - lift the sample and contact the PMMA/graphene from below on oneedge,
   - remove under gentle nitrogen flow the residual dH_2O.
4. Annealing of CVD graphene to substrate
   - start Sawatec 300, hot plate,
   - run annealing program as shown in Fig. 3.
5. Remove PMMA protective layer
   - UV exposure (power 500 W, lamp current 25 V) for 30 min,
   - immerse the chip in acetone for 30 min at 30 °C,
   - remove the chip under continuous rinsing with acetone to avoid contamination,
   - rinse the substrate with isopropanol thoroughly,
   - gently dry under nitrogen flow.

The CVD graphene is now deposited onto the electrodes and ready for electrical characterization or bio-functionalization.

2.1.5 Fabrication of reduced graphene oxide FET

Following the cleaning and silanization of the surface with APTES, 20 μL of aqueous GO solution with a final concentration of 12.5 μg/mL are drop

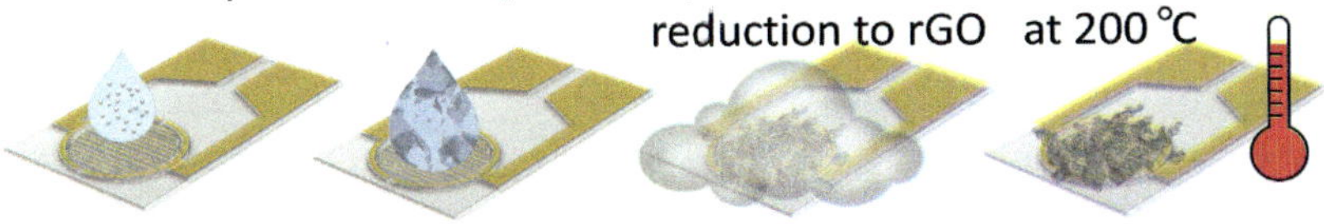

Fig. 4 The deposition of GO onto IDE starts with a self-assembly monolayer of APTES on the glass interlayer to insure good adhesion of the negatively charged GO suspension. Subsequently, an aqueous GO solution with 12.5 µg/mL is drop casted onto the IDE. After a rinsing and baking step for 1 h at 120 °C, the GO is reduced to rGO in hydrazine vapor for 4 h. A thermal reduction afterward increases the stability of the rGO gFET.

casted onto the IDE. The deposition of GO and the chemical reduction process is shown in Fig. 4.

2.1.5.1 Preparation of GO solutions

GO used for the deposition on the interdigitated electrodes was provided from the University of Bayreuth (Feicht et al., 2017). The stock concentration of the GO solution is 4.28 mg/mL. Four different dilutions were prepared to determine the influence of the GO concentration on the transfer of IDE: 50, 25, 12.5 and 6.25 µg/mL in deionized H_2O. After the measurements with the produced gFETs, the optimized concentration of 12.5 µg/mL was used to produce gFETs.

2.1.5.2 GO drop casting

The GO suspensions are applied by a drop cast method onto the chips. For this, a drop of 15 µL of GO suspension is pipetted onto the interdigitated electrodes of the chips, allowing the GO flakes to attach to the APTES layer for 2 h precisely. The time is very crucial to form a uniform layer on the sensing area. Subsequently, the excess GO suspension is rinsed off with deionized H_2O, not directing the jet onto the array area and carefully blow drying with compressed air.

2.1.5.3 Reduction of graphene oxide

2.1.5.3.1 Hydrazine reduction All GO-modified IDE chips are placed in the center of a glass Petri dish for chemical reduction. Because of the carcinogenic properties of hydrazine, it is important to work under the fume hood at all times and treat the waste with special care. 1 mL of hydrazine is pipetted into the corners of the glass dish, while the chips are all placed in the center. Immediately, the lid is sealed air-tight with Kapton tape right afterward and the dish is placed in the oven inside the fume hood at 80 °C for 4 h.

After reduction, the Petri dish remains in the fume hood without lid for 1 h to evaporate the remaining hydrazine before rinsing each chip with dH_2O and subsequently with isopropanol to remove hydrazine residues. Gently blow-dry and store the chips in a desiccator or directly perform the thermal reduction.

2.1.5.3.2 Thermal reduction For the thermal reduction, IDE chips are placed in a pre-heated oven at 200 °C under vacuum for 2 h. This leads to a more stable and reliable electrical chip performance. The chips are taken out of the oven after 2 h and cooled down to room temperature. The resistance of the chips is measured with a Fluke multimeter immediately after reduction. The chips are stored in a desiccator under vacuum until further functionalization and measurement.

2.1.5.3.3 Optimize GO concentration GO concentration of 4.28 mg/mL in dH_2O is too high for the deposition of monolayer GO on the surface, therefore it needs to be diluted. To optimize the properties of the rGO field-effect transistors, different GO solutions were deposited on the substrate. The criteria for a good field-effect transistor for bio-sensing are manifold:

- Flakes need to connect the drain- with the source electrodes
- Presence of semiconductive properties (transfer characteristic)
- Single- or few-layered GO deposition
- High mobility of the finally produced gFET
- Linear dependency to environmental changes (pH, ionic strength)
- Low ohmic resistance to avoid gate leakage current

Four different concentrations were drop cast on the gFET and judged according to these criteria.

2.2 Scanning electron microscope images of graphene

SEM images were recorded in order to (i) determine the nanostructure of the graphene flakes, (ii) to verify the coverage of the drain-source channel with graphene, and (iii) to observe the covered sensing area. Fig. 5 shows the drain-source channels of the IDE covered with CVD graphene.

For comparison, rGO flakes forming the sensor channels are shown in Fig. 6. The SEM images of a gFET after 16 h of hydrazine reduction are shown in Fig. 7, where defects in the rGO substrate can be observed. This observation led to an investigation of the minimum time needed to reduce graphene oxide used later in the electronic sensing protocol.

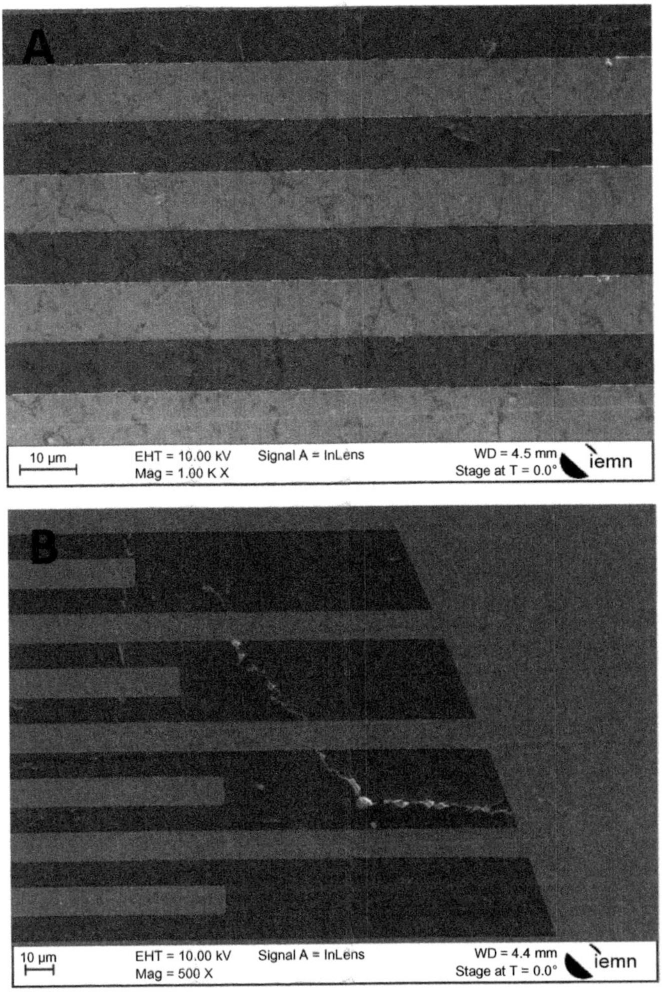

Fig. 5 SEM images of CVD graphene on IDE chips. (A) Full coverage of drain-source channel at magnification of 1000 ×. (B) Crack in CVD graphene at the edge of electrode array, magnification 500 ×.

The SEM images in Figs. 5 and 6 allow for a comparison of CVD graphene and rGO. Both graphene sources are forming a conductive drain-source channel but exhibit significant morphological differences.

The deposited CVD graphene consists of uniform graphene sheet, only interrupted by minor cracks. The rGO consists of flakes with dimension between 3 and 30 µm due to the exfoliation process of GO. The concentration of rGO flakes needs to be above the percolation threshold to guarantee a

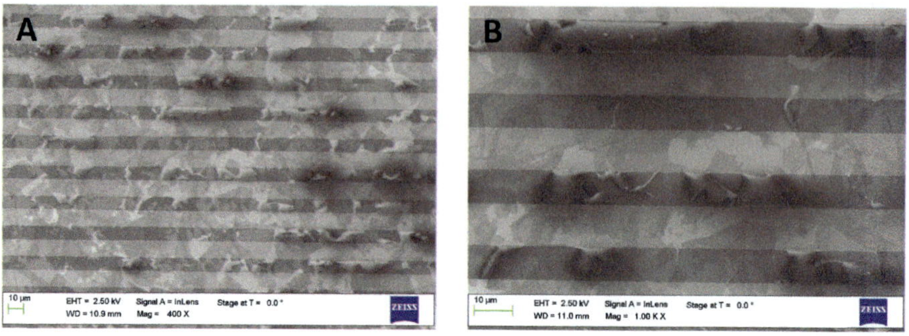

Fig. 6 SEM images of IDE chips modified with drop casted GO at 12.5 μg/mL in aqueous solution, followed by 4 h of reduction in hydrazine vapor. (A) Spread coverage of IDE, magnification 400 ×. (B) Visible multilayers of rGO flakes, magnification 1000 × (bright areas correspond to less rGO layers than dark ones).

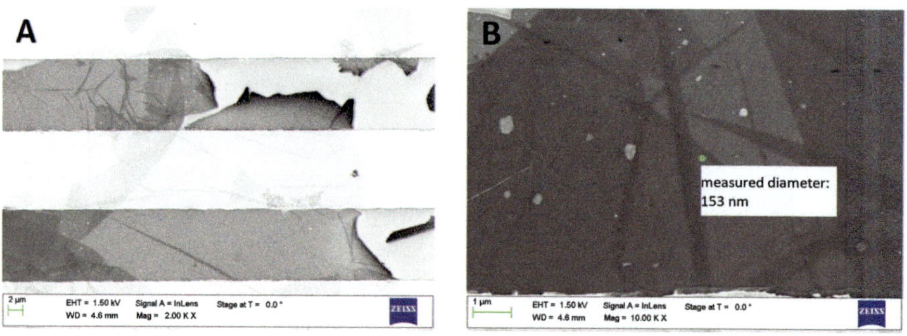

Fig. 7 SEM images of GO coated IDE after 12 h reduction in hydrazine vapor. Defect size approximately 100–200 nm. (A) rGO flakes covering the IDE 2000 ×. (B) Defect holes in rGO flakes, magnification 10,000 ×.

drain-source connection. The rGO images were taken after the reduction of 4 h in hydrazine vapor. After the reduction in hydrazine vapor for more than 16 h, the graphene sheets suffered from damages in the rGO lattice, leading to the formation of holes in the rGO flakes, as shown in Fig. 7.

2.3 Real-time GO reduction measurement

The reduction of the GO to rGO is a crucial component in the fabrication of rGO derived gFETs. Only rGO flakes connecting the drain and source electrodes as semiconducting material will be effective in electronic bio-sensing. The reduction of GO was carried out in two steps: firstly hydrazine vapor reduction (Larisika, Huang, Tok, Knoll, & Nowak, 2012) and subsequently thermal reduction (Chen, Yan, & Bangal, 2010). The change of conductivity is an indicator for the success of the reduction and was measured in real-time during the reduction process. For that purpose, a measurement set-up was created with a source measure unit (SMU). Two wires were soldered to the two electrodes of the interdigitated Micrux chip and a constant voltage of 10 mV was applied. The drain-source current I_{DS} was measured throughout the entire reduction procedure and should increase in case of the reduction of the graphene oxide. The measurement was carried out during the hydrazine vapor reduction and thermal reduction. The conductance of the rGO, which is the reciprocal of Ohm's law is calculated with Eq. (1),

$$G_{DS} = \frac{I_{DS}}{U_{DS}} \tag{1}$$

where G_{DS} is the calculated drain-source conductance, in S or $1/\Omega$, I_{DS} the measured drain-source current in A, and U_{DS} the applied drain-source voltage in V.

The measured drain-source current I_{DS} was logged during the reduction of graphene oxide. The resulting increase of conductance is shown in Fig. 8. The hydrazine vapor reduction and thermal reduction are recorded subsequently on the same IDE chip with an increasing conductivity, indicating

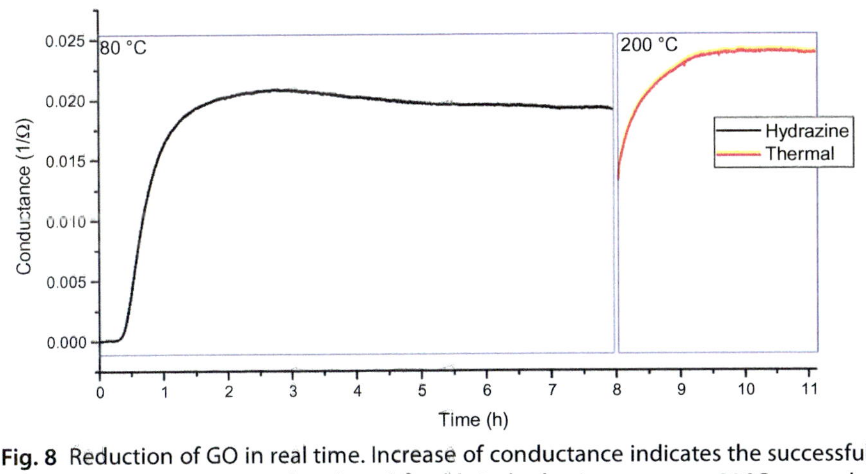

Fig. 8 Reduction of GO in real time. Increase of conductance indicates the successful reduction of GO to rGO. GO reduced for 8 h in hydrazine vapor at 80°C was subsequently thermally annealed at 200°C under vacuum for 3 h to obtain best performing rGO.

successful GO reduction. The hydrazine vapor reduces the GO within the first hour to approximately 75% of the final conductance. After 3 h, the reduction reaches its maximum and decreases thereafter. This is either due to a decrease of hydrazine in the vapor chamber or the formation of damages in rGO nanosheets.

Due to the decrease of temperature and the absence of hydrazine, the conductance dropped significantly before the thermal reduction, but during the thermal reduction, the conductance increased again and remained constant after ~1 h thermal reduction. The hydrazine and thermal reduction sequence is important. Protocols with the inversed order lead to a dramatically lower conductance values and were therefore not further investigated.

2.4 SEM analysis and conductance evaluation for different rGO coverages

Scanning electron microscope (SEM) images were recorded to obtain the coverage and dispersion of rGO flakes on the interdigitated electrodes after the reduction. Four different GO concentrations were prepared and drop cast onto the electrodes. The corresponding SEM images are shown in Fig. 9.

The SEM images show, that the rGO covers almost the entire surface at GO concentrations of 50 and 25 μg/mL. There are many overlapping sheets, leading to multilayer formation. The dilution of 12.5 μg/mL covers the drain-source channel, but has little overlapped sheets. Dilution of 6.25 μg/mL is loosely packed on the surface, leading to a poor connection between the 10 μm distant electrodes.

The SEM microscopic analysis indicates that interdigitated electrodes of a 10 μm gap size, covered with a drop casted GO solution of 12.5 μg/mL are optimal. However, electrical measurements need to be carried out to correlate the different surface coverages with their electrical performance in a FET device. To this end, transfer characteristic, or $I_D V_G$ curves, were measured, as shown in Fig. 10.

The gFET formed from GO suspensions of 12.5 μg/mL has the steepest slope (I_{DS}/V_{GS}-ratio) after hydrazine vapor and thermal reduction. The slope increases for each gFET from before to after thermal reduction. It can be observed, that the thermal reduction leads to the highest increase in slope for low GO concentrations, but the sweet spot for GO coverage is at a concentration of 12.5 μg/mL with the steepest total slope.

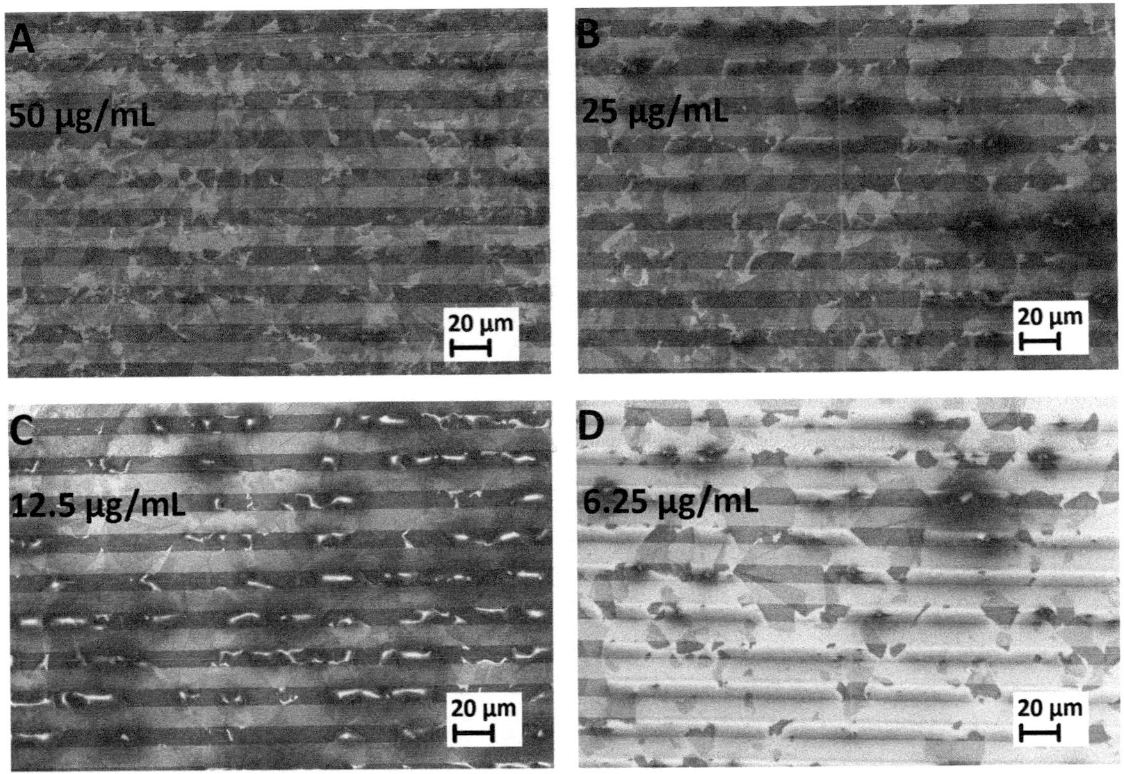

Fig. 9 The SEM images of the IDE chips formed by drop casting different GO solutions to the interfaces. All SEM images were recorded at a magnification of 400× (A and B) densely packed rGO flakes. (C) rGO flakes connect the drain and source electrodes, but minor overlap of different GO sheets. (D) The rGO concentration is very low and connects the electrodes partially.

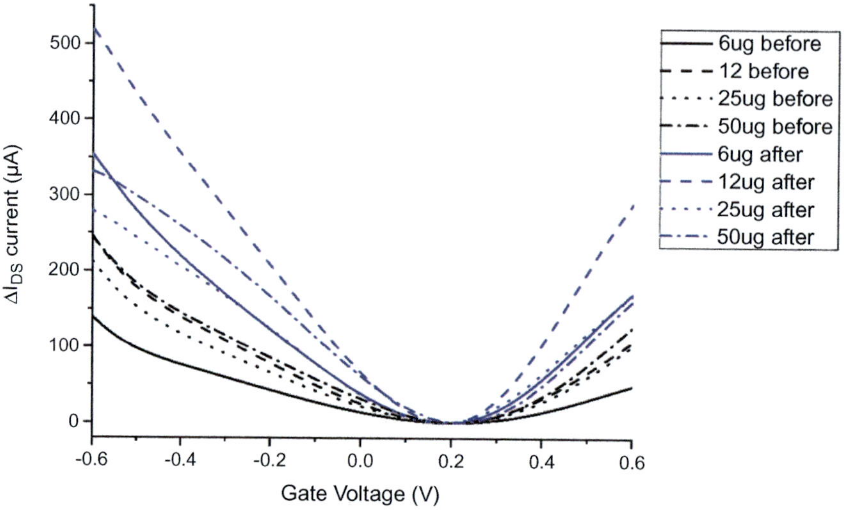

Fig. 10 $I_{DS}V_G$ curves of IDE electrodes formed by drop casting different GO concentrations onto the chips. The black curves were measured after the reduction in hydrazine vapor at 80 °C at ambient pressure. The blue curves were measured on the same gFET after an additional thermal reduction at 200 °C in a vacuum oven. There is a significant increase for all slopes after thermal reduction. The gFET with a deposited GO concentration of 12.5 μg/mL shows the highest mobility.

2.5 Comparison of CVD graphene and rGO

For any sensor application, the electrical characteristics of the graphene substrates are essential; in particular, the transfer characteristic and baseline drift over time are parameters influencing the quality of the bio-sensing performance. The comparisons of gFET with CVD graphene and rGO are shown in Fig. 11.

The linear fit of the slopes, negative and positive to the Dirac point, allows for the calculation of the mobility μ of the fabricated graphene field-effect transistor. Graphene has remarkable electron and hole mobilities at room temperature and this parameter can be calculated for different architectures of the graphene field-effect transistors and the applied drain-source voltage V_{DS}, using Eq. (2):

$$\mu_{eff} = m_{lin} \cdot \frac{L}{W \cdot V_{DS} \cdot C_i} \tag{2}$$

where m_{lin} is the slope from the linear fit (change of I_{DS} vs. V_G), L and W the channel length and width, respectively, V_{DS} the applied drain-source voltage, and C_i the gate capacitance (Tanaka et al., 2010). The evaluated value of the gate capacitance for this system is 3.3 μF.

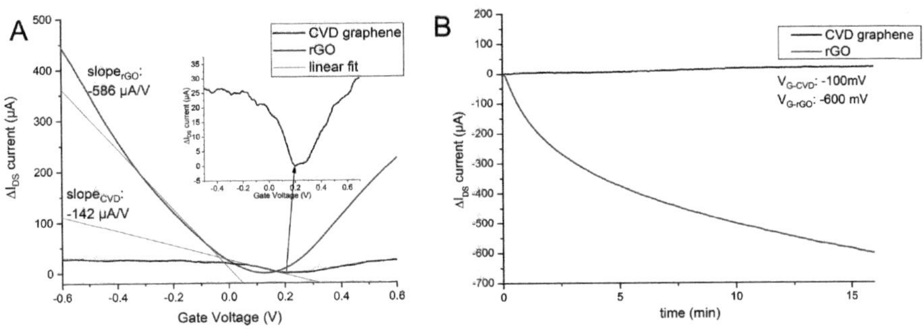

Fig. 11 Comparison of electrical transfer characteristics of CVD graphene and rGO-FETs. Plot A shows the comparison of the transfer characteristics of different fabricated gFETs. Blue is rGO and black CVD graphene as drain-source channel substrate. The insert is a close-up to the CVD graphene I_DV_G curve. Plot B illustrates the baseline drift of the rGO-FET (blue) and the CVD graphene FET (black) at an applied gate voltage of −600 mV for rGO and −100 mV for CVD graphene, to adjust the working point. The CVD graphene FET has only 30% responsiveness compared to the rGO-FET, but a very stable baseline.

The linear fits of rGO and CVD graphene FETs are displayed in Fig. 12 and the values of the linear fit m_{lin} are used to calculate the mobility:

$\mu_{rGO} = 725\ ^{cm^2}/_{Vs}\ \pm 10\%.$

$\mu_{CVD} = 176\ ^{cm^2}/_{Vs}\ \pm 10\%.$

The values are low, compared to previous reports (5000 cm^2/Vs (Lemme et al., 2008) to 200,000 cm^2/Vs (Bolotin et al., 2008)), while others reported low mobilities of 1 cm^2/Vs (Castro et al., 2010); however, the absolute changes in current are by three orders higher than other reports. While usually low μA are recorded in order to measure the changes in the current, the gFETs described here exhibit currents up to mA, improving the signal to noise ratio due to lower amplifier requirements.

3. Surface functionalization of gFETs

In order for gFETs to be sensitive and selective as sensors for the detection of odorants and other analytes, receptors that recognize these ligands must be attached to the graphene channel. The used surface modification strategy must meet the following requirements: (a) biorecognition elements should be (sufficiently) firmly attached to the graphene surface, (b) and have to preserve their biological activity, (c) the modification process should be controlled and highly reproducible, (d) should allow for tuning into the optimal surface density of biorecognition elements.

Several strategies can be employed in the case of CVD graphene or rGO. They are based on the use of π-π stacking interactions with other aromatic

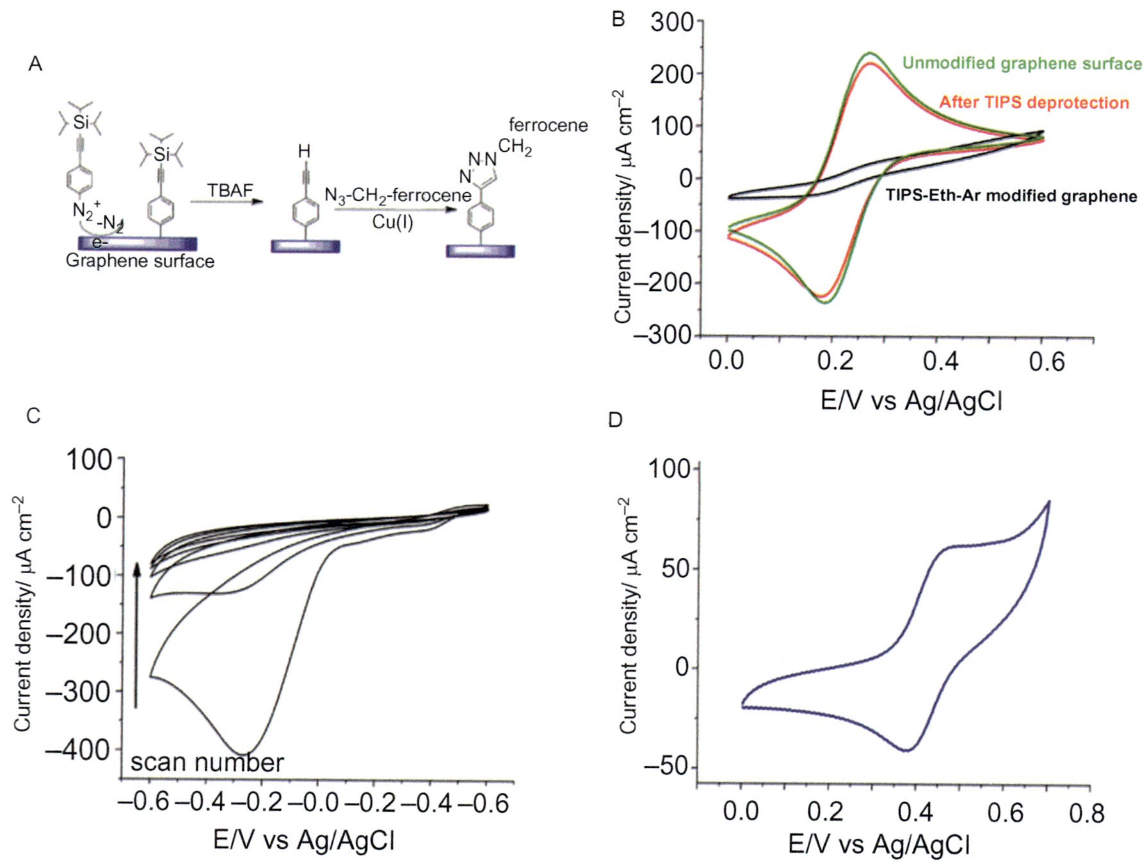

Fig. 12 Surface modification of gFET: (A) Schematic illustration of the covalent modification strategy using silyl-protected diazonium salts as ligands, followed by tetrabutylammonium fluoride (TBAF)-based deprotection and "click" reaction using azidomethylferrocene as a model compound. (B) Cyclic voltammograms recorded on unmodified graphene surface (green), TIPS-Eth-Ar modified graphene surface (black) and after triisopropylsilyl (TIPS) deprotection in 0.1 M tetrabutylammonium fluoride (TBAF) using ferrocene methanol (1 mM)/KCl (0.1 M), scan rate = 100 mV s^{-1}. (C) Change of the current density as a function of scan number upon reduction of the diazonium salt and the formation of a TIPS-Eth-Ar modified graphene interface. (D) Cyclic voltammogram of the interface after "click" chemistry of the ferrocenemethanol derivative in acetonitrile/NBu_4PF_6 0.1 M, scan rate 100 mV s^{-1}.

molecules such as pyrene, dopamine and others. For example, the use of 1-pyrenecarboxylic acid allows for subsequent coupling of amino-containing receptors to activated carboxyl groups through the formation of amide bond following NHS/EDC coupling strategy. This protocol will be described below.

While simple and easy to perform, the interaction is rather weak making it difficult to achieve control over the number of immobilized receptors and achieving high stability. The other used graphene modification strategy is based on the formation of a covalent linkage between the graphene transducer and the ligand. This can be achieved via the electrochemical reduction of diazonium salts, (Allongue et al., 1997; Mishyn et al., 2019) a popular approach for carbon-based materials. The major drawback of this approach is the difficulty to control the extent of the reaction—notably to limit the reaction to the formation of a functional monolayer, important for gFET applications. The highly reactive nature of the formed aryl radical results in the formation of disordered poly-aryl multilayers, which in the case of a gFET sensor might limit the dynamic sensing range. The introduction of bulky substitutions on the ArN_2^+ moieties is known to limit radical addition reactions and allows the formation of ultrathin functional layers. On the basis of this concept, the precursor 4-[(triisopropylsilyl)-ethylenyl]benzenediazonium tetrafluoroborate (TIPS-Eth-ArN_2^+) was used. (Mishyn et al., 2019) After covalent attachment of this molecule to the graphene and subsequent cleavage of the protective group, this "anchor" allows to take advantages of "click" chemistry reactions, such as a high reaction rate and high yield, to attach an azide terminated biorecognition element to the alkyne group on the surface.

4. Modification of the graphene surface using diazonium electrografting

4.1 Electrochemical surface functionalization by TIPS-eth-ArN_2^+

A classic three-electrode electrochemical cell was used with platinum wire and Ag/AgCl electrodes as the counter and the reference electrodes, respectively. The graphene working electrode was cycled between +0.6V to 0.6V vs. the reference electrode at a scan rate 50 mV s^{-1} in a solution containing 1 mM of TIPS-Eth-ArN_2^+ in 0.1 M tetrabutylammonium hexafluorophosphate ($TBAPF_6$) in acetonitrile. After five cycles, the electrode was withdrawn from the solution and copiously rinsed in pure acetonitrile.

4.2 Deprotection of TIPS

In order to cleave the triisopropylsilyl (TIPS) groups and get access to the alkyne group on the surface, the modified graphene surface was immersed in 0.1 M tetrabutylammonium fluoride (TBAF) in tetrahydrofuran (THF) for 1 h under constant agitation at 100 rpm. After the reaction was completed, the interface was copiously rinsed with THF and left in pure THF for an additional hour under the same agitation speed to remove any residues of TBAF. Then the surface was dried at ambient conditions.

4.3 "Click" chemistry

The deprotected substrate was further treated by "click" chemistry (Huisgen 1,3-dipolar cyclization) using azidomethylferrocene as a model compound to demonstrate the feasibility of the protocol and to estimate the surface coverage by electrochemical means. The interface was immersed into an aqueous solution of $CuSO_4$ (10 mM) and L-ascorbic acid (20 mM) in the presence of azidomethylferrocene (0.83 mM in THF) and left for 1 h under an argon atmosphere. The interface was then treated with an aqueous solution of 10 mM EDTA for 10 min to chelate any remaining Cu^{2+} residues and finally washed copiously with acetone and water and left to dry.

5. Surface characterization

5.1 Electrochemical characterization

To validate TIPS-Eth-ArN_2^+ deposition and removal of the protective group (Fig. 12A), cyclic voltammograms (CVs) were recorded after each step using 1 mM ferrocenemethanol (Fc) as a redox couple. It is important to mention that all parameters of the electrochemical measurements (such as the surface of the working electrode, the concentration of the redox couple, scanning rate, etc.) remained unchanged for all measurements. In other words, the peak intensity in the voltammogram exclusively depends on the surface conductivity, i.e., the ability to exchange electrons with an external redox couple. Since graphene is a good conductor, the unmodified interface shows a reversible electrochemical behavior (Fig. 12B).

The electrochemical grafting step was monitored by cyclic voltammetry. An intense cathodic peak at −0.2 V was present in the first cycle, which indicates the reduction of the diazonium compound (Fig. 12C). During this process, the diazonium molecule accepts one electron from the surface of the working electrode and cleaves a neutral N_2 molecule. The resulting radical from this reaction attacks the surface of graphene and forms a covalent bond with the graphene surface (Fig. 12A). In this case, sp^2 hybridized graphene carbon is transformed into sp^3 state. In subsequent cycles, this peak disappears, as the deposition occurs quickly and almost complete during the first cycle. The formed ultrathin layer of the organic molecules on the surface of graphene is an insulator and prevents electron exchange with the external redox couple. To verify the completeness of the surface coating, a cyclic voltammogram of the interface was recorded after this modification (Fig. 12B). A decrease in the cathode and anode signals indicates that the surface is completely covered by this ultrathin non-conductive layer.

The deprotection step removes the sterically bulky TIPS groups and gives access to electron exchange with the external redox couple. As a result, after removal of the protective group, the cathodic and anodic peaks recover their intensity close to unmodified graphene (Fig. 12B). After the deprotection, alkyne groups are accessible on the surface of graphene for "click" chemistry. The success of the integration of ferrocene units onto the graphene surface was seen from the presence of the one-electron reversible oxidation and reduction peaks on the cyclic voltammogram of the ferrocene-modified electrode (Fig. 12D) where no external redox couple was used. A surface concentration of $\Gamma = 4.37 \times 10^{-10}\,\mathrm{mol\,cm^{-2}}$ of bound ferrocene groups was derived from these measurements using Eq. (3):

$$\Gamma = Q/_{nFA} \qquad (3)$$

where Q is the passed charge (3.67×10^{-6} C), n is the number of exchanged electrons ($n = 1$), F is the Faraday constant and A is the electroactive surface of the electrode determined as $0.087\,\mathrm{cm^2}$. Considering the ferrocene molecules as spheres with a diameter of 6.6 Å, the theoretical maximum coverage for an idealized ferrocene monolayer can be estimated as $\Gamma = 4.8 \times 10^{-10}\,\mathrm{mol\,cm^{-2}}$. In our case, more than 90% of coverage was achieved.

5.2 SEM imaging and XPS analysis

SEM images were recorded for both rGO and CVD graphene-based devices to verify the stability of the graphene layers after the modification and to assess the surface morphology for defects. Comparing devices with rGO and CVD graphene (Fig. 13A and B, respectively), it can be observed that after the integration of ferrocene, no defects are visible on rGO and the surface morphology remained the same as it was on the initial interface. In contrast to rGO, macro defects with an approximate size of several micrometers appeared on the surface of CVD graphene, indicating an increase in the number of defects in the graphene structure.

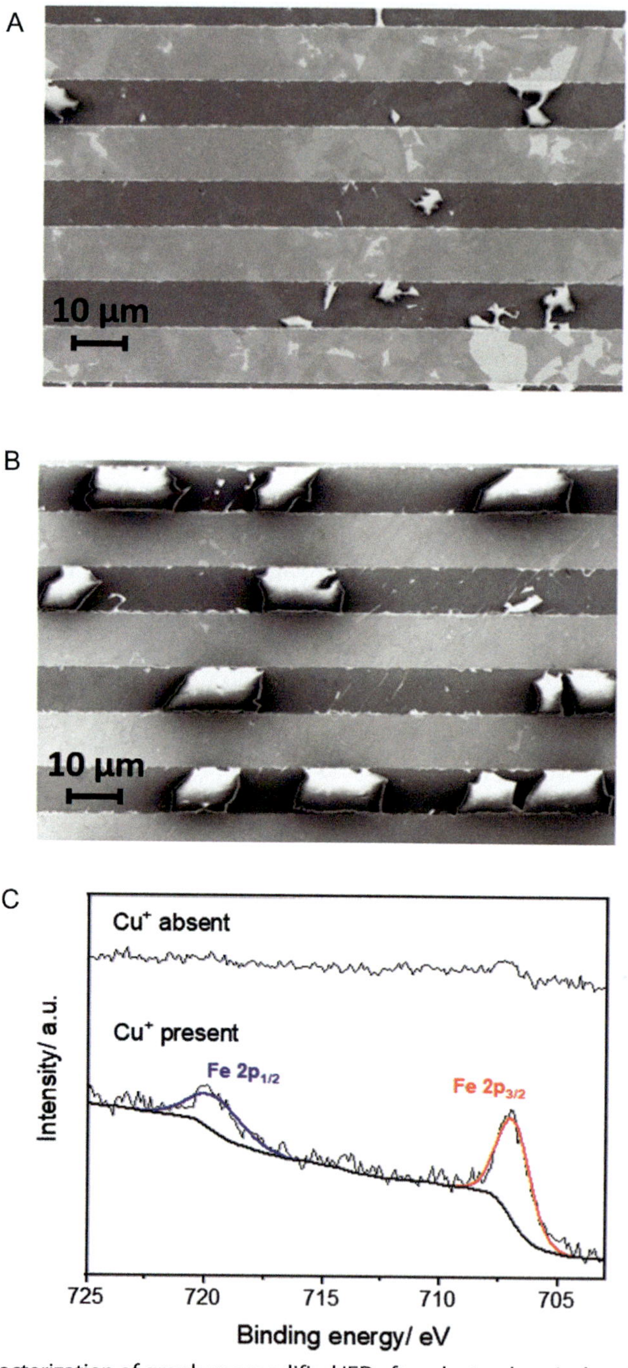

Fig. 13 Characterization of graphene modified IED after electrochemical grafting of diazonium salts followed by clicking of ferrocene units: SEM images of (A) rGO and (B) CVD graphene on IDE after the integration of ferrocene at magnification of 1000×. The white spots on (A) correspond to uncovered by rGO glass substrate and on (B) correspond to defects caused by diazonium treatment. (C) High-resolution Fe_{2p} X-ray photoelectron spectroscopy (XPS) core-level spectrum of ferrocene modified interfaces.

The success of the integration of ferrocene units onto graphene surface was confirmed from the presence of the Fe_{2p} component in the XPS survey spectrum (Fig. 13C). In the absence of copper catalyst, no click reaction took place and no Fe_{2p} component could be recorded.

6. Modification of the graphene surface using π-π—Stacking of pyrene derivatives

A second approach, that we briefly introduce for the attachment of receptor proteins to the graphene sensing area, is based on a protocol by which the graphene surface is chemically modified by a bi-functional linker, 1-pyrenebutanoic acid succinimidyl ester (PBSE). On one end, the linker firmly attaches to the graphene surface through π-π interactions with the pyrene group and on the other end it covalently reacts with the amino group of the protein to form an amide bond. This concept is schematically displayed in Fig. 14.

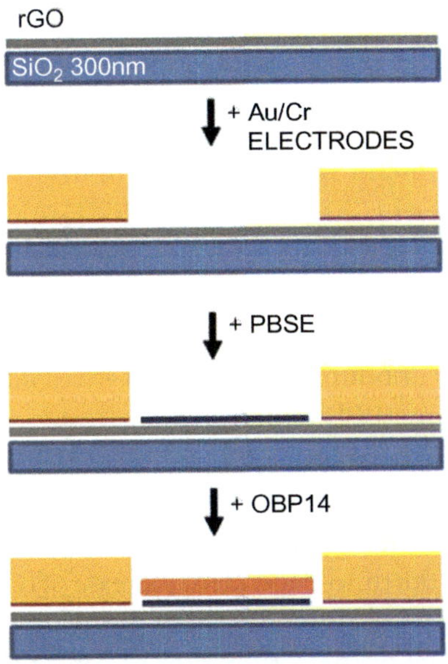

Fig. 14 Schematics of the fabrication and functionalization of rGO-FETs: a silicon wafer with 300 nm SiO_2 and reduced graphene oxide (rGO) flakes was coated with Au/Cr electrodes (5 nm Cr, 70 nm Au); next, the linker, 1-pyrenebutanoic acid succinimidyl ester (PBSE), was coupled to the substrate, followed by protein (here: OBP14) immobilization (randomly) by active ester chemistry.

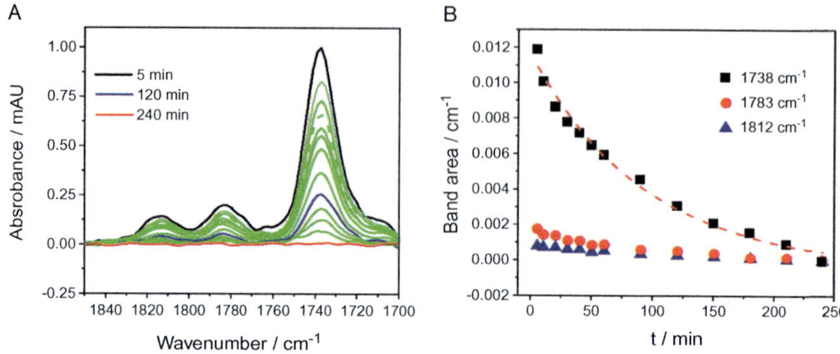

Fig. 15 (A) Hydrolysis of the active ester in PBS leads to a decrease of the bands at 1738, 1781 and 1815 cm^{-1}. (B) The area of the 1738, 1783, and 1812 cm^{-1} bands decrease as a function of time. The dashed red curve guides the eye, indicating a time constant of hours.

Firstly, we tested the chemical stability of the linker system, applied as 20 µL of a 5 µM PBSE solution in tetrahydrofuran was placed onto the rGO-FET channel, in the absence of any protein (coupling reaction). Fig. 15 shows FTIR measurements of the immobilized PBSE linker, incubated in PBS buffer (A). The bands of the active ester (1738, 1781, and 1815 cm^{-1}, respectively) decrease as a function of time, because of spontaneous hydrolysis. Fig. 15 B depicts the exponential decrease of all three bands, with a time constant in the order of hours.

In the next experiment, 20 µL of a 5 µM PBSE solution in tetrahydrofuran was placed onto the rGO-FET channel; the fast evaporation of tetrahydrofuran allowed immediate deposition of 30 µL of a 10 µM OBP14 solution in PBS buffer (1 mM; pH 8.0) onto the substrate, which was incubated for 2 h at 4 °C. Loosely or unbound linker/OBP14 was removed by rinsing the detection area with same PBS buffer.

IR spectra of the linker monolayer with the OBP solution, taken at different times during the assembly (binding) to the gate, are given in Fig. 16A, while Fig. 16B summarizes the quantitative analysis of the functionalization process by monitoring the time-dependent increase of the amide I and II bands of the protein and the corresponding decrease of the band at 1738 cm^{-1}, assigned to the cleavage of the active ester group of the linker molecules. The whole functionalization process ending with the covalent attachment of the OBP (here OBP 14 from the honey bee) as odorant receptor is completed within 30 min.

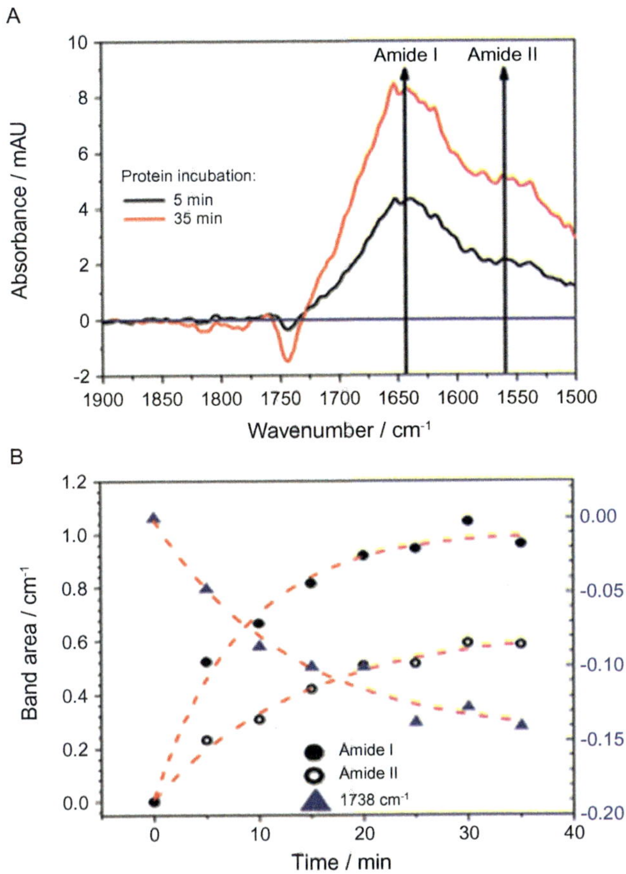

Fig. 16 (A) Infrared spectra of the PBSE linker attached to gate area of the rGO surface, and OBP14 immobilized for 5 or 35 min, as indicated (spectra measured in ATR configuration and water corrected). (B) Time-dependent increase of the amide I and II bands, upon binding of OBP14 to the linker molecules at the gate surface, and the corresponding decrease of the band @ 1738 cm^{-1} upon cleavage of the active ester of the linker molecules during the protein immobilization. The dashed red curves are guides to the eye, qualitatively indicating a time constant of the coupling reaction of tens of minutes.

7. Control measurements

Fig. 17 shows different types of control measurements that were performed in order to demonstrate that the response signals obtained by the olfactory biosensor are specific and thus clearly assignable as protein-ligand

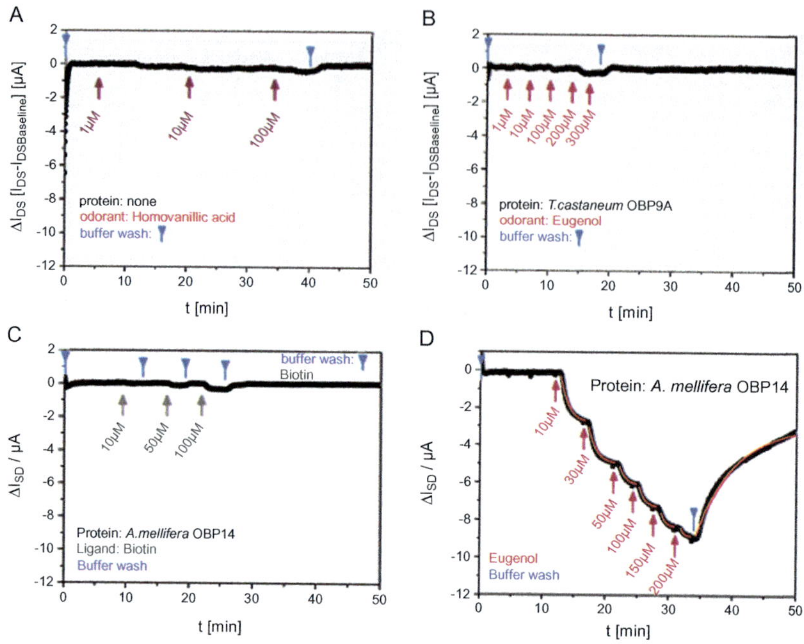

Fig. 17 (A) Real-time response of a PBSE-functionalized rGO-FET to different concentrations of Homovanillic acid. (B) Real-time response of a FET biosensor functionalized with OBP 9 from the beetle *Tribolium castaneum* to different Eugenol concentrations, a non-ligand for this beetle. (C) Real-time response of OBP14-functionalized biosensor to different concentrations of a nontarget ligand, i.e., biotin. (D) ΔI_{SD} vs. time measurement of an OBP 14-functionalized rGO-FET to different concentrations of Eugenol, a ligand for the honey bee with $K_d = 20–40\ \mu M$.

interactions rather than non-specific interaction of the tested odorants with the rGO surface. Firstly, rGO-FET devices were fabricated and functionalized by the linker PBSE as described above, but without any odorant binding protein immobilized. Fig. 17A demonstrates that after addition of different concentrations of a phenolic-structured odorant like Homovanillic acid, no sensor signal could be observed. Similar observations were made for other ligands (not shown). Fig. 17B demonstrates that after immobilizing an odorant binding protein, OBP9 from *Tribolium castaneum*, to the rGO surface, different concentrations of Eugenol, a non-ligand to this species, do not cause a detectable change in the source drain-current. Introducing a small but unnatural ligand like biotin to an OBP functionalized sensor surface, does not show a distinct response signal, here shown for the OBP 14 from the honey bee (C). For comparison, Fig. 17D demonstrates that Eugenol is an extremely strong ligand to OBP14 from the honeybee with a $K_d = 20–40\ \mu M$.

In conclusion, all negative control experiments have clearly demonstrated that although it seems likely that π-orbitals of phenolic compounds like Eugenol or Homovanillic acid can be easily stacked on the rGO surface similar to pyrene derivatives (e.g., PBSE), the sensor response signals clearly reflect the affinity between OBPs and their ligands and is not caused by unspecific interactions of odorants with the rGO surface.

8. Binding kinetic evaluation upon molecule adsorption

In order to describe the kinetics of binding reactions, such as antigen to antibody or protein to ligand, the Langmuir adsorption model has been widely used for biosensor applications. (O'Shannessy & Winzor, 1996; Yang, Yau, & Chan, 1998) This model describes the adsorption of ligands on a solid surface with a finite number of binding sites at a certain pressure, volume and temperature. (Langmuir, 1916) To be able to apply the Langmuir adsorption model, the parameters K_A and K_d have to be introduced. The application of the Langmuir adsorption model requires three assumptions:

1. only monolayer coverage is possible,
2. homogenous surface,
3. neighboring binding sites do not influence the adsorption of one binding site and
4. equivalence of all binding sites

8.1 Equilibrium binding affinity

If all conditions are met, the binding equilibrium can be written as:

$$A + B \underset{k_{off}}{\overset{k_{on}}{\rightleftharpoons}} AB \tag{4}$$

where A is the analyte and B is the immobilized receptor (Schasfoort & Tudos, 2008). The forward reaction, the adsorption, is described by the association rate constant k_{on} in $M^{-1}s^{-1}$, while the reverse reaction, the desorption, is associated with the dissociation rate constant k_{off} in s^{-1}. At a dynamic equilibrium, the rates of association and dissociation are equal, (Eq. 5):

$$k_{on}[A][B] = k_{off}[AB] \tag{5}$$

Under these assumptions, the measured data of different analyte concentrations can be used to calculate the association and dissociation rate constants for comparison of measurement to data from literature. Furthermore, the affinity and dissociation constants K_A and K_d can be evaluated using Eqs. (6) and (7) (Schasfoort & Tudos, 2008),

$$K_A = \frac{[AB]}{[A][B]} = \frac{k_{on}}{k_{off}} \tag{6}$$

$$K_d = \frac{[A][B]}{[AB]} = \frac{k_{off}}{k_{on}} \tag{7}$$

where K_A in M^{-1} is the inverse of K_d in M.

The Langmuir adsorption model describes the surface coverage, where θ is the percentage of formed complexes AB with respect to the total amount of binding sites B. Therefore, $1-\theta$ is the number of free binding sites and equal to [B]. Eq. (8) describes the fraction of the concentration of the complex AB to the concentration of B.

$$\frac{[A][B]}{[B]} = \frac{\theta}{1-\theta} \tag{8}$$

Regarding Eq. (6), Eq. (8) is equal to:

$$K_A \cdot [A] = \frac{\theta}{1-\theta} \tag{9}$$

Under the assumption of an infinite reservoir of [A], due to a constant inflow of the analyte A, [A] can be expressed as a constant c. Substitution into Eq. (9) and rearrangement yield θ according to Eq. (10). This equation is called Langmuir isotherm and is only valid for a system with constant temperature, pressure and [A].

$$\theta = \frac{c \cdot K_A}{1 + c \cdot K_A} \tag{10}$$

Fig. 18 shows the lin-lin and lin-log plots of the Langmuir isotherms, observed for the binding of Eugenol to OBP14 from the honey bee. The dissociation constant K_d can be determined at 50% surface coverage.

The Langmuir isotherm fit can be used for biosensor calibration by reference measurements with a known analyte concentration. Furthermore, the saturation of the surface can be evaluated, as well as the surface density of receptors. Beside the quantification of analytes or receptors binding to a surface, the Langmuir isotherm fit is an important tool for comparison of

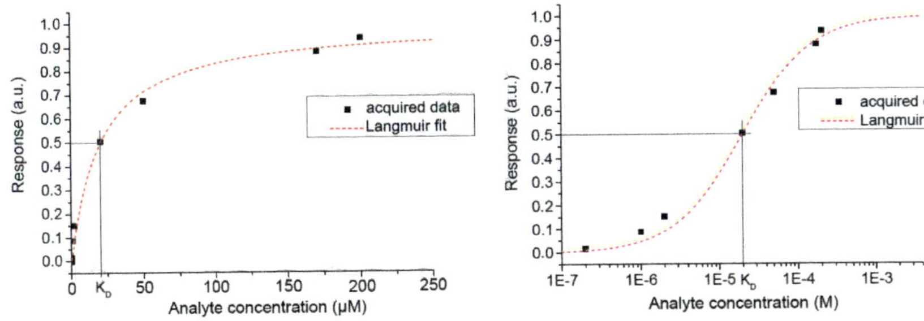

Fig. 18 Langmuir fit applied to acquired data points to determine the dissociation constant K_d. K_d can be observed at 50% of surface coverage. At this value, 50% of all binding sites B formed a complex with the ligand A.

different biosensors. It can be also used to optimize specific sensor types. Due to the evaluation of surface coverage θ with the Langmuir isotherm equation, the results of different measurement systems can be compared. The surface coverage at a certain concentration should be the same for electrical, optical or any other measurement method.

8.2 Linear equation for k_{obs}

The biomolecular interaction of an analyte A with an immobilized receptor B, forms a complex AB under the association and dissociation rate constants k_{on} and k_{off}. A new parameter for the binding kinetics is introduced, called k_{obs}. This parameter is given by the linear Eq. (11), determined by k_{on}, k_{off} and the ligand concentration [A] (Schasfoort & Tudos, 2008),

$$k_{obs} = k_{on}[\mathrm{A}] + k_{off} \tag{11}$$

where k_{on} represents the slope and k_{off} the intercept. A change of [AB] over time is the result of association of [A] with [B] to [AB] and dissociation of [AB]. The derivative of [AB] with respect to the time is given by Eq. (12):

$$\frac{d[AB]}{dt} = k_{on}[\mathrm{A}][\mathrm{B}] - k_{off}[\mathrm{AB}] \tag{12}$$

where the association with the rate constant k_{on} leads to an increase of [AB] and the dissociation with the rate constant k_{off} decreases the concentration of the AB complexes (Schasfoort & Tudos, 2008). Due to an infinite reservoir of analyte A, the concentration [A] remains the same as initially: $[\mathrm{A}] = [\mathrm{A}_0]$. The concentration of the formed complex AB decreases if more binding sites B are free, while B_0 is the total number of binding sites:$[AB] = [B_0] - [B]$. Substituting these relationships in Eq. (12) and integrate it over time, Eq. (13) is obtained.

$$\int_{t_0}^{t} \frac{d[AB]}{dt} dt = \int_{t_0}^{t} \left([B] \cdot \left(k_{on}[\mathrm{A}] - k_{off}[\mathrm{B}]\right) - k_{off}[\mathrm{B}_0]\right)\mathrm{dt} \tag{13}$$

After introducing k_{obs}, Eq. (14) can be obtained.

$$\int_{t_0}^{t} \frac{d[AB]}{dt} dt = \int_{t_0}^{t} \left([B] \cdot k_{obs} - k_{off}[\mathrm{B}_0]\right)\mathrm{dt} \tag{14}$$

Under the assumption of $k_{off} \ll k_{obs}$, it can be simplified to:

$$\int_{t_0}^{t} \frac{d[AB]}{dt} dt = \int_{t_0}^{t} [B] \cdot k_{obs}\,\mathrm{dt} \tag{15}$$

On the other hand, the derivative of [AB] can be expressed as the derivation of $[B_0] - [B]$:

$$\frac{d[AB]}{dt} = \frac{d[\mathrm{B}_0]}{dt} - \frac{d[B]}{dt} \tag{16}$$

while the concentration of B_0 remains constant over time. Substituting this result in Eq. (15), Eq. (17) can be obtained:

$$\int_{t_0}^{t} -\frac{d[B]}{dt} dt = \int_{t_0}^{t} [B] \cdot k_{obs}\,\mathrm{dt} \tag{17}$$

resulting in Eq. (18) for the exponential fit of binding kinetics for a specific concentration.

$$[\mathrm{B}] = [\mathrm{B}_0] \cdot e^{-k_{obs} dt} \tag{18}$$

According to Eq. (18), k_{obs} can be obtained through an exponential fit to the data and afterward the concentration [A] can be extracted due to its linear dependence with k_{obs}, shown in Fig. 19A. The observed parameter k_{obs} for a certain concentration is resulting from this exponential fit and can be plotted against the concentration of the analyte, see Fig. 19B. The association and dissociation rate constants k_{on} and k_{off} can be calculated from k_{obs} with Eq. (11). K_A and K_d are determined using Eqs. (6) and (7). The values for the calculated affinity and dissociation constants should be in the same order as the one evaluated with the Langmuir isotherm Eq. (10).

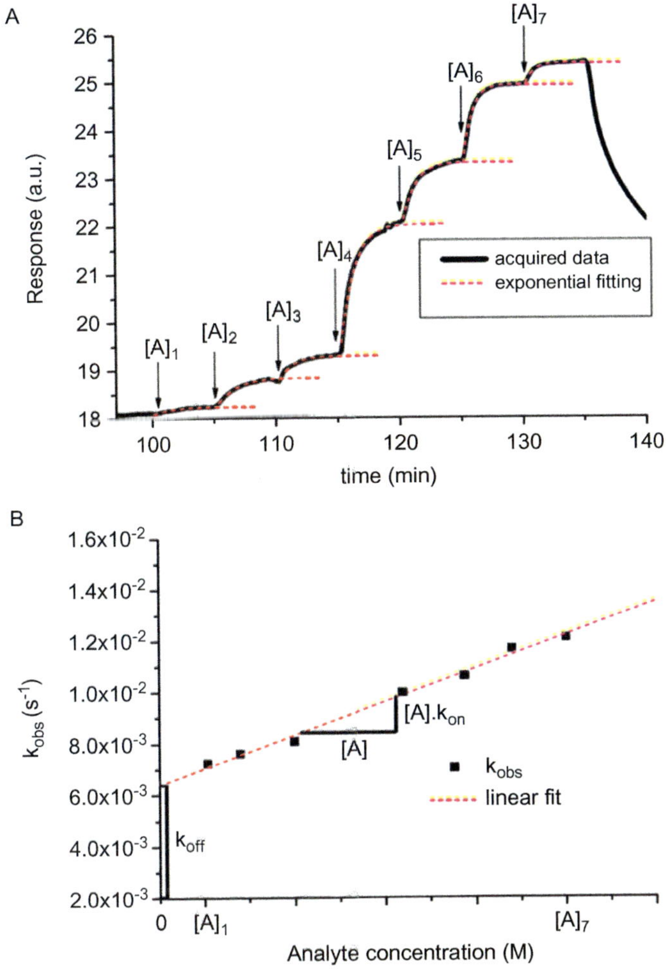

Fig. 19 A time resolved sensor response is used to evaluate the binding kinetics at different concentrations, here the current I_{DS} of a field-effect transistor. The exponential fit to the acquired data results in the parameter k_{obs} in the exponent. In (A) the exponential fit of the sensor response to every new concentration was performed. (B) Shows the corresponding k_{obs} to every concentration. It obeys to Eq. (11), which relates [A] linearly to k_{obs}.

9. Discussion and conclusion

The better understanding of the biochemical basis and the biomolecular mechanisms underlying olfaction in nature has led to a revival in research also on the technical emulation of the sense of smell in various strategies aimed at developing an artificial nose. The new and promising ansatz in this biomimetic approach is the use of functional modules also used in nature, e.g., odorant binding proteins. Their robustness, stability toward chemical and physical stress, and their cheap availability in huge amounts attracted increased interest in complementing our advanced technical solutions for physical, i.e., optical, acoustical, and mechanical sensors by the still two missing chemical senses, i.e., for taste and, in particular, for smell.

Among the different transduction concepts reported in the literature and also covered in this book, electronic biosensor platforms hold great promise, complementing electrochemical and optical detection formats for the quantitative monitoring of bioanalytes. They offer highly sensitive and label-free detection schemes, disposable and cheap devices, with signals that are easy to analyze and interpret, suitable for multiplexed operation and for remote control, compatible with NFC technology, etc.

In particular, the use of graphene as the conductive gate material in the preparation of thin film transistors as the sensing device offers a tremendous advantage compared to the use of organic semiconducting materials or compared to Si-based transistors, which both require far more demanding preparation protocols. This has been demonstrated convincingly in many examples reported in the literature.

As it is demonstrated in this chapter, graphene-based smell sensors, functionalized by OBPs for the sensitive and selective recognition and binding of odorants as ligands, offer an enormous potential for fundamental and quantitative studies of the molecular interaction of odorants with their receptors, i.e., OBPs, complementing solution-based methods.

For the technical realization of a biomimetic olfactory biosensor based on graphene field-effect transistors functionalized by odorant-binding proteins, the described protocols lead to very reproducible sensors. Their performance was demonstrated for the functionalization by OBP14 from the honey bee (*Apis mellifera*) and for the operation in situ and in real-time by monitoring a broad spectrum of odorants known to be attractants for bees, binding from solution. The electrical measurements of the binding of a test odorant were shown to follow the Langmuir model for ligand–receptor interactions. The results demonstrate that such devices can bind odorants even after immobilizing the receptor OBP on the graphene channel and are able to discriminate between ligands binding within a broad range of dissociation constants.

References

Allongue, P., Delamar, M., Desbat, B., Fagebaume, O., Hitmi, R., Pinson, J., et al. (1997). Covalent modification of carbon surfaces by aryl radicals generated from the electrochemical reduction of Diazonium salts. *Journal of the American Chemical Society*, *119*(1), 201–207. https://doi.org/10.1021/ja963354s.

Bolotin, K. I., Sikes, K. J., Jiang, Z., Klima, M., Fudenberg, G., Hone, J., et al. (2008). Ultrahigh electron mobility in suspended graphene. *Solid State Communications*, *146*(9), 351–355. https://doi.org/10.1016/j.ssc.2008.02.024.

Bushdid, C., Magnasco, M. O., Vosshall, L. B., & Keller, A. (2014). Humans can discriminate more than 1 trillion olfactory stimuli. *Science*, *343*(6177), 1370–1372. https://doi.org/10.1126/science.1249168.

Castro, E. V., Ochoa, H., Katsnelson, M. I., Gorbachev, R. V., Elias, D. C., Novoselov, K. S., et al. (2010). Limits on charge carrier mobility in suspended graphene due to flexural phonons. *Physical Review Letters*, *105*(26), 266601. https://doi.org/10.1103/PhysRevLett.105.266601.

Chen, W., Yan, L., & Bangal, P. R. (2010). Preparation of graphene by the rapid and mild thermal reduction of graphene oxide induced by microwaves. *Carbon*, *48*(4), 1146–1152. https://doi.org/10.1016/j.carbon.2009.11.037.

Feicht, P., Siegel, R., Thurn, H., Neubauer, J. W., Seuss, M., Szabó, T., et al. (2017). Systematic evaluation of different types of graphene oxide in respect to variations in their in-plane modulus. *Carbon*, *114*, 700–705. https://doi.org/10.1016/j.carbon.2016.12.065.

Kern, W. (1970). Cleaning solution based on hydrogen peroxide for use in silicon semiconductor technology. *RCA Review*, *31*, 187–205.

Langmuir, I. (1916). The constitution and fundamental properties of solids and liquids. Part I. Solids. *Journal of the American Chemical Society*, *38*(11), 2221–2295. https://doi.org/10.1021/ja02268a002.

Larisika, M., Huang, J., Tok, A., Knoll, W., & Nowak, C. (2012). An improved synthesis route to graphene for molecular sensor applications. *Materials Chemistry and Physics*, *136*(2), 304–308. https://doi.org/10.1016/j.matchemphys.2012.08.003.

Lemme, M. C., Echtermeyer, T. J., Baus, M., Szafranek, B. N., Bolten, J., Schmidt, M., et al. (2008). Mobility in graphene double gate field effect transistors. *Solid-State Electronics*, *52*(4), 514–518.

Mishyn, V., Aspermair, P., Leroux, Y., Happy, H., Knoll, W., Boukherroub, R., et al. (2019). "Click" chemistry on gold electrodes modified with reduced graphene oxide by electrophoretic deposition. *Surfaces*, *2*(1), 193–204. https://doi.org/10.3390/surfaces2010015.

O'Shannessy, D. J., & Winzor, D. J. (1996). Interpretation of deviations from pseudo-first-order kinetic behavior in the characterization of ligand binding by biosensor technology. *Analytical Biochemistry*, *236*(2), 275–283. https://doi.org/10.1006/abio.1996.0167.

Schasfoort, R. B. M., & Tudos, A. J. (2008). *Handbook of surface plasmon resonance*. Royal Society of Chemistry.

Tanaka, S., Goto, H., Tomori, H., Ootuka, Y., Tsukagoshi, K., & Kanda, A. (2010). Effect of current annealing on electronic properties of multilayer graphene. *Journal of Physics: Conference Series*, *232*, 012015. https://doi.org/10.1088/1742-6596/232/1/012015.

Yang, M., Yau, H. C. M., & Chan, H. L. (1998). Adsorption kinetics and ligand-binding properties of thiol-modified double-stranded DNA on a gold surface. *Langmuir*, *14*(21), 6121–6129. https://doi.org/10.1021/la980577i.

CHAPTER SEVENTEEN

Electronic biosensors based on EGOFETs

Mohammad Y. Mulla[a], Luisa Torsi[b,c,d], Kyriaki Manoli[b,c,*]

[a]Research Institutes of Sweden (RISE), Norrköping, Sweden
[b]Dipartimento di Chimica - Università degli Studi di Bari "A. Moro", Bari, Italy
[c]CSGI (Centre for Colloid and Surface Science), Unit of Bari, Bari, Italy
[d]Center for Functional materials, The Faculty of Science and Engineering, Åbo Akademi University, Turku, Finland
*Corresponding author: e-mail address: kyriaki.manoli@uniba.it

Contents

Abstract

There is an increasing interest for low cost, ultrasensitive, time saving yet reliable, point-of-care bioelectronic sensors. Electrolyte gated organic field effect transistors (EGOFETs) are proven compelling transducers for various sensing applications, offering direct electronic, label-free transduction of bio-recognition events along with miniaturization, fast data handling and processing. Given that field effect transistors act as intrinsically signal amplifiers, even a small change of a chemical or biological quantity may significantly alter the output electronic signal. In EGOFETs selectivity can be guaranteed by the immobilization of bioreceptors able to bind specifically a target analyte. The layer of receptors can be linked to one of the electronic active interfaces of the transistor, and the interactions with a target molecule affect the electronic properties of the device. The present chapter discusses main aspects of EGOFETs transducers along with detailed examples of how to tailor the device interfaces with desired functionality. The development of an "electronic tongue" based on an EGOFET device coupled to odorant binding proteins (OBPs) for enantiomers differentiation is presented.

Methods in Enzymology, Volume 642
ISSN 0076-6879
https://doi.org/10.1016/bs.mie.2020.07.003

1. Introduction

A transistor is a three-terminal device where the current flowing in the region between the source (S) and the drain (D) electrodes, called the channel, is controlled by a third electrode, addressed as the gate electrode (G). The control of the channel conductivity is achieved by capacitively coupling the gate electrode and the channel by applying an electric field across an insulator. This is known as field-effect-transistor (FET) (Sze & Ng, 2006). At variance, the controlling electrode can be capacitively coupled to the channel of the device via an electrolyte and such a device is called electrolyte gated field effect transistor (EGFET). To meet the requirements of large area, printable and flexible electronic devices, most channel materials are based on organic polymer based semiconductors (OSCs), thus is common to use the acronym EGOFETs for electrolyte gated organic field effect transistors (Kim et al., 2013).

Organic transistors gated via ionically conducting and electronically insulating electrolytes, have been the focus of an intensive research activity in the past two decades as bioelectronic sensors. Taking advantage of the high electrical double layer capacitance (EDLC), EGOFETs can be operated in sub-volt regime (i.e., <0.5 V) and can be employed as transducers for biosensing applications since analyte molecules can be detected in liquid media. The sensing in liquid media is primarily important because of the necessity of sensing biomolecules of interest in aqueous media such as saliva, blood serum, mucus, potable water etc. in-vivo and ex-vivo. The use of electrolyte gated transistors as biosensors for the detection of various biomarkers is growing fast, mainly due to their label-free operation mode, their ultra-sensitivity and high-specificity. Several efforts have been made to develop new electronic biosensor architectures and to find innovative materials that ensure the integration of active bioreceptors in the device while maintaining high performance (Casalini et al., 2013; Bortolotti, Berto, Sensi, Di Lauro, & Biscarini, 2019).

In EGOFET biosensors, a recognition element (e.g., an antibody or an aptamer) must be either immobilized at one of the relevant interfaces; gate/electrolyte or semiconductor/electrolyte. Upon binding, the capacitive coupling enables the amplification of subtle electrostatic potential, so even a minor change can be potentially translated into a large change of transistor's figures of merit such as drain current, threshold voltage (Macchia, Manoli, Di Franco, Scamarcio, & Torsi, 2020; Manoli et al., 2015; Picca et al., 2019).

Although many methods for biomolecules immobilization on metal and polymer substrates exist, integrating bio-receptors in an electronic device is a major challenge. The interfacing of biomaterials to electronic devices is one of the most challenging research field that has relevance for both fundamental studies and for the development of highly performing biosensors.

The first attempts of integrating biomolecules on the EGOFETs were focused on the biofunctionalization of the organic semiconductor (Buth, Donner, Sachsenhauser, Stutzmann, & Garrido, 2012; Kergoat et al., 2012; Magliulo, Mallardi, et al., 2013a; Mulla, Seshadri, et al., 2015). Soon after, this approach was abandoned since most organic semiconductors are inherently hydrophobic to host hydrophilic biomolecules and amenable to deterioration. Therefore, most recent examples report EGOFET biosensors with recognition moieties linked directly to the gate terminal, which is usually a gold electrode (Casalini et al., 2013; Mulla, Tuccori, et al., 2015; Macchia, Manoli, et al., 2018; Macchia, Romele, et al., 2018; Macchia et al., 2019).

In this chapter, important aspects related to the field of EGOFET biosensors are discussed. Most relevant strategies for biomolecules integration into the EGOFET device are presented. The transduction mechanism of EGOFET based biosensing platforms is also argued.

2. Biofunctionalization routes

In general, immobilized biomolecules such as enzymes or proteins are attached to a substrate or entrapped within a matrix. These immobilized enzymes or proteins offer several potential advantages over soluble proteins. Immobilized enzymes are typically macroscopic catalysts that are retained in the reactor; therefore, continuous replacement of enzyme is not necessary, and separation of the enzyme from other components in the reaction mixture is simplified (Clark & Blanch, 1997). For bio-sensor technology, fundamental studies of cell biology, diagnostic biosensors, and tissue engineering; bio-receptors such as proteins, peptides and cells need to be patterned in a predefined geometry. For example, spatially controlled growth of neurons or cells will allow the study of neuronal response under direct visualization of cells and synapses. This would lead to a better understanding of basic mechanisms, circuitry transactions for learning, cognition, and information storage in the brain, cell-array based biosensors and artificially designed neuronal networks (Csucs, Michel, Lussi, Textor, & Danuser, 2003; Scholl et al., 2000). Thus, organizing multiple biomolecules on surfaces with micron to nanometer-scale resolution is crucial in the development of biosensors.

The integration of biomolecules in an electronic device is a key step in the fabrication process of a biosensor. Several factors determine the adopted immobilization strategy, such as the properties of the bio- system and of the supporting material, the type of transducer used, and more importantly the application for which the sensor is designed. Experimental conditions during the immobilization and operation, as temperature, ionic strength, pH, should be carefully set for guaranteeing the vitality and functionality of the biomolecules. Moreover, a successful immobilization is characterized by the number of anchored bioprobes in preferred orientation so as to establish a high affinity with ligand of interest and limited nonspecific adsorption (Pavlickova, Schneider, & Hug, 2004).

In the following paragraph, current immobilization techniques, i.e. physical adsorption, covalent binding, and biochemical affinity reactions will be introduced. Strategies that can lead to highly oriented structures, homogeneous covering of the surface and provide well exposed active sites for direct coupling with the target component are discussed.

2.1 Physical adsorption

Biomolecules can be adsorbed physically onto different surfaces via hydrophobic, ionic and/or Van der Waals interactions. Adsorption is a simple and inexpensive technique that is commonly employed for biomolecules immobilization. For instance, proteins like antibodies can be directly adsorbed onto noble metals (Au, Ag, Pt), carbon materials (glassy carbon, carbon paste, nanotubes), metal oxides and conductive or semiconductive materials (Pavlickova et al., 2004; Seshadri et al., 2018). The immobilization protocol consists of exposing the proteins in solution to the support under the appropriate conditions where mutual attraction between the solid surface and the protein results in coverage of the surface. The negatively charged backbone of the DNA can easily adsorbed on wide range of solid substrates that includes carbonaceous substrates, charged polymers, metal surfaces, etc. (Pan et al., 2009; Schreiner et al., 2010). The simplicity of adsorption, however, can be offset by the ease of desorption, depending on the various conditions. But this disadvantage can serve an important function, as easy removal of proteins from the substrate is useful in its regeneration; that is, once the immobilized enzyme has lost its activity, the inactive adsorbed proteins can often be stripped from the substrate and subsequently replaced with new proteins (Clark & Blanch, 1997). Variety of molecular surface to biochip surface interactions can lead to large variations in protein concentrations as well as possibly introducing the reduced protein bioactivity or protein denaturation. Furthermore, adsorption may not provide surface density and mobility of immobilized bio-molecules that can be obtained by other immobilization techniques (Ulbrich, Golbik, & Schellenberger, 1991). Thus, more robust strategies that involve covalent attachment to a functionalized surface or immobilization through affinity reactions allow to control better the density and the conformational freedom of the probes.

2.2 Covalent binding

A comparatively stable method of protein immobilization is to link a protein to the substrate via a covalent bond. Functional groups, i.e., carboxylic -COOH, hydroxyl -OH, amine $-NH_2$ and thiol -SH groups, are usually involved to covalently bind biomolecules on a solid support. The functional groups present on the surface of the material usually form bonds with groups present in the exterior of the biomolecules (Rusmini, Zhong, & Feijen, 2007). For instance, in peptides and proteins covalent binding occurs via the terminal groups in the side chains of amino acids. The attachment of biomolecules can be site specific or random. Unless the surface of the substrate consists of reactive groups in its native form, intermediate components can be introduced and act as linkers between the surface and the biomolecule. Moreover, intermediate linkers are often used to achieve better orientation, conformation freedom and limit steric hindrance effects. A covalent bond can be formed in one or more steps. Several chemical routes exist for covalently coupling functional groups. Some examples of the most utilized covalent immobilization protocols involve:

(i) carbodiimide and N-hydroxysuccinimide esters (EDC/NHS) chemistry for conjugation of carboxylates to primary amines

(ii) sulfhydryl-reactive crosslinker chemistry, where maleimide reagents are used for conjugating to a thiol group

(iii) hydrazide-activated crosslinkers for conjugation of carbonyls (aldehydes, ketones),

(iv) aryl azides, aryl diazirines and benzophenone can serve as photo-crosslinkers for covalent binding of biomolecules

The main advantage of covalent linking is the strength of binding, and detachment during use is usually negligible. However, this also makes regeneration not possible and the reagents involved increase the cost of the immobilization procedure (Clark & Blanch, 1997; Kalantar-zadeh & Fry, 2007). Moreover, covalent binding can but cannot guarantee the proper orientation of the recognition molecule.

2.3 Affinity coupling

Affinity binding approaches offer more precise control of the orientation and density of biomolecules on a surface. Avidin-biotin interaction is a strong non-covalent interaction and is employed for biofunctionalization of different surfaces in bio-electronic detection systems (Green, 1975; Wilchek & Bayer, 1988). It allows biotinylated molecules to be bound to avidin modified surfaces or vice versa. Avidin analogs, such as streptavidin and neutravidin, can be used since they present similar affinity to biotin. Biotin can covalently bind to other biomolecules (e.g. antibodies, peptides, DNA). Biotinylated reagents bearing NHS, maleimide, EDC, hydrazide groups can react with primary amines, thiols, carboxyls and carbohydrates. Photo-reactive groups can also be incorporated to biotin. Histidine based affinity method (His Tag system) offers site specific immobilization as well. Protein A and protein G directly coupled to the substrate surface are commonly used to bind the Fc region of antibodies. In this case, a significantly higher fraction of active, well-oriented antibodies is achieved, but usually suffers from a lower surface density of immobilized antibodies and increased variability since two immobilization steps are required (Pavlickova et al., 2004). Biomolecules can be also immobilized by DNA hybridization. Single stranded (ssDNA)-protein conjugates can be formed by direct covalent attachment, avidin-biotin affinity system, bifunctional linkers etc. (Kalantar-zadeh & Fry, 2007).

2.4 Self-assembled monolayers (SAM)

Another widely employed method of bio-molecules immobilization is using self-assembled monolayers (SAMs). Self-assembly is the phenomenon in which molecules are self-organized in a systematic way on a surface (metal-gold, silver or oxidic surfaces). SAMs are formed by self-aligning of certain molecules in two dimensions on surfaces forming a quasi-crystalline structure. Different systems of self-assembled monolayers (SAMs) exist based on various organic molecular components, which either inhibit or promote proteins immobilization (Ulman, 1996). Most widely used and perfectly developed are SAMs of alkanethiols on gold surface. The sulfur atoms of alkanethiols adhere to gold surface and tightly packed alkyl chains are tilted by an angle of 30° to the gold surface (Xia & Whitesides, 1998).

Such molecules should present two different head and tail functional groups where one end binds to the surface of the sensor and the other end anchors the biomolecules. For the formation of SAMs, the substrate is usually immersed into a dilute solution of the SAM compounds. The duration of SAM formation can last few minutes up to several hours. Parameters like chain length, concentration of the solution and nature of the end groups affect the formation of SAMs (Matharu, Bandodkar, Gupta, & Malhotra, 2012). SAMs can be attached to biomolecules by physisorption or by chemical modification of the end group. Modified substrates with SAMs having different reactive terminal groups can be used for multicomponent immobilization, hence can be used for detecting several analytes. SAMs with non- reactive end groups or different chain lengths can be used as spacers. Spacer between SAMs with reactive groups are also beneficial for immobilizing large biomolecules. For instance, proteins can attach to the surface without undergoing conformational changes due to steric interactions.

Plethora of heterobifunctional thiol-SAMs of different chain lengths and functional end groups are commercially available. These functional groups like -COOH, -NH2, etc. can covalently bind to biomolecules via crosslinkers such EDC/NHS (Matharu et al., 2009), aldehyde (Mantzila, Maipa, & Prodromidis, 2008), etc. Avidin modified thiol-SAMs have also been reported for anchoring biotinylated antibodies or DNA strands (Matharu et al., 2012). Also SAM of Ni^{2+} chelates have been employed to bind His-tagged A/G protein for oriented antibody immobilization (Patrie & Mrksich, 2007). In general, the above protocols of Au biofunctionalization require two or more stages to be completed. Alternatively, one step SAM formation using thiolated biomolecules like antibodies is possible (Park & Kim, 1998; Zhang et al., 2008). The latter approach has been proven to be rather effective for providing sufficient surface coverage and well-oriented antibodies. The thiolation of antibodies can be done using amine and sulfhydryl reactive heterobifunctional crosslinkers. Moreover, antibody fragments can directly assemble on the gold surface by Au—S bonds. SAMs of alkanethiols on gold offer many advantages such as: high packing density, dense monolayers simple procedure of preparation, lower defect density, better stability in ambient conditions, practicality in applications and most important, possibility to control the interfacial properties of systems (Xia & Whitesides, 1998).

3. Electrolyte gated organic transistors

3.1 Main features and operation principle of EGOFETs

Polymer electrolyte gated organic field effect transistors (OFET) can be traced back to 2005, where a pentacene based OFET was demonstrated using a solid electrolyte as insulator (Panzer & Frisbie, 2005). In 2007 Misra et al. used ionic liquids to gate an inorganic based FET (Misra, McCarthy, & Hebard, 2007). While few years later, a water droplet was used as gating medium showing field effect performance on thin film organic transistors (Kergoat et al., 2010) opening new era of aqueous electrolyte gated OFETs. In all cases, the concept is rather simple. The organic semiconductor is being deposited on a pre-patterned substrate with the source and drain electrodes, that define the active channel. The channel is exposed to an electrolyte solution and a gate electrode is immersed in the solution. A water or buffer droplet is enough to act as an insulator between gate electrode and the organic semiconductor (OSC).

In a standard operation, a voltage (V_G) is applied to the gate electrode with respect to the grounded source and a second bias is applied between source and drain contacts. When operated within a regime of electrostatic accumulation, upon gate bias, a potential difference is established at the interface between the electrolyte solution and the channel. The resulting potential difference controls the current flow in the channel but depends on the gate voltage and the processes occurring at the gate interface with electrolyte. If a polarizable electrode is used as gate, a second a potential difference is built between the gate and the electrolyte. It is worth noting that this ideal state is not achieved, and hence a gate leakage current is expected. This not true for a non-polarizable gate electrode, where all the potential drops across the interface with the channel. Field induced carriers on the semiconductor (holes or electrons) are confined at the top few monolayers of the semiconducting film. This is generally referred as 2D carrier transport that does not depend on the film thickness.

No redox reactions should occur and the response to changes in potential is rapid without diffusion limitations. Water being an electrolyte is electrochemically active. Therefore, attention should be paid to avoid any occurrence of electrochemical phenomena. Considerable faradaic current can be observed due to electrolysis of water if the potential difference between source and drain electrodes exceeds 1.0 V limit.

From the above it is clear that by triggering the interfaces formed at the two sides of the electrolyte, gate/electrolyte and OSC/electrolyte, we can affect the conductivity of the channel. To do so a proper design of the geometrical features is a prerequisite. Hence, an EGOFET device can act as biosensing platform (Picca et al., 2019).

3.2 Bio-functionalization of the organic semiconductor via PE-CVD

During the past years, different protocols have been reported to introduce a biolayer on the surface of the semiconductors. Most polymer-based semiconductors are hydrophobic by nature, thus chemical modification is required to render them hostile for biomolecules immobilization. There are also hydrophilic organic semiconductors, but not that common (Belyaeva, van Deursen, Barbetsea, & Schneider, 2018). Initially, the idea was to synthesize monomers bearing biologically relevant side chains or groups. After polymerization, films displaying the desired functionality on the surface can be obtained (Kergoat et al., 2012). However, the presence of functional groups on the backbone or on the side chains of the polymer impact on the electrical properties leading to devices of inferior performance. Besides, the bioprobes are not homogeneously distributed in the bulk of the polymer layer and they are not confined or properly oriented at the surface, preventing effective binding with the target analyte.

To overcome such issues, alternative methods to immobilize bioprobes solely on the surface, without affecting the bulk properties of the semiconductor, have been reported. For instance, Buth et al. (Buth et al., 2012) studied different surface treatments, including UV-oxidation and silanization with (3-aminopropyl)triethoxysilane (APTES) to biofunctionalize α-sexithiophene (α6T). Most interesting approach is the use of well-established surface treatments that can be applied in most of surfaces independently of the nature of the material. As such we should consider plasma treatment, that can be used to render a surface with the desired functionality. Plasma enhanced chemical vapor deposition (PE-CVD) of thin films, treatments at low and atmospheric pressure are widely utilized in Material Science and Technology to modify the surface of materials, generally within a thickness of hundreds of nanometers, with no change of bulk properties. Other advantages of this technology are the easy integration in existing industrial lines and the minimal impact to the environment, since the amount of gas/vapor chemicals used is negligible, and solvents are not used at all.

Herein, we discuss in more detail such an example. In particular, carboxylic groups were coated on the surface of poly(3-hexylthipohene) (P3HT) surface by using PE-CVD. The modified P3HT surface further served as sensing stage in an EGOFET biosensor (Magliulo, Pistillo, et al., 2013). Functionalizing surfaces with carboxylic groups (-COOH) to enhance biocompatibility of materials is widely cited. The presence of carboxylic groups not only helps in achieving hydrophilic surfaces but can also serve as an anchor sites for the covalent immobilization of biomolecules. Biomolecules can be easily conjugated to these carboxylic groups using EDC/NHS chemistry. Such an approach has been successfully employed for anchoring vesicles containing NH_2-functionalized phospholipids. Proteins can then attach to the bilayer surface through amide, disulfide, thioether covalent conjugation or by means of biotin/streptavidin binding) (Magliulo, Mallardi, et al., 2013b).

3.2.1 *Materials, equipment, and reagents*

3.2.1.1 Equipment

1. Home-build low pressure RF (13.56 MHz) plasma reactor
2. Spin-coater
3. Bath-sonicator
4. Analytical balance
5. Thermal incubator
6. Soxhlet extractor
7. Filters (0.2 μm PTFE and 100 nm PC)
8. Avanti mini extruder (Avanti Polar)
9. X-ray photoelectron spectrometer (Theta Probe Thermo-VG Scientific Instrument)
10. Semiconductor parameter analyzer (Agilent 4155C)
11. Home-build probe station

3.2.1.2 Materials and reagents

1. Poly(3-hexylthiophene) (P3HT) from BASF (RR > 98%, Sepiolid P200)
2. Photolithographically patterned with interdigitated S & D electrodes ($L = 10\,\mu m$, $W = 10\,mm$) Si/SiO_2 wafers
3. Solvents: Isopropanol (IPA), ethanol (EtOH), methanol (MeOH), hexane, chloroform ($CHCl_3$), HPLC water (all purchased from Sigma Ardich)
4. Vapors/Gases: Acrylic acid vapors (AA 99%, Aldrich, USA), Ethylene (Eth, Air Liquide), Argon (Ar, Air Liquide), Nitrogen (N_2, Air Liquide)
5. Soy bean lecithin (EPIKURON 200 from CARGILL)

6. N-((6-(biotinoyl)amino)hexanoyl)-1,2-dihexadecanoyl-sn-glycero-3-phosphoethanolamine (Biotin-X DHPE from INVITROGEN)
7. Phosphatidylethanolamine (Avanti Polar)
8. Phosphate buffer solution (10mM, KCl 2.7mM, 137mM NaCl, pH=7.4, pellets from Sigma Aldrich)
9. Ethyl (dimethylaminopropyl) carbodiimide (EDC from Sigma Aldrich)
10. N-Hydroxysuccinimide (NHS from Sigma Aldrich)
11. Streptavidin (SA) from Streptomyces avidinii (Sigma Aldrich)

3.2.2 *Protocol*

3.2.2.1 Substrates cleaning and deposition of the organic semiconductors

1. Cut with a diamond tip pen the prepatterned Si/SiO_2 substrates
2. Immerse each substrate in beakers of 20 ml filled with solvents of increasing polarity (IPA, EtOH and MeOH) and sonicate for 10 min each
3. Rinse with HPLC water
4. Dry under N_2 flow
5. Purify P3HT by successive Soxhlet extractions with methanol and hexane
6. Weight 2.6 mg of dried purified P3HT and dissolve in 1 ml of $CHCl_3$
7. The P3HT solution is sonicated until the polymer is completely dissolved
8. Filter the solution through a 0.2 um PTFE filter
9. Place the clean substrate on the spin coater holder. Drop cast with micropipette the filtered P3HT solution (~60–90 uL to cover the surface) and spin at 2000 rpm for 30s. The thickness obtained is around 30 ± 5 nm
10. Anneal at 75°C for 1 h in the thermal incubator

Tip: The solution of P3HT should be always prepared in a dark amber vial since P3HT is photosensitive and degrades the electronic performance of final device upon exposure to ambient light, especially UV wavelengths.

3.2.2.2 PE-CVD process

1. Acrylic acid vapors were passed from a needle valve; ethylene and argon were passed from two electronic mass flow meters. Gases were mixed before the reactor. Conditions: Radio frequency power (50 W), pressure (250 mTorr) and flow rate of precursors (acrylic acid flow 2.5 sccm, ethylene flow 7.5 sccm, argon flow 2.5 sccm).

2. P3HT surfaces are inserted in the reactor and they are plasma processed with the vapors of a mixture of acrylic acid and ethylene giving plasma deposited Ethylene/Acrylic acid (pdEthAA).

Tip: Ethylene is added as a cross-linker, to stabilize the coating in aqueous media and eliminate problems such as leaching of compounds and delamination from substrates.

Tip: A deposition rate $33 \pm 1\,\text{nm}\,\text{min}^{-1}$ was measured in such conditions. The deposition time was 3 s.

Tip: PE-CVD conditions were optimized in order to satisfy the following requirements: acquire stable -COOH-functionalized coatings in water media, achieve optimum electrical performance of the transistor and have satisfactory percentage of -COOH functionality on the surface of P3HT to bind the bioreceptor's layer.

3.2.2.3 Bioconjugation of phospholipid (PL) layer with P3HT-pdEthAA

1. Soybean lecithin (4.5 mg), proper amount (50–150 μg) of biotin-X DHPE and phosphatidylethanolamine (0.5 mg) are dissolved in chloroform
2. Dry the solution under N2 flow and keep under vacuum for 60 min
3. Rehydrate the PLs in 1 mL of PBS
4. Sonicate for 10 min on ice
5. Extrude through a polycarbonate filter having 100 nm pore sizes
6. The substrates with plasma-treated P3HT are incubated in a freshly prepared 100 mM ethyl (dimethylaminopropyl) carbodiimide (EDC) and 100 mM N-Hydroxysuccinimide (NHS) solution (both in 10 mM PBS pH 7.4) for 1 h at room temperature with mild stirring
7. After activation, the substrates rinse three times in PBS buffer
8. Immerse them in in the vesicles suspension (biotinylated or non-biotinylated) overnight under mild agitation at room the temperature
9. Rinse three times in PBS buffer
10. Prepare the standard solution of SA (in PBS) for carrying the sensing measurements

Tip: Phospholipid layer are chosen because of their versatility, being the main components of biological membranes, they are amphipathic molecules, i.e. endowed with both hydrophilic and hydrophobic moieties, that spontaneously self-assemble in water forming bilayers facing the outer aqueous domains with their hydrophilic moiety (the PL head) and secluding in the interior of the bilayer the two hydrophobic alcyl chains (the PL tails).

The EGOFET biological functionalization is shown in Fig. 1

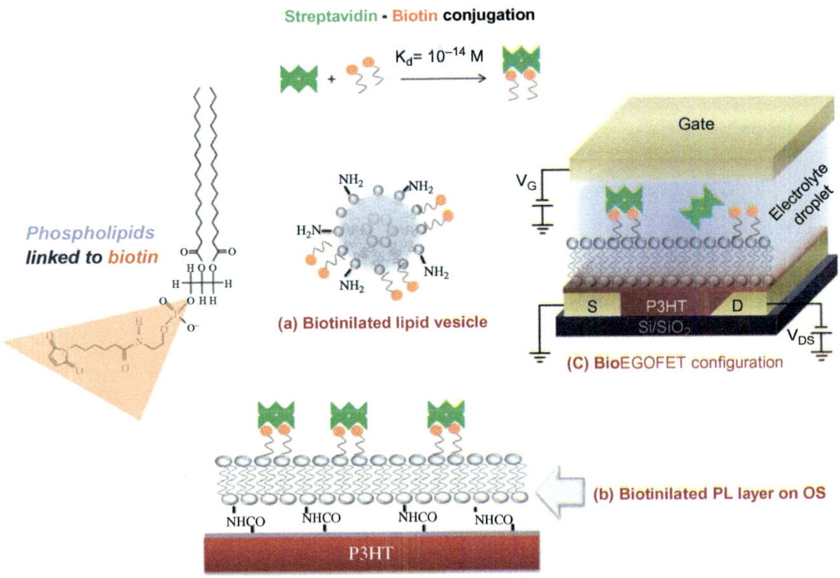

Fig. 1 Schematic illustration of the process of phospholipid layer formation and attachment to pdEthAA layer deposited on the OS of EGOFET.

3.2.2.4 XPS analysis

1. Surface chemical composition of P3HT before and after surface treatment was measured by X-ray photo-electron spectroscopy (XPS). Survey spectra (0–1200 eV binding energy, BE) and high-resolution spectra were acquired at 150 and 100 eV pass energy, respectively
2. Data was analyzed using the Thermo Avantage software, version 4.75. C1s spectra were best-fitted with the following peak components, representing the different functional groups: C0 (C—C, C-H; BE 284.8 ± 0.1 eV, BE reference), C1 (C—O, C—S, C-COOH; BE 286.5 ± 0.3 eV), C2 (C=O, O—C—O; BE 287 ± 0.3 eV), and C3 (-COOH/R; BE 289.4 ± 0.2 eV). Peak components were normalized to the highest peak
3. Samples used to investigate the aging were kept in petri dish sealed with the paraffin film and stored at ambient conditions

Tip: X-ray photoelectron spectroscopy (XPS) measurements were performed with a Theta Probe Thermo-VG Scientific instrument using an Al Ka monochromatic source (15 kV with 300 μm X-ray spot size and take-off angle of 37 degree, the base pressure was maintained at 10 = −9 mbar).

Tip: Three distinct points were analyzed per sample; the resulting atomic percentage values were reported as mean value with associated standard deviation for each set of measurements.

3.2.2.5 Electric measurement of the EGOFETs

1. The EGOFET electrical measurements were carried out by using a 2 μL droplet of water (HPLC grade) or phosphate buffered saline (PBS 10 mM, pH 7.2) as gate dielectric, produced with a micropipette and deposited on the surface of the organic semiconductor. A gold lamina roughly of 3 mm^2 size in contact with the electrolyte solution was used as a gate electrode
2. Bio-Sensing measurements are performed by measuring the transfer characteristics of bio functionalized EGOFET first using a drop of PBS (baseline)
3. Streptavidin is drop-casted on the active area (i.e., S-D IDE region) for 15 min
4. The channel is rinsed with buffer to remove the unbound proteins
5. A transfer curve is obtained after exposure to SA solution
6. Repeat the same procedure for the remaining SA solutions
7. Prepare the calibration curve by plotting the drain current at maximum gate voltage versus the SA concentrations

Tip: For the output characteristics, the drain current (I_{DS}) was measured as function of the drain voltage (V_{DS}) at different gate voltages applied (V_{GS}) from 0 to −0.5 V, steps of −0.1 V. For the transfer characteristics, I_{DS} was measured as function of V_{GS} (from 0 to −0.5 V) at a constant V_{DS} of −0.5 V. In all I–V measurements, the voltage was swept back and forth to evaluate the occurrence of any hysteresis.

Tip: Electrical characterization of the EGOFETs, before and after plasma treatment are carried out using the semiconductor analyzer. Standard I_{DS} vs V_{DS} output and I_{DS} vs V_{GS} transfer characteristics were recorded and compared.

Tip: During the measurements, a water vapor saturated environment was kept around the devices to avoid evaporation of water.

Tip: The transistor electrical figures of merit, namely the field-effect mobility (μ_{FET}), the threshold voltage (V_{th}, transistor turn-on bias) and the I_{ON}/I_{OFF} current ratio were extracted from the characteristics in the saturation regime.

3.2.3 Results and discussion

Satisfactory transistor performance exhibiting good field effect modulation, as shown in Fig. 2 from the output and transfer characteristics, and comparable to that of the EGOFET with the bare P3HT is observed after the device has undergone 3 s plasma exposure. Three seconds lead to formation of 2 nm of pdEthAA layer. The electrical figures of merit such as field-effect mobility and threshold voltage of the P3HT/pdEthAA device were similar to those of bare P3HT device. Mobility was found in the order of 10^{-3}–10^{-4} cm^2/Vs and threshold voltage was equal to $-0.03\,V \pm 0.02$. However, a reduction in I_{ON}/I_{OFF} ratio ($\sim 10^2$) was observed and is attributed to the increase in hydrophilicity of the P3HT surface after pdEthAA layer deposition, that leads to an increase of the doping of the OSC. XPS spectra of and 3 s plasma deposited pdEthAA chip (green) and bare P3HT sample (red) is shown Fig. 2. An accurate curve fitting was applied to C1s spectra from both pristine and plasma treated samples, to investigate any surface chemical changes. The distinct characteristic peak of -COOH/R group at binding energy $289.4 \pm 0.2\,eV$ confirms the presence of -COOH/R

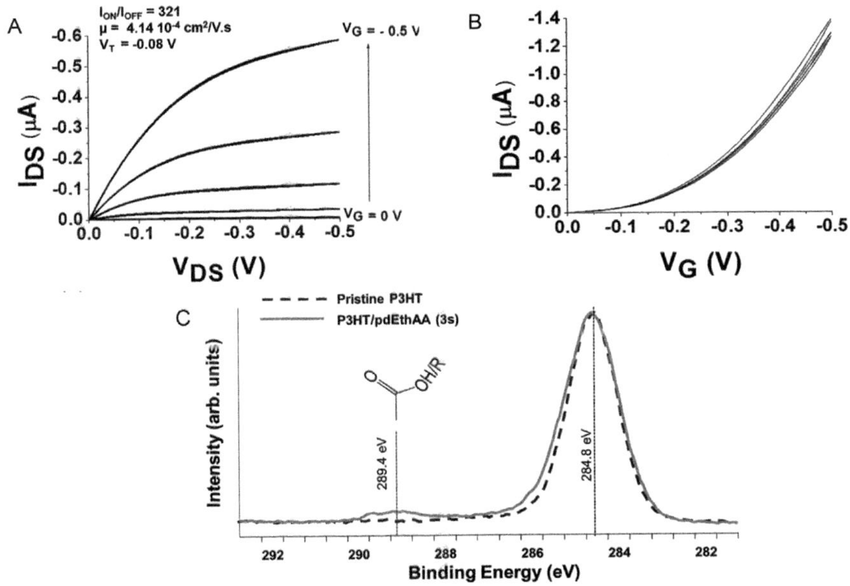

Fig. 2 (A) Output and (B) transfer characteristics of the P3HT/pdEthAA EGOFET device (C) High resolution spectra of C*1*s XPS peak for a P3HT-pdEthAA(*3* s) (gray solid line) and bare P3HT (dashed line).

functionality on the surface of plasma deposited P3HT sample. No peak shape changes were observed on the aged samples (after 136 days) confirming the stability of the deposited layer.

After bio-functionalization of active area, the field effect is retained quite well. The incorporation of the phospholipid layer in the architecture of the EGOFET improves the electrical performance of the device in terms of on/off ratio. Thus, previously reported issue of doping and hysteresis effect are satisfactorily suppressed since the PL layer acts as ion penetration inhibitor. We should note here that hysteretic phenomena in EGOFETs are often indicative of electrochemical phenomena but can occur due to the charge polarization/ion movement in the electrolyte as well. The EGOFET is further investigated as a sensor. For the sensing measurements, streptavidin–biotin couple was selected as prototype. Accordingly, the electrical performances of EGOFET devices prepared with biotinylated phospholipid layer bilayers exposed to increasing concentrations of streptavidin (SA) solutions were investigated. An example of sensor response to 100 μg/mL (1.6 μM in 10 mM PBS, pH = 7.4) streptavidin concentration when incubated with biotinylated phospholipid layer functionalized EGOFET is shown in Fig. 3A. It is clearly seen that as compared to device transfer characteristics exposed to PBS, after streptavidin exposure considerable increase in I_{DS} is observed. SA holds a negative charge at pH = 7.4 (isoelectric point ~5), thus when binds to the biotinylated layer is considered to act as an extra (negative) gating inducing more holes in the P3HT channel. Streptavidin solutions at different concentrations, from 1 μg/mL to

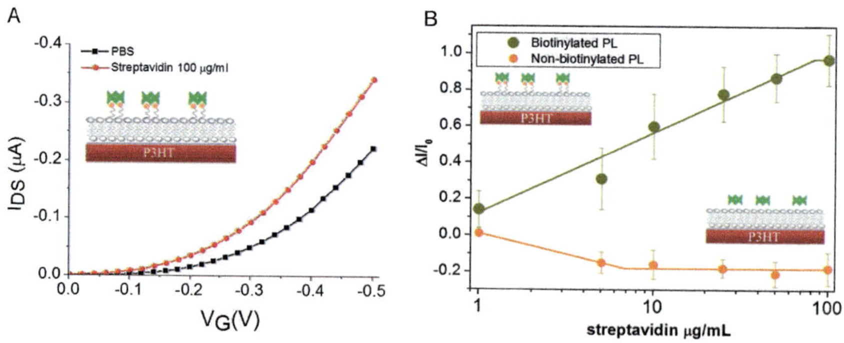

Fig. 3 (A) Transfer characteristics of EGOFET with biotinylated phospholipid layer before and after exposing to streptavidin concentration of 100 μg/mL. (B) Streptavidin sensing calibration curve of biosensing upon exposure to increasing concentrations of streptavidin solutions to P3HT-pdEthAA based EGOFET devices integrated with biotinylated phospholipid layer and non-biotinylated phospholipid layer devices.

100 μg/mL, were tested to evaluate the sensitivity and the detection limit of the EGOFET devices. Fractional increase in I_{DS} at $V_{GS} = -0.5$ V, i.e., $\Delta I/I_0$ vs increasing concentration of streptavidin is plotted as a calibration curve shown in Fig. 3B. For each streptavidin concentration, the average values were obtained by measuring 15 distinct transistors and the average value and standard error (SE) are reported in the graphs. As a control, same streptavidin concentrations were measured with devices integrated with non-biotinylated phospholipid layer.

3.2.4 Alternative methods/procedures

3.2.4.1 UV-crosslinking of polyacrylic acid (PAA) on the organic semiconductor

More recently, Mulla, Seshadri, et al. (2015) functionalized the surface of the OSC with carboxylic functionality by introducing an ultrathin UV-cross-linked polyacrylic acid (PAA) layer. It has been demonstrated that PAA can be cross-linked by low wavelength (254 nm) UV-C radiations can be used to cross-linked ultrathin layer of PAA without employing any heat treatment or chemical cross-linker. More specifically, polyacrylic acid with average Mw around 450 K was dissolved in HPLC water to form 3% by weight solution. The solution was stirred overnight to dissolve the PAA completely in the water. The PAA solution was drop-casted onto the semiconductor allowed to rest for at least 30 mins. The substrate was then spin coated at low rpm (around 500 rpm) for few minutes until the excess of PAA is removed. After spinning, the substrate is inserted into custom made UV-C exposure chamber. A commercially available germicidal mercury lamp (254 nm wavelength) is used to expose the OS-PAA substrate for 5 min. Finally, the substrate is rinsed with DI water. The XPS investigation confirmed the presence and stability of the carboxylic groups' functionality on the surface of P3HT. Furthermore, it can be noted from high percentage (around 12%) of available -COOH functional groups. Such an amount is much higher than what is obtained with the plasma treatment, i.e., 2–3%. The PAA UV-crosslinked surface was further used to immobilize biotinylated PLs using standard EDC/NHS activation chemistry and detect SA.

3.2.5 Summary

Selective label free detection of streptavidin was demonstrated by properly modifying the surface of the organic semiconductor. The P3HT-pdEthAA based EGOFET was able to detect SA within a range from 10 nM to 1 μM with a limit of detection (LOD) equal to 10 nM. The PAA UV- crosslinked

EGOFET sensor detected SA down to 10 pM while the dynamic range was from 0.01 nM to 1.6 μM. As seen, the LOD obtained using the EGOFET platform is higher than the reported biotin-SA binding affinity. Biotin and streptavidin is known to have a strong binding affinity with a femtomolar dissociation constant. However, dissociation constant of the streptavidin-biotin complex commonly cited in the literature can be used as an approximation. It usually refers to pure D-biotin and tetrameric streptavidin complexes formed in solution. Binding constants of streptavidin complexed with immobilized biotin have been found to range from 10^{-6} to 10^{-11} M (D'Agata, Grasso, & Spoto, 2008; Qureshi, Yeung, Wu, & Wong, 2001). Moreover, it has been shown that detection limit and sensitivity may depend on the density of the probes, the surface area and the sample volume used (De Avila et al., 2013; Jung et al., 1999). The performance of the EGOFET biosensors compete the performances of traditional optical techniques.

From an applications point of view, the EGOFET with integrated phospholipid layer approach opens several perspectives since antibodies and proteins can be also bound to phospholipid bilayers supported on solid surfaces without altering the membrane properties. PLs are the main components of biological membranes. They are amphipathic molecules, bearing both hydrophilic and hydrophobic moieties. They can spontaneously self-assemble in water forming bilayers facing the outer aqueous domains with their hydrophilic moiety (head) and isolating in the interior of the bilayer the two hydrophobic acyl chains (the PL tails). In this way, the non-polar inner part minimizes ionic diffusion through the membrane, thus limiting the OSC doping, while the PL head provides the binding sites for capturing the analyte.

Indeed, biotin PL/SA layer has been used to further functionalize antibodies for detection of clinically relevant biomarkers. The plasma treated EGOFET has been investigated as a biosensor for (C-reactive protein) CRP detection (Palazzo et al., 2014). In this case, a biotinylated CRP antibody (Ab) is initially bound to the SA/PL layer. The latter binds specifically CRP protein. The binding of CRP to the PL/SA/Ab gives rise to a three-stacked protein system, labeled as: PL/SA(AV)/Ab/CRP. Although, the binding of CRP occurs far from the OSC/electrolyte interface at distances much larger than Debye screening length (λ), the EGOFET sensor was able to detect the analyte. The electronic transduction was dominated by the outermost protein in the stacked layer and not by those residing underneath, even though they are closer to the device electronic channel.

The mechanism of sensing was correlated to changes in the capacitance of the gating system. To identify the origin of capacitance modulated transduction, it is necessary to describe in more detail the gate capacitance.

As mentioned, upon gate biasing EDLs form at each interface of the electrolyte medium. The total capacitance (C_{TOT}) of the gate-electrolyte-semiconductor system is given by the series of the gate-electrolyte capacitance (C_{GE}) and electrolyte-semiconductor capacitance (C_{ES}). If the two EDL capacitances are equal ($C_{GE}=C_{ES}=C$), the total capacitance is $C_{TOT}=C/2$, otherwise the smallest one will dominate. According to the Stern model, each EDL is composed of two layers, the Helmholtz double layer (HDL) and the diffusion layer. All the applied voltage is dropped across the two HDL generating large capacitances ($> 10\,\mu F/cm^2$). This is true for high ions concentration solution, like ionic liquids and solid polyelectrolytes. The situation changes when conventional buffer solutions are employed as gating medium. In that case, only a small fraction of charges is in the HDL, while the rest form the Gouy-Chapman diffuse double layer (GCDL). Moreover, the bound proteins are charged and the space between them is analogous to an ionic gel where the fixed polyelectrolyte ions establish a continuous background charge, and the mobile counter-ions can move freely within the layer. The counter-ions cannot diffuse too far into the solution because they form a net opposite charge in the protein layer leading to a Donnan's equilibrium. An electrical double layer is formed at the outer protein layer by the excess of counter-ions and the fixed charges.

Hence, the overall capacitance (C) of the EGOFET has a contribution from a series of capacitors: a fixed capacitance$=C_{fix}$ accounting for the HDL and the PL bilayer capacitance, the C_{GCDL} and, in the presence of protein layer, the Donnan's capacitance that is proportional to the ionic strength ($C_{DON} \propto i_s$). A capacitance that is in parallel and will be denoted by C_{par} is also foreseen to fully account for the stacked proteins:

$$C=\left(\frac{1}{C_{FIX}}+\frac{1}{C_{GC}}+\frac{1}{C_{DON}}\right)^{-1}+C_{par.}$$

The CRP binding adds a new layer that changes the capacitance impacting the $1/C_{DON}$ term and to a lesser extent the C_{PAR} term. According to this model, the bio EG-OFET response is indeed governed by a capacitive effect that is independent from the position at which the Donnan's equilibrium is set and therefore does not depend on Debye's length. The latter dominates in electrostatic, rather than capacitive, detections. Therefore, another major advantage of proposed research is that the proposed EGOFET sensors are operated even in high ionic strength electrolytic solutions in highly effective manner. This will open the doors for its applications in real life biosensors.

3.3 Biofunctionalization of the gate electrode

This paragraph describes the functionalization of the second interface of the EGOFETs, i.e., the gate electrode. The metal gate electrode (Gold, Au) is functionalized with Odorant Binding Proteins from porcine nasal mucus (pOBP) using thiol-SAMs with carboxylic end groups (-COOH), as anchoring functional group. Carvone enantiomers dissolved in water, were used as analyte. For the first time, chiral discrimination between two enantiomers by OBPs is demonstrated. The results obtained by the EGOFET platform are supported by competitive fluorescent binding assay.

3.3.1 Materials, equipment, and reagents

3.3.1.1 Equipment

1. Electrochemical analyzer (Model CH-1140B) from CH instruments
2. Novascan bench top UV/O3 cleaner
3. Emission fluorescence spectrometer (Perkin Elmer LS55)
4. Centrifuge
5. Magnetic stirrer with hotplate
6. Vacuum desiccator
7. Spin-coater
8. Bath-sonicator
9. Analytical balance
10. Thermal incubator
11. Filters (0.2 μm PTFE)
12. Pyrex glass electrochemical cell
13. X-ray photoelectron spectrometer (Theta Probe Thermo-VG Scientific instrument)
14. Semiconductor parameter analyzer (Agilent 4155C)
15. Home-build probe station

3.3.1.2 Materials and reagents

1. Gold plate (99.99 % purity, 0.5 mm thickness from Kurt. J. Lesker)
2. polyethylene naphthalate (PEN) flexible substrates
3. Home made Ag/AgCl (KCl sat.) reference electrode (RE)
4. Platinum (Pt) plate as counter electrode (CE)
5. Sulfuric acid(H_2SO_4(97%, v/v))

6. Hydrogen peroxide (H_2O_2(30%, v/v))
7. HPLC water
8. Ethanol
9. Acetone
10. Methanol
11. Potassium chloride (KCl)
12. 3-mercaptopropionic acid (3MPA)
13. Acetic acid
14. Ethyl(dimethylaminopropyl)carbodiimide (EDC)
15. N-hydroxysuccinimide (NHS)
16. Sodium phosphate buffer (20 mM, pH 8)
17. Tris-HCl (50 mM)
18. 1-Aminoanthracene (1AMA) Fluorescence Probe
19. Poly[2,5-bis(3-tetradecylthiophen-2-yl)thieno[3,2-*b*]thiophene] (PBTTT-C14 by Ossila)
20. 1,2-dichlorobenzene (1,2 DCB)
21. Chloroform ($CHCl_3$)
22. D (+) Carvone and L (−) Carvone Enantiomers
23. 2-Phenylethanol
24. Odorant Binding Proteins from porcine nasal mucus (pOBP pOBP-mutant-F88W) The odorant binding proteins were supplied by the group of Prof. Krishna Persaud (University of Manchester, UK)

3.3.2 *Protocol*

3.3.2.1 Gate cleaning

1. Cut the gold lamina and bend in L shape. The area of gate electrode encountering the electrolyte is approximately 3 mm^2
2. The gold gates were first cleaned sequentially by rinsing with HPLC grade water, acetone, isopropanol, HPLC grade water and ethanol
3. Dry well under nitrogen flow
4. Prepare Piranha solution by mixing sulfuric acid and hydrogen peroxide in a ratio of 3:1
5. Immerse the gate electrode in freshly prepared Piranha solution. The piranha solution was heated on hot plate at 80 °C for 10 min and then put in warm ultra-sonic bath for 10 min
6. Rinse well with water and immerse gates in boiling water for 10 Min
7. Dry with N_2 blow and put the gates in Novascan bench top UV/O3 cleaner for 20 min with heating 150 °C
8. Rinse with HPLC water, ethanol and dried under N2 stream

9. Electrochemical polishing of the gates using a three-electrode configuration (Au gate is working electrode (WE), Ag/AgCl is RE and Pt as CE, 0.5 M sulfuric acid as electrolyte, scan the potential from 0 to 1.5 V for 30 cycles, scan rate 0.1 V/s).

10. Rinse rigorously HPLC grade water and then with ethanol and dry under N_2 stream

Tip: Before cleaning the gold gates, all glass-wares were carefully pre-cleaned by piranha solution, then by rinsing with ethanol and dried under nitrogen blow. The beakers were immediately sealed by sealing film to avoid contamination from ambient atmosphere to re-deposit on the inner surface of glass-wares.

Tip: piranha reaction is extremely dangerous; care should be taken while handling and disposing off this solution.

Tip: Piranha cleaning was performed at least two times, between which UV/ozone cleaning was performed.

Tip: The cleaned substrates are immediately used for next process or stored under vacuum in vacuum desiccator connected to lab scale vacuum pump.

3.3.2.2 Bioconjugation of odorant binding proteins using SAMs

1 Prepare the SAM solution (50 mM solution of 3-mercaptopropionic acid (3MPA) in ethanol containing 5% acetic acid).

2 Rinse SAM functionalized gold gate electrode with ethanol

3 Activate the carboxylic terminated 3MPA SAMs by immersion into a 100 mM EDC and 200 mM NHS aqueous solution for 1 h at 25 °C

4 Rinse well with water and incubate the activated gates in the pOBP solution (0.7 mg/mL in 20 mM sodium phosphate buffer, pH 8.00) for 2 h

5 Rinse with sodium phosphate buffer and dry under gently blown nitrogen stream

6 Prior to immersion of gold electrode into the solutions, nitrogen was bubbled through the ethanol for 10 min and beaker was kept under nitrogen atmosphere

Tip: Prior to immersion of gold electrode into the solution, purge N_2 for 10 min.

Tip: The functionalized electrodes are stored in petri dish sealed with paraffin film. The electrodes could be stored in lab refrigerator at around 4 °C for 1 week. No major difference in sensing experiments was observed in sensing measurements with functionalized gates stored up to 2–3 weeks.

This also confirms the stability of pOBP. But in all the experimental results shown here, the functionalized gates were freshly prepared or stored maximum up to 1 or 2 days.

3.3.2.3 Competitive fluorescent binding assay (Determination of Kd)

1. Affinity of pOBP toward fluorescent 1-AMA: Prepare 1 μM pOBP-F88W solution in 10 mM phosphate buffer, pH 7.4, and titrated with aliquots of 1 mM 1-AMA in methanol to final concentrations of 0.25, 0.5, 1, 2, 3, 4, 6, 8, 12, 16, 20 μM
2. The affinity of pOBP toward carvone enantiomers was measured in competitive binding assays, using both pOBP and probe at 1 μM as the final concentration
3. The ligand solutions, with concentrations ranging from 2 to 40 μM, were prepared in 10 mM phosphate buffer pH 7.4 and used for the binding experiments
4. Measure the emission fluorescence spectra of each solution. Spectrums were recorded at room temperature, with a 1 cm light path quartz cuvette and 5 nm slits for both excitation and emission. The probe was excited at 295 nm and emission spectra were recorded between 300–570 nm (pOPB/1-AMA complex $\lambda = 484$ nm).

Tip: The dissociation constant (Kd) of the pOBP/1-AMA complex ($Kd_{1\text{-AMA}}$) was calculated from the binding curve by non-linear squares fit of the experimental data using the equation: $y = B_{max} \times [1\text{-AMA}]/(Kd_{1\text{-AMA}} + [1\text{-AMA}])$ where, [1-AMA] is the concentration of the free probe, y is the specific binding derived by measuring fluorescence intensity and B_{max} is the maximum amount of complex formed at saturation. The computer program used to fit the data was SigmaPlot 12 (Systat software, Inc. USA).

Tip: Dissociation constants of competitors were calculated from the corresponding IC50 values, using the equation: $K_{Sol} = [IC50]/1 + [1\text{-AMA}]/K_{1\text{-AMA}}$, where [1-AMA] is the free concentration of the fluorescence probe and $K_{1\text{-AMA}}$ is the dissociation constant of the complex OBP/1-AMA.

3.3.2.4 EGOFET device fabrication

1. Gold source (S) and drain (D) electrodes were photo-lithographically patterned using titanium as gold adhesion promoting layer to form interdigitated geometry. The channel length and width were 5 μm and 10 mm respectively

2. The substrates are cleaned by sonication in acetone and isopropanol for 5 min sequentially and rinsed with small quantity of chloroform and 1,2-dichlorobenzene
3. PBTTT-C14 is dissolved in mixture of two solvents, 1,2-DCB and $CHCl_3$ in the ratio of (9:1) at the concentration of 7 mg/mL by heating at 80°C until clear solution is observed
4. The OSC solution is filtered through the 0.45 μm PTFE filtered and kept warm until spinning
5. PBTTT-C14 solution is spin coated on this substrate at 7000 rpm for 60 s
6. Annealing is performed at 120°C for 10 min on hot plate and allowed to cool gradually

Tip: PBTTT-C14 being light sensitive, the prepared devices were stored in dark and under vacuum (in lab vacuum desiccator).

3.3.2.5 X-ray photo-electron spectroscopic XPS analysis

XPS analysis was performed on functionalized gate electrodes at different steps of fabrication process to confirm the functionalization process. Measurements settings and other details are described in Section 3.2.2.5.

3.3.2.6 Sensing protocol of chiral EGOFET sensor

1. The EGOFET electrical measurements were carried out by using a 3 μL droplet of water (HPLC grade) gate dielectric, produced with a micropipette and deposited on the surface of the organic semiconductor
2. Measure the transfer curve of the pOBP functionalized gate electrode but without any analyte bound to the electrode. Repeat three times with a pause of 5 min. This is the baseline (blank measurements)
3. Prepare the standard solutions of carvone enantiomers and 2-Phenylethanol (control) (range 1–10^6 pM in 10 mM phosphate buffer pH7.4)
4. For sensing measurements, gate electrode was incubated for 5 min in 200 μL analyte solution in a small container (specifically, the cap of 2 mL vial was used)
5. Rinse the gate very gently by dipping in HPLC grade water beaker and directly brought in contact with the water droplet (3 μL) already dispensed over the device channel area
6. Measure the Transfer after the Exposure to the Analyte
7. Repeat the steps 4–6 for all concentrations (starting from lower to higher concentration).

8. The data are plotted as the $\Delta I/I_0$ average values measured on three different devices for each curve

Tip: For the output characteristics, the drain current (I_{DS}) was measured as function of the drain voltage (V_{DS}) at different gate voltages applied (V_{GS} from 0 to −0.5 V, steps of −0.1 V). For the transfer characteristics, I_{DS} was measured as function of V_{GS} (+0.2 V to −0.5 V) at a constant $V_{DS}=-0.5$ V.

Tip: The room temperature was controlled in the range of 16–20 °C while humidity was between 38% and 45%.

Tip: Before each sensing measurement process, the device was allowed to stabilize until overlapping IDS vs VGS characteristics are observed.

Tip: The (D(+)Carvone or L( )Carvone enantiomers are perceived as spearmint or caraway odors with human threshold for detection of 30 and 420 nM, respectively (Pelosi & Viti, 1978).

3.3.3 Results and discussion

3.3.3.1 Determination of dissociation constant (Kd) by binding assay titration

In the study of odorant binding proteins and its binding affinities with target analytes, competitive fluorescent binding assay is very widely used technique (Capone et al., 2009). Herein, the non-fluorescent analyte, enantiomers of carvone (D(+) Carvone or L(−) Carvone), competed for binding sites in an OBP with a fluorescent ligand that is used as a probe (1-AMA). Depending on the binding affinity between protein and analyte, the degree of decrease in bound (1-AMA) probe to the protein percentage varies, as shown in Fig. 4 Thus, the dissociation constant K_D is inversely proportional to the binding

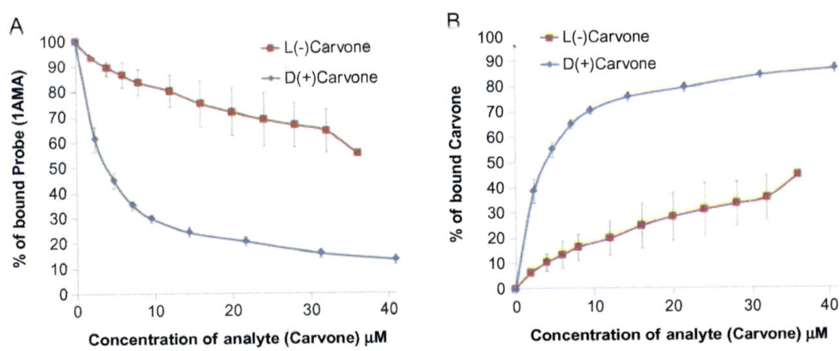

Fig. 4 Percentage (A) of bound *1*AMA to pOBP and (B) of bound analyte (Carvone enantiomers) to pOBP in binding assay titration.

affinity between protein and analyte. The Kd obtained from the binding assay between non-delipidated pOBP and carvone enantiomers is: K_d for D(+) Carvone was 21.38 μM and for L(−) Carvone was 1.76 μM.

There is one order of magnitude difference between the affinities of carvone enantiomers with the pOBP. Although, in general proteins are known to be enantioselective, there are no references about the demonstration of enantioselectivity by OBPs. This is the first demonstration of enantioselectivity by odorant binding proteins. The dissociation constants of pOBPs also provide an indication that they are binding with carvone enantiomers with affinity in micro-molar concentration range. Hence, as a general thumb rule, for actual sensing measurements with EGOFET, concentration range chosen was two orders of magnitude below and above the concentrations given by binding assay measurements.

3.3.3.2 Carvone enantiomers differentiation using pOBP functionalized gate in an EGOFET device

The electrical performance of the EGOFET device using the pOBP functionalized gate is depicted in Fig. 5. The gold gate electrode was modified with SAMs bearing carboxylic tail groups. The porcine odorant binding proteins (pOBPs) were covalently attached to the carboxylic groups by standard EDC/NHS chemistry. Good field effect modulation and no hysteresis is observed. As previous stated, this strategy keeps the OS undisturbed, enabling optimum performance of the EGOFET device. The calibration curves for thc (D)-(+)-carvone, (L)-(−) carvone along with the control experiment with 2-phenylethanol ligands, that binds very weakly to pOBP, are reported in Fig. 5 as $\Delta I/I$ versus the ligand concentration. The curves are distinguishable down to few tens of picomolar concentration. As it can be seen, enantiomeric differentiation at extremely low concentrations is demonstrated. The behavior observed is correlated to larger binding affinity upon pOBP-(D)-(+)-carvone interaction compared to the interaction with the (L)-(−)-carvone that is confirmed by the fluorescent binding assay titration. As a control, non-specific analyte, 2-phenylethanol (PE) was used for sensing measurements. In this case the response is much lower.

An interesting phenomenon is observed if we see the graph of the normalized response with one-site binding algorithm as shown in Fig. 5D. The response curve for D(+)ccarvone and L(−) carvone are overlapping. This infers that the K_d calculated from the EGOFET sensor response is almost the same. The K_d for D(+) carvone was found to be

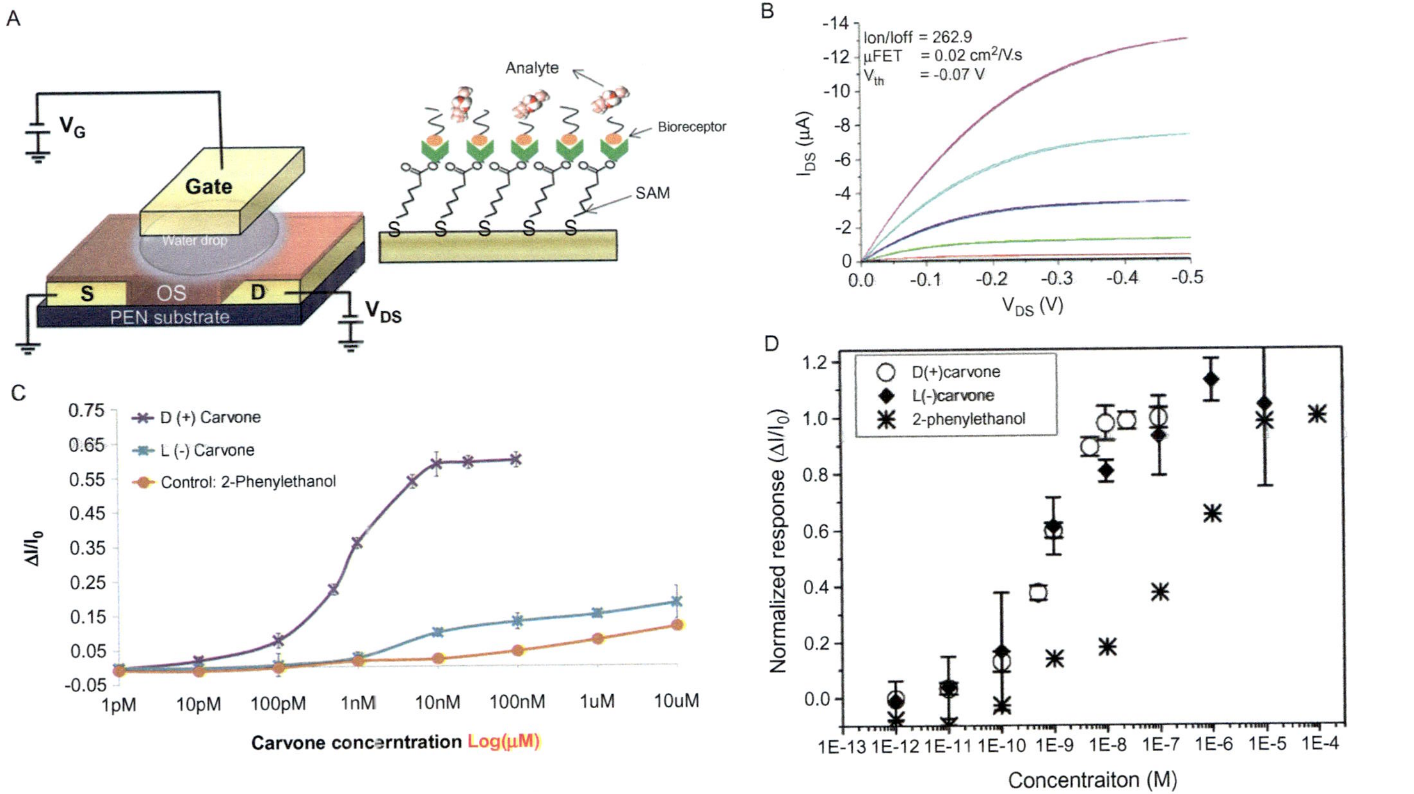

Fig. 5 (A) Illustration of the Au gate functionalization using -COOH terminated SAMs and of EGOFET device configuration, (B) Output characteristics of the pOBP functionalized gated EGOFET, (D) Calibration curve s of D(+)Carvone, L(−)Carvone and non-specific analyte (2-phenylethanol), (D) Normalized sensor response to three analytes.

$7.6 \pm 0.4 \times 10^{-10}$ M while that for the L(−) carvone was $7 \pm 2 \times 10^{-10}$ M. Thus, it can be fairly concluded that the lower response of L(−)Carvone is inherently due to transduction mechanism and not an artifact.

On the other hand, is really promising the ability of the EGOFET device to detect concentrations in the pM level. The mechanism behind the sensitive response of the transistor is attributed to changes in the capacitance of the protein layer. Consider that a third and lower capacitance, due to the protein layer, is added to the total capacitance of the gating system. The three capacitances are in series. Hence, the smallest capacitance is the dominant one and modulates the transistor. Even minute changes in the protein layer capacitance due to the ligand binding are enough to give rise to a large variation in the signal (Mulla, Tuccori, et al., 2015). The chiral EGOFET sensor allowed reaching dissociation constants ($K_{FET} \sim$ nM range) at least three orders of magnitude lower than those measured in solution (K_{Sol}). This has been explained observing that the binding at the pOBP-SAM is favored compared with pOBP in solution, since a lower entropy change is associated with ligands binding to the proteins orderly confined on the gate surface.

4. Summary

It was shown that odorant binding proteins can be effectively linked to the gold gate surface of an EGOFET device through SAM linkers. Such a protocol can be used for the development of EGOFET immunosensors since antibodies or aptamers can be also bound to the electrode in similar manner. It was also demonstrated that by integrating mammalian odorant binding proteins, i.e., porcine odorant binding proteins (pOBP) to the gate electrode of an electrolyte gated organic field effect transistor enantiomeric differentiation of carvone is possible in the pM level. Most importantly this is the first demonstration of enantioselectivity by odorant binding proteins using an EGOFET bioelectronic platform. The results are further confirmed by data using fluorescence spectroscopy. These results could shed more light on the odor discrimination by mammals and insects as well.

References

Belyaeva, L. A., van Deursen, P. M., Barbetsea, K. I., & Schneider, G. F. (2018). Hydrophilicity of graphene in water through transparency to polar and dispersive interactions. *Advanced Materials*, *30*, 1703274.

Bortolotti, C. A., Berto, M., Sensi, M., Di Lauro, M., & Biscarini, F. (2019). Biosensing with electrolyte gated organic field effect transistors. In *Organic bioelectronics for life science and healthcare* (pp. 71–96). Materials Research Foundations.

Buth, F., Donner, A., Sachsenhauser, M., Stutzmann, M., & Garrido, J. A. (2012). Biofunctional electrolyte-gated organic field-effect transistors. *Advanced Materials*, *24*, 4511–4517.

Capone, S., De Pascali, C., Francioso, L., Siciliano, P., Persaud, K., & Pisanelli, A. (2009). Electrical characterization of a pig odorant binding protein by impedance spectroscopy. In *SENSORS, 2009 IEEE*. IEEE.

Casalini, S., Leonardi, F., Cramer, T., & Biscarini, F. (2013). Organic field-effect transistor for label-free dopamine sensing. *Organic Electronics*, *14*, 156–163.

Clark, D. S., & Blanch, H. W. (1997). *Biochemical engineering* (4th). CRC Press.

Csucs, G., Michel, R., Lussi, J. W., Textor, M., & Danuser, G. (2003). Microcontact printing of novel co-polymers in combination with proteins for cell-biological applications. *Biomaterials*, *24*, 1713–1720.

D'Agata, R., Grasso, G., & Spoto, G. (2008). Real-time binding kinetics monitored with surface plasmon resonance imaging in a diffusion-free environment. *The Open Spectroscopy Journal*, *2*, 1–9.

De Avila, B., Watkins, H. M., Pingarrón, J. M., Plaxco, K. W., Palleschi, G., & Ricci, F. (2013). Determinants of the detection limit and specificity of surface-based biosensors. *Analytical Chemistry*, *85*, 6593–6597.

Green, N. M. (1975). Avidin. *Advances in Protein Chemistry*, *29*, 85–133.

Jung, L. S., Nelson, K. E., Campbell, C. T., Stayton, P. S., Yee, S. S., Pérez-Luna, V., et al. (1999). Surface plasmon resonance measurement of binding and dissociation of wild-type and mutant streptavidin on mixed biotin-containing alkylthiolate monolayers. *Sensors and Actuators B: Chemical*, *54*, 137–144.

Kalantar-zadeh, K., & Fry, B. (2007). *Nanotechnology-enabled sensors*. Springer Science & Business Media.

Kergoat, L., Herlogsson, L., Braga, D., Piro, B., Pham, M. C., Crispin, X., et al. (2010). A water-gate organic field-effect transistor. *Advanced Materials*, *22*, 2565–2569.

Kergoat, L., Piro, B., Berggren, M., Pham, M.-C., Yassar, A., & Horowitz, G. (2012). DNA detection with a water-gated organic field-effect transistor. *Organic Electronics*, *13*, 1–6.

Kim, S. H., Hong, K., Xie, W., Lee, K. H., Zhang, S., Lodge, T. P., et al. (2013). Electrolyte-gated transistors for organic and printed electronics. *Advanced Materials*, *25*, 1822–1846.

Macchia, E., Manoli, K., Di Franco, C., Scamarcio, G., & Torsi, L. (2020). New trends in single-molecule bioanalytical detection. *Analytical and Bioanalytical Chemistry*, *412*, 5005–5014.

Macchia, E., Manoli, K., Holzer, B., Di Franco, C., Ghittorelli, M., Torricelli, F., et al. (2018). Single-molecule detection with a millimetre sized transistor. Nature Communications, 9, 3223.

Macchia, E., Manoli, K., Holzer, B., Di Franco, C., Picca, R. A., Cioffi, N., et al. (2019). Selective single-molecule analytical detection of C-reactive protein in saliva with an organic transistor. *Analytical and Bioanalytical Chemistry*, *411*, 4899–4908.

Macchia, E., Romele, P., Manoli, K., Ghittorelli, M., Magliulo, M., Kovács-Vajna, Z. M., et al. (2018). Ultra-sensitive protein detection with organic electrochemical transistors printed on plastic substrates. *Flexible and Printed Electronics*, *3*, 034002.

Magliulo, M., Mallardi, A., Mulla, M. Y., Cotrone, S., Pistillo, B. R., Favia, P., et al. (2013a). Electrolyte-gated organic field-effect transistor sensors based on supported biotinylated phospholipid bilayer. *Advanced Materials*, *25*, 2090–2094.

Magliulo, M., Mallardi, A., Mulla, M. Y., Cotrone, S., Pistillo, B. R., Favia, P., et al. (2013b). Electrolyte-gated organic field-effect transistor sensors based on supported biotinylated phospholipid bilayer. *Advanced Materials*, *25*, 2090–2094.

Magliulo, M., Pistillo, B. R., Mulla, M. Y., Cotrone, S., Ditaranto, N., Cioffi, N., et al. (2013). PE-CVD of hydrophilic-COOH functionalized coatings on electrolyte gated field-effect transistor electronic layers. *Plasma Processes and Polymers*, *10*, 102.

Manoli, K., Magliulo, M., Mulla, M. Y., Singh, M., Sabbatini, L., Palazzo, G., et al. (2015). Printable bioelectronics to investigate functional biological interfaces. *Angewandte Chemie-International Edition*, *54*, 12562–12576.

Mantzila, A. G., Maipa, V., & Prodromidis, M. I. (2008). Development of a faradic impedimetric immunosensor for the detection of Salmonella typhimurium in milk. *Analytical Chemistry*, *80*, 1169–1175.

Matharu, Z., Bandodkar, A. J., Gupta, V., & Malhotra, B. D. (2012). Fundamentals and application of ordered molecular assemblies to affinity biosensing. *Chemical Society Reviews*, *41*, 1363–1402.

Matharu, Z., Bandodkar, A. J., Sumana, G., Solanki, P. R., Ekanayake, E. M., Kaneto, K., et al. (2009). Low density lipoprotein detection based on antibody immobilized self-assembled monolayer: Investigations of kinetic and thermodynamic properties. *The Journal of Physical Chemistry B*, *113*, 14405–14412.

Misra, R., McCarthy, M., & Hebard, A. F. (2007). Electric field gating with ionic liquids. *Applied Physics Letters*, *90*, 052905.

Mulla, M., Seshadri, P., Torsi, L., Manoli, K., Mallardi, A., Ditaranto, N., et al. (2015). UV crosslinked poly (acrylic acid): A simple method to bio-functionalize electrolyte-gated OFET biosensors. *Journal of Materials Chemistry B*, *3*, 5049–5057.

Mulla, M. Y., Tuccori, E., Magliulo, M., Lattanzi, G., Palazzo, G., Persaud, K., et al. (2015). Capacitance-modulated transistor detects odorant binding protein chiral interactions. *Nature Communications*, *6*, 1–9.

Palazzo, G., De Tullio, D., Magliulo, M., Mallardi, A., Intranuovo, F., Mulla, M. Y., et al. (2014). Detection beyond Debye's length with an electrolyte-gated organic field-effect transistor. *Advanced Materials*, *27*, 911–916.

Pan, F., Zhao, X., Perumal, S., Waigh, T. A., Lu, J. R., & Webster, J. R. (2009). Interfacial dynamic adsorption and structure of molecular layers of peptide surfactants. *Langmuir*, *26*, 5690–5696.

Panzer, M. J., & Frisbie, C. D. (2005). Polymer electrolyte gate dielectric reveals finite windows of high conductivity in organic thin film transistors at high charge carrier densities. *Journal of the American Chemical Society*, *127*, 6960–6961.

Park, I.-S., & Kim, N. (1998). Thiolated *Salmonella* antibody immobilization onto the gold surface of piezoelectric quartz crystal. *Biosensors and Bioelectronics*, *13*, 1091–1097.

Patrie, S. M., & Mrksich, M. (2007). Self-assembled monolayers for MALDI-TOF mass spectrometry for immunoassays of human protein antigens. *Analytical Chemistry*, *79*, 5878–5887.

Pavlickova, P., Schneider, E. M., & Hug, H. (2004). Advances in recombinant antibody microarrays. *Clinica Chimica Acta*, *343*, 17–35.

Pelosi, P., & Viti, R. (1978). Specific anosmia to l-carvone: The minty primary odour. *Chemical Senses*, *3*, 331–337.

Picca, R. A., Manoli, K., Macchia, E., Sarcina, L., Di Franco, C., Cioffi, N., et al. (2019). Ultimately sensitive organic bioelectronic transistor sensors by materials and device structures' design. *Advanced Functional Materials*, *30*, 1904513.

Qureshi, M. H., Yeung, J. C., Wu, S.-C., & Wong, S.-L. (2001). Development and characterization of a series of soluble tetrameric and monomeric streptavidin muteins with differential biotin binding affinities. *Journal of Biological Chemistry*, *276*, 46422–46428.

Rusmini, F., Zhong, Z., & Feijen, J. (2007). Protein immobilization strategies for protein biochips. *Biomacromolecules*, *8*, 1775–1789.

Scholl, M., Sprössler, C., Denyer, M., Krause, M., Nakajima, K., Maelicke, A., et al. (2000). Ordered networks of rat hippocampal neurons attached to silicon oxide surfaces. *Journal of Neuroscience Methods*, *104*, 65–75.

Schreiner, S. M., Shudy, D. F., Hatch, A. L., Opdahl, A., Whitman, L. J., & Petrovykh, D. Y. (2010). Controlled and efficient hybridization achieved with DNA probes immobilized solely through preferential DNA-substrate interactions. *Analytical Chemistry*, *82*, 2803–2810.

Seshadri, P., Manoli, K., Schneiderhan-Marra, N., Anthes, U., Wierzchowiec, P., Bonrad, K., et al. (2018). Low-picomolar, label-free procalcitonin analytical detection with an electrolyte-gated organic field-effect transistor based electronic immunosensor. *Biosensors and Bioelectronics*, *104*, 113–119.

Sze, S. M., & Ng, K. K. (2006). *Physics of semiconductor devices*. John Wiley & Sons.

Ulbrich, R., Golbik, R., & Schellenberger, A. (1991). Protein adsorption and leakage in carrier–enzyme systems. *Biotechnology and Bioengineering*, *37*, 280–287.

Ulman, A. (1996). Formation and structure of self-assembled monolayers. *Chemical Reviews*, *96*, 1533–1554.

Wilchek, M., & Bayer, E. A. (1988). The avidin–biotin complex in bioanalytical applications. *Analytical Biochemistry*, *171*, 1–32.

Xia, Y., & Whitesides, G. M. (1998). Soft lithography. *Angewandte Chemie International Edition*, *37*, 550–575.

Zhang, B., Jiang, Y., Kuang, H., Yao, C., Huang, Q., Xu, S., et al. (2008). Development of a spiral piezoelectric immunosensor based on thiol self-assembled monolayers for the detection of insulin. *Journal of Immunological Methods*, *338*, 7–13.

CHAPTER EIGHTEEN

Gravimetric biosensors

Khasim Cali[a], Elena Tuccori[b], Krishna C. Persaud[a,*]

[a]Department of Instrumentation and Analytical Science, The University of Manchester, Manchester, United Kingdom

[b]Pall Corporation, Portsmouth, United Kingdom

*Corresponding author: e-mail address: krishna.persaud@manchester.ac.uk

Contents

Methods in Enzymology, Volume 642
ISSN 0076-6879
https://doi.org/10.1016/bs.mie.2020.05.010

Abstract

Gravimetric transducers produce a signal based on a change in mass. These transducers can be used to construct gas sensors or biosensors using odorant binding proteins (OBPs) as recognition elements for small volatile organic compounds. The methods described in this chapter are based on the immobilization of the OBPs onto functionalized (activated) self-assembled monolayer (SAMs) on gold and on nanocrystalline diamond surfaces. Depending on the surface immobilization methods used to fabricate the biosensor, recombinant proteins can be engineered to express six histidine tags either on the N-terminal or C-terminal of the proteins and these can also be used to facilitate protein immobilization. These methods are used to produce functional sensors based on quartz crystal microbalances or surface acoustic wave devices and are also applicable to other types of gravimetric transducers.

Abbreviations

CMUT capacitive micromachined ultra-sonic transducer
DCM dichloromethane
ddH$_2$O double distilled water
DMF dimethylformamide
EDC 1-ethyl-3-[3-dimethylaminopropyl] carbodiimide hydrochloride
FBAR film bulk acoustic resonator
LB Langmuir-Blodgett
LIFT laser induced forward transfer
MEMS micro-electromechanical system
MUP major urinary protein
NHS N-hydroxysuccinimide
NTA nitrilotriacetic acid, Nα, Nα-Bis-(carboxy-methyl)-L-lysine hydrate
OBP odorant binding protein
PFA perfluoroalkoxy
PTFE polytetrafluoroethylene
QCM quartz crystal microbalance
Q-Factor quality factor
SAM self-assembled monolayer
SAW surface acoustic wave
SVP saturated vapor pressure
TA thioctic acid

1. Introduction

A gravimetric transducer produces a signal based on a change in mass (Fanget et al., 2011). For gas sensors or biosensors, a sensitive coating is utilized. Mass loading of the transducer is related to parameters describing the efficiency of analyte adsorption onto the sensitive coating. The partition coefficient K describes the adsorbing properties of the layer at equilibrium. K depends on chemical, vapor solubility properties, polarizability, dipolarity, hydrogen bond acidity, hydrogen bond basicity interactions, between analyte molecules and the sorbent layer. This chapter focusses on the use of gravimetric transducers to construct gas sensors or biosensors using odorant binding proteins (OBPs) as the sorbent elements for different classes of chemical species. Incorporating the sensing components of biological olfaction systems can increase the selectivity and sensitivity of the resulting sensors and has promise to overcome the limitations of materials normally used in sorptive based gas sensors or biosensors. Persaud and Tuccori demonstrated that OBPs can be successfully immobilized on quartz crystal microbalances to produce viable gas sensors capable of sensing pheromones (Persaud & Tuccori, 2014). Subsequently it was demonstrated that OBPs and the closely related family of major urinary proteins (MUPs) could be immobilized on micro-cantilever devices or surface acoustic wave transducers to produce robust and sensitive gas sensor arrays (Cali, Scorsone, & Persaud, 2019; Manai et al., 2014; Scorsone et al., 2016).

The methods described in this chapter are based on the immobilization of the OBPs onto functionalized (activated) self-assembled monolayer (SAMs) on gold (for quartz crystal microbalance (QCM) biosensors) and on diamond (for surface acoustic wave (SAW) biosensors). Depending on the surface immobilization methods used to fabricate the biosensor, recombinant proteins can be engineered to express six histidine tags either on the N-terminal or C-terminal of the proteins and these can also be used to facilitate protein immobilization.

1.1 The piezoelectric effect

The discovery of the piezoelectric effect by Jaques and Pierre Curie in 1880 (Curie & Curie, 1880) opened a new era of research relating mechanical states and electrical states in crystals. They demonstrated that when crystals such as Quartz or Rochelle salt (sodium potassium tartrate tetrahydrate) were compressed, an electrical potential could be measured between the faces of the crystals. Lippmann predicted the converse effect should also occur (Lippmann, 1881) which was also verified by the Curie brothers. In 1917, Langevin suggested the design for the first piezoelectric transducer based on these studies, employing quartz plates to serve as emitters and receivers of high-frequency waves under water that led to the development of sonar, Rutherford also being instrumental in employing piezoelectric devices as hydrophones (Katzir, 2012).

The nature of piezoelectricity is linked to the number of electric dipoles present in the material. A dipole is a vector having a direction and a value dependent on the other electrical charges in the vicinity. These dipoles tend to have the same direction when next to each other forming regions called Weiss domains (Weiss, 1906) that are generally randomly oriented. With some materials these domains can be aligned by poling which is a process by which a strong electric field is applied across the material. The reason why piezoelectric material creates a voltage is because when a mechanical stress is applied, the crystalline structure is disturbed changing the direction of the polarization of the electric dipoles. Depending on the nature of the dipole this change in the polarization might either be caused by a reconfiguration of the ions within the crystalline structure or by a reorientation of molecular groups. The resulting change in dimensions from an applied stress can shift the centers of mass for the positive and negative ions thus producing a dipole throughout the material (Fig. 1).

The dipoles inside the material cancel each other out, but on the surface of the material this cannot happen, producing a polarity on the surface due to the surface charge density. If electrodes are connected to the faces of a thin slab of this material and these are connected to a current measurement device, a current can be measured through the external circuit when stress is applied to the crystal. Releasing the stress causes a transient current flow in the opposite direction. If the converse effect is used and an alternating potential difference applied, mechanical oscillations occur within the crystal lattice. Stable oscillations occur at the natural resonant frequency of the crystal and at that frequency the crystal presents a low impedance to the exciting voltage. If the crystal is incorporated into the feedback loop of an oscillating circuit, it becomes the frequency determining element of the circuit, as its Q (quality factor—defined as the ratio of the energy stored in the oscillating resonator to the energy dissipated per cycle by damping processes) is very high, typically several thousand. Pioneering researchers such as Cady (Cady, 1921, 1946), Pierce (Pierce, 1923) and others designed crystal controlled oscillators with high stability for use as tuning devices and crystal filters for communication lines and radio that are now embedded in a multitude of modern electronic devices. A quartz crystal has a hexagonal prism structure as shown in Fig. 1C. The z-axis runs along the length of the crystal and is called the optical axis. The x-axis or the electric axis is parallel to a line bisecting the angles between adjacent prism faces. The y-axis, which is also known as the mechanical axis, is at right angles to the face of prism and also to the x-axis. Initially oscillator crystals were cut in the y-z plane with the electric field applied along the x-axis. These crystals were rather unstable and

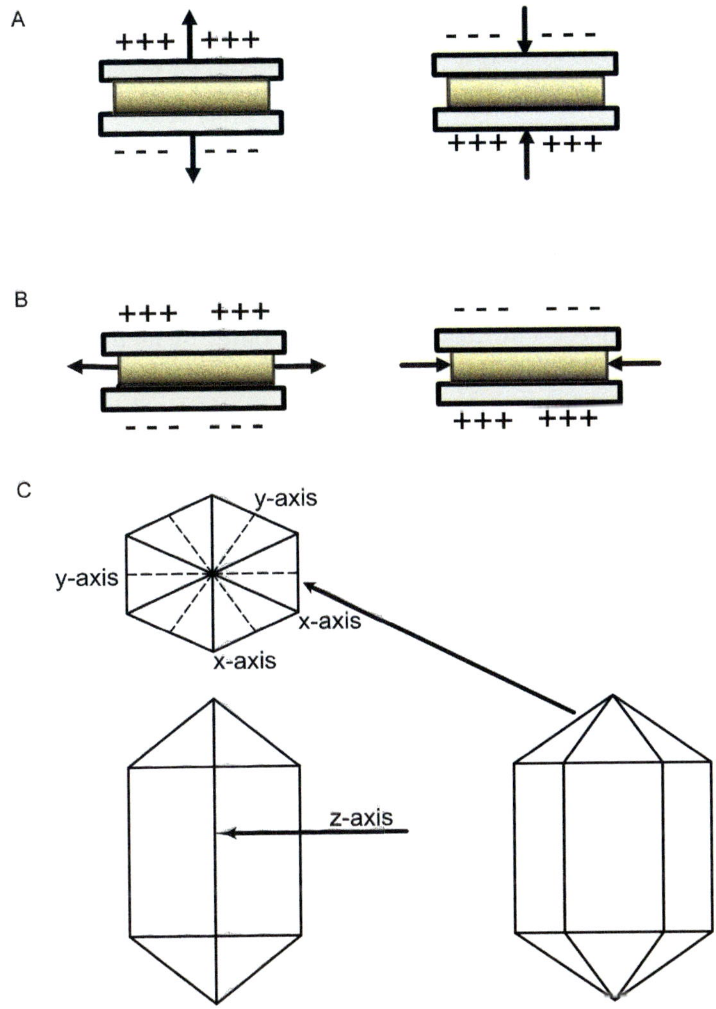

Fig. 1 The piezoelectric effect is illustrated; mechanical tension or compression on a piezoelectric material will induce a current depending type of stress and direction shown by the arrows shown in (A) and (B). When a stress is applied to a piezoelectric material, the dimensions of the material changes. Depending on the direction the stress is applied, the resulting change in dimensions can shift the centers of mass for the positive and negative ions; this produces a dipole throughout the material. The dipoles inside the material cancel each other out, but on the surface of the material the dipoles are not canceled out, producing a polarity between the faces of the material. (C) A quartz crystal has a hexagonal prism structure shown on the right. As shown on the left, the z-axis runs along the length of the crystal and is called the optical axis. The x-axis or the electric axis is parallel to a line bisecting the angles between adjacent prism faces. The y-axis, which is also known as the mechanical axis, shown in dotted lines, is at right angles to the face of prism and also to the x-axis.

could only work at low frequencies. Lack, Willard, and Fair (1930) showed that cutting crystals at various angles changed the temperature coefficients and frequency constants. Experiments with different cuts of crystals yielded two cuts with zero temperature coefficients $+35°15'$ (AT cut), and $-49°00'$ (BT cut) and these have been widely adopted since then.

1.2 Quartz crystal microbalances and chemical sensing

In 1959, Sauerbrey established a relationship between mass absorption on the surface of an AT-cut quartz crystal and offset of its frequency in gas phase, which established the foundation for quartz crystal microbalance (QCM) research in the area of chemical sensing (Sauerbrey, 1959). This relationship is summarized in Eq. (1)

$$\Delta f = \frac{-2 f_0^{\,2}}{A\sqrt{\rho_q \mu_q}} \Delta m, \tag{1}$$

where

Δf = the measured resonant frequency change (Hz)
f_0 = the intrinsic frequency of the quartz crystal (Hz)
A = the piezoelectrically active crystal area (electrodes area, cm^2)
ρ_q = the density of quartz ($2.643\,g/cm^2$)
μ_q = the shear modulus of quartz for the AT-cut crystal ($2.947 \times 10^{11}\,g/cm\,s^2$)
Δm = the elastic mass change (g)

This simplifies to

$$\Delta f = -2.26.10^{-6} \frac{f_0^2}{A} \Delta m \tag{2}$$

King (1972) described a piezoelectric sorption detector and demonstrated its ability to measure extremely small mass changes in an air environment. He also described an approach which should be taken to extend this concept to liquid systems. For quartz crystals coated with various materials the Sauerbrey equation only applies to the elastic subjects such as metallic coatings, metal oxides, and thin adsorbed layers that do not dissipate energy during the oscillation and there may be substantial deviations due to viscoelastic effects of coatings (Vogt, Lin, Wu, & White, 2004).

QCMs have gained widespread use in creating arrays of gas sensors (Di Natale et al., 2008; Muñoz-Aguirre, Nakamoto, & Moriizumi, 2005; Nakamoto, Fukunishi, & Moriizumi, 1990), and have been widely adopted as transducers for "electronic nose" technologies. Fig. 2 shows a picture of

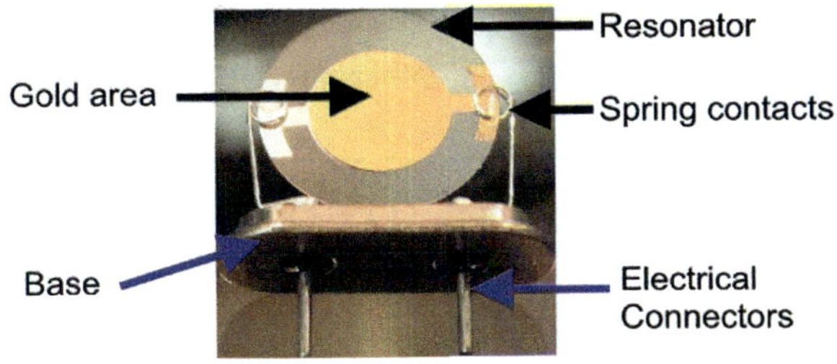

Fig. 2 An AT-cut 20 MHz quartz crystal is shown. The resonator is supported by two spring contacts at the edges. These make electrical and mechanical contact with evaporated gold disks deposited on both sides of the QCM and are connected to the electrical contacts on the crystal base. The crystal sets the base frequency of an oscillator circuit that produces a train of electrical pulses at a frequency set by the resonant frequency of the crystal.

a quartz crystal commonly used as a transducer for gas sensors. It has a resonance frequency of 20 MHz and is 7.95 mm in diameter. On both sides of the crystal is a thin layer of gold (4.9 mm diameter) which is evaporated over a titanium adhesion layer and this is used for immobilization of sorbent materials such as polymers, proteins or other materials. Electrical connections are made via spring contacts that are clamped to the gold connectors on the edges of the quartz resonator.

1.3 Surface acoustic wave transducers

Surface acoustic waves (SAWs) are elastic waves, first described by Rayleigh (1885), that can travel along the surface of a solid piezoelectric—the amplitude decaying with depth into the material proportionally to the wavelength. Interdigitated electrodes were utilized by White and Voltmer to generate an acoustic wave across a piezoelectric material (White & Voltmer, 1965) and this concept was used to develop a host of resonant devices that have become commonly used in filters and tuning elements in the electronic industry (e.g., smartphones). Because the mechanical vibrations are a strong function of their environment, including both their frequency and amplitude, changes in the environment can lead to changes in the SAW itself. This forms the basis for a SAW chemical sensor; target analytes change the SAW environment and changes in the SAW are subsequently detected. The physical structure of such a device is shown in Fig. 3.

The most general mode used for gas sensing used is the Rayleigh wave, in which the SAW propagates at the speed of sound of the crystal and the surface displacement has two components—one normal to the SAW propagation and one parallel (shear) to the substrate—resulting in an elliptical

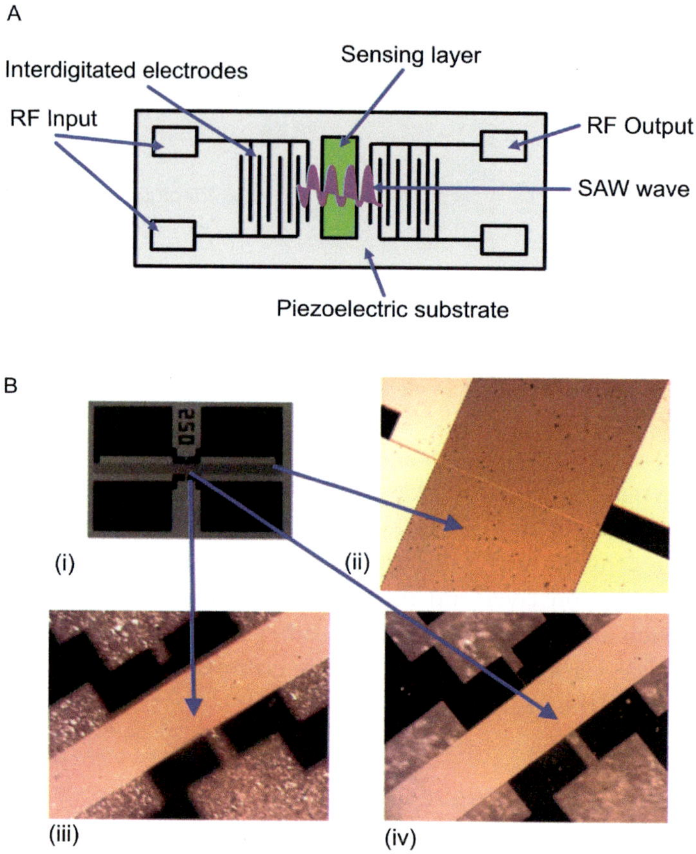

Fig. 3 (A) A schematic of a simple SAW resonator is shown. Two Interdigitated electrode structures (IDT) are deposited on a piezoelectric substrate. This makes up a SAW delay line. One IDT acts as the input or the transmitting IDT which converts radiofrequency (RF) electrical signals to acoustic energy that propagates on the surface of the piezoelectric substrate to the output (IDT) transducer which converts acoustic energy back to an electric signal creating a delay equivalent to the time taken by the surface acoustic wave to travel between the two IDTs. If an adsorbent material is placed between the two IDTs, then adsorption of molecules from the vapor phase on to the material will case a change in the propagation delay of the surface acoustic wave. (B) (i) a SAW transducer manufactured by KIT (Karlsruhe Institute of Technology, Germany)—the band in the center running from left to right constitutes interdigitated electrodes shown in more detail in (ii) the gap between each electrode sets the resonant frequency of the structure. (iii) and (iv) show views of the SAW transducer modified by deposition of nanocrystalline diamond (CEA, France) which allows attachment of OBPs via NTA modified diamond surfaces. *Panel (A) adapted from Voiculescu, I., & Nordin, A. N. (2012). Acoustic wave based MEMS devices for biosensing applications.* Biosensors and Bioelectronics, 33*(1), 1–9. https://doi.org/10.1016/j.bios.2011.12.041.*

motion. The Lamb wave (Lamb, 1917) resembles the Rayleigh wave and is typically generated using very thin substrates that are only a few wavelengths thick. These devices can be used also for sensing in liquid media. For sensing in liquids, a thin film called a guiding layer is often applied to the surface of the device to activate a Love wave mode (Love, 1911), which also inhibits radiation of acoustic pressure into the liquid. The interdigitated transducer (IDT) electrode pattern shown in Fig. 2 determines the resonant frequency of the device, where in a typical IDT the electrode spacing is an integer multiple or 1/2, 1/3, or 1/4 of the wavelength $\lambda = 2\pi/k$, with k being the wavenumber. For most devices, the wavelengths are between 0.01 and 1.0 mm. The most basic design is planar where the IDT consists of simple linear electrode "fingers" with equal spacing between them, but many more complex designs have been implemented. SAW devices operate at much higher frequencies than quartz crystals—typical devices are made to resonate at 433 MHz which is within a useful operational waveband for many communications devices.

In general, a SAW chemical sensing system consists of a sensing layer on the SAW substrate that is designed to react or bind with the analytical target, and electrical components designed for read out and recording the sensor output. Any perturbation will alter the delay in propagation of an acoustic wave across the surface of the device. This can be sensed as a phase shift, frequency-shift, or time-delay in the output electrical signal. The sensing layer interacts with the target to cause detectable changes in the characteristics of the acoustic wave, such as the velocity ν or amplitude, which results in a frequency-shift f or phase shift ϕ (Eq. 3).

$$\frac{\Delta \nu}{\nu} = \frac{\Delta f}{f} = -\frac{\Delta \phi}{\phi} \tag{3}$$

The sensing layer is a critical component of sensor design, and a wide variety of sensing layer materials have been explored for both chemical and biological SAW sensors. These include polymers, metal oxides, nanoparticles, or for biological applications proteins such as antibodies. Comprehensive reviews of SAW chemical sensors include Länge, Rapp, and Rapp (2008) and Go, Atashbar, Ramshani, and Chang (2017).

1.4 Film bulk acoustic resonator (FBAR)

The film bulk acoustic resonator is a CMOS compatible sensor based on QCM theory and have a mass sensitivity three orders of magnitude higher than QCM. This is due to the thin films achieved in the structure as compared to the diced quartz used in QCM. The FBAR membrane structure is well known in MEMS fabrication. A bulk wave propagating inside a piezoelectric thin film is generated by putting the piezoelectric thin film between electrodes and applying a high-frequency signal (Fig. 4). Standing acoustic waves with a defined resonant frequency, f_r, are generated by the rf signal due to the inverse piezoelectric effect. Addition of a mass on the surface of the resonator results in a change in f_r. By tracking changes in f_r, mass changes on the resonators can be detected. Chemical transducers operating at much higher frequencies than SAW devices are now being exploited (Chang et al., 2016; Mastromatteo & Villa, 2013; Tukkiniemi et al., 2009; Zhang, Du, Fang, & Zhao, 2017; Zhang, Luo, Flewitt, Cai, & Zhao, 2018).

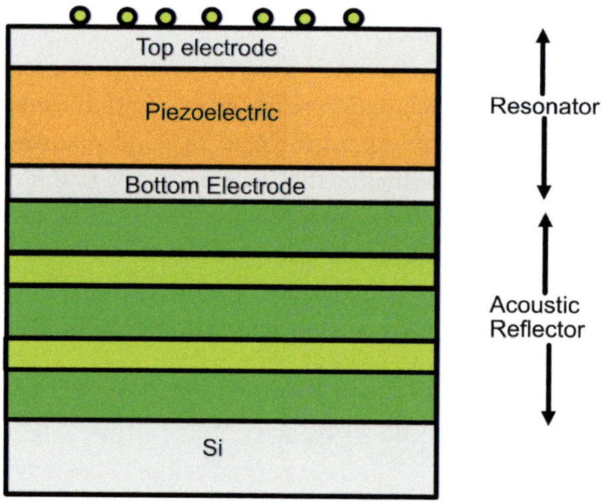

Fig. 4 The structure of a film bulk acoustic resonator (FBAR) is shown. Piezoelectric films fabricated using MEMS technology are thinner, can reach higher frequencies and offers the advantage of monolithic integration with advanced CMOS technology for peripheral readout and signal processing circuitry. FBARs are fabricated from a thin central piezo-electric layer with metal electrodes micropatterned on both sides. The FBAR's Q-factor can be increased by placing it on top of an acoustic mirror. The acoustic mirror is formed using alternating oxide and metallic layers. Acoustic isolation from the substrate is achieved when a Bragg reflector, comprised of layer pairs with contrasting acoustic impedance is placed underneath the FBAR. The exposed electrode at the top of the FBAR can be functionalized to immobilize proteins or other materials to create biosensors. *Adapted from Voiculescu, I., & Nordin, A. N. (2012). Acoustic wave based MEMS devices for biosensing applications.* Biosensors and Bioelectronics, 33*(1), 1–9. https://doi.org/10.1016/j.bios.2011.12.041.*

The wave resonates at a discrete frequency according to the thickness of the thin film. Like SAW the frequency f can be expressed as in Eq. (3) in terms of wavelength λ and acoustic velocity ν:

$$\lambda f_0 = \nu \tag{4}$$

where the wavelength λ is related to the thickness h of the piezoelectric film

$$\lambda = 2h \tag{5}$$

These devices typically operate at around 2 GHz.

1.5 Micro-electromechanical systems (MEMS)

Microcantilevers are small beams which are anchored at one end. Molecular adsorption on a cantilever surface can generate bending as a result of adsorption-induced changes in surface stress. In addition, mass loading due to molecular adsorption results in changes in the resonance frequency of the cantilever (Fig. 5). Chemical or biological selectivity can be achieved

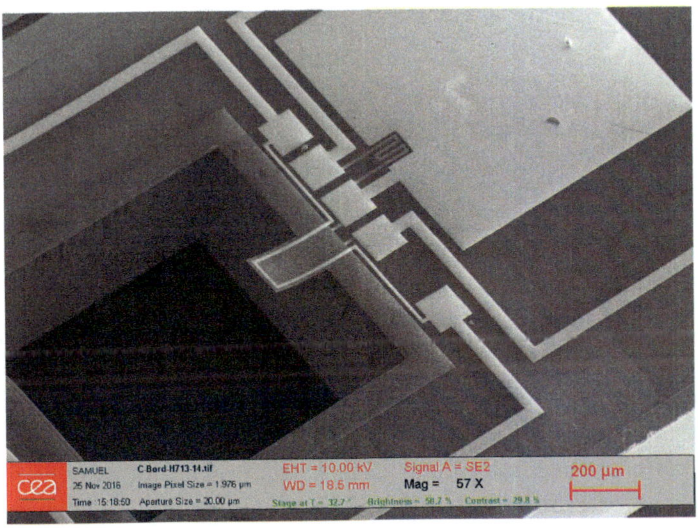

Fig. 5 Photograph of a microcantilever device fabricated by CEA, France. Microcantilever sensors operate by detecting changes in resonance response or deflection caused by mass loading, surface stress variation, or changes in damping conditions. Four resonance response parameters (resonance frequency, amplitude, Q-factor, and deflection) can be detected simultaneously. Surface stress produced as molecules adsorb on a cantilever can be observed as changes in deflections. *Photograph—courtesy of Dr. Emmanuel Scorsone, CEA, France.*

via immobilized sorptive materials on the cantilever surface (Thundat, Oden, & Warmack, 1997). Resonant micro cantilevers with different transduction techniques have been commonly used for gas concentration measurement. Piezoresistive layers are often used to detect the cantilever bending that can be produced by different ways (e.g., electrostatic or magnetic actuation forces). These devices are comprehensively reviewed by Voiculescu and Nordin (2012). Possas-Abreau et al. described the development of diamond and silicon MEMS sensor arrays for vapor detection (Fig. 5) (Possas-Abreu, Ghassemi, et al., 2017).

1.6 Capacitive micromachined ultra-sonic transducer (CMUT)

Capacitive micromachined ultrasonic transducers (CMUT) are membrane micro-electromechanical system (MEMS) devices that use capacitive transduction for mass load detection. They have very small capacitances and it is necessary to connect many CMUT devices into a parallel array. Using the group capacitance, this gives extremely high mass sensitivity afforded by the array of devices and allows for a range of operating frequencies dependent on the thickness of the resonating membrane (Fanget et al., 2011; Nazemi, Joseph, Park, & Emadi, 2019).

2. Protein immobilization

All of the gravimetric transducers described previously are appropriate for producing biosensors or gas sensors based on OBPs. However, care needs to be taken in the immobilization of biomolecules on the transducer surface to ensure optimum sensor performance. The selective immobilization of proteins is a key step in developing useful biosensors, because both protein stability and functionality require special control over protein orientation at the sensitive layers of devices. The techniques used have great effects on the specificity, sensitivity, reproducibility and reversibility of the resulting sensor. Some of the requirements of an immobilization process include: (1) retention of biological activity of biomolecules; (2) achievement of reproducible and stable attachment with the substrate against variations of pH, temperature, ionic strength, chemical nature of the microenvironment; and (3) uniform, dense, and oriented localization of the biomolecules. Bioconjugation, where biomolecules are linked together or to solid supports, is an important aspect of the biological sciences. Noncovalent methods of protein immobilization are widely employed and involve either passive adsorption on to hydrophobic surfaces or electrostatic interactions with charged surfaces. However, this method often induces conformational changes in the proteins that reduce their binding activity.

For biosensor fabrication it is much more efficient to covalently immobilize the protein via a functional group that is naturally present (Wong, Khan, & Micklefield, 2009). It is possible to couple proteins via exposed amines, e.g., from lysine groups, using an active ester, an aldehyde or epoxides (Fig. 6). The cysteine residue may also be used as the thiol group readily undergoes conjugate addition with α,β-unsaturated carbonyls (e.g., maleimides) to form stable thioether bonds. Similarly carboxylic groups from acidic moieties such as aspartic and glutamic residues, can be immobilized by conversion to their corresponding active esters in situ with a carbodiimide coupling agent such as 1-ethyl-3-(3-dimethylaminopropyl)carbodiimide (EDC) with an auxiliary nucleophile such as N-hydro-succinimide (NHS) (Fig. 7).

Apart from covalent methods it is also possible to use non-covalent but biologically mediated methods of immobilization. The protein can be expressed with a genetically encoded polyhistidine tag (Hochuli, Bannwarth, Dobeli, Gentz, & Stuber, 1988). This tag usually consists of six sequential His residues, and has the ability to chelate transition metals including Cu(II), Co(II), Zn(II), or Ni(II). Thus, a biosensor surface can be modified with a chelating moiety such as nitrilotriacetic acid (NTA) or iminodiacetic acid and is then treated with a solution of the relevant metal salt (commonly nickel) to produce a support containing the metal ions. This metal-activated support is then used for protein immobilization through chelation with the His residues of the tag (Fig. 8).

There are also developments in materials coating the transducer that can improve covalent immobilization of proteins. Coating surface acoustic wave devices or microcantilevers with diamond nanoparticles confers attractive properties for chemical detection. It is a chemically inert material with a high Q-factor producing a stable sensor platform, yet the carbon terminated surface offers wide possibilities for covalent attachment of specific receptors including odorant binding proteins (Chevallier, Scorsone, & Bergonzo, 2009; Manai et al., 2014; Possas-Abreu, Rousseau, et al., 2017).

2.1 Self-assembled monolayers

There has been much research effort dedicated to optimizing the conditions for covalent binding and self-assembled monolayer (SAM) technology has

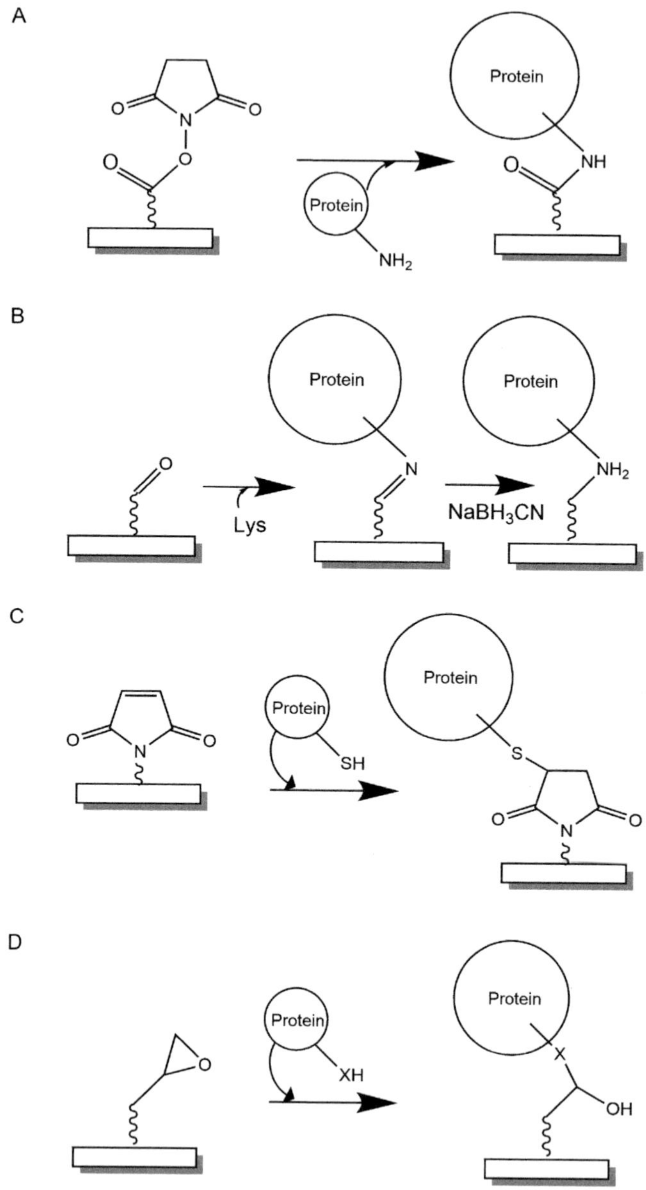

Fig. 6 Conventional covalent methods for immobilization with nucleophilic residues of proteins. Covalent bonds are formed through reaction with functional groups present on the protein surface. These methods can be used on unmodified proteins since they rely only on naturally present functional groups. (A) Reaction of lysine residues to NHS esters or aldehydes (B) and cysteine residue bonding to maleimide groups (C). Epoxides may react with either of the nucleophilic residues (D). *Adapted from Wong, L. S., Khan, F., & Micklefield, J. (2009). Selective covalent protein immobilization: Strategies and applications.* Chemical Reviews, 109*(9), 4025–4053. https://doi.org/10.1021/cr8004668.*

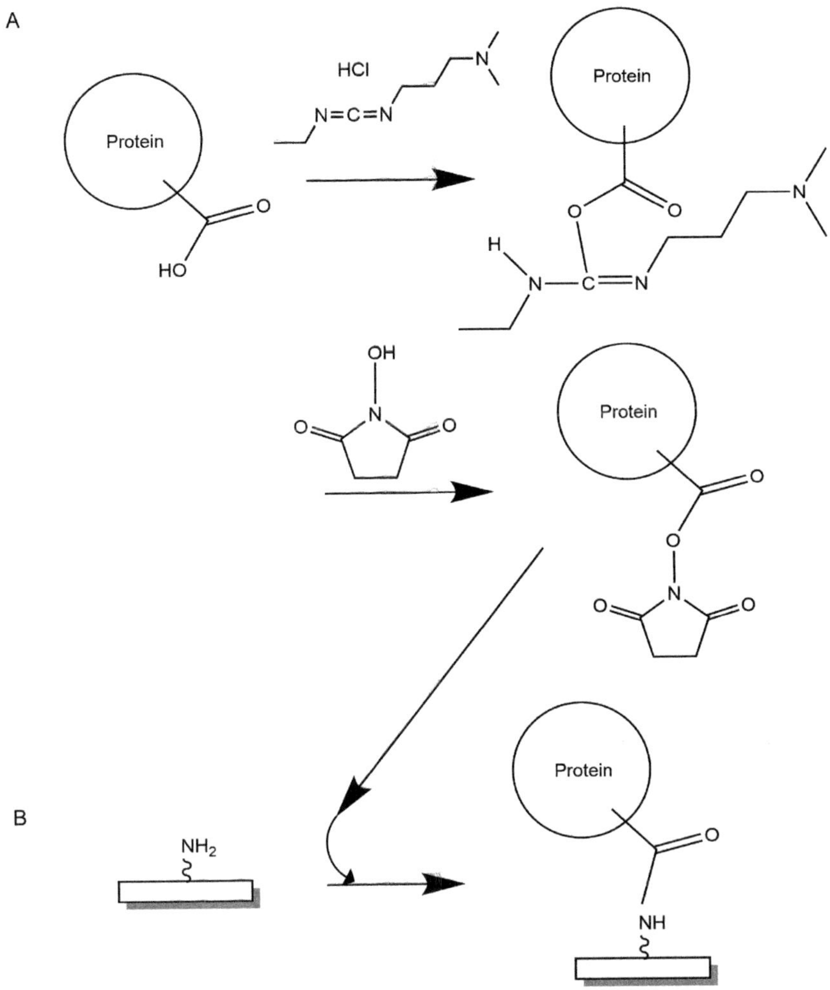

Fig. 7 Immobilization through acidic amino acid residues such as Glu and Asp. (A) By conversion to their corresponding active esters in situ with a carbodiimide coupling agent (EDC) 1-ethyl-3-(3-dimethylaminopropyl) carbodiimide followed by in situ generation of an NHS active ester (B) that is used to couple the protein.

come to the forefront (Briand et al., 2006; Ferretti, Paynter, Russell, Sapsford, & Richardson, 2000; Limbut, Kanatharana, Mattiasson, Asawatreratanakul, & Thavarungkul, 2006; Mauriz, Calle, Montoya, & Lechuga, 2006; Minamiki, Sasaki, Tokito, & Minami, 2017; Vaughan, O'Sullivan, & Guilbault, 1999). SAM is the generic name for techniques for generation of monomolecular layers of biological molecules on a variety

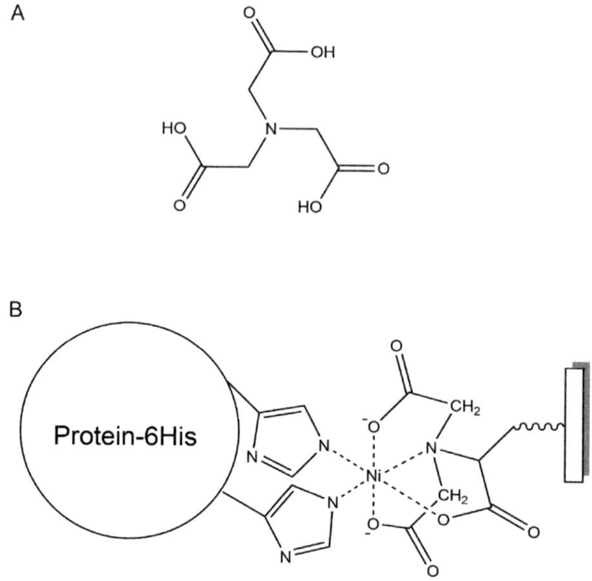

Fig. 8 Coupling via 6-histidine tagged proteins. (A) Nitrolotriacetic acid (NTA) used to functionalize the support. The chelating NTA moiety is first "activated" with the addition of a hexacoordinate metal ion (e.g., Ni^{2+}) followed by addition of the tagged protein. (B) The protein is immobilized on the support through chelation of the metal ion by the imidazole moieties of the His residues.

of substrates. Immobilization of proteins on SAMs allows reliable control over both the packing density and the environment of an immobilized protein at a substrate surface. In general, SAMs form by the spontaneous adsorption of amphiphilic adsorbates onto an appropriate substrate (Fig. 9) that is initially driven by chemical affinity between the adsorbates and the substrate. They can be formed either by gas phase deposition or from the liquid phase. They take advantage of the fact that many organic compounds have the capacity for self-assembly including long chain carboxylic acids or alcohols (RCOOH, ROH), where R is an alkyl chain, reacting with metal oxide substrates; organosilane species ($RSiX_3$, R_2SiX_2 or R_3SiX), where X is a chlorine atom or an alkoxy group, reacting with hydroxylated substrates (glass, silicon and aluminium oxide, etc.); and organosulfur-based species reacting with noble metal (gold, silver) surfaces. The latter is of great interest for biosensor construction due to their unique attractive features. Sulphur-containing compounds (alkanethiols, dialkyl disulfides and dialkyl sulfides) have a strong affinity for noble metal surfaces as they are spontaneously chemisorbed, with a regular organization and high thermal, mechanical

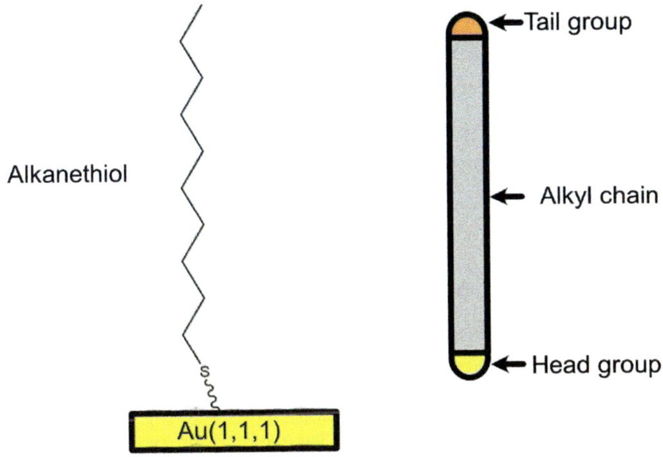

Fig. 9 Thiol-based SAMs. There are three important parts. The thiol moiety (head group) and chemisorption of this part to the substrate is the most important process in monolayer formation. An alkyl chain (methylene spacer) allows Van der Waals interaction between chains and provide a significant driving force for the adsorbate organization. The third part consists of a terminal functional group (tail group) to which a protein may be attached.

and chemical stability. These SAMs are easy to generate, they are densely packed and precisely oriented, and the thickness and the surface properties of the films can be easily adjusted. In thiol based SAMs the structure of the adsorbate molecule is divided into three parts (Fig. 9).

The first part is the thiol moiety (head group) and chemisorption of this part to the substrate is the most energetic and most important process in monolayer formation. The second part is an alkyl chain (methylene spacer), where Van der Waals interaction between chains provide a significant driving force for the adsorbate organization. The third part consists of a terminal functional group (tail group). In alkanethiol-based SAMs tail groups are methyl moieties that are exposed at the outermost surface of the monolayers and exert the most direct influence on the interfacial properties of the films. Because of their stability, orientation and ability to functionalize the terminal groups on the molecules, SAMs can offer a very convenient vehicle for covalent immobilization of biomolecules on gold surfaces for biosensor development. Often mixed SAMs formed by co-adsorption of mixtures of two thiols are used. One of these may be a thiol containing a functional group like a carboxylic acid that allows covalent attachment of a protein, the other being a diluting thiol without the functional group. This is used to reduce the surface concentration of functional groups, thus minimizing steric hindrance, partial denaturation of the potential immobilized protein and nonspecific interactions that could produce interference signals. This can be very useful when immobilizing large proteins like antibodies but is less useful with small proteins such as OBPs.

3. Methods

3.1 Preparation of an OBP based sensor

This section describes immobilization of OBPs on quartz crystal microbalances (QCMs, Fig. 2) using a activated self-assembled monolayer of thioctic acid (TA) (Cali & Persaud, 2020; Persaud & Tuccori, 2014; Tuccori & Persaud, 2019). For this method the immobilized protein could be with or without a 6-Histidine tag.

3.2 Rationale

Thioctic acid (alpha-lipoic acid) is a molecule with a large disulfide-containing base, a short alkyl chain with four CH_2 units, and a carboxyl termination (Fig. 10). Carboxyl terminated SAMs of thioctic acid adsorbed on a gold surface evaporated on a quartz crystal are utilized for covalent immobilization of OBPs. Thioctic acid has distinct advantages for surface modification. The disulfide-containing base gives added stability and yields two gold-sulfur bonds per molecule in surface-attached species.

The widely utilized covalent protein immobilization method through amide bond formation is facilitated by reaction of amines on exposed lysine with N-hydroxysuccinimide (NHS) activated esters on the gold surface (Fig. 6). A carboxylic acid terminated self-assembled monolayer (SAM—in this case lipoic acid) is prepared by dipping the gold substrate (in this case QCM) into the thioctic acid. The SAM modified gold surface is then activated (functionalized) by reacting with 1-ethyl-3-[3-dimethylaminopropyl] carbodiimide hydrochloride (EDC). EDC reacts with a carboxyl group to form an active ester, an O-acylisourea. In general, covalent attachment of protein to the EDC reactive surface is possible. However, the EDC active surface is unstable to hydrolysis. To overcome this limitation, the use of N-hydroxysuccinimide (NHS) allows the conversion of the O-acylisourea intermediate into a relatively stable amine reactive NHS ester (Fig. 10). This NHS ester can then cross-link with amine groups on the protein through a covalent amide bond (Tan et al., 2011).

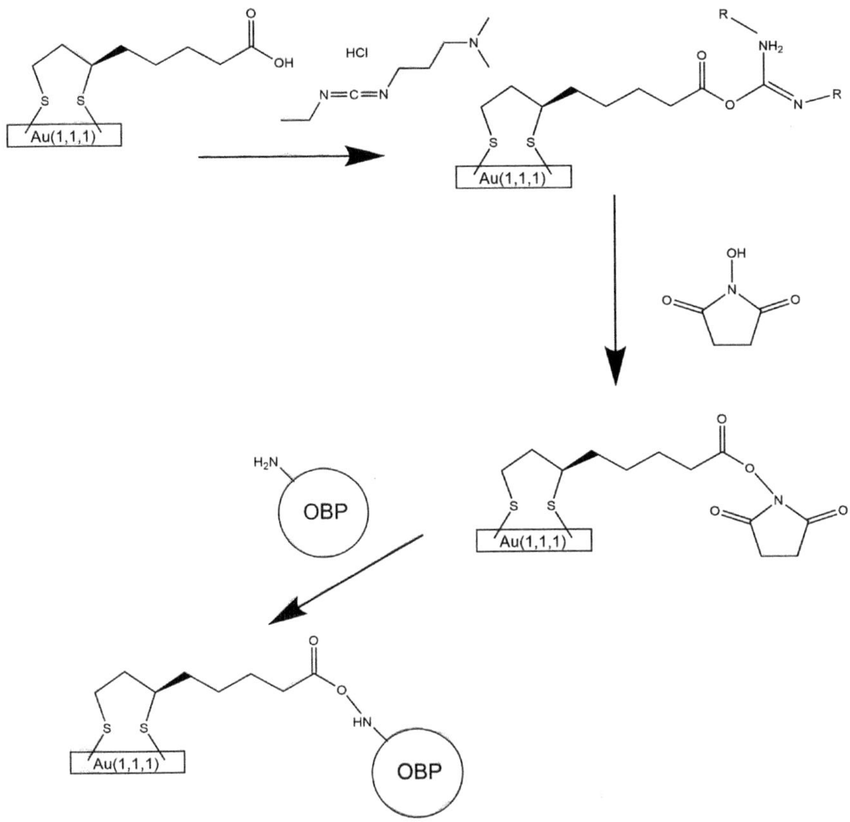

Fig. 10 Immobilization of OBPs on a gold surface using a SAM formed from thioctic acid. The SAM modified gold surface is activated by reacting with 1-ethyl-3-[3-dimethylaminopropyl] carbodiimide hydrochloride (EDC). EDC reacts with a carboxyl group to form an active ester, an O-acylisourea. N-hydroxysuccinimide (NHS) allows the conversion of the O-acylisourea intermediate into a relatively stable amine reactive NHS ester. This NHS ester can then cross-link with an amine group on the protein through a covalent amide bond.

3.3 Materials, equipment and reagents

Recombinant OBPs were produced in the laboratory standard procedures for the expression and purification of OBPs with or without a 6-Histidine tag (Cali & Persaud, 2020). All reagents used were of analytical grade purchased from Sigma-Aldrich, Fisher Scientific, or VWR. Genes were custom synthesized at Eurofins MWG GmbH, Ebensburg, Germany. The JLMQ USB interface for 4 QMB sensors (JLM Innovation GmbH, Tuebingen, Germany) was used as a readout system for 20 MHz quartz crystal microbalances with gold coated electrodes that were fabricated by IMM-CNR, Italy. The SAW readout device was from Karlsruhe Institute of Technology (KIT), Germany. SAW transducers with the structure shown in Fig. 3B are utilized. Nanocrystalline diamond coated surface acoustic wave (SAW) sensors were produced using strategies reported by Manai et al. (2014) and Scorsone et al. (2016).

4. Protocols

4.1 Preparation of SAM onto QCMs

This procedure also applies to any gravimetric transducer that has a gold active surface.

This procedure should be carried out in a fume cupboard

1. Clean glass vials (of 2–4 mL capacity) by incubating with Piranha solution (1:3: 30% H_2O_2:H_2SO_4) for a few minutes. Remove the Piranha solution and rinse the vials with plenty of distilled water. Dry vials in an oven.
2. Place required number of QCMs into clean glass Petri dish. Add about 10 μL of Piranha solution to the gold surface of a QCM crystal. Leave it for a few minutes to remove any organic residues from the surface and to obtain hydrophilic gold surface. Make sure the Piranha solution does not touch the spring contacts at the edges of the crystal as these are made of steel and will corrode. Rinse the QCM with ddH_2O and let it dry at air. Repeat the whole procedure on the other side of the QCM. It is possible to remove the quartz resonators from the supports to carry out the procedure, but this can risk breaking the fragile quartz crystals.
3. To each cleaned vial add about 2 mL of a solution of 10 mM thioctic acid (TA) in ethanol, and then immerse one QCM into the vial. Then place the vials into a container with a loosely closed lid. Leave the reaction to continue for 20 h under slow nitrogen flow (20sccm/min). However, the procedure can be also be done without nitrogen flow in a dessicator under inert atmosphere. While the interaction of thioctic acid with gold is very rapid, the formation of a stable SAM takes many hours.
4. Rinse the QCMs with absolute ethanol to remove unbound molecules of TA and leave to dry in air. Place the QCMs into a clean Petri dish or glass vial for storage.

4.2 Activation of the SAM

This procedure is carried out on the lab bench

1. Place the Petri dish (es) containing the QCMs on your lab working bench. To activate the carboxylic acid groups of the TA SAM add 20 μL of a solution of 180 mM ethyl (dimethylaminopropyl) carbodiimide (EDC) and 180 mM of N-hydroxysuccinimide (NHS) in 10 mM sodium phosphate buffer (pH 7) to the gold surface of the QCM crystal. Close the Petri dish to stop the activation solution from evaporating. Leave the reaction to continue for 2 h at room temperature.
2. Rinse with ddH_2O and dry the crystal at air.
3. Repeat step 1 and 2 on the other side of the QCM.

4.3 Immobilization the OBP onto activated TA SAM

This procedure is carried out on the lab bench. Concentration of OBPs needed is between 0.5 and 1 mg/mL.

1. The immobilization of the proteins is performed by pipetting 10 μL of the OBP solution onto the gold surface containing the activated SAM and leaving it to react for an hour at room temperature.
2. Gently rinse with ddH_2O and then dry the crystal at air.
3. Repeat steps 1 and 2 on the other side of the QCM.

While some researchers block unreacted groups on the SAM using a solution of ethanolamine, it is simpler just to wash the surface with distilled water as the EDC/NHS activation is not stable.

5. Immobilization of OBPs on to nanocrystalline diamond coated transducers

5.1 Rationale

Diamond can be activated to produce hydrogen-terminated diamond surfaces and functionalized using the chemistry normally used for carbon. A suitable approach is use of hexanoic acid which is covalently attached to hydrogen-terminated diamond surfaces followed by ethyl(dimethylaminopropyl) carbodiimide (EDC)/N-hydroxysuccinimide (NHS) activation and coupling with OBPs. The first step of this method is the formation of hexanoic acid radical covalently onto the hydrogen-terminated diamond surface on SAW sensors. This covalent immobilization of aliphatic primary amines on hydrogen-terminated diamond is a nucleophilic substitution involving the formation of carbanion at the diamond surface in the presence of OH^- ions in solution.

The reaction of the protonated amine with the labile carbanion results in the release of ammonia and carbon-carbon covalent grafting of the aliphatic compound over the diamond surface.

5.2 Protocol

It is advisable to prepare replicate sensors for each protein. While the SAW devices are not as fragile as QCMs, care needs to be taken in handling them—use soft forceps such as those used for handling surface mount electronic devices and avoid touching the electrode surfaces.

1. Immerse the hydrogen-terminated diamond substrate SAW sensor into 2 mL of 1 mM solution of hexanoic acid (caproic acid-ACA) in 0.2 M phosphate buffer (pH 7.0) for 10 min.
2. Thoroughly rinse the sensor with ddH_2O and then dry under gentle N_2 flow.
3. Activate the carboxylic acid groups of the resulting SAM with a solution of 30 mM ethyl (dimethylaminopropyl) carbodiimide (EDC)/60 mM of N-hydroxysuccinimide (NHS) in 10 mM sodium phosphate buffer (pH 7) for 2 h at room temperature.
4. Rinse with ddH_2O and dry the sensor at air.
5. Expose the activated diamond substrate to a small quantity of OBP solution (0.5–1 mg/mL in 50 mM $NaHCO_3$ buffer, pH 8.0) for 2 h.
6. Rinse the sensor thoroughly with ddH_2O and store in phosphate buffer saline (PBS) solution at 4 °C or dry in air and store in a vial at 4 °C.

6. Immobilization of histidine tagged OBPs

6.1 Rationale

The purpose of this method is to control the orientation of the proteins immobilized on the surface Here NTA forms a chemical linker between the diamond surface and the protein. This approach involves formation of NTA SAM onto diamond surface. NTA has the ability to bind nickel (II), and the resulting NTA-Ni^{2+} complex has a higher affinity to histidine thus enabling the immobilization of 6-Histidine tagged OBPs (Fig. 8).

6.2 Protocol

1. Immerse the hydrogen-terminated diamond substrate SAW sensor in a solution of 1 mM of Nα, Nα-Bis-(carboxy-methyl)-L-lysine hydrate (≥97%) (NTA) in 0.2 M phosphate buffer, pH 7.0, for 10 min.
2. Thoroughly rinse the sensor with ddH_2O and then dry under gentle N_2 flow.

3. To activate the NTA SAM, immerse the diamond surface in nickel (II) chloride hexahydrate (99.99%) solution for 2h at room temperature.
4. Immerse the NTA-Ni2+ complex terminated diamond surfaces in the 6-Histidine tagged OBP solution (1 mg/mL in 50mM phosphate buffer saline (PBS), pH 8.0) for 2h.
5. Rinse the sensor thoroughly with ddH_2O and store in phosphate buffer saline (PBS) solution.

7. Analyte vapor generation

OBP based gas sensors can respond to different analytes from ppb to ppm levels and test apparatus used very much depends on the nature of the analyte of interest, the saturated vapor pressure and the whether complex mixtures or single components are used. Burlachenko et al. review appropriate sampling methods for electronic noses (Burlachenko, Kruglenko, Snopok, & Persaud, 2016).

For simple systems an automated flow system consisting of electronically controlled solenoid valves and mass flow controllers is used to produce and deliver selected concentrations of solvent vapors to the detectors. To obtain the desired analyte concentration, a stream of carrier gas is passed through a bubbler that is filled with an analyte or solvent of interest. Saturation of the carrier gas with the solvent vapor can be verified through measurement of the rate of mass loss of the solvent in the bubbler. The vapor-saturated carrier gas is then diluted with pure carrier gas (which in this case can be normal room air, or air from a cylinder that has added humidity) through the use of mass flow controllers.

For simple tests—generation of a saturated vapor headspace can suffice for testing operation of the OBP sensor, especially when dealing with substances of low volatility. The substance can be placed in closed 40mL vial. Prepare each analyte sample at least in replicate. As an example, between 1 and 20 μL of the liquid analytes (e.g., acetone) or 1–30mg of the solid analytes (e.g., tobacco) into closed 40mL vials (Fig. 11) and left for 1h (for highly volatile compounds) to few hours (for less volatile compounds) at room temperature to allow the formation of the saturated vapors.

To generate very low concentrations of vapors permeation tubes may be used. They provide a stable concentration of a specific trace chemical, including those with low vapor pressures. When held at a constant temperature, these devices emit a small stable flow of chemical vapor that is measured in nanograms or nanoliters per minute (Mitchell, 2000).

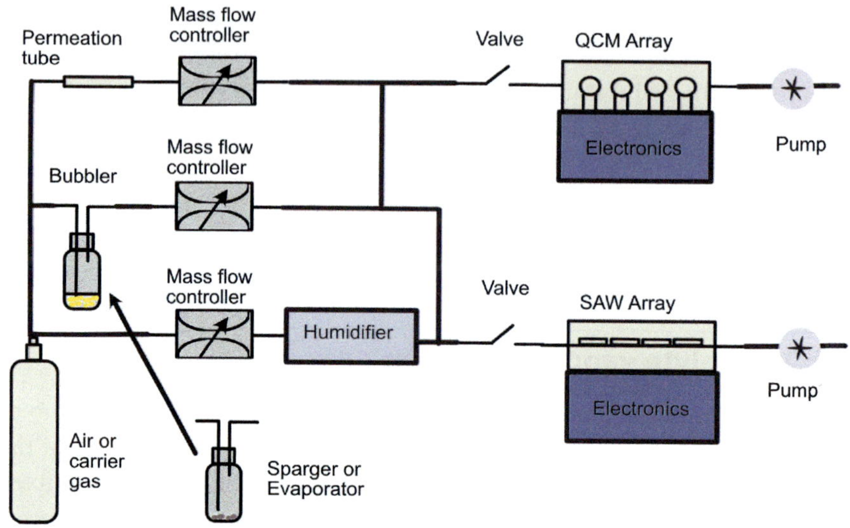

Fig. 11 Gas Rig for testing OBP sensors using either QMB or SAW platforms. Analyte generation may be through bubblers, permeation tubes, spargers, headspace stripping devices or others. The generated vapor is diluted with clean humidified air via mass flow controllers to produce desired concentration ranges of vapors to be applied to the sensors. A small pump aspirates vapor across the sensor arrays.

7.1 Testing of sensors with analyte vapors

This procedure is carried out in the fume cupboard. The QCM measurements are carried out at 22% RH or above in air at room temperature. Extremely low relative humidity should be avoided as the protein requires a shell of water molecules to retain the active tertiary structure and active binding pocket.

1. Device set up: Mount (insert) the QCM biosensors developed above into the QCM reader sensing chamber. Then place the device into a fume cupboard. Connect the device to the computer and a pump (unless the device has an embedded pump). The device should have inlet and outlet tubing connectors. Different QCM readers have different numbers of sensors that they can accommodate. The JLMQ device can hold up to 4 sensors while others can accommodate up to 12 sensors. For newly made biosensors, it is highly recommended to allow them to stabilize by passing clean air over the sensors for several hours. An example set up is shown in Fig. 11.
2. Device operation: Start the software in use and the pump to pass clean air (approximately 0.1 L/min—flow rate depends on the sampling system utilized) through to the sensors to allow the establishment of the baseline. Small rotary or diaphragm pumps for gases are available from a range of suppliers.

3. Detection of the vapor from target analytes: Any tubing used should be inert to minimize adsorption/absorption of analyte vapor. Suitable materials are polytetrafluoroethylene (PTFE) or perfluoroalkoxy (PFA). For the flow rates normally used, tubing can be 3.2 or 4mm diameter dependent on connectors that are used. The sample vial containing the analyte saturated vapor should have tubing that will be connected to the measurement device via a connector. The pump will pull the analyte sample through to the device sensing chamber. Allow the exposure of the analyte vapor to the sensors for around 10–20s (but this is up to the operator to decide according to the quality of the signal obtained). Save/export the results according to the appropriate data format (e.g., text file, CSV file, Excel file, etc.).

8. Safety considerations and standards

Piranha solution is highly corrosive and adequate person protection should be implemented.

When using gases and volatile analytes a fume hood with adequate ventilation should be used.

9. Analysis

Fig. 12 shows the raw data recorded from an OBP-QMB sensor to a saturated headspace of cocaine hydrochloride. This has an extremely low vapor pressure (SVP 1.84211×10^{-11} atm at 20 °C), but the device produces a small but useful signal, illustrating the potential sensitivity of such devices. A decrease in frequency is observed due to the increased mass loading of the QMB and this returns to normal frequency as the analyte desorbs from the sensor.

Fig. 13 shows raw data recorded from an array of eight SAW-OBP sensors (S1–S8) with immobilized OBPs from a range of insect and mammalian OBPs. Like QMBs the actual change in frequency is a decrease, but by convention the data have been normalized with respect to sign and are shown as positive changes in frequency. Responses are shown to pulses of (A) alpha-pinene vapor and (B) tobacco vapor. The frequency changes above are much higher than those observed with QMBs due to the higher operating frequency of the devices (433MHz for SAW devices as opposed to 20MHz for QMBs). It is seen that different OBPs in the array respond very differently to the different samples presented to them. Using such an array allows traditional "electronic nose" approaches to be used to discriminate different classes of volatile chemicals or indeed complex mixtures of volatile compounds.

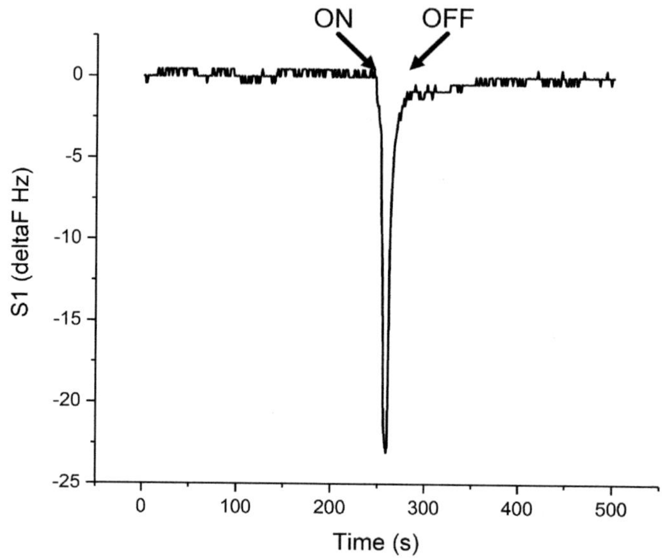

Fig. 12 Raw data response of an OBP-QMB sensor to cocaine-HCl vapor. A saturated vapor headspace was aspirated across the sensor array and the frequency change of each sensor was recorded. A decrease in frequency is observed as the mass of the QMB increased due to the loading of the adsorbed analyte, and this is reversible when clean air is applied to the sensor.

The appropriate data processing methods for gas sensor arrays are reviewed in detail by Persaud (2013). Pattern recognition in an "electronic nose" system may be regarded as a branch of artificial intelligence that involves the mimicking of human intelligence to solve chemical problems. Two main approaches to pattern recognition are commonly used: parametric and non-parametric. Parametric methods rely upon obtaining or estimating the probability density function of the parameters used to characterize the response of a system. Conversely, non-parametric methods require no assumption about the fundamental statistical distributions of data. Two types of non-parametric learning or classification methods are available: supervised and non-supervised (De Vito, Fattoruso, Pardo, Tortorella, & Di Francia, 2012; Hines, Llobet, & Gardner, 1999; Pardo & Sberveglieri, 2005). Supervised methods involve the learning of data based on advance knowledge of the classification, whereas unsupervised methods make no prior assumption about the sample classes but try to separate groups or clusters. This is a large and complex field that is evolving very rapidly with developments in advanced algorithms as well as dedicated integrated circuits capable of encoding neural networks.

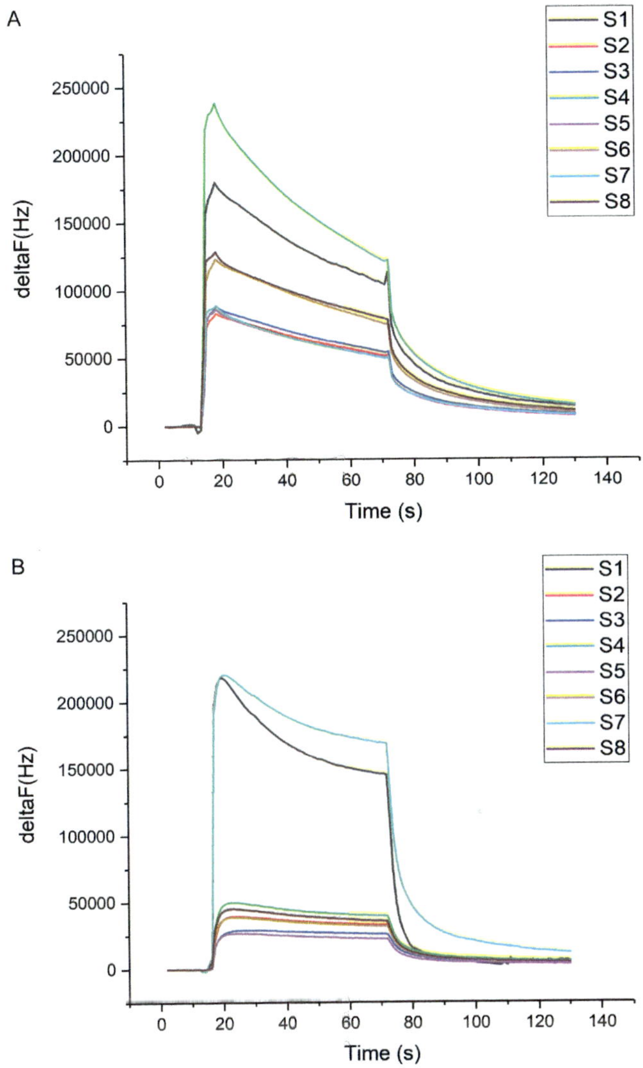

Fig. 13 Responses of an OBP-SAW array comprising eight OBPs from insect and mammalian origin. S1–S8 are individual SAW devices with different immobilized proteins (A) responses to a saturated vapor of alpha-pinene (B) responses to tobacco headspace vapor. The responses of individual OBPs are very different dependent of the analyte components reflecting the diversity of interactions that may be possible. The generated pattern of responses can be used to discriminate a range of different volatile chemicals or complex mixtures such as tobacco.

10. Related techniques

Apart from self-assembled monolayers, another major technique for the fabrication of supramolecular structures on surfaces is the Langmuir-Blodgett (LB) method. Langmuir films are composed of an organized monolayer of amphiphilic molecules at a liquid/gas interface. These molecules, by virtue of possessing a hydrophobic tail and a hydrophilic head, are spontaneously oriented at a liquid/gas interface. In LB amphiphilic molecules and macromolecules with amphiphilic segments are transferred onto a solid support by vertically lifting the support through the film.

SAMs can either be formed directly from solution (as reported here) or by gas phase deposition (Shuturminska, O'Malley, Collis, Conde, & Azevedo, 2018). While gold is the preferred noble substrate for SAM formation for protein immobilization, other metals have been used including copper, iron and silver. Copper is a widely used due to its electric conduction and thermal properties. Iron has been investigated due to its magnetic hyperthermia properties. The geometry and lattice structure of the metal surface can be utilized further by evenly distributing specific ligands (such as aromatic peptides and imidazoles with alkyl tails) over the surface to enhance and stabilize the metal properties. Silicate glass is also being used since unlike a metal surface; it is relatively chemically inert due to the strength of Si—O bond in the silica network. Glass surfaces need to be modified by etching (using extreme pH to break Si—O bonds) to form free hydroxyl groups on the surface which provide linkage points where other molecules can be chemically grafted. Alternatives to silicates and metals for SAMs are the carbon-based substrates known as nanocarbons. Examples of these nanocarbons include fullerenes, carbon nanotubes (CNTs) and graphene. Apart from SAMs with terminal carboxylic acid and SAMs with terminal NTA, other SAMs used for protein immobilization include SAMs with terminal amino groups (glutaraldehyde reaction to link with an amino group of the objective substance), SAMs with terminal biotin (biotin immobilization combined with streptavidin labeled objective substance), and SAMs with latent aldehyde (Hahn et al., 2007).

The formation of SAMs is not restricted to single types of molecules, and so mixed SAMs are also used. There are two ways to prepare mixed SAMs. One is to prepare by mixing two different solutions of alkane thiol derivatives, and the other is to use an asymmetric disulfide derivatives. Nonspecific adsorption caused by electrostatic interaction is not un-common in single SAM, but occurrence of nonspecific adsorption can be prevented significantly by using mixed SAM, particularly when dealing with large proteins. One example of mixed SAM used for biosensor applications is the use of alkane thiol derivative introduced oligoethyleneglycol moiety (Kyo, Usui-Aoki, & Koga, 2005; Laibinis, Fox, Folkers, & Whitesides, 1991).

Other methods of immobilization of OBPs to form sensors have also been reported. These range from simple drop casting of OBPs on to SAW transducers (Di Pietrantonio et al., 2015) to the use of laser-induced forward transfer (LIFT) (Duocastella, Colina, Fernández-Pradas, Serra, & Morenza, 2007) by the same authors for depositing OBPs on to SAW transducers.

11. Pros and cons

Pros	Cons
One of the important features of SAM is simple preparation as no special equipment is required to prepare SAM layers. After chemisorption, the SAMs are stable in ambient temperature and pressure for prolonged periods of time that allows subsequent chemical functionalization of the assembled molecules with biological entities such as proteins. The stability of SAMs during chemical functionalization further expands their applications, such as in microelectronics.	Even though the preparation of SAMs from solution is relatively easy, care needs to be taken to ensure their formation is optimal under certain conditions. E.g., Selection of appropriate solvents—although most polar solvents including dimethylformamide (DMF), dichloromethane (DCM), and acetone can assist the formation of SAMS, preparation of SAMs of alkyl thiols is typically carried out in pure ethanol which is optimal to guarantee production of defect-free SAMs.
QCM and SAW sensors can measure intermolecular interaction without labeling measuring objects by monitoring the changes of frequencies of the sensors. Once immobilized on QCMs and SAWs OBP biosensors remains stable and be able to continue to sensitively detect the target analytes vapor up to 1 year. Having affinity to a great number of organic molecules (odorants and pheromones) makes OBPs attractive to build reliable biosensors for different applications.	There could be small daily variations in response due to variations in the concentrations of saturated vapor generated at ambient temperatures. Having affinity to a great number of organic molecules makes OBP biosensors susceptible to cross-sensitivity to a range of substances. Hence, they are best implemented as arrays of sensors containing different OBPs with different binding affinities to a range of vapors.

Continued

—cont'd

Pros	Cons
OBPs have binding pockets capable of binding certain groups of chemicals with high specificity.	Binding pockets are not completely selective so cross-sensitivity will be observed across a range of analytes. Nonspecific binding is also possible on the outside of the proteins as well as on the supports used to immobilize the proteins.

12. Troubleshooting and optimization

Problem	Solution
OBP biosensors are sensitive to the humidity of the environment. If too dry they will stop working completely.	Never expose the OBP biosensors to completely dry gases. If that happens for any reason exposure to water vapor often leads to complete recovery. It is recommended to expose the sensors to a pulse of water vapor on a regular basis.
Occasionally OBP sensors may lose activity suddenly—especially if exposed to high concentrations of certain vapors.	If the sensor does lose activity suddenly exposure to water vapor often leads to recovery.

13. Summary

Methods are described enabling recombinant OBPs with or without a 6-Histidine tag to be immobilized using self-assembled monolayers on gravimetric transducers to create functional sensors capable of sensing a range of volatile organic compounds with selectivity dependent on the binding pockets of the OBPs. SAMs allow well oriented covalent attachment of proteins on to transducers surfaces. NTA immobilization utilizing 6-histidine tags attached on to the N-terminal or the C-terminal of the OBPs can produce well ordered stable films without affecting access of ligands to the protein binding sites.

Acknowledgments

Khasim Cali was supported though the CURSOR project which has received funding from the European Union's Horizon 2020 research and innovation programme under grant agreement No 832790 and from the Japan Science and Technology Agency.

Elena Tuccori was supported through a Marie Curie Early Stage Researcher grant (FLEXSMELL grant agreement No 2009-238454).

The opinions expressed in this document reflect only the author's view and reflects in no way the European Commission's opinions. The European Commission is not responsible for any use that may be made of the information it contains.

References

Briand, E., Salmain, M., Herry, J. M., Perrot, H., Compère, C., & Pradier, C. M. (2006). Building of an immunosensor: How can the composition and structure of the thiol attachment layer affect the immunosensor efficiency? *Biosensors and Bioelectronics, 22*(3), 440–448. https://doi.org/10.1016/j.bios.2006.05.018.

Burlachenko, J., Kruglenko, I., Snopok, B., & Persaud, K. (2016). Sample handling for electronic nose technology: State of the art and future trends. *TrAC Trends in Analytical Chemistry, 82*, 222–236. https://doi.org/10.1016/j.trac.2016.06.007.

Cady, W. G. (1921). The piezo-electric resonator. *Proceedings Institute of Radio Engineers, 6*, 83–114.

Cady, W. G. (1946). *Piezoelectricity* (1st ed.). New York and London: McGraw-Hill.

Cali, K., & Persaud, K. C. (2020). Modification of an Anopheles gambiae odorant binding protein to create an array of chemical sensors for detection of drugs. *Scientific Reports, 10*(1), 1–13. https://doi.org/10.1038/s41598-020-60824-7.

Cali, K., Scorsone, E., & Persaud, K. C. (2019). Odour sensing using arrays of odorant binding proteins. In *Chemical senses, Vol. 44* (p. E49). Oxford, England: Oxford University Press.

Chang, Y., Tang, N., Qu, H., Liu, J., Zhang, D., Zhang, H., et al. (2016). Detection of volatile organic compounds by self-assembled monolayer coated sensor array with concentration-independent fingerprints. *Scientific Reports, 6*, 1–12. https://doi.org/10.1038/srep23970.

Chevallier, E., Scorsone, E., & Bergonzo, P. (2009). Modified diamond nanoparticles as sensitive coatings for chemical SAW sensors. *Procedia Chemistry, 1*(1), 943–946. https://doi.org/10.1016/j.proche.2009.07.235.

Curie, J., & Curie, P. (1880). Comptes rendus des séances de l'Académie des Sciences. *Journal of Chemical Information and Modeling*, 1–30. https://doi.org/10.1017/CBO9781107415324.004.

De Vito, S., Fattoruso, G., Pardo, M., Tortorella, F., & Di Francia, G. (2012). Semi-supervised learning techniques in artificial olfaction: A novel approach to classification problems and drift counteraction. *IEEE Sensors Journal, 12*, 3215–3224. https://doi.org/10.1109/JSEN.2012.2192425.

Di Natale, C., Martinelli, E., Paolesse, R., D'Amico, A., Filippini, D., & Lundström, I. (2008). An experimental biomimetic platform for artificial olfaction. *PLoS One, 3*(9), e3139. https://doi.org/10.1371/journal.pone.0003139.

Di Pietrantonio, F., Benetti, M., Cannatà, D., Verona, E., Palla-Papavlu, A., Fernández-Pradas, J. M., et al. (2015). A surface acoustic wave bio-electronic nose for detection of volatile odorant molecules. *Biosensors and Bioelectronics, 67*(October 2016), 516–523. https://doi.org/10.1016/j.bios.2014.09.027.

Duocastella, M., Colina, M., Fernández-Pradas, J. M., Serra, P., & Morenza, J. L. (2007). Study of the laser-induced forward transfer of liquids for laser bioprinting. *Applied Surface Science, 253*(19), 7855–7859. https://doi.org/10.1016/j.apsusc.2007.02.097.

Fanget, S., Hentz, S., Puget, P., Arcamone, J., Matheron, M., Colinet, E., et al. (2011). Gas sensors based on gravimetric detection—A review. *Sensors and Actuators, B: Chemical, 160*(1), 804–821. https://doi.org/10.1016/j.snb.2011.08.066.

Ferretti, S., Paynter, S., Russell, D. A., Sapsford, K. E., & Richardson, D. J. (2000). Self-assembled monolayers: A versatile tool for the formulation of bio- surfaces. *TrAC Trends in Analytical Chemistry, 19*, 530–540. https://doi.org/10.1016/S0165-9936(00)00032-7.

Go, D. B., Atashbar, M. Z., Ramshani, Z., & Chang, H. C. (2017). Surface acoustic wave devices for chemical sensing and microfluidics: A review and perspective. *Analytical Methods Royal Society of Chemistry*, *9*, 4112–4134. https://doi.org/10.1039/c7ay00690j.

Hahn, C. D., Leitner, C., Weinbrenner, T., Schlapak, R., Tinazli, A., Tampé, R., et al. (2007). Self-assembled monolayers with latent aldehydes for protein immobilization. *Bioconjugate Chemistry*, *18*(1), 247–253. https://doi.org/10.1021/bc060292e.

Hines, E. L., Llobet, E., & Gardner, J. W. (1999). Electronic noses: A review of signal processing techniques. *IEE Proceedings—Circuits, Devices and Systems*, 146(6), 297. https://doi.org/10.1049/ip-cds:19990670.

Hochuli, E., Bannwarth, W., Dobeli, H., Gentz, R., & Stuber, D. (1988). Genric approach to facilitate purification of recombinant proteins with a novel metal chelate adsorbent. *Nature Biotechnology*, *6*, 1321–1325. (November). https://doi.org/10.1038/nbt1188-1321.

Katzir, S. (2012). Who knew piezoelectricity? Rutherford and Langevin on submarine detection and the invention of sonar. *Notes and Records of the Royal Society Royal Society*, *66*(2), 141–157. https://doi.org/10.1098/rsnr.2011.0049.

King, W. H., Jr. (1972). The use of resonating devices to make small mass measurements. *Bulletin of the New York Academy of Medicine*, *48*(2), 459–464.

Kyo, M., Usui-Aoki, K., & Koga, H. (2005). Label-free detection of proteins in crude cell lysate with antibody arrays by a surface plasmon resonance imaging technique. *Analytical Chemistry*, 77(22), 7115–7121. https://doi.org/10.1021/ac050884a.

Lack, F. R., Willard, G. W., & Fair, I. E. (1930). Some improvements in quartz crystal circuit elements. *Bell SystemTechnical Journal*, *453*(1), 453–463.

Laibinis, P. E., Fox, M. A., Folkers, J. P., & Whitesides, G. M. (1991). Comparisons of self-assembled monolayers on silver and gold: Mixed monolayers derived from HS(CH2) 21X and HS(CH2)10Y (X, Y = CH3, CH2OH) have similar properties. *Langmuir*, 7(12), 3167–3173.

Lamb, H. (1917). On waves in an elastic plate. *Proceedings of the Royal Society of London, Series A*, *93*(648), 114–128. https://doi.org/10.1098/rspa.1917.0008.

Länge, K., Rapp, B. E., & Rapp, M. (2008). Surface acoustic wave biosensors: A review. *Analytical and Bioanalytical Chemistry*, *391*(5), 1509–1519. https://doi.org/10.1007/s00216-008-1911-5.

Limbut, W., Kanatharana, P., Mattiasson, B., Asawatreratanakul, P., & Thavarungkul, P. (2006). A comparative study of capacitive immunosensors based on self-assembled monolayers formed from thiourea, thioctic acid, and 3-mercaptopropionic acid. *Biosensors and Bioelectronics*, *22*(2), 233–240. https://doi.org/10.1016/j.bios.2005.12.025.

Lippmann, M. G. (1881). On the principle of the conservation of electricity. *The London, Edinburgh, and Dublin Philosophical Magazine and Journal of Science*, *12*, 151–154. https://doi.org/10.1080/14786448108627081.

Love, A. E. H. (1911). *Some problems of geodynamics*. Cambridge: University Press.

Manai, R., Scorsone, E., Rousseau, L., Ghassemi, F., Possas Abreu, M., Lissorgues, G., et al. (2014). Grafting odorant binding proteins on diamond bio-MEMS. *Biosensors and Bioelectronics*, *60*, 311–317. https://doi.org/10.1016/j.bios.2014.04.020.

Mastromatteo, U., & Villa, F. F. (2013). High sensitivity acoustic wave AlN/Si mass detectors arrays for artificial olfactory and biosensing applications: A review. *Sensors and Actuators B: Chemical*, *179*, 319–327. https://doi.org/10.1016/j.snb.2012.10.033.

Mauriz, E., Calle, A., Montoya, A., & Lechuga, L. M. (2006). Determination of environmental organic pollutants with a portable optical immunosensor. *Talanta*, *69*(2), 359–364. https://doi.org/10.1016/j.talanta.2005.09.049.

Minamiki, T., Sasaki, Y., Tokito, S., & Minami, T. (2017). Label-free direct electrical detection of a histidine-rich protein with sub-femtomolar sensitivity using an organic field-effect transistor. *ChemistryOpen*, *6*(4), 472–475. https://doi.org/10.1002/open.201700070.

Mitchell, G. D. (2000). A review of permeation tubes and permeators. *Separation and Purification Methods*, *29*(**1**), 119–128. Taylor & Francis Group. https://doi.org/10.1081/SPM-100100005.

Muñoz-Aguirre, S., Nakamoto, T., & Moriizumi, T. (2005). Study of deposition of gas sensing films on quartz crystal microbalance using an ultrasonic atomizer. *Sensors and Actuators B: Chemical*, *105*(2), 144–149. https://doi.org/10.1016/j.snb.2004.05.063.

Nakamoto, T., Fukunishi, K., & Moriizumi, T. (1990). Identification capability of odor sensor using quartz-resonator array and neural-network pattern recognition. *Sensors and Actuators B: Chemical*, *1*(1–6), 473–476. https://doi.org/10.1016/0925-4005(90)80252-U.

Nazemi, H., Joseph, A., Park, J., & Emadi, A. (2019). Advanced micro-and nano-gas sensor technology: A review. *Sensors (Switzerland)*, *19*(6), 1285. https://doi.org/10.3390/s19061285.

Pardo, M., & Sberveglieri, G. (2005). Classification of electronic nose data with support vector machines. *Sensors and Actuators, B: Chemical*, *107*(2), 730–737. https://doi.org/10.1016/j.snb.2004.12.005.

Persaud, K. C. (2013). Engineering aspects of olfaction. In *Neuromorphic olfaction* (pp. 1–58). Dordrecht: Springer. https://doi.org/NBK298822 [bookaccession].

Persaud, K. C., & Tuccori, E. (2014). Biosensors based on odorant binding proteins. Vol. 9789401786133, Bioelectronic nose: Integration of biotechnology and nanotechnology (pp. 171–190). Netherlands: Springer. https://doi.org/10.1007/978-94-017-8613-3_10.

Pierce, G. W. (1923). Piezoelectric crystal resonators and crystal oscillators applied to the precision calibration of wavemeters. *Proceedings of the American Academy of Arts and Sciences*, *59*(4), 81. https://doi.org/10.2307/20026061.

Possas-Abreu, M., Ghassemi, F., Rousseau, L., Scorsone, E., Descours, E., & Lissorgues, G. (2017). Development of diamond and silicon MEMS sensor arrays with integrated readout for vapor detection. *Sensors (Basel, Switzerland)*, *17*(6), 1163. https://doi.org/10.3390/s17061163.

Possas-Abreu, M., Rousseau, L., Ghassemi, F., Lissorgues, G., Habchi, M., Scorsone, E., et al. (2017). Biomimetic diamond MEMS sensors based on odorant-binding proteins: Sensors validation through an autonomous electronic system. In *ISOEN 2017—ISOCS/IEEE international symposium on olfaction and electronic nose, proceedings*. https://doi.org/10.1109/ISOEN.2017.7968909.

Rayleigh, L. (1885). On waves propagated along the plane surface of an elastic solid. *Proceedings of the London Mathematical Society*, *s1–17*(1), 4–11.

Sauerbrey, G. (1959). Verwendung von Schwingquarzen zur Wägung dünner Schichten und zur Mikrowägung. *Zeitschrift für Physik*, *155*, 206–222. https://doi.org/10.1007/BF01337937.

Scorsone, E., Manai, R., Ricatti, M. J., Redaelli, M., Bergonzo, P., Persaud, K. C., et al. (2016). Major urinary proteins on nanodiamond-based resonators toward artificial olfaction. *IEEE Sensors Journal*, *16*(17), 6543–6550. https://doi.org/10.1109/JSEN.2016.2585187.

Shuturminska, K., O'Malley, C., Collis, D. W. P., Conde, J., & Azevedo, H. S. (2018). Displaying biofunctionality on materials through templated self-assembly. In *Self-assembling biomaterials: Molecular design, characterization and application in biology and medicine* (pp. 341–370). Elsevier. https://doi.org/10.1016/B978-0-08-102015-9.00018-6.

Tan, Y. H., Schallom, J. R., Ganesh, N. V., Fujikawa, K., Demchenko, A. V., & Stine, K. J. (2011). Characterization of protein immobilization on nanoporous gold using atomic force microscopy and scanning electron microscopy. *Nanoscale*, *3*(8), 3395–3407. https://doi.org/10.1039/c1nr10427f.

Thundat, T., Oden, P. I., & Warmack, R. J. (1997). Microcantilever sensors. *Microscale Thermophysical Engineering*, *1*(3), 185–199. https://doi.org/10.1080/108939597200214.

Tuccori, E., & Persaud, K. C. (2019). Pheromone detection using odorant binding protein sensors. In *ISOEN 2019—18th international symposium on olfaction and electronic nose, proceedings*. https://doi.org/10.1109/ISOEN.2019.8823345.

Tukkiniemi, K., Rantala, A., Nirschl, M., Pitzer, D., Huber, T., & Schreiter, M. (2009). Fully integrated FBAR sensor matrix for mass detection. In *Vol. 1. Procedia chemistry* (pp. 1051–1054). Elsevier. https://doi.org/10.1016/j.proche.2009.07.262.

Vaughan, R. D., O'Sullivan, C. K., & Guilbault, G. G. (1999). Sulfur based self-assembled monolayers (SAM's) on piezoelectric crystals for immunosensor development. *Fresenius' Journal of Analytical Chemistry*, *364*(1), 54–57. https://doi.org/10.1007/s002160051300.

Vogt, B. D., Lin, E. K., Wu, W.-L., & White, C. C. (2004). Effect of film thickness on the validity of the Sauerbrey equation for hydrated polyelectrolyte films. *The Journal of Physical Chemistry B*, *108*, 12685–12690. https://doi.org/10.1021/jp0481005.

Voiculescu, I., & Nordin, A. N. (2012). Acoustic wave based MEMS devices for biosensing applications. *Biosensors and Bioelectronics*, *33*(1), 1–9. https://doi.org/10.1016/j.bios.2011.12.041.

Weiss, P. (1906). La variation du ferromagnetisme du temperature. *Comptes Rendus*, *143*, 1136–1149.

White, R. M., & Voltmer, F. W. (1965). Direct piezoelectric coupling to surface elastic waves. *Applied Physics Letters*, 7(12), 314–316. https://doi.org/10.1063/1.1754276.

Wong, L. S., Khan, F., & Micklefield, J. (2009). Selective covalent protein immobilization: Strategies and applications. *Chemical Reviews*, *109*(9), 4025–4053. https://doi.org/10.1021/cr8004668.

Zhang, M., Du, L., Fang, Z., & Zhao, Z. (2017). A sensitivity-enhanced film bulk acoustic resonator gas sensor with an oscillator circuit and its detection application. *Micromachines*, *8*(1), 25. https://doi.org/10.3390/mi8010025.

Zhang, Y., Luo, J., Flewitt, A. J., Cai, Z., & Zhao, X. (2018). Film bulk acoustic resonators (FBARs) as biosensors: A review. *Biosensors and Bioelectronics*, *116*, 1–15. https://doi.org/10.1016/j.bios.2018.05.028.

Further reading

Pinnaduwage, L. A., Hawk, J. E., Boiadjiev, V., Yi, D., & Thundat, T. (2003). Use of microcantilevers for the monitoring of molecular binding to self-assembled monolayers. *Langmuir*, *19*, 7841–7844. https://doi.org/10.1021/LA034969N.

CHAPTER NINETEEN

Plasmonic biosensors relying on biomolecular conformational changes: Case of odorant binding proteins

Anil Bozdogan[a,b], **Simone Hageneder**[a], **Jakub Dostalek**[a,*]
[a]AIT-Austrian Institute of Technology GmbH, Tulln, Austria
[b]CEST Kompetenzzentrum für elektrochemische Oberflächentechnologie GmbH TFZ, Wiener Neustadt, Austria
*Corresponding author: e-mail address: jakub.dostalek@ait.ac.at

Contents

Abstract

Plasmonic nanostructures serve in a range of analytical techniques that were developed for the analysis of chemical and biological species. Among others, they have been pursued for the investigation of odorant binding proteins (OBP) and their interaction with

Methods in Enzymology, Volume 642
ISSN 0076-6879
https://doi.org/10.1016/bs.mie.2020.04.068

odorant molecules. These compounds are low molecular weight agents, which makes their direct detection with conventional surface plasmon resonance (SPR) challenging. Therefore, other plasmonic sensor modalities need to be implemented for the detection and interaction analysis of OBPs. This chapter provides a guide for carrying out such experiments based on two techniques that take advantage of conformation changes of OBPs occurring upon specific interaction with their affinity partners. First, there is discussed SPR monitoring of conformation changes of biomolecules that are not accompanied by a strong increase in the surface mass density but rather with its re-distribution perpendicular to the surface. Second, the implementation of surface plasmon-enhanced fluorescence energy transfer is presented for the sensitive monitoring of conformational changes of biomolecules tagged with a fluorophore at its defined part. Examples from our and other laboratories illustrate the performance of these concepts and their applicability for the detection of low molecular weight odorant molecules by the use of OBPs attached to the sensor surface is discussed.

Abbreviations

ATP	adenosine triphosphate
ATR	attenuated total internal reflection
n_s	aqueous sample refractive index
n_b	biolayer refractive index
BRE	biomolecular recognition elements
S_b	bulk refractive index sensitivity
D	diameter
d	distance/thickness
λ_{em}	emission wavelength
EF	electric field enhancement factor
λ_{ex}	excitation wavelength
h	height
HEPES	4-(2-hydroxyethyl)-piperazine-1-ethanesulfonic acid
LSPR	localized surface plasmon resonance
LSPs	localized surface plasmons
OBP	odorant binding protein
Λ	period
PEF	plasmon-enhance fluorescence
PEG	oligoethylene glycol-thiol
L_p	probing depth
PSPs	propagating surface plasmons
RIU	refractive index unit
SAM	self-assembled monolayer
λ_{SPR}	SPR wavelength
SERS	surface plasmon-enhanced Raman spectroscopy
SEIRA	surface-enhanced infrared absorption
SPR	surface plasmon resonance
Γ	surface mass density
TCEP	tris(2-carboxyethyl)phosphine hydrochloride solution
UV-LIL	UV-laser interference lithography
λ	wavelength

1. Introduction

Plasmonics represents a recently established interdisciplinary research domain that concerns tight confinement of light at metallic nanostructures. It exploits resonant excitation of surface plasmons that originate from collective oscillations of electron density. The associated electromagnetic field probes the close proximity of the metallic (typically gold) nanostructures to short sub-wavelength distances. This probing depth can be tuned at a length-scale comparable to the size of a variety of chemical and biological species by the design of the metallic nanostructures. Such tuneable confinement of light energy makes the plasmonic nanostructures ideally suited for optical observation of biomolecules and their interactions at solid surfaces, which have been implemented in a range of biosensor modalities. The measurement of refractive index changes induced by the binding of biomolecules at noble metals is employed in surface plasmon resonance (SPR) biosensors (Homola, 2008; Liedberg, Nylander, & Lundstrom, 1983). The increased intensity of the probing surface plasmon field allows the efficient fingerprinting of molecular analytes in surface plasmon-enhanced Raman spectroscopy (SERS) (Jeanmaire & Van Duyne, 1977; Langer et al., 2019) and it opens doors for in-situ surface-enhanced infrared absorption spectroscopy (SEIRA) (Adato & Altug, 2013) observations of species that are otherwise masked by the absorption of water. Moreover, the plasmonic management of light at the metallic surfaces is utilized for the plasmon-enhance fluorescence (PEF) assays to enhance their sensitivity and suppress the background signal (Bauch, Toma, Toma, Zhang, & Dostalek, 2014).

In the majority of plasmonic biosensors, the detection of target species is performed by the use of biomolecular recognition elements (BRE) that affinity capture target analyte from a liquid sample at the sensor surface. The specific binding is directly monitored through the increased surface mass density (SPR), by the occurrence of specific vibrational bands (SERS, SEIRA) or by using labels (PEF, SERS). In general, all these techniques have been implemented to affinity biosensors for the detection of analytes with medium or large molecular weight. However, their utilization of low molecular weight analytes can be complicated. The SPR approach often does not allow their detection as the small compounds do not induce sufficiently large and thus measurable refractive index change. Likewise, for direct SERS-based analysis, many important low molecular weight analytes are not Raman-active, and the SEIRA approach typically relies on the amplification of the signal only in a narrow spectral window. Moreover, assays with a fluorophore and Raman-active tags (PEF, SERS) typically rely on a sandwich format that is not suitable for low molecular weight analytes as they do not present two regions to which (typically bulky BREs like antibodies) can simultaneously bind.

To overcome this limitation, BREs that are attached at the sensor surface and undergo a conformational change upon the specific interaction with the target analyte can be utilized. Then, these binding events can be accompanied by a change that is substantially stronger than that induced by the presence of the target analyte itself. For the SPR readout, BRE conformation change can manifest itself as a spatial re-distribution of the surface mass density perpendicular to the surface and be translated to measurable refractive index variations (Hall et al., 2011; Jonsson, Jonsson, Dahlin, & Hook, 2007). Besides, the BRE conformation change can lead to the shifts of its Raman or infrared absorption bands (Jones, Balakrishnan, Squier, & Spiro, 2014; Neumann et al., 2009) and it can be conjugated with a reporting molecule (e.g., fluorophore). Then, the conformation change can be converted to an optical signal based on distance-amended fluorescence quenching (Zheng, Seferos, Giljohann, Patel, & Mirkin, 2009) or SERS signal strength (Jung et al., 2007).

This chapter aims at providing a practical guide to the implementation of two plasmonic sensor modalities for the measurement of conformation changes of biomolecules. In particular, we introduce SPR and PEF approaches relying on the probing of the sensor surface with localized surface plasmons on gold nanoparticles and by propagating surface plasmons traveling along continuous gold films. A possible implementation of such experiments is described by the use of specific protocols. A possible interpretation of the measured optical response due to the refractive index re-distribution on the surface (SPR readout) and due to the plasmon-enhanced fluorescence energy transfer (quenching of PEF) is explained. Lastly, using these techniques in the context of odorant binding proteins (OBP) is discussed. It is worth noting that the analysis of such interactions is a challenge and plasmonic biosensors hold the potential to provide new insights to these studies and can serve in translating this knowledge toward practical applications.

2. Probing of biomolecules at interfaces with surface plasmons

In plasmonic biosensors, the surface of metallic nanostructures carrying analyzed species is probed by the resonantly excited surface

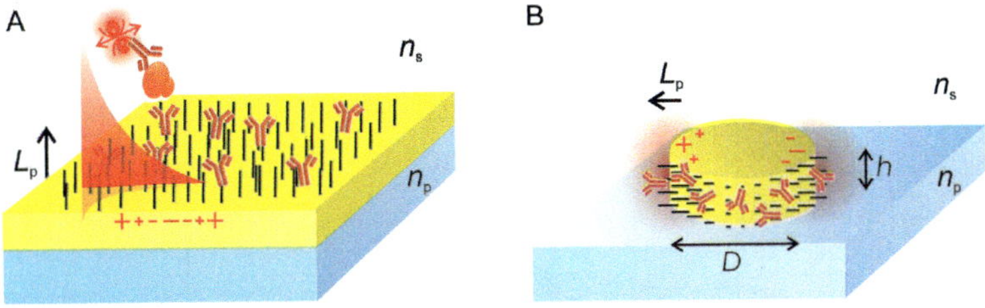

Fig. 1 Schematics of the probing of biomolecular binding events with (A) propagating surface plasmons (PSPs) traveling along a metallic film and (B) localized surface plasmons (LSPs) supported by a metallic nanoparticle with indicated probing depth L_p and refractive index above the structure n_s and below the structure n_p.

plasmons. As illustrated in Fig. 1, two types of surface plasmons can be utilized: (a) propagating surface plasmons (PSPs) that travel along continuous metallic films and (b) localized surface plasmons (LSPs) supported by metallic nanoparticles.

On the surface of gold structures (which are preferably used for their chemical stability and range of surface chemistries available), optical waves at a wavelength in the red and near-infrared parts of the spectrum are often used for the resonant excitation of PSPs and LSPs. For the optical excitation of PSPs, typically, periodically corrugated metallic films or so-called Kretschmann configuration of attenuated total internal reflection (ATR) method are employed. The profile of field for the ATR excited PSPs on a 50 nm thick gold film in contact with water is showed in Fig. 2A. It reveals that at a wavelength of $\lambda = 633$ nm, the PSP electric field intensity peaks at the gold surface (field intensity enhancement of $EF > 30$) and exponentially decays into the adjacent water medium as:

$$|E(z)|^2 - EF \cdot \exp\left(-2d/L_p\right), \tag{1}$$

where EF is the electric field intensity enhancement, d is the distance from the surface and $L_p/2$ is the probing depth. For the PSP field shown, the probing depth of $L_p/2 = 93$ nm was determined.

Significantly tighter confinement of light energy is accompanied by the excitation of LSPs. Fig. 2B shows an example of a cross-section of the LSP electric field intensity for arrays of gold disk nanoparticles (diameter $D = 140$ nm and height $h = 50$ nm) in contact with water. For the geometry with the bottom (substrate) dielectric refractive index of $n_p = 1.46$ that is different from that of water $n_s = 1.33$ on the top, the LSPs are excited at a wavelength of $\lambda = 708$ nm for the period $\Lambda = 360$ nm. Compared to PSPs, a higher enhancement factor of about $EF \sim 100$ occurs due to the

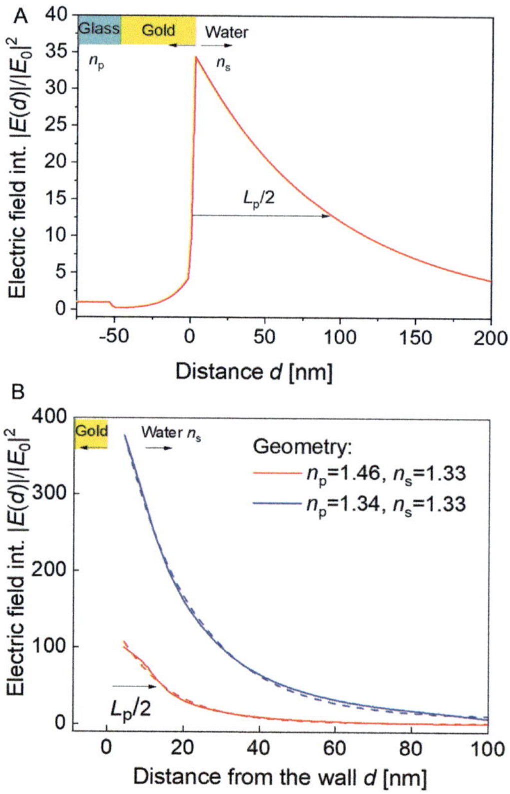

Fig. 2 The profile of the electric field intensity $|E|^2$ normalized with the electric intensity of the excitation optical beam $|E_0|^2$ for (A) resonantly excited PSPs by ATR configuration with a 50 nm thick gold film in contact with water (refractive index of $n_s = 1.33$) at a wavelength of 633 nm and (B) LSPs on arrays of cylindrical nanoparticles with a diameter of $D = 140$ nm and height $h = 50$ nm in the refractive index asymmetrical (refractive index of the substrate of $n_p = 1.45$, period of $\Lambda = 360$ nm) and nearly symmetrical geometry (refractive index of the substrate of $n_p = 1.34$, period of $\Lambda = 525$ nm). Solid lines correspond to finite difference time domain simulations and dashed lines are exponential fits.

tighter confinement of the field featuring $L_p/2 = 13$ nm. Interestingly, the probing depth L_p can be substantially increased by adjusting the nanoparticle arrays period to $\Lambda = 525$ nm and setting the refractive index of the bottom dielectric to $n_p = 1.34$ (so it is close to that above water with $n_s = 1.33$). In such refractive index symmetrical geometry, so-called lattice LSPs can be excited and the presented example in Fig. 2B reveals an increased electric field intensity enhancement factor of $EF \sim 400$ and a less localized field with a probing depth of $L_p/2 = 18$ nm.

The optical coupling to PSPs and LSPs is typically manifested as a dip in the reflectivity or transmission wavelength spectrum centered at a wavelength λ_{SPR}, where part of the light energy is resonantly transferred to surface plasmon modes. This energy is partially dissipated in the metal due to the Ohmic losses, which, as Fig. 3 illustrates, affects the spectral width of the resonant dip $\Delta\lambda$. An example of the SPR dip in the reflectivity spectrum for the PSP excitation on a gold film is showed in Fig. 3A; an SPR dip in the transmission spectrum for LSPs excitation on arrays of gold cylindrical nanoparticles can be seen in Fig. 3B.

Importantly, the spectral position of the SPR dip λ_{SPR} in reflectivity or transmission spectra shifts with the changes in refractive index n_s of the

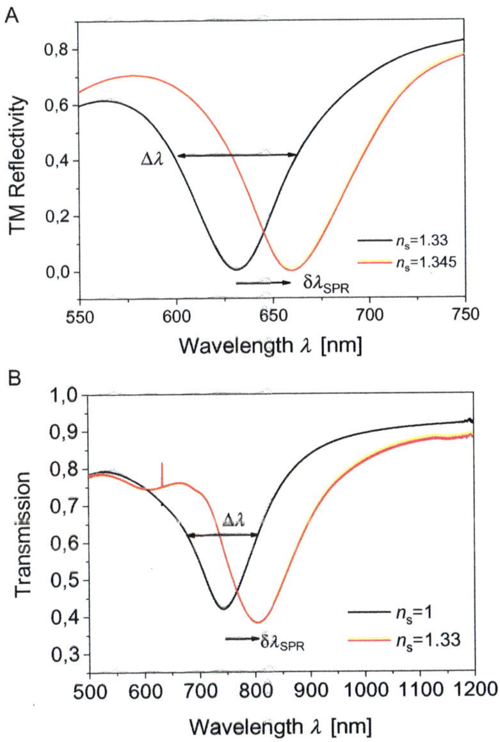

Fig. 3 Refractive index sensitivity for (A) ATR excitation of PSP modes with wavelength interrogation and (B) for the transmission measurement of resonant excitation of disk nanoparticle arrays. The excitation of PSP is simulated in transversally magnetic (TM) reflectivity for a prism with $n_p = 1.845$, 50 nm thick Au layer, water on the top, and angle of incidence of $\theta = 51.7$ deg. (B) Measured LSPR spectra for the resonant coupling to LSPs for normal angle of incidence of $\theta = 0$ on gold arrays of gold disk nanoparticles with a diameter $D = 175$ nm, height $h = 50$ nm, and period $\Lambda = 460$ nm on a glass surface with $n_p = 1.5$.

dielectric adjacent to the metal. These spectral shifts are linear for small changes δn_s and can be quantified by a bulk refractive index sensitivity S_b. For the examples shown, the bulk refractive index sensitivity for the wavelength interrogation of resonantly excited PSPs in ATR configuration is $S_b = \delta\lambda_{SPR}/\delta n_s = 2000\,\text{nm/RIU}$ (unitless RIU states for refractive index unit). For the resonant coupling to LSPs, the bulk refractive index is one order of magnitude smaller and $S_b = 190\,\text{nm/RIU}$ was measured in the presented experiment.

The metallic structures for the resonant excitation of PSPs and LSPs are typically prepared on glass substrates by a combination of thin-film deposition and lithography. The following protocol describes the use of thermal evaporation and UV-laser interference lithography (UV-LIL) for the preparation of continuous thin gold films (supporting PSPs) and arrays of cylindrical gold nanoparticles (supporting LSPs). For illustration, the procedure is summarized in Fig. 4.

2.1 Definition

Preparation of plasmonic nanostructures by UV-interference lithography.

2.2 Rationale

Metallic nanostructures and metallic thin films are deployed on the surface of plasmonic sensors to optically probe the reactions of biomolecules occurring on their top when contacted with analyzed liquid samples.

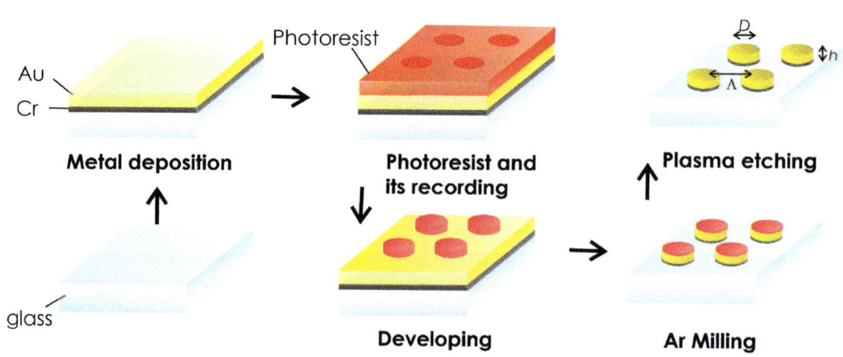

Fig. 4 Schematics of steps in the preparation of arrays of cylindrical gold nanoparticles on a glass substrate.

2.3 Materials, equipment, and reagents

- Microposit S1805 G2 positive photoresist (Micro Resist Technology, Germany)
- Propylene glycol monomethyl ether acetate (PGMEA, Sigma Aldrich, Germany)
- Developer AZ 303 (Micro Resist Technology, Germany)
- Au granules and Cr powder (Mateck, Germany)
  - Vacuum thermal evaporation instrument (HHV AUTO 306 from HHV LTD, UK)
  - Spincoater G3P-8 (Pi-Chem, UK)
  - UV-interference lithographer with a 4 mW HeCd laser (model IK 3031 R-C from Kimmon, Japan) emitting at a wavelength $\lambda = 325$ nm coherent beam and Lloyds mirror configuration, details can be seen in publications (Khan, Keshmiri, Kolb, Dimopoulos, & List-Kratochvil, 2016; Quilis et al., 2018)
  - Argon milling system IonSys 500 (Roth & Rau, Germany)
  - Oxygen plasma etcher

2.4 Protocol

(1) Subsequent deposition of 2 nm Cr and 50 nm Au layers in vacuum better than 10^{-6} mBar from tungsten and molybdenum boats, respectively, deposition rate 0.1 Å/s.

(2) Deposition of S1805 G2 photoresist solution diluted 1:2 with propylene glycol monomethyl ether acetate by the use of spin-coating at 4500 rpm for 45 s, followed by a soft baking step at 100 °C on a hot plate for 120 s.

(3) Recording of the UV-interference pattern to prepared 100 nm thick photoresist layer, irradiation power 30 μW/cm^2, irradiation dose of 6.75 mJ/cm^2.

(4) Developing the structure recorded to the photoresist layer by AZ 303 developer diluted with distilled water at a ratio 1:15. Typical developing time is between 90 and 240 s.

(5) Directional dry etching with an argon milling is applied to transfer the pattern from the resist into the underlying Au layer (etching time 450 s, 70 degrees).

(6) Finally, the remaining resist is removed using oxygen plasma at 1 mbar and 40 W applied for 5 min.

Pros	Cons
Structuring over the large surface area	The geometry of metallic nanoparticles is restricted to cylindrical and elliptical shapes
Well-ordered period arrays of metallic nanoparticles	
Tuneable geometry characteristics—period and diameter of the arrays	

2.5 Alternative methods/procedures

Plasmonic nanostructures can be alternatively prepared by electron beam and focused ion beam lithography, which allow for the fabrication of an almost arbitrary shape of nanoparticles. However, these methods are time-consuming and not suitable for routine use in analytical technologies as structuring areas larger than 100 μm is complicated and complex infrastructure is required.

Moreover, colloidal lithography can be employed for facile structuring of large areas. Wet chemical synthesis of metallic nanoparticles also provides efficient routes for the cost-efficient preparation of metallic nanostructures. However, these methods lack the possibility to arrange the structures in perfect lattices.

2.6 Troubleshooting and optimization

Problem	Solution
Metallic nanoparticles are connected	Increase the irradiation dose or development time
Metallic nanoparticles are not present	Often due to the over-etching during the ion milling step

2.7 Summary

UV-interference lithography—UV-LIL—is a method able to combine the advantages of large-area structuring and relatively high throughput and it is suitable for the routine preparation of plasmonic nanostructures whenever perfect lattice arrangement is required. If needed, it can be employed for the preparation of master structures for subsequent UV-nanoimprint lithography when scaling up of the production is important.

3. Monitoring of biomolecule conformation changes by SPR

In SPR biosensors, the kinetics of surface reactions can be monitored by the tracking of variations of SPR wavelength λ_{SPR} due to the binding-induced refractive index changes. Series of transmission or reflectivity spectra are acquired in time, and by their analysis, the SPR wavelength detuning is determined as a shift of the SPR dip minimum. Typically, it is determined by fitting the spectrum with an analytical function or by using a centroid approach. These methods provide accuracy in the λ_{SPR} baseline measurement with the noise of $\sigma(\lambda_{SPR}) = 10^{-2} - 10^{-3}$ nm (Dahlin, Tegenfeldt, & Hook, 2006; Reiner et al., 2017).

In regular SPR measurements, changes in λ_{SPR} due to the increase of the surface mass density Γ that is ascribed to the capture of target analyte on the sensor surface are monitored:

$$\Gamma = d(n_b - n_s)\frac{\partial c}{\partial n}, \tag{2}$$

where n_b states the refractive index of a biolayer (representing the BRE and the captured analyte), d represents its thickness, and the coefficient $\partial n/\partial c = 0.2\,\text{mm}^3\,\text{mg}^{-1}$ relates the changes in refractive index with the concentration of biomolecules bound to the surface. (Stenberg, Persson, Roos, & Urbaniczky, 1991)

The SPR wavelength shift $\delta\lambda_{SPR}$ due to refractive index changes of the biolayer n_b depends on its thickness as the probing field exponentially decreases with a characteristic probing depth $L_p/2$ (see Fig. 2). The SPR wavelength shift induced by the formation of a biolayer with a thickness d and refractive index n_b can be mathematically expressed based on perturbation theory (Snyder & Love, 1983) as:

$$\delta\lambda_{SPR} = S_b \int_0^d (n_b - n_s)|E(z)|^2 dz \Big/ \int_{-\infty}^{\infty} |E(z)|^2 dz. \tag{3}$$

This equation can be further simplified when using the penetration depth of the probing field L_p as follows:

$$\delta\lambda_{SPR} \sim S_b(n_b - n_s)\left(1 - e^{-2d/L_p}\right) = S_b\Gamma\frac{\partial n}{\partial c}\frac{1 - e^{-2d/L_p}}{d}. \tag{4}$$

In general, the SPR wavelength shifts not only when the surface mass density Γ is increased due to the capture of biomolecules on the metallic surface, but also when the biolayer is redistributed in a way that its overall surface mass density Γ stays constant (e.g., by increasing its thickness d and decreasing its refractive index n_p, so the $d/(n_b-n_s)$ is not changed). If the limit of the biolayer thickness much lower than the surface plasmon probing depth $d \ll L_p$, Eq. (4) yields:

$$\delta\lambda_{SPR} = S_b \Gamma \frac{\partial n}{\partial c} \frac{2}{L_p}, \tag{5}$$

and for the opposite situation when $d \gg L_p$ this equation holds

$$\delta\lambda_{SPR} = S_b \Gamma \frac{\partial n}{\partial c} \frac{1}{d}. \tag{6}$$

The sensitivity of SPR wavelength λ_{SPR} to the re-distribution of the biolayer mass in the direction perpendicular to the metal surface can be exploited for the observation of conformational changes of biomolecules. In principle, the conformational change can manifest itself as a change of respective biolayer refractive index n_p and thickness d. If the distance range at which this re-distribution occurs (represented by the effective thickness of the biolayer d) is larger or comparable with the surface plasmon probing depth $d \sim L_p$, such change can translate to a measurable shift in λ_{SPR}. Fig. 5A shows simulations that elucidate such an experiment. In this example, a shift of SPR wavelength λ_{SPR} due to the formation of a biolayer with a thickness of $d_0 = 10\,\text{nm}$ was determined for the probing depth of $L_p/2 = 100$, 10, and 1 nm by using Eq. (4). Then, the surface mass density of the layer Γ was assumed constant and the changes in the SPR wavelength were determined for its thickness d varied between 5 and 15 nm (and refractive index n_b concluded from Eq. 2). The simulated SPR wavelength $\lambda_{SPR}(d)$ was normalized to that determined for the original thickness $\lambda_{SPR}(d_0)$. These data predict that for the probing depth of $L_p/2 = 100\,\text{nm}$ (representing PSP—see Fig. 2A) and $L_p/2 = 10$ and 1 nm (representing LSPs—see Fig. 2B), a 10% change in the biolayer thickness translates to a change in the SPR wavelength λ_{SPR} of about 0.5%, 5%, and 10%, respectively, when compared to that corresponding to the initial biolayer.

To illustrate the above theory, let us relate it to the results reported by the van Duyne laboratory on the conformational change of calmodulin. It is a protein with a prolonged shape with a characteristic length of about 6.5 nm that increases by about 1 nm (corresponding to about 15%) upon its

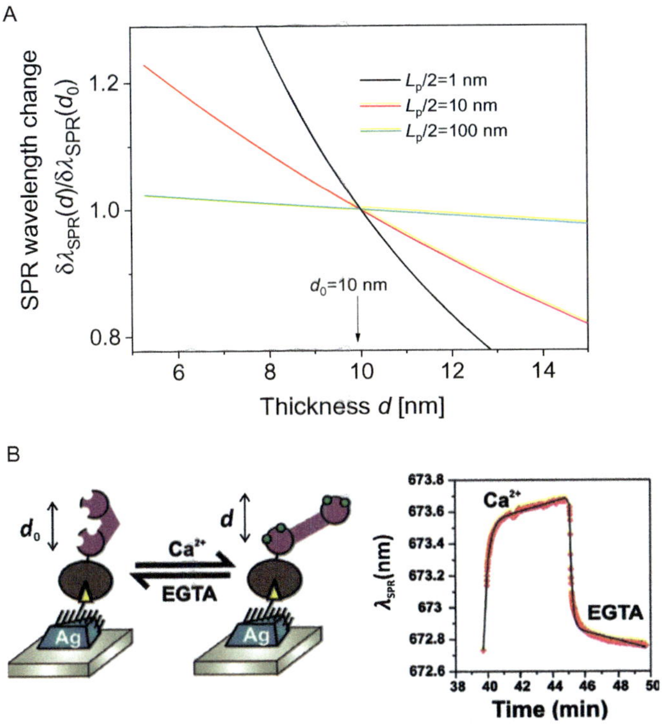

Fig. 5 (A) Simulations of the SPR wavelength changes induced by a conformational change of a biomolecule represented by variations of a biolayer characteristic thickness of $d_0 = 10$ nm in the range $d = 5$–25 nm for three surface plasmon probing depths $L_p/2$ stated in the graph. (B) Example of a conformation change of a protein calmodulin observed by tracking of LSPR wavelength changes. Adapted with permission from (Hall et al., 2011). Copyright 2011 American Chemical Society.

interaction with calcium ions (Hall et al., 2011). This conformation change was observed on silver triangular nanoparticles with the LSPR probing depth of $L_p/2 = 5$–10 nm. The immobilization of calmodulin protein on the sensor carrying a thiol SAM linker layer induced a shift in the LSPR wavelength of $\delta\lambda_{SPR}(d_0) = 3.8$ nm. Upon the interaction with Ca^{2+}, a change up to 0.9 nm was observed in $\delta\lambda_{SPR}$ (corresponding to 23%). These observations are in qualitative agreement with the above-presented model, taking into account the uncertainties in the conformation change on the surface and in bulk as well as in the probing field depth $L_p/2$. It is worth noting that these observations were made possible by an instrument that allowed for accurate tracking of the LSPR wavelength changes with a baseline standard deviation of $\sigma(\lambda_{SPR}) = 2 \times 10^{-3}$ nm.

4. Monitoring of biomolecule conformation changes by PEF

An alternative approach to the monitoring of conformational changes can be utilized based on the fluorescence energy transfer that occurs at close proximity to metals. This phenomenon leads to strong quenching of emitted fluorescence light at distances below or comparable with a Förster distance of 10–15 nm. It is worthy of noting that such distance-dependent quenching of fluorescence depends on the orientation of the fluorophore (see Fig. 6) and that in most common experiments the measured signal is averaged over all possible orientations.

On a gold surface, the distance-dependent fluorescence quenching of a fluorophore attached to a BRE can be implemented for the monitoring of its conformation changes induced by the affinity interaction with a target analyte, see Fig. 7. The additional interaction of the emitter with surface plasmons at the excitation wavelength λ_{ex} (close to the spectral position of the fluorophore absorption band) and emission wavelength λ_{em} offers means to enhance the strength of the detected fluorescence signal and to control the angular distribution of the emitted light. In the example presented in Fig. 7, an aptamer hairpin was used as BRE for a low molecular weight analyte and it was designed with an Alexa Fluor 647 fluorophore attached to its 5′ end with a thiol group at its 3′ end (Sergelen et al., 2017). The aptamer was composed of two key sequences that partially

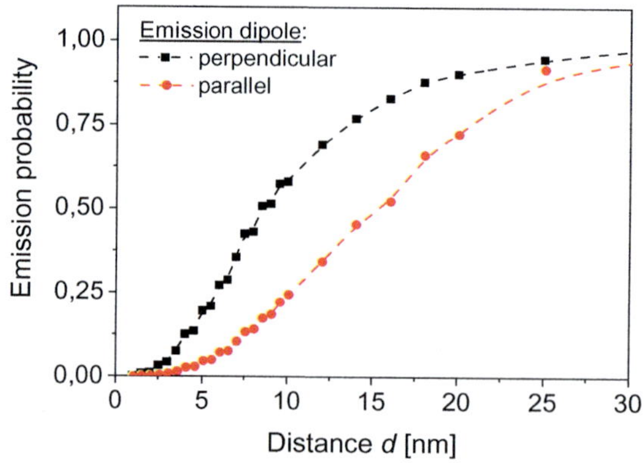

Fig. 6 Simulations of emission probability of a fluorophore represented as a dipole radiating at a wavelength of $\lambda = 670$ nm above a gold surface in contact with water. Emission probability to the far-field radiation propagating in water was simulated.

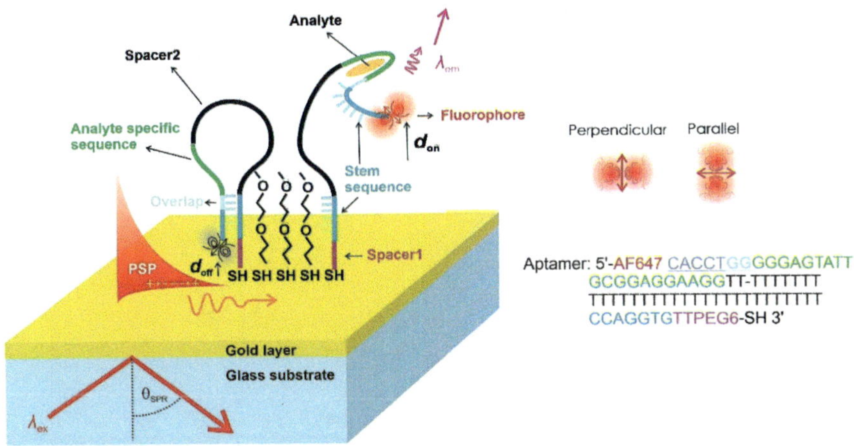

Fig. 7 Schematics of the aptamer-based assay on the gold sensor surface that utilizes a fluorophore coupled with propagating surface plasmons (PSP). The hairpin aptamer was anchored at the gold surface via a thiol group at the 3′ end and the surface density was diluted by using polyethylene glycol-thiol (PEG) for the affinity capture of target analyte—adenosine triphosphate. *Reproduced with permission from Sergelen, K., Fossati, S., Turupcu, A., Oostenbrink, C., Liedberg, B., Knoll, W., et al. (2017). Plasmon field-enhanced fluorescence energy transfer for hairpin aptamer assay readout.* ACS Sensors, *2(7), 916–923. Copyright 2017 American Chemical Society.*

overlap: a stem sequence and analyte-specific sequence. The stem sequence leads to a formation of a hairpin of the aptamer oligonucleotide chain in the absence of the target analyte. When the aptamer is anchored to a gold surface via the 3′ end with the thiol group, the fluorophore attached to the opposite aptamer 5′ end is present in the close proximity to the surface and thus is strongly quenched (see the rapidly decreasing emission probability when lowering the distance d in Fig. 6). Upon the affinity binding of the target analyte (low molecular weight adenosine triphosphate—ATP—was used in the discussed example), the stem sequence interaction is disrupted and the aptamer takes an open-loop structure. As can be seen in Fig. 8, the fluorescence signal then strongly increases as the fluorophore is lifted from the gold surface up. These measurements were carried out with the ATR configuration for the excitation of PSP modes with a monochromatic optical beam at a wavelength of $\lambda_{ex} = 633$ nm coinciding with the absorption band of the used fluorophore Alexa Fluor 647. As can be seen in Fig. 8B, the affinity binding of ATP analyte at the gold sensor surface with aptamer hairpin leads to an occurrence of a strong fluorescence peak measured at the fluorophore emission wavelength $\lambda_{em} = 670$ nm. The peak occurs at an angle of

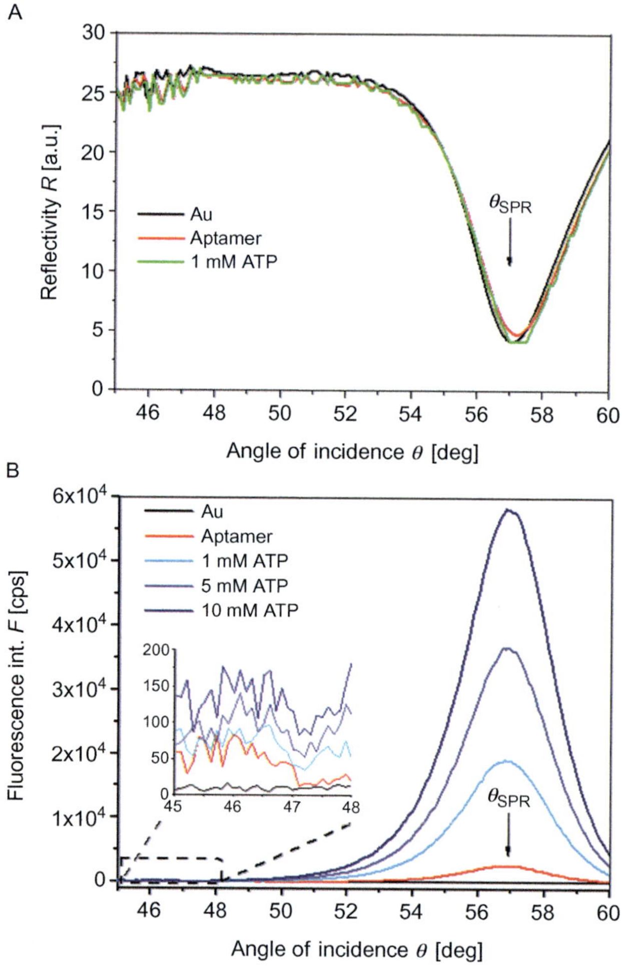

Fig. 8 (A) Angular SPR reflectivity spectra $R(\theta)$ and (B) respective fluorescence intensity spectra $F(\vartheta)$ measured for a hairpin aptamer BRE attached at the sensor surface that was brought in contact with a sample spiked with concentrations of a target analyte—ATP—of 0, 1, 5, and 10 mM. The gold sensor surface was functionalized with a long loop ATP hairpin aptamer (see schematics in Fig. 7). *Reproduced with permission from Sergelen, K., Fossati, S., Turupcu, A., Oostenbrink, C., Liedberg, B., Knoll, W., et al. (2017). Plasmon field-enhanced fluorescence energy transfer for hairpin aptamer assay readout.* ACS Sensors, *2(7), 916–923. Copyright 2017 American Chemical Society.*

incidence of the excitation beam with λ_{ex} at which PSP modes amplify the fluorescence signal emitted away from the gold sensor surface. When increasing the concentration of target analyte, the peak fluorescence signal is increasing as a higher fraction of the aptamer BREs on the surface is switched to the open-loop configuration. Interestingly, the regular SPR response presented in Fig. 8A shows no measurable change due to the ATP induced conformation change of the aptamer (at a distance range $d < 10\,\text{nm}$), as the probing depth is much larger ($L_p/2 = 93\,\text{nm}$).

4.1 Definition

Surface plasmon resonance—SPR—combined with plasmon-enhanced fluorescence—PEF—spectroscopy.

4.2 Rationale

The combination of SPR and PEF techniques can serve for in-situ characterization of biointerfaces and for the monitoring of affinity binding events of high and low molecular analytes. In conjunction with appropriate models, the kinetic data can be analyzed in order determine binding rates and equilibrium affinity constants.

4.3 Materials, equipment, and reagents

- (11-Mercaptoundecyl)triethylene glycol (PEG-thiol, SPT-0011) from SensoPath Technologies (USA)
- DNA ATP hairpin aptamer with a fluorophore at one and anchoring thiol group at the other end. An example is AlexaFluor 647N 5′-CA CCT GGG GGA GTA TTG CGG AGG AAGGTT TTT TTT TTT TTT TTT TTT TTT TTT TTT TTT CCAGGT G TT PEG6-SH 3′ synthesized by IDT-DNA (Leuven, Belgium)
- All additional chemicals are by Sigma Aldrich, Austria:
  - o Tris buffer (10 mM Tris hydrochloride + 6 mM $MgCl_2$, pH 7.4)
  - o HEPES buffer (10 mM 4-(2-hydroxyethyl)-piperazine-1-ethanesulfonic acid containing 150 mM NaCl, 5 mM MgC12, and 0.005% Tween 20, pH 7.4)
  - o Tris(2-carboxyethyl)phosphine hydrochloride solution (TCEP)
  - o Adenosine, guanosine and adenosine 5′-triphosphate (ATP) disodium salt hydrate
- Optical setup relying on Kretschmann configuration of ATR for combined SPR and PEF experiments: A He-Ne laser (2 mW) emitting at $\lambda_{ex} = 633\,\text{nm}$ is coupled to a prism made of high refractive index glass LASFN9 (exhibiting $n_p = 1.845$) with optically matched glass substrate carrying gold layer. The enhanced electromagnetic field intensity associated with the resonant excitation of PSPs at λ_{ex} excites Alexa Fluor 647 fluorophores and the emitted fluorescence light at the emission wavelength $\lambda_{em} = 670\,\text{nm}$ is collected from the direction perpendicular to

the surface. The emitted fluorescence light beam is spectrally cleaned by using a notch filter (XNF-632.8-25.0M from CVI Melles Griot) and bandpass filter (670FS10–25 from Andover Corporation Optical Filter) and its intensity F is detected by using an avalanche photodiode (Count-200-FC, Laser Components, Germany) or photomultiplier tube. More details can be found in the publication (Sergelen et al., 2017) and Fig. 9.

4.4 Protocol

(1) Freshly prepared borosilicate or LASFN9 glass substrates with a layer of 2 nm chromium and 50 nm gold or with arrays of gold nanoparticles (see protocol above).

(2) Incubation of the substrates in a 1 μM solution of ATP aptamer in Tris buffer.

(3) Optically matching the substrate to a LASFN9 glass prism by using refractive index matching oil (from Cargille, USA), clamping a flow cell on top and connecting the cell to a pump. Mounting into the SPR/PEF optical setup.

(4) Incubation of the substrates in TCEP (100 μM) for 4 h for reduction of the aptamer.

(5) Addition of 10 μM PEG-thiol to the solution and incubation for 2 h form a self-assembled monolayer (SAM).

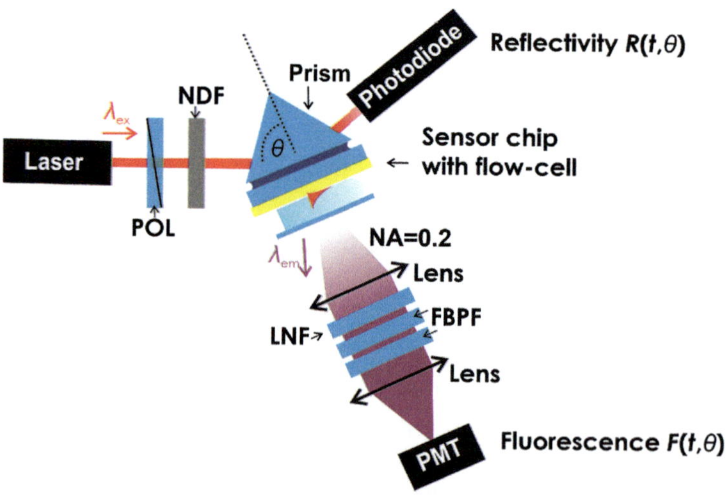

Fig. 9 Optical setup for combined SPR and SPFS measurement: Polarizer (POL), laser notch filter (LNF), fluorescence bandpass filter (FBPF), and photomultiplier (PMT). The systems can measure reflectivity R and fluorescence intensity F depending on the angle of incidence θ and time t.

(6) Washing of the sensor surface for 15 min with Tris or HEPES buffer.

(7) Sequential injection of the target analyte (ATP) and reference analytes for 3 min in either Tris or HEPES buffer with real-time monitoring of fluorescence signal F emitted upon the resonant excitation of PSPs at a wavelength λ_{ex} that is close to the absorption band of the used fluorophore.

Pros	Cons
Real-time observation of biomolecular interaction kinetics	Often unspecific interactions complicate the measurements
Possible monitoring of both low and high molecular weight analytes	Identification of affinity captured species can be done only in combination with other techniques such as mass spectrometry
Parallel screening of multiple interactions available by multichannel devices or in combination with microarrays	

4.5 Alternative methods/procedures

Direct monitoring of affinity surface reaction kinetics can be performed by quartz crystal microbalance (QCM) or optically by methods including reflectometric interference spectroscopy (RIFS) and integrated waveguide Mach Zehnder interferometry.

4.6 Troubleshooting and optimization

Problem	Solution
Analysis of complex samples often complicates the blocking of the sensor surface with abundant components present in complex analyzed samples	Design of biointerface by using polymer surface architectures that exhibit antifouling properties. Using buffers with blocking agents at a high concentration such as BSA or Tween 20
Need of parallel analysis of multiple reactions on the surface	SPR and PEF can be combined with microarray detection format, and a range of multichannel sensors is commercially available based on, e.g., surface plasmon imaging (SPRI)

4.7 Summary

The combination of SPR and PEF provides a versatile platform for the monitoring of affinity interaction of a wide spectrum of analytes ranging from high to low molecular weight biomolecules. Besides, it can be employed as a quality control for the development of biointerface preparation protocols as well as to support the overall process of assay development for the identifying of a suitable biomolecular recognition element, implementing biointerface architecture, and optimizing concrete assay steps.

5. Context of odorant binding protein interaction

Odorants represent an important class of low molecular weight analytes (up to 300 Da) that recently attracted attention of the biosensor research community aiming at mimicking the smell functionality of insects and vertebrates (Pelosi, Mastrogiacomo, Iovinella, Tuccori, & Persaud, 2014). They are highly hydrophobic molecules and they are transferred across the mucosa in vertebrates and trigger a cascade of reactions including those with olfactory receptors on a lipid membrane of nasal epithelial cells. Odorant binding proteins—OBPs—are present in the mucosa to capture hydrophobic odorants and to facilitate their transport to the olfactory receptors. OBPs are medium-sized proteins between 10 and 30 kDa and they have been proposed to serve as a BREs for the low molecular weight odorants (Larisika et al., 2015) based on their conformational change. The OBPs of vertebrates exhibit a structure of eight antiparallel β-sheets and a short α-helical segment. Their conformation change upon the affinity capture of odorants has been subjected of numerous studies including molecular dynamics simulations of rat and pig OBPs (Hajjar, Perahia, Debat, Nespoulous, & Robert, 2006) combined with circular dichroism measurements (Nespoulous, Briand, Delage, Tran, & Pernollet, 2004) and protein crystallography (White, Briand, Scott, & Borysik, 2009). These studies indicate that the size of the OBPs is several nanometres and that their overall structure is still preserved upon the capture of odorant molecules in their binding pocket. The distance between different domains can vary subtly, still, according to our knowledge, this change has not yet been exactly quantified in terms of, e.g., hydrodynamic radius or mass distribution change.

Hurot et al. reported a biosensor for odorant molecules using rat OBPs and SPR sensor configuration relying on ATR to probe its gold surface with

PSP modes at a wavelength of 663 nm (Hurot et al., 2019). Three genetically mutated rat OBP proteins were modified to carry a cysteine group at their N-terminus allowing the formation of a self-assembled monolayer on the gold sensor surface with a surface mass density of about $\Gamma \sim 2000\,\text{pg/mm}^2$ (deduced from the reported surface density of 6×10^{12} proteins/cm^2 and molecular weight of 20 kDa). These OBPs were spotted in a microarray format, including an OBP3-w natural protein (recognizing odorant molecules with a benzene ring structure), mutated OBP3-a (designed for interaction with aldehydes with increased affinity), and control OBP3-c. Afterward, the sensor surface was exposed to aqueous solutions with the dissolved odorant molecules β-ionone, hexanal, and control hexanoic acid. The affinity binding of these compounds was monitored from detuning of the SPR signal with an instrument providing the resolution of $\delta\Gamma = 5$–$10\,\text{pg/mm}^2$. The binding of the odorant molecules gave a specific response for even sub nM concentrations. The authors speculate that this can be ascribed to an amplification mechanism related to the conformational change of OBP. For the probing depth of the PSP field of about $L_p/2$–100 nm and assuming the model presented in Fig. 5, the measurable conformation change with the instrument would need to be above 5% of the effective thickness change of the OBP protein layer. It should be noted that the SPR change due to the mass increase associated with the capture of the target itself analyte is not measurable (corresponding to $\Gamma = 5$–$10\,\text{pg/mm}^2$ when all OBPs capture one odorant molecule).

An alternative approach to detect odorant molecules by SPR was reported by (Cennamo et al., 2015). This work also employed an SPR detection principle based on the probing of the gold sensor surface with PSP modes at a wavelength of about 640 nm with a miniature fiber optic probe. Such a platform does not typically allow resolving as small changes in surface mass density $\delta\Gamma$ as the previous ATR-based example; thus a competitive assay was developed. The gold surface was chemically modified with thiol self-assembled monolayer carrying carboxyl groups that were subsequently used for attaching of butyric acid by amine coupling. This compound has a similar chemical structure to the odorant molecule butanal and can be recognized by porcine OBP dissolved in an aqueous solution in contact with the sensor. By spiking the analyzed sample containing target butanal with the porcine OPB, the butanal concentration-dependent response is obtained. It is based on the blocking of the binding pocket of the OBP with butanal that hinders its affinity interaction with butyric acid on the surface. According to the obtained data, a limit of detection of 25 μM was achieved by such an indirect measurement.

The third example of the detection of odorant molecules by a plasmonic biosensor relies on the use of gold nanostructures that support more confined LSP modes for direct detection. Zhang et al. developed a plasmonic nanostructure consisting of a thin gold film that was perforated with arrays of nanoholes (Zhang et al., 2015). The gold forms small metallic nanoclusters with a characteristic size of about 20 nm. The nanoholes allow the excitation of LSPs with a tightly confined probing depth $L_p/2$ of less than 10 nm. On the surface of this structure honey bee Acer-ASP2 OBP was immobilized by a thiol SAM through amine coupling, leading to a shift of the LSPR wavelength $\delta\lambda_{SPR}$ of 23 nm. To directly detect the target odorant—β-ionone—the changes in an LSPR wavelength $\delta\lambda_{SP}$ associated with the conformation change of the used OBP were monitored. The β-ionone interaction with the OBP serving as a BRE at the sensor surface was measured by the analysis of transmission intensity and LSPR wavelength shifts. The observed shift in the LSPR wavelength λ_{SPR} of 10–20 nm was observed upon increasing odorant concentration from 10 nM to 1 mM. With respect to the model presented in Fig. 5, such a strong response cannot be solely explained by the conformational change of the OBP (when compared to that for the immobilization of the OPB itself which reached similar value). In order to fully explain these measurements, more detailed information needs to be acquired, e.g., measuring the affinity binding kinetics by accurately tracking the respective changes in the LSPR wavelength λ_{SPR} to be able to distinguish the specific response from other effects.

6. Summary

This chapter presents an implementation of two plasmonic biosensor techniques for the investigation of affinity interaction of low molecular weight molecules with their ligands attached at a solid surface. The possible utilization of conformational changes occurring upon such interaction for sensitive target analyte detection is discussed. In the field of odorant biosensors relying on odorant binding proteins, there has been used SPR-based monitoring of conformational changes that are accompanied with the re-distribution of the surface mass density perpendicular to the sensor surface. We discuss the accuracy of this approach and how it depends on the probing depth of the surface plasmon field, which can be controlled by the choice of the metallic nanostructures that support surface plasmon modes (either propagating or localized). A practical guide for the preparation of metallic nanostructures with controlled plasmonic properties and possible means to modify their surface with biorecognition elements are given and the performance of plasmonic biosensors relying on the OBP conformational changes is illustrated by selected examples from literature. In addition, an approach based on plasmon-enhanced fluorescence energy transfer is introduced and its utilization presented for biomolecular recognition elements, which can be selectively tagged with a fluorophore at their defined parts. The conformation change due to the capture of a low molecular weight target analyte is then accompanied by a change in the distance between the fluorophore and the metal, leading to strong fluorescence signal variations. This sensitive approach was not yet explored for detection of odorant molecules, but may become an attractive modality in conjunction with recently explored alternatives to OBPs based on synthetically made DNA and peptide hairpin biomolecular recognition elements capable to specifically capture odorant molecules (Gaggiotti et al., 2020).

Acknowledgment

S.H. and J.D. received support from the European Union's Horizon 2020 research and the innovation program under Grant Agreement No. 633937, project ULTRAPLACAD, the Austrian Research Promotion Agency (FFG) with grant agreement No. 861578 (ERANET project PLABAN), and by Lower Austria EFRE project IKTHEUAP (WST3-F-5030820/010-2019).

References

Adato, R., & Altug, H. (2013). In-situ ultra-sensitive infrared absorption spectroscopy of biomolecule interactions in real time with plasmonic nanoantennas. *Nature Communications*, *4*, 2154. https://doi.org/10.1038/ncomms3154.

Bauch, M., Toma, K., Toma, M., Zhang, Q., & Dostalek, J. (2014). Plasmon-enhanced fluorescence biosensors: A review. *Plasmonics*, *9*, 781–799.

Cennamo, N., Di Giovanni, S., Varriale, A., Staiano, M., Di Pietrantonio, F., Notargiacomo, A., et al. (2015). Easy to use plastic optical fiber-based biosensor for detection of Butanal. *PLoS One*, *10*(3), e0116770. https://doi.org/10.1371/journal.pone.0116770.

Dahlin, A. B., Tegenfeldt, J. O., & Hook, F. (2006). Improving the instrumental resolution of sensors based on localized surface plasmon resonance. *Analytical Chemistry*, *78*(13), 4416–4423. https://doi.org/10.1021/ac0601967.

Gaggiotti, S., Hurot, C., Weerakkody, J. S., Mathey, R., Buhot, A., Mascini, M., et al. (2020). Development of an optoelectronic nose based on surface plasmon resonance imaging with peptide and hairpin DNA for sensing volatile organic compounds. *Sensors and Actuators B: Chemical*, *303*, 127188. https://doi.org/10.1016/j.snb.2019.127188.

Hajjar, E., Perahia, D., Debat, H., Nespoulous, C., & Robert, C. H. (2006). Odorant binding and conformational dynamics in the odorant-binding protein. *Journal of Biological Chemistry*, *281*(40), 29929–29937. https://doi.org/10.1074/jbc.M604869200.

Hall, W. P., Modica, J., Anker, J., Lin, Y., Mrksich, M., & Van Duynet, R. P. (2011). A conformation-and ion-sensitive plasmonic biosensor. *Nano Letters*, *11*(3), 1098–1105. https://doi.org/10.1021/nl103994w.

Homola, J. (2008). Surface plasmon resonance sensors for detection of chemical and biological species. *Chemical Reviews*, *108*(2), 462–493. https://doi.org/10.1021/cr068107d.

Hurot, C., Brenet, S., Buhot, A., Barou, E., Belloir, C., Briand, L., et al. (2019). Highly sensitive olfactory biosensors for the detection of volatile organic compounds by surface plasmon resonance imaging. *Biosensors & Bioelectronics*, *123*, 230–236. https://doi.org/10.1016/j.bios.2018.08.072.

Jeanmaire, D. L., & Van Duyne, R. P. (1977). Surface Raman electrochemistry part I. Heterocyclic, aromatic and aliphatic amines adsorbed on the anodized silver electrode. *Journal of Electroanalytical Chemistry*, *84*, 1–20.

Jones, E. M., Balakrishnan, G., Squier, T. C., & Spiro, T. G. (2014). Distinguishing unfolding and functional conformational transitions of calmodulin using ultraviolet resonance Raman spectroscopy. *Protein Science*, *23*(8), 1094–1101. https://doi.org/10.1002/pro.2495.

Jonsson, M. P., Jonsson, P., Dahlin, A. B., & Hook, F. (2007). Supported lipid bilayer formation and lipid-membrane-mediated biorecognition reactions studied with a new nanoplasmonic sensor template. *Nano Letters*, 7(11), 3462–3468. https://doi.org/10.1021/nl072006t.

Jung, J., Chen, L. X., Lee, S., Kim, S., Seong, G. H., Choo, J., et al. (2007). Fast and sensitive DNA analysis using changes in the FRET signals of molecular beacons in a PDMS microfluidic channel. *Analytical and Bioanalytical Chemistry*, *387*(8), 2609–2615. https://doi.org/10.1007/s00216-007-1158-6.

Khan, I., Keshmiri, H., Kolb, F., Dimopoulos, T., & List-Kratochvil, E. (2016). Plasmonic absorber based on multi-diffractive grating. *Advanced Optical Materials*, *4*(3), 435–443.

Langer, J., De Aberasturi, D. J., Aizpurua, J., Alvarez-Puebla, R. A., Auguié, B., Baumberg, J. J., et al. (2019). Present and future of surface-enhanced Raman scattering. *ACS Nano*, *14*(1), 28–117.

Larisika, M., Kotlowski, C., Steininger, C., Mastrogiacomo, R., Pelosi, P., Schutz, S., et al. (2015). Electronic olfactory sensor based on A. mellifera odorant-binding protein 14 on a reduced graphene oxide field-effect transistor. *Angewandte Chemie, International Edition*, *54*(45), 13245–13248. https://doi.org/10.1002/anie.201505712.

Liedberg, B., Nylander, C., & Lundstrom, I. (1983). Surface-plasmon resonance for gas-detection and biosensing. *Sensors and Actuators*, *4*(2), 299–304.

Nespoulous, C., Briand, L., Delage, M. M., Tran, V., & Pernollet, J. C. (2004). Odorant binding and conformational changes of a rat odorant-binding protein. *Chemical Senses*, *29*(3), 189–198. https://doi.org/10.1093/chemse/bjh017.

Neumann, O., Zhang, D. M., Tam, F., Lal, S., Wittung-Stafshede, P., & Halas, N. J. (2009). Direct optical detection of aptamer conformational changes induced by target molecules. *Analytical Chemistry*, *81*(24), 10002–10006. https://doi.org/10.1021/ac901849k.

Pelosi, P., Mastrogiacomo, R., Iovinella, I., Tuccori, E., & Persaud, K. C. (2014). Structure and biotechnological applications of odorant-binding proteins. *Applied Microbiology and Biotechnology*, *98*(1), 61–70. https://doi.org/10.1007/s00253-013-5383-y.

Quilis, N. G., Leveque, M., Khan, I., Knoll, W., Boujday, S., Lamy De La Chapelle, M., et al. (2018). Tunable laser interference lithography preparation of plasmonic nanoparticle arrays tailored for SERS. *Nanscale*, *10*, 10268–10276.

Reiner, A. T., Ferrer, N. G., Venugopalan, P., Lai, R. C., Lim, S. K., & Dostalek, J. (2017). Magnetic nanoparticle-enhanced surface plasmon resonance biosensor for extracellular vesicle analysis. *Analyst*, *142*(20), 3913–3921. https://doi.org/10.1039/c7an00469a.

Sergelen, K., Fossati, S., Turupcu, A., Oostenbrink, C., Liedberg, B., Knoll, W., et al. (2017). Plasmon field-enhanced fluorescence energy transfer for hairpin aptamer assay readout. *ACS Sensors*, *2*(7), 916–923.

Snyder, A. W., & Love, J. (1983). *Optical waveguide theory*. Springer.

Stenberg, E., Persson, B., Roos, H., & Urbaniczky, C. (1991). Quantitative-determination of surface concentration of protein with surface-plasmon resonance using radiolabeled proteins. *Journal of Colloid and Interface Science, 143*(2), 513–526. https://doi.org/10.1016/0021-9797(91)90284-f.

White, S. A., Briand, L., Scott, D. J., & Borysik, A. J. (2009). Structure of rat odorant-binding protein OBP1 at 1.6 angstrom resolution. *Acta Crystallographica Section D. Structural Biology, 65*, 403–410. https://doi.org/10.1107/s090744490900420x.

Zhang, D. M., Lu, Y. L., Zhang, Q., Yao, Y., Li, S., Li, H. L., et al. (2015). Nanoplasmonic monitoring of odorants binding to olfactory proteins from honeybee as biosensor for chemical detection. *Sensors and Actuators B: Chemical, 221*, 341–349. https://doi.org/10.1016/j.snb.2015.06.091.

Zheng, D., Seferos, D. S., Giljohann, D. A., Patel, P. C., & Mirkin, C. A. (2009). Aptamer nano-flares for molecular detection in living cells. *Nano Letters, 9*(9), 3258–3261. https://doi.org/10.1021/nl901517b.

9780128211571